Materials for Additive Manufacturing

This book introduces the theory and technology of the preparation and application of materials for additive manufacturing (AM) in a comprehensive and systematic manner. Its main contents can be divided into three parts: overview of AM technology and materials; introduction of AM polymer materials (such as powder materials, liquid materials and wire materials), AM metal materials (such as powder and wire materials) and AM ceramic materials; application cases of AM. This book is featured by its systematization, comprehensiveness, universality and novelty, with both illustrations and texts. It includes both theoretical research and practical application, which is a reference book of practical value for the research and application of AM materials.

The intended readers of this book include both engineers & technicians in the field of AM, and teachers & students of related majors (as a reference book).

3D Printing Technology Series

Materials for Additive Manufacturing

Yusheng Shi
Huazhong University of Science and Technology, Wuhan, China

Chunze Yan
Huazhong University of Science and Technology, Wuhan, China

Yan Zhou
China University of Geosciences (Wuhan), Wuhan, China

Jiamin Wu
Huazhong University of Science and Technology, Wuhan, China

Yan Wang
Wuhan Institute of Technology, Wuhan, China

Shengfu Yu
Huazhong University of Science and Technology, Wuhan, China

Ying Chen
Huazhong University of Science and Technology, Wuhan, China

Academic Press is an imprint of Elsevier
125 London Wall, London EC2Y 5AS, United Kingdom
525 B Street, Suite 1650, San Diego, CA 92101, United States
50 Hampshire Street, 5th Floor, Cambridge, MA 02139, United States
The Boulevard, Langford Lane, Kidlington, Oxford OX5 1GB, United Kingdom

Copyright © 2021 Huazhong University of Science and Technology Press.
Published by Elsevier Ltd. All rights reserved.

No part of this publication may be reproduced or transmitted in any form or by any means, electronic or mechanical, including photocopying, recording, or any information storage and retrieval system, without permission in writing from the publisher. Details on how to seek permission, further information about the Publisher's permissions policies and our arrangements with organizations such as the Copyright Clearance Center and the Copyright Licensing Agency, can be found at our website: www.elsevier.com/permissions.

This book and the individual contributions contained in it are protected under copyright by the Publisher (other than as may be noted herein).

Notices
Knowledge and best practice in this field are constantly changing. As new research and experience broaden our understanding, changes in research methods, professional practices, or medical treatment may become necessary.

Practitioners and researchers must always rely on their own experience and knowledge in evaluating and using any information, methods, compounds, or experiments described herein. In using such information or methods they should be mindful of their own safety and the safety of others, including parties for whom they have a professional responsibility.

To the fullest extent of the law, neither the Publisher nor the authors, contributors, or editors, assume any liability for any injury and/or damage to persons or property as a matter of products liability, negligence or otherwise, or from any use or operation of any methods, products, instructions, or ideas contained in the material herein.

British Library Cataloguing-in-Publication Data
A catalogue record for this book is available from the British Library

Library of Congress Cataloging-in-Publication Data
A catalog record for this book is available from the Library of Congress

ISBN: 978-0-12-819302-0

For Information on all Academic Press publications
visit our website at https://www.elsevier.com/books-and-journals

Publisher: Matthew Deans
Acquisitions Editor: Christina Gifford
Editorial Project Manager: Emily Thomson
Production Project Manager: Anitha Sivaraj
Cover Designer: Miles Hitchen

Typeset by MPS Limited, Chennai, India

Contents

Foreword   xi

1. **Overview of additive manufacturing technology and materials**   1
   1.1 Research and development status of additive manufacturing technology   1
   1.2 Research and development status of additive manufacturing materials   3
       1.2.1 Polymer materials   3
       1.2.2 Metal materials   5
       1.2.3 Ceramic materials   7
   References   7

2. **Polymer materials for additive manufacturing—powder materials**   9
   2.1 Selective laser sintering processing mechanism of polymer and its composite powder materials   9
       2.1.1 Heating process of polymer powder materials by laser   11
       2.1.2 Mechanism of selective laser sintering of polymer powder materials   19
       2.1.3 Influence of properties of polymer and its composite powder materials on selective laser sintering processing   24
   2.2 Preparation, composition, and characterization of polymers and their composite powder materials   49
       2.2.1 Preparation of polymer powder materials   49
       2.2.2 Composition of selective laser sintering polymer materials   53
       2.2.3 Characterization of selective laser sintering polymer materials   59
   2.3 Nylon 12 powder materials   61
       2.3.1 Preparation process of nylon 12 powder materials   62
       2.3.2 Selective laser sintering process characteristics of the nylon 12 powder   73
       2.3.3 Performance of nylon 12 powder selective laser sintering parts   89

2.4 Composite powder material of nylon 12 91
    2.4.1 Composite powder material of nylon 12/rectorite 92
    2.4.2 Nanosilica/nylon 12 composite powder materials 98
    2.4.3 Composite powder material of nylon 12 coated aluminum 111
    2.4.4 Nylon 12/copper composite powder material 130
    2.4.5 Composite powder material of nylon 12/potassium titanate whisker 138
    2.4.6 Carbon fiber/nylon 12 composite powder material 150

2.5 Styrene-based amorphous polymer powder materials 162
    2.5.1 Polystyrene powder materials 163
    2.5.2 Styrene–acrylonitrile copolymer powder material 163
    2.5.3 High impact polystyrene powder material 168

2.6 Polycarbonate powder material 176
    2.6.1 Selective laser sintering process and properties of polycarbonate powder material 178
    2.6.2 Effects of postprocessing on the properties of polycarbonate sintered parts 182

2.7 Acrylonitrile–butadiene–styrene powder material 184
    2.7.1 Basic characteristics of acrylonitrile–butadiene–styrene materials 184
    2.7.2 Sintering performance of acrylonitrile–butadiene–styrene powder 186

References 189

3. Polymer materials for additive manufacturing: liquid materials 191

3.1 Overview of stereolithography apparatus formed photopolymer 191
    3.1.1 Stereolithography apparatus material 195
    3.1.2 Stereolithography apparatus reaction mechanism 203
    3.1.3 Characteristic parameters of photopolymer and ultraviolet light source 215
    3.1.4 Characteristics of photopolymer materials and their stereolithography apparatus formability 218

3.2 Research on stereolithography apparatus solid materials 228
    3.2.1 Study on benzyl alcohol accelerator in cationic stereolithography apparatus system 230
    3.2.2 Study on iodonium salt and its photoinitiators 240
    3.2.3 Study on stereolithography apparatus kinetics of trimethylene oxide 260
    3.2.4 Study on the preparation and properties of solid materials in cationic systems 264
    3.2.5 Study on the preparation and properties of hybrid system solid materials 279
    3.2.6 Study on preparation and properties of solid materials in free-radical system 284

	3.3	Oligomers in stereolithography apparatus solid materials	288
		3.3.1 Polypropylene glycol diglycidylether diacrylate	288
		3.3.2 Low-viscosity urethane acrylate	293
		3.3.3 Oligomer of stereolithography apparatus support material	305
		3.3.4 Synthesis and properties of waterborne urethane acrylate	306
		3.3.5 Synthesis and properties of polyethylene glycol diacrylate	320
		3.3.6 Study on the preparation of support materials by oligomers	323
	3.4	Modified stereolithography apparatus forming materials	329
		3.4.1 Nano-SiO_2 modified stereolithography apparatus forming material	330
		3.4.2 Toughened photopolymer material of epoxy acrylate	340
		3.4.3 Synthesis and application of a novel alicyclic epoxy acrylate	349
	References		359
4.	**Polymer material for additive manufacturing—filament materials**		**361**
	4.1	Fused deposition modeling principle and process of polymer filament materials	361
		4.1.1 Fused deposition modeling principle	361
		4.1.2 Analysis of material modeling process	361
		4.1.3 Thermodynamic transformation of polymer processing	366
		4.1.4 Performance requirements for fuse deposition modeling polymer materials	371
	4.2	Modeling materials in fuse deposition modeling	373
		4.2.1 ABS filament	373
		4.2.2 Polylactic acid filament	378
		4.2.3 Polycarbonate and its composites filaments	383
		4.2.4 Nylon filaments	386
	4.3	Support materials in fuse deposition modeling	387
		4.3.1 Overview of support materials	387
		4.3.2 Break-away support materials	389
		4.3.3 Water-soluble support materials	395
	References		400
5.	**Metal materials for additive manufacturing**		**403**
	5.1	Additive manufacturing technologies for metal materials and the principles	403
		5.1.1 Selective laser melting technology	403
		5.1.2 Wire and arc additive manufacture	407

5.2 Forming mechanisms of metal materials — 409
- 5.2.1 Laser energy transfer — 409
- 5.2.2 Absorption of laser energy by metal — 412
- 5.2.3 Absorption of laser by metal powder — 415
- 5.2.4 Temperature, stress and strain fields in selective laser melting forming process — 419
- 5.2.5 Dynamics and stability of melting pool — 436

5.3 Metal powder for selective laser melting — 439
- 5.3.1 Effects of powder particle size on formability — 439
- 5.3.2 Effects of powder sphericity on formability — 450
- 5.3.3 Effects of powder oxygen content on formability — 452
- 5.3.4 Common metal and alloy powder materials for additive manufacturing — 454

5.4 Properties and microstructure characteristics of metal powder for additive manufacturing — 455
- 5.4.1 Metallurgical characteristics of selective laser melting metal powder — 455
- 5.4.2 Surface roughness and dimensional accuracy of formed parts by the selective laser melting technology — 507
- 5.4.3 Microstructure characteristics and mechanical properties of typical metal materials for additive manufacturing — 545

5.5 Metal wire for arc fuse deposition forming — 575
- 5.5.1 Design and preparation technology of wire materials — 575
- 5.5.2 Characterization of metal wire properties — 578

5.6 Microstructure and properties of wire and arc additive manufacture — 579
- 5.6.1 Microstructure and properties of multiaxial pipe joint component — 579
- 5.6.2 Microstructure and performance of typical repaired components — 588

References — 594

6. Ceramic materials for additive manufacturing — 597

6.1 Additive manufacturing technology and principle of ceramic materials — 597
- 6.1.1 Stereolithography technology and principles of ceramic slurry — 599
- 6.1.2 Three-dimensional printing technology and principles of ceramic powders — 600
- 6.1.3 Selective laser melting technology and principles of ceramic powders — 601
- 6.1.4 Laminated objected manufacturing technology and principles of ceramic sheets — 602
- 6.1.5 Fused deposition modeling technology and principles of ceramic filaments — 604

		6.1.6	Selective laser sintering technology and principles of ceramic powders	605
	6.2	Forming mechanism of the ceramics prepared by selective laser sintering		606
	6.3	Preparation of ceramic materials for selective laser sintering		609
		6.3.1	Ceramic powders and binders for selective laser sintering	609
		6.3.2	Preparation of composite ceramic powders for selective laser sintering	612
		6.3.3	Properties of ceramic composite powders and ceramics prepared by selective laser sintering	617
	References			645
7.	Application cases of additive manufacturing materials			649
	7.1	Application case 1 of additive manufacturing polymer powder material		649
	7.2	Application case 2 of additive manufacturing polymer powder material		650
	7.3	Application case 3 of additive manufacturing polymer powder material		653
	7.4	Application case 1 of additive manufacturing polymer wire material		654
	7.5	Application case 1 of polymer liquid materials for additive manufacturing		656
	7.6	Application case 1 of metal powder materials for additive manufacturing		657
		7.6.1	Drum lid	657
		7.6.2	Lid and box	660
	7.7	Application case 2 of metal powder materials for additive manufacturing		662
	7.8	Application case 1 of metal wire materials for additive manufacturing		665
		7.8.1	Printing of ultralarge (thin wall) parts	666
		7.8.2	Multimaterial forging die forming	667
		7.8.3	Hot forging die remanufacturing	668
		7.8.4	Part remanufacturing	669
	7.9	Application case 2 of metal wire materials for additive manufacturing		670
	7.10	Application case 3 of metal wire materials for additive manufacturing		672
	7.11	Application case 1 of ceramic powder materials for additive manufacturing		673
	7.12	Application case 2 of ceramic powder materials for additive manufacturing		674
	7.13	Application case 3 of ceramic powder materials for additive materials		675

		7.13.1	Manufacturing of cordierite ceramic parts	675
		7.13.2	Manufacturing of Al_2O_3 ceramic parts	676
		7.13.3	Manufacturing of SiC ceramic parts	677
8.	Materials for four-dimensional printing			679
	8.1	Definition of four-dimensional printing		679
	8.2	Research and development status of four-dimensional printing materials at home and abroad		680
		8.2.1	Polymers and their composite materials	680
		8.2.2	Metals and their composite materials	683
		8.2.3	Ceramics and their composite materials	685
	8.3	Research progress of our team on four-dimensional printing materials		686
		8.3.1	Cu-Al-Ni-based shape memory alloys	686
		8.3.2	Cu-Zn-Al-based shape memory alloys	702
		8.3.3	Double network hydrogel reinforced by carbon nanotubes	710
		8.3.4	Acrylate-based shape memory polymer	721
	References			735
Index				741

Foreword

Additive manufacturing (AM) is a new digital manufacturing technology intergrating machinery, computer, numerical control, materials, and other disciplines. It has a history of more than 30 years in the field of advanced manufacturing in the world. It transforms the traditional, part-design oriented to manufacturing process into a new design oriented to performance, revolutionizing today's manufacturing industry.

Since 1991, Huazhong University of Science and Technology began theoretical and applied research on AM technology. It is one of the earliest of the research carried out with AM technology in China. Up to now, it has developed series of equipment and its supporting materials for four types of AM processes, including lamella material layering, stereolithography, selective laser sintering, and selective laser melting, and realized industrialization, which have been widely used at home and abroad. It has won the second prize of National Science and Technology Progress Award and the second prize of State Science and Technology Invention Award, five first prizes of provincial and ministerial level, five second prizes of provincial and ministerial level, and more than 40 invention patents. Relevant research results were evaluated by the academicians of the Chinese Academy of Sciences and Chinese Academy of Engineering as the Chinese top 10 scientific and technological progress in 2011.

The core of AM technology is equipment and materials. After more than 10 years of development, it has formed a relatively mature series of products of AM equipment in China, and some indicators have reached the international advanced level. However, with the development of AM technology and expansion in its application fields, the shortage of AM materials becomes increasingly prominent. At present, AM materials include polymer, metal, ceramic, and composite materials, but only more than 100 kinds of materials are applied practically with limited performance and expensive price. Therefore, the existing AM materials have been unable to meet the needs of practical applications, and have become one of the main bottlenecks restricting the development of AM technology. Aiming at the problems of the above AM materials, the rapid manufacturing team of Huazhong University of Science and Technology, supported by the domestic and international major cooperative projects, such as Key Project of National Natural Science Foundation, National Key Research and Development Plan, Ultra

Large Scale Integration Circuit Manufacturing Technology and Packaged Process (02) and Advanced CNC Machine Tools and Basic Manufacturing Equipment (04) of National Science and Technology Major Project of the Ministry of Science and Technology of China, National Science and Technology Support Program and the 7th Framework Program, carried out basic and applied research for a long time, has developed a variety of new AM materials and their preparation methods, obtaining more than 20 national invention patents. Some materials have been industrialized and widely used at home and abroad. For example, under the support of "AM Polymer Composite Materials R&D and Industrialization Innovation Team in Guangdong Province" project, the AM polymer materials developed in this project have been industrialized in Guangdong Silver Age Sci. & Tech. Co. Ltd., formed multiple production lines of powder materials and wire materials for AM. In addition to meeting the domestic market demands, the products are exported to the United States, Germany, and other countries in a large scale. To cultivate scientific and technological talents in related fields and further study AM materials technology, the rapid manufacturing team of Huazhong University of Science and Technology condensed and summarized the team's research results of in this field, and published the monograph *AM Materials*.

The preparation, forming mechanism and process of AM polymers, metals, ceramics, and their composite materials are discussed in this book. The book is divided into seven chapters. The first chapter discusses the overview of AM technology and material, mainly including the development of AM technology and materials. The second chapter discusses polymer powder materials for AM, mainly including nylon (PA), polycarbonate (PC), polystyrene (PS), and their composite materials; Chapter 3, Polymer materials for additive manufacturing: liquid materials discusses polymer liquid materials for AM, mainly including photosensitive resin and its composite materials; Chapter 4, Polymer material for additive manufacturing—filament materials discusses polymer wire materials for AM, mainly including ABS and its composite materials; Chapter 5, Metal materials for additive manufacturing discusses metal powder materials and wire materials for AM, mainly including metal powder materials (Ti, Ni, Fe-based alloy) and wire materials; Chapter 6, Ceramic materials for additive manufacturing discusses ceramic materials for AM, mainly including Al_2O_3, ZrO_2, Kaolinite, and their composite materials; Chapter 7, Application cases of additive manufacturing materials discusses the application of AM, mainly including the typical applications of polymers and metal materials in industry and medicine; and Chapter 8, Materials for four-dimensional printing discusses materials for 4D printing.

In the process of writing this book, the authors, based on the scientific research achievements made in AM materials technology for decades, considered the requirements of readers with different knowledge backgrounds, which not only guaranteed the novel contents and reflected the latest research results but also discussed theoretical knowledge and practical application cases. Therefore, this book can be read by engineering and technical personnel in different fields and used as a reference book by teachers and students in related fields.

This book focuses on the achievements of the rapid manufacturing team of Huazhong University of Science and Technology. These achievements were obtained by the research team consisting of hundreds of people after persistent study for decades. The listed authors, Shi Yusheng, Yan Chunze, Zhou Yan, Wu Jiamin, Wang Yan, Yu Shengfu, and Chen Ying, are only the representatives of the research team. First of all, I would like to sincerely thank Professor Huang Shuhuai, founder of the rapid manufacturing team of Huazhong University of Science and Technology, for his guidance and establishing a good research platform for the latecomers. In the process of writing this book, I referred to the papers and achievements of some graduate students of this team, they are Wei Qingsong, Liu Jie, Wen Shifeng, Li Xiangsheng, Lin Liulan, Gan Zhiwei, Huang Biwu, Liu Jinhui, Lu Zhongliang, Yang Jingsong, Zhang Wenxian, Li Ruidi, Liu Haitao, Wang Li, Zhang Sheng, Liu Kai, Zhao Xiao, Zhu Wei, Li Wei, Zhang Linlin, Tang Fulan, Xia Sijie, Xu Wenwu, Tang Ping, Cheng Di, He Wenting, Jin Jiabing, Xing Shule, etc., I would like to express my heartfelt thanks to these graduate students! In addition, Post-doctor Cheng Lijin, Liu Rongzhen, Li Zhaoqing, Graduate Student Ji Xiantai, Ma Gao, Wu Xueliang, Chen Annan, Chen Jingyan, Dai Yili, Sun Yan, Chen Peng, Liu Shanshan, Xiao Huan, Chen Shuang, Ma Yixin, Fu Hua, Liu Yang, Chen Keyu, Wang Chong, Hu Hui, Tang Mingchen, Zhou Wangbing, and Xu Zhongfeng participated in the preparation work.

Since this book is the first academic monograph on AM materials as the theme, it involves extensive contents, some of which derived from our latest research results, while some research work is still ongoing, so we have been deepening our understanding of this technology. Due to an insufficient understanding of some issues and the limited academic level and knowledge of the authors, some mistakes and defects are inevitable in the book, and I look forward to the criticism and corrections from peer experts and readers.

Shi Yusheng
June 2019 in Wuhan

Chapter 1

Overview of additive manufacturing technology and materials

1.1 Research and development status of additive manufacturing technology

AM technology is mainly concentrated in Europe and the United States, of which the United States is the origin of AM technology, and the country with the widest research on this technology. The Laboratory for Freeform Fabrication at the University of Texas at Austin is one of the world's first established AM research center, covering all aspects of AM technology. W. M. Keck Center for 3D Innovation, established by the University of Texas at El Paso, cooperated with the University of New Mexico, Youngstown State University, Lockheed Martin Company, Northrop Grumman Company, RP + m, and Stratasys Company to be dedicated to the development of AM technology for aerospace systems. The University of Pennsylvania, in conjunction with the Battelle Memorial Institute and Sciaky Corporation, established the Center for Innovative Materials Processing through Direct Digital Deposition to focus on the design of materials such as metals and polymers and study on industrial application. In addition, scientific research units in other developed countries in Europe and the United States have set up AM research centers. For example, the University of Sheffield in the United Kingdom has set up the Centre for Advanced Additive Manufacturing, focusing on the study of structural design of AM parts, inkjet printing, laser forming of biomaterials, selective laser melting of aerospace materials, and new materials for laser sintering. The University of Nottingham in the United Kingdom has established the EPSRC Centre for Innovative Manufacturing in Additive Manufacturing to make innovative breakthrough for multifunctional AM technology and AM material system design. The Centre for Additive Layer Manufacturing, established by the University of Exeter in the United Kingdom is devoted to addressing the challenges of combining AM with industrial applications. The Fraunhofer Laser Research Institute in Germany has established the Fraunhofer Additive Manufacturing Alliance to focus on AM forming of metals, polymers, ceramics and biomaterials. Its 11 research centers are located throughout Germany. France has set up the Center for Technology Transfers in Ceramics (CTTC) to form brittle materials that are difficult to process using AM techniques such as inkjet printing, bond jetting, and ceramic direct deposition. The School of Mechanical Engineering at the University of Leuven in Belgium has conducted in-depth research on the types of AM technology and applied it to actual production. In addition to the aforementioned countries in Europe and America, the Monash University in Australia established the Monash Centre for Additive Manufacturing, which has the world's largest selective laser melting facility: Concept Laser X-Line 1000, and manufactured the world's first all-metal aero engine in 2015. In Asia, Singapore also established an AM center to study future manufacturing, marine applications, biomedical, and architectural printing. It covers AM equipment in various fields such as metal and biology, and is committed to becoming the pivot of AM center in Southeast Asia.

China has also actively established a number of AM technology innovation centers and service centers. For example, Academician Lu Bingheng in the field of AM took the lead to gather leading figures in AM from Tsinghua University, Beijing University of Aeronautics and Astronautics, Xi'an Jiaotong University, Northwestern University of Technology and Huazhong University of Science and Technology, and integrate domestic and foreign research and development forces to establish an AM research institute in Nanjing with the focus on the development of industrial technologies such as AM processes, equipment, materials, and applications in the fields of aerospace manufacturing, aerospace technology, automobile research and development (R&D), biomanufacturing, and medical rehabilitation, and gradually realize the transformation and incubation of technology. The AM innovation centers centered on local industry characteristics have been established in Tianjin and Qingdao successively to create an exhibition and experience center, a processing service center as well as a technology R&D center, displaying and disseminating global AM and the cutting-edge technologies and dynamics of industrial development, and promoting industrial development. Beijing set up the Collaborative Innovation Alliance of Digital Medical AM to mainly make breakthroughs in key technologies such as digital medical AM materials, processes and equipment, and tool software, and establish the first domestic and world-class "Collaborative Innovation Center of Digital Medical AM" and "Service Platform." Shanghai allied with five R&D teams to set up an AM technology innovation center, which combines local industry application needs to promote the application of AM technology in many fields. Guizhou Province in the southwestern region officially established an AM technology center in the Guiyang Economic and Technological Development Zone to provide industrial design and technology innovation platform and rapid manufacturing services for equipment manufacturing enterprises in the Zone. Changsha established an AM technology industry base that relies on equipment manufacturing enterprises to tackle key technologies in AM materials and equipment, so as to promote the improvement of the AM industry chain, and to drive the continuous breakthrough and development of AM technology [1,2].

1.2 Research and development status of additive manufacturing materials

AM technology has completely changed the way and principle of traditional manufacturing, which is a subversion of the traditional manufacturing model. At present, AM materials have become the main bottleneck restricting the development of AM, as well as the key and difficult point of making breakthrough in AM technology. Only by developing more new materials can the application field of AM technology be expanded. Currently, AM materials mainly include polymer materials, metal materials, ceramic materials and composite materials. AM materials are an important foundation for the development of AM technology. To some extent, the degree of material development determines the application range of AM. The polymer materials, metal materials, and ceramic materials will be introduced separately later in the chapter.

1.2.1 Polymer materials

The polymer materials used for AM are mainly thermoplastic polymers and composite materials thereof, and thermoplastic polymers can be divided into amorphous and crystalline states. Among them, amorphous polymers are commonly used to prepare parts with low strength requirements but high dimensional accuracy, mainly including polycarbonate (PC), polystyrene (PS), high impact PS (HIPS) and poly (methyl methacrylate) (PMMA). In 1993, DTM Company in the United States used PC powder for the formation of investment casting parts for the first time. Shi Yusheng, Wang Yan et al. from Huazhong University of Science and Technology used epoxy resin system to postprocess PC AM parts. The mechanical properties of postprocessed PC AM parts had great improvement, which can be used as a functional component with less demanding performance [3]. EOS and 3D System introduced the commercial powder materials for AM PrimeCast and CastForm based on PS in 1998 and 1999 respectively. Compared with PC, this AM material, characterized by lower sintering temperature, less sintering deformation, and excellent formability, is more suitable for investment casting process, so PS powder gradually replaces the application of PC powder in investment casting. The University of Hong Kong and Nanchang Hangkong University enhanced the performance of PS AM parts by preparing composite particles. Shi Yusheng et al. [4] from Huazhong

University of Science and Technology has successively improved the strength of PS AM parts by postprocessing impregnating epoxy resin and preparing PS/polyamide (PA) composite materials. The final formed parts can meet the requirements of general functional parts. Although the forming temperature of the PS is low and the precision is high, the strength of the AM parts is low, and it is difficult to form complicated, thin-walled parts. Therefore, Yang Jingsong et al. proposed to use HIPS powder materials to prepare resin molds for precision casting. The mechanical properties of the AM parts are much higher than PS sintered parts, and can be used to form parts with complicated, thin-walled structures. Shi Yusheng et al. from Huazhong University of Science and Technology also prepared the HIPS initial parison by AM, and then processed the HIPS functional parts with high precision and mechanical properties to meet the general requirements by postprocessing of impregnated epoxy resin.

AM parts by crystalline polymer have higher strength. PA is the most commonly used crystalline polymer in AM. It can be used to produce AM parts with high density and high strength, and can be used as functional part directly. Therefore, it has received wide attention, accounting for more than 95% of the current market of AM polymer materials. Gibson from the University of Hong Kong, Childs and Tontowi from the University of Leeds in the United Kingdom, Das et al. from the University of Michigan, Lin Liulan et al. from Huazhong University of Science and Technology, Xu Chao et al. from China Academy of Engineering Physics, and Zarringhalam and Ajoku et al. from Loughborough University in UK have carried out research on AM process and properties of pure nylon powder materials such as nylon 6, nylon 11, nylon 12, nylon 1010, and nylon 1212 respectively in early stage, which proved that nylon is the best material for directly preparing plastic functional parts by AM technology. However, the AM parts by nylon composite material have more outstanding performance than by the pure nylon AM parts, so that it can meet the performance requirements of plastic functional parts for different occasions and applications [5]. Nylon composite powder material has become an AM material mainly developed by 3D System, EOS and CRP, and new products are emerging. 3D System has introduced a series of nylon composite powder materials such as DuraForm GF, Copper PA, DuraForm AF, DuraForm HST, etc., among them, DuraForm GF is a nylon powder filled with glass beads, which has good forming precision and appearance quality; Copper PA is a mixture powder material of copper powder and nylon powder. With high heat resistance and thermal conductivity, it can be used to form injection molds directly by AM for small scale production of general plastic products such as polyethylene (PE), polypropylene (PP), and PS, with a production batch of hundreds of pieces; DuraForm AF is a mixed powder material of aluminum powder and nylon powder, and its AM part has metal appearance and high hardness and modulus. EOS also produces glass beads/nylon composite powder, PA3200GF, aluminum powder/nylon composite powder, Alumide, and the carbon fiber/nylon composite powder, CarbonMide, newly launched in 2008. CRP also introduced the glass beads/nylon composite powder, WindForm GF, aluminum powder/glass beads/nylon composite powder, WindForm Pro, and carbon fiber/nylon composite powder, WindForm XT.

In addition, the outstanding advantage of AM technology lies in that it can be used to manufacture complex and individualized products [6], which is very compatible with the needs of the biomedical field. Therefore, some bioactive or biocompatible polymer composites have recently become the forefront of academic research. These composite materials are usually made of bioactive bioceramics such as hydroxyapatite (HA) and biocompatible thermoplastic polymer materials such as polyvinyl alcohol (PVA), polylactide-glycolide (PLAGA), Poly-L-lactic acid (PLLA), polyetheretherketone (PEK), PE, and other components.

1.2.2 Metal materials

AM metal powder, as the most important part of the AM industry chain of metal parts, lies in the greatest value [7,8]. AM metal powder needs to have good plasticity, and must meet the requirements of fine powder particle size, narrow particle size distribution, high sphericity, good fluidity and high bulk density. At present, the AM consumable metal powder is mainly prepared through atomization method. The atomization method mainly includes water atomization and gas atomization. The powder prepared by gas atomization has the advantages of high purity, low oxygen content, controllable powder particle size, low production cost and high sphericity, and has become the main development direction of high performance and special alloy powder preparation technology.

Metal materials used for AM include iron-based alloys, nickel-based alloys, aluminum-based alloys, and titanium-based alloys [9,10]. Iron-based alloys are rich in resources, which are widely used in industrial production and life. They can be subdivided into three categories: stainless steel, high-strength steel, and die steel; 304 and 316 austenitic stainless steel powders (and their low carbon steel grades) were the first stainless steel materials developed for AM, and are now the typical processing materials in the AM market. After that, ASTM321 austenitic stainless steel was introduced to the market. Aermet 100 steel is a secondary hardening ultrahigh-strength steel. This type of alloy is widely used in the aerospace industry, but its melting and forming processes are complex. Now it has been successfully applied to AM technology. Research on high-strength steels such as 300 M, 30CrMnSiA and 40CrMnSiMoVA is also being carried out step by step. At present, Paderbon University in Germany uses AM technology to directly form H13 hot work die steel. IFW Dresden in Germany studied the AM FeCrMoVC tool steel, the formed tool steel has high density and no crack, which can be used as a knife mold. The AISI420 die steel for AM technology developed by Huazhong University of Science and Technology has achieved the forging level.

Nickel-based alloy refers to a kind of alloy of comprehensive performance with high strength and certain antioxidation and corrosion resistance at 650°C–1000°C, and is widely used in aerospace, petrochemical and other fields. Inconel 718, Inconel 625, and Inconel 738 alloys have been used as a typical material for AM. Inconel 600, Inconel 690, and Inconel 713 have also been used for AM. The Rene series is a nickel-based alloy material that GE has independently developed for its high-temperature parts. At the beginning of the 21st century, the mechanical properties and strength index of Rene 95 alloy AM parts is close to the Grade C standard of powder metallurgy, and the plasticity index exceeds the Grade A standard of powder metallurgy. In addition, in the Chinese high-temperature alloy grades, PGH 95 is the first high-temperature alloy powder developed in the early 1980s with its composition similar to Rene 95, whose room-temperature mechanical properties of AM parts are very close to the technical standards of powder metallurgy.

Aluminum-based alloys have been widely used in cooling and lightweight parts in the automotive, aerospace and other fields, and are also receiving attention from the AM industry. Professor Huang Weidong's team used AlSi12 alloy powder to repair ZL104 alloy and 7050 aluminum alloy by AM. The mechanical properties of the repaired parts even exceeded that of the base alloy. Lore Thijs et al. used AM technology to study the formation of AlSi10Mg alloy powder, and obtained aluminum alloy parts with good structure. Other aluminum alloy materials such as AlSi7Mg, AlSi9Cu3, AlMg4.5Mn4 and 6061 have also been studied and applied. Recently, EOS's new product, AlSi10Mg Speed 1.0, has been able to obtain fully dense parts by AM.

Titanium-based alloy is a new research hotspot toward the metal materials in the field of AM. Titanium alloy has the advantages of high-temperature resistance, corrosion resistance, high strength, low density and good biocompatibility. It has been widely used in aerospace, chemical industry, nuclear industry, sports equipment and medical equipment. TiAl6V4 (TC4) is one of the first titanium alloys used in the AM industry. American AeroMet company used AM technology to realize the installation of titanium alloy components. However, its performance does not meet the forging standard and cannot be used as the main bearing component. Professor Wang Huaming from Beihang University broke through the key technology of AM and successfully manufactured TC4 titanium alloy. Its mechanical properties such as room-temperature and high-temperature tensile, high-temperature creep and high-temperature durability are significantly higher than forgings. The structural parts have been installed on the aircraft. After studying the TC4 AM parts, Professor Huang Weidong from Northwestern Polytechnical University showed that the mechanical properties of the parts both in the

AM and the heat-treatment states were better than those annealed forged standard parts. Arcam's CP2 and Ti6Al4V have been successfully used in biomedical applications. In addition, M. Spiers et al. formed a Ti-13Nb-13Zr alloy stent using AM technology. Kyoto University in Japan formed Ni-Ti alloy by AM and successfully transplanted different artificial bones for four patients with cervical disk herniation.

1.2.3 Ceramic materials

Ceramic materials have the advantages of high-temperature resistance and high strength, and are widely used in industrial manufacturing, biomedical, aerospace and other fields. At present, few types of ceramic materials can be used for AM, which has become an important factor restricting the development of AM ceramic technology. At this stage, the ceramic materials for AM mainly include aluminum oxide (Al_2O_3), tricalcium phosphate (TCP), porous silicon nitride (Si_3N_4), and titanium silicide (Ti_3SiC_2) [11]. Ceramic materials combined with an organic precursor for AM mainly include SiC, Si_3N_4, SiOC, and SiNC. Schaedler et al. used an AM technology based on ultraviolet curing to realize a polymer precursor of three-dimensional structure with organic macromolecules. After pyrolysis at 1000°C, a high-density ceramic with nearly shrinkage uniformity was obtained. Weng Zuohai et al. from Northwestern Polytechnical University have successfully obtained silicon nitride ceramics with high porosity by AM. W. Sun et al. combined with AM and cold isostatic pressing technology to prepare a dense Ti_3SiC_2 ceramic. The University of the West of England has developed a new ceramic material "Viri-Clay" that forms personalized ceramic tableware by AM. The HRL Labs in Malibu, California, developed a preceramic resin that enables forming complex structural ceramics by AM.

The development and application of AM technology depends on the development and industrialization of basic materials. Based on this, this book will mainly address the forming mechanism, the preparation methods of materials and the properties of formed parts of three types of AM materials: polymers, metals, and ceramics.

References

[1] S. Kumar, Selective laser sintering: a qualitative and objective approach, JOM 55 (10) (2003) 43−47.
[2] D.T. Pham, S. Dimov, F. Lacan, Selective laser sintering: applications and technological capabilities, Proc. Inst. Mech. Eng. Part. B J. Eng. Manuf. 213 (5) (1999) 435−449.
[3] Y. Shi, J. Chen, Y. Wang, et al., Study of the selective laser sintering of polycarbonate and post process for parts reinforcement, Proc. Inst. Mech. Eng. Part. L J. Mater. Des. Appl. 221 (1) (2007) 37−42.
[4] Y.S. Shi, J.S. Yang, C.Z. Yan, et al., An organically modified montmorillonite/nylon-12 composite powder for selective laser sintering, Rapid Prototyp. J. 17 (1) (2011) 28−36.

[5] R.D. Goodridge, M.L. Shofner, R.J.M. Hague, et al., Processing of a Polyamide-12/carbon nanofiber composite by laser sintering, Polym. Test. 30 (1) (2011) 94−100.
[6] F.E. Wiria, C.K. Chua, K.F. Leong, et al., Improved bio composite development of poly (vinyl alcohol) and hydroxyapatite for tissue engineering scaffold fabrication using selective laser sintering, J. Mater. Sci.: Mater. Med. 19 (3) (2008) 989−996.
[7] W. Yanqing, S. Jingxing, W. Haiquan, Application and research status of AM materials, J. Aeronaut. Mater. 36 (4) (2016) 89−98.
[8] H. Jie, L. Wenjun, D. Liuliu, et al., Research progress of metal materials in additive manufacturing technology, Mater. Rev. (s2) (2014) 459−462.
[9] Z. Zeng, W. Lianfeng, Y. Biao, Research progress of AM metal materials, Shanghai Nonferrous Met. 37 (1) (2016) 57−60.
[10] Y. Nina, P. Xionghou, Method of preparation of AM metal powder, Sichuan Nonferrous Met. 4 (2013) 48−51.
[11] B. Yue, Z. Le, W. Shuai, et al., Research progress of AM ceramic materials, Mater. Rev. 30 (21) (2016) 109−118.

Chapter 2

Polymer materials for additive manufacturing—powder materials

Polymer materials are widely used in various fields of the national economy and rapidly developed with their excellent properties. The appearance of emerging technologies such as additive manufacturing (AM) has put forward higher requirements for polymer materials, and promoted its development toward high-performance and functionalization. The development of polymers and composite materials for AM has received more and more attention. In AM technology, the polymers and its composite powder materials are mostly used consumables in the market, which are mainly used in selective laser sintering (SLS) technology, that is, the bond is formed with the thermal part of laser irradiation by polymer powder materials. This chapter, centered on the SLS-forming polymers and their composite powder materials, mainly introduces the preparation of nylon and its composite powder materials, polystyrene (PS) powder materials, polycarbonate powder materials, and their SLS process.

2.1 Selective laser sintering processing mechanism of polymer and its composite powder materials

The technological process of SLS is shown in Fig. 2.1. First, the 3D solid model file of the part is sliced in the Z direction, and the cross-sectional information of the part body is stored in the STL file; then a layer of powder material is laid on the workbench with powder, and the laser beam emitted by the CO_2 laser selectively scans the powder layer according to the computer-aided design data of the cross section of each layer under the control of the computer. In the area scanned by the laser, the powder material is sintered together, and the powder not irradiated by the laser is still loose as a formed part and the support of the next powder layer; after the first layer is sintered, the workbench is lowered by the height of one layer of cross section to carry out powder spreading and sintering for the next layer, and the new layer and the previous layer are naturally sintered together; thus, when all the cross sections are sintered, the excess powder not sintered is removed to obtain the

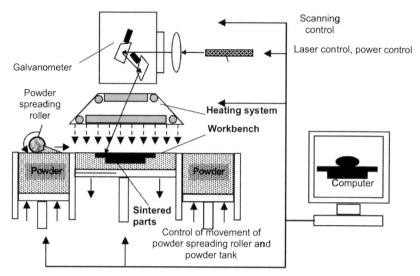

FIGURE 2.1 Schematic diagram of the technological process of SLS [1].

designed three-dimensional (3D) solid part. As shown in Fig. 2.1, the laser scanning process, laser switch-on/off and power, temperature preheating, and movement of powder spreading roller and powder tank are all performed under the precise control of the computer system.

Compared to other 3D printing technologies, the characteristics of SLS technology are as follows:

1. Extensive processing materials. In theory, any powder material that absorbs laser energy and has a reduced viscosity can be used for SLS. These materials can be polymer, metal, and ceramic powder materials.
2. Wide applications. The various forming materials determine that SLS technology can use different powder materials to form complex parts that meet different applications. SLS can form plastic prototypes and functional parts for structural verification and functional testing, and can be used to form metal or functional ceramic parts by direct or indirect methods. At present, SLS parts have been widely used in automotive, aerospace, medical biology, and other fields.
3. High material utilization rate. In the SLS process, the powder material not scanned by the laser is still in a loose state and can be reused. Therefore, SLS technology has a high material utilization rate.
4. No need of support. Since the unsintered powder can support the cavity and the cantilever portion of the formed part, it is not necessary to additionally design the support structure like stereolithography and fused deposition modeling.

Only small laser power is required for sintering, so the material used for SLS is mainly polymer powder. Since a certain number of voids and gaps are often present inside the sintered structure, the mechanical properties of the sintered material of the polymer powder material are generally lower than that of the molded part. The composite material is composed by two or more materials which have different physical and chemical properties. The submicroscopic or microscopic uneven dispersion of the second phase in the composite materials can often produce amazing improvement in the properties of composite materials. Therefore, it is particularly important to compensate for the defects of material properties with the use of composite materials. Composite materials are more commonly used as 3D printing materials due to their ability to cooperate with the functions of the various components.

In the currently used resin-based composite materials, one phase serves as a reinforcing part, and the other phase of resin condenses and adheres to the reinforcing material to form a whole body, and the components of the formed composite material maintain the identity of the original material. Moreover, the excellent performance lacked in the original single component material can be obtained through the complementary performance of each component in the composite material. The resin mainly plays the role of partially carrying and transferring the stress, while protecting the filler and fiber from the erosion and corrosion of the surrounding medium. Although short-fiber and powder fillers cannot be used as load-bearing materials, they can improve the mechanical properties of composite materials.

The SLS processing of the polymer material is to form a spot having very high energy density and small size of light beam emitted by the CO_2 laser on the working surface through the focusing lens, and the spot sinters the polymer powder material laid on the workbench. This forming method includes two basic processes of heating the polymer powder material by laser and sintering the polymer powder material. A correct understanding of these two basic processes is the basis for the successful application of SLS technology. In this part, the two basic processes are theoretically analyzed to reveal various factors and their interactions related, so as to provide a basis for the development of high-performance SLS polymer materials and optimization of sintering process.

2.1.1 Heating process of polymer powder materials by laser

2.1.1.1 Characteristics of laser input energy

The laser beam in the SLS-forming system is a Gaussian beam. Since the working plane is on the focal plane of the laser beam, the light intensity distribution of the laser beam is:

$$I(r) = I_0 \exp\left(\frac{-2r^2}{\omega^2}\right) \tag{2.1}$$

where $I(r)$ is the light intensity; I_0 is the maximum light intensity at the center of the spot; w is the characteristic radius of spot, here the light intensity I is $e^{-2}I_0$; r is the distance from the inspection point to the spot.

The size of I_0 is related to the laser power (P):

$$I_0 = \frac{2P}{(\pi w^2)} \tag{2.2}$$

Eq. (2.1) shows that the powder under the center of the laser scanning line receives a larger amount of energy, while the energy at the edge is lower, but when there is a certain overlap of the scanning lines, the laser on the entire scanning area can reach a more uniform level due to the packing energy. The CO_2 laser can be operated in a pulsed or continuous manner. When the repetition rate is high, the output is quasicontinuous and can be processed in a continuous manner. The cross-sectional energy intensity distribution of the continuous laser scanning line is:

$$E(y) = \sqrt{\frac{2}{\pi}} \frac{P}{w\nu} \exp\left(\frac{-2y^2}{w^2}\right) \tag{2.3}$$

where ν is the movement rate of the scanning laser beam. Eq. (2.3) shows the cross-sectional energy distribution of a single scanning line. For a plurality of overlapping scanning lines, the cross-sectional energy density distribution is related to parameters such as scanning pitch.

In the SLS process, the laser scanning speed is fast, and the laser energy can be linearly superposed in successive scanning processes. Set the scanning pitch as S, assuming that the equation of a certain starting scanning line is $y = 0$, then the equation of Ith scanning line after this is $y = IS$. The distance from a point $P(x, y)$ to Ith scanning line is $y - IS$, and the influence of Ith scanning line on point P is:

$$E(y) = \sqrt{\frac{2}{\pi}} \frac{P}{w\nu} \exp\left[\frac{-2(y-IS)^2}{w^2}\right] \tag{2.4}$$

The packing energy of multiple scanning lines is:

$$E_s(y) = \sum_{I=0}^{n} \left\{ \sqrt{\frac{2}{\pi}} \frac{P}{w\nu} \exp\left[\frac{-2(y-IS)^2}{w^2}\right] \right\} \tag{2.5}$$

Fig. 2.2 is a graph of laser energy distribution calculated according to Eq. (2.5) when the laser spot diameter is 0.4 mm, the moving speed (ν) of the scanning laser beam is 1500 mm/s, the laser power (P) is 10 W, and the scanning pitch is 0.3, 0.2, 0.15, and 0.1 mm, respectively.

As shown in Fig. 2.2, as the scanning pitch increases, the uniformity and maximum value of the laser energy distribution change accordingly. The laser energy increases with the decrease of the scanning pitch. For a

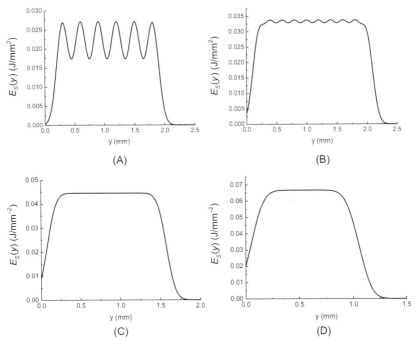

FIGURE 2.2 Laser energy distribution of multiple overlapped scanning lines: (A) $d_{sp} = 0.3$ mm; (B) $d_{sp} = 0.2$ mm; (C) $d_{sp} = 0.15$ mm; (D) $d_{sp} = 0.1$ mm.

laser beam with a spot diameter of 0.4 mm, when the scanning pitch exceeds 0.2 mm, the distribution of scanning laser energy is extremely uneven, showing a wave peak and wave trough (see Fig. 2.2A and B). The uneven energy distribution will result in uneven quality of the sintered parts. Therefore, during the laser sintering process, the scanning pitch should be less than 0.2 mm, that is, the scanning pitch should be smaller than the radius of laser spot.

2.1.1.2 Interaction between laser and polymer powder materials

The laser will be reflected, transmitted, and absorbed when it is incident on the surface of the powder material, during which the energy changes following the law of energy conservation:

$$E = E_{\text{reflection}} + E_{\text{tranmission}} + E_{\text{absorption}} \qquad (2.6)$$

where E is the laser energy incident on the surface of the powder material; $E_{\text{reflection}}$ is the energy reflected by the surface of the powder; $E_{\text{tranmission}}$ is the energy of the laser after it is transmitted through the powder; and $E_{\text{absorption}}$ is the energy absorbed by the powder material.

The above equation can be converted into:

$$R + \varepsilon + \alpha r = 1 \qquad (2.7)$$

where R is the reflection coefficient; ε is the transmission coefficient; αr is the absorption coefficient.

For polymer powder, the transmittance of CO_2 laser with a wavelength of 10.6 μm is very low. Therefore, the amount of laser energy absorbed by the powder material is mainly determined by the absorption coefficient and the reflection coefficient. The larger the reflection coefficient, the smaller the absorption coefficient, and the smaller the laser energy absorbed by the powder material; on the contrary, the laser energy absorbed by the powder material is larger.

The absorption of laser energy by the material is related to the laser wavelength and the surface state of the material. The CO_2 laser with a wavelength of 10.6 μm is easily absorbed by the polymer powder material. Due to the large surface roughness of the polymer powder material, the laser beam generates multiple reflections on the peak-trough side wall, and even has interference with it, thereby generating strong absorption. Therefore, the absorption coefficient of the polymer powder material to the CO_2 laser beam is large enough to reach 0.95–0.98.

The laser energy absorbed by the surface of the powder material collides with the basic energy particles in the polymer material through laser photons, and the energy is instantaneously converted into heat energy, and the heat energy is expressed in the form of an increase in the temperature of the material. As the temperature of the material increases, thermal radiation on the surface of the material returns the energy.

$$\Delta E = E_{in} - E_{out} \qquad (2.8)$$

The temperature change on the surface of the material conforms to the following rules:

1. Under the condition of same time of laser action, the larger the ΔE, the faster the temperature increase of the material.
2. Under the condition that ΔE is same, the smaller the specific heat of the material, the higher the temperature.
3. Under the same condition of laser irradiation, the smaller the thermal conductivity of the material, the greater the temperature gradient between the laser active region and its adjacent region.

The thermal conductivity of the polymer solid material is about 0.2 W/(m K), and the thermal conductivity of the powder is related to the thermal conductivity (λ_s) of the solid, the thermal conductivity (λ_g) of the air, and the void fraction (ε) of the powder.

The thermal conductivity λ_g of air can be calculated using an empirical equation:

$$\lambda_g = 0.004372 + 7.384 \times 10^{-5} T \qquad (2.9)$$

Polymer materials for additive manufacturing—powder materials **Chapter | 2** 15

The void fraction (ε) indicates the content of the void volume in the powder, and can be expressed by the powder density (ρ) and the solid density (ρ_s) of the material:

$$\varepsilon = \frac{(\rho_s - \rho)}{\rho_s} \quad (2.10)$$

The bulk density of spherical powder can be calculated by the following equation:

$$\rho = \frac{\pi \rho_s}{6} \quad (2.11)$$

Then the relative density of the powder material is:

$$\rho_R = \frac{\rho}{\rho_s} = 0.523 \quad (2.12)$$

$$\text{Void ratio } \varepsilon = 1 - \rho_R = 0.477 \quad (2.13)$$

The shape of the polymer powder prepared by different methods is different, and the bulk density of the powder is different, but the void ratio (ε) of most powders is about 0.5.

The thermal conductivity (λ) of the powder can be calculated using the Yagi-Kun model:

$$\text{When } \lambda \leq 673K, \ \lambda = \frac{\lambda_s(1-\varepsilon)}{(1 + \varphi \lambda_s / \lambda_g)} \quad (2.14)$$

where $\varphi = 0.02 \times 10^{2(\varepsilon - 0.3)}$.

It can be calculated from Eq. (2.14) that the thermal conductivity of the polymer powder material at room temperature is about 0.07 W/(m K). Since the thermal conductivity of the polymer powder material is very low, the temperature gradient between the laser active region and its adjacent region is large during the laser sintering process, and the sintering parts are prone to warping deformation. Therefore, the polymer powder material should be suitably preheated in the laser sintering process to reduce the laser power and the temperature gradient, thus preventing warping deformation.

2.1.1.3 Transfer of heat in polymer powder materials

When the laser beam scans and sinters the powder layer, the intensity of the laser light incident inside the powder material decreases geometrically with the increase of the depth. Therefore, the inside of the powder is heated in a conductive manner. The powder directly irradiated by the laser beam is heated and exchanges heat with the surrounding powder and the environment. Although the heat transfer under the action of laser also obeys the basic laws of thermodynamics, including three heat transfer modes of conduction, radiation and convection, it has many special characteristics. Since the laser beam is a heat source for motion, its action time with the powder generally ranges from several milliseconds to dozens of milliseconds, so the powder-heating speed is fast and the temperature gradient is large; during the heating process, the physical properties of the powder materials such as the absorption rate, thermal conductivity and specific heat capacity to the laser light change with the increase of temperature. The powder material also transits from solid state to liquid state then to solid state, with changes of density, which also affects the thermal conductivity and specific heat capacity, and the temperature values of internal points of the powder materials are also changing. Therefore, the laser sintering process of the powder material is an extremely complex and unstable heat conduction process.

Because the temperature field changes rapidly during laser sintering, and is affected by many factors, it is impossible to directly measure the change of temperature distribution by experiment; while grasping the variation law of temperature field under laser heating condition has practical guiding significance to control the processing technology reasonably, thus achieving the best processing results. Therefore, many scholars have proposed various heat transfer models for simulating the changes of temperature field during laser sintering. However, there are some errors in the models compared with the actual situation. So far, there is not a perfect heat transfer model for laser heating, and it is quite difficult to solve the equation of the heat transfer model.

SLS polymer materials are divided into crystalline polymer and amorphous polymer. During the sintering process, the crystalline polymer has latent heat, while the amorphous polymer has no latent heat. Therefore, different mathematical models must be used.

2.1.1.4 Selective laser sintering mathematical model of amorphous polymers

In principle, the temperature field for SLS-forming of the amorphous polymer powder involves 3D nonlinear heat conduction problem, but under certain conditions, the problem can be simplified. Festa et al. compared the temperature field models of different laser surface processing and found that when the parameter $N > 3.9$, a one-dimensional model can be used for approximation. Where

$$N = \frac{\nu(\omega)}{\alpha} \qquad (2.15)$$

ν is the laser scanning speed; ω is the radius of laser spot; α is the thermal diffusion coefficient of the material. Since the thermal diffusion coefficient of the polymer powder is very small, $N > 1000$, the one-dimensional approximation model can be used to calculate the SLS-forming temperature field of the polymer powder. The heat transfer control equation is

Polymer materials for additive manufacturing—powder materials **Chapter | 2** 17

$$\rho C_p \left(\frac{\partial T}{\partial t}\right) = \partial \frac{\left(\lambda \left(\frac{\partial T}{\partial z}\right)\right)}{\partial z} \tag{2.16}$$

On the upper surface of the powder layer, the boundary conditions of heat flow are given: during the illumination, there is heat flow from laser heating (q), heat dissipation heat flow from radiation and convection; after the illumination, the heat is dissipated through radiation and convection,

$$-\lambda \frac{\partial T}{\partial z}\bigg|_{z=0} = q - U(T - T_a) \tag{2.17}$$

In Eqs. (2.16) and (2.17), ρ is the density of the powder layer at z, C_p is the specific heat of the material, λ is the local effective thermal conductivity of the powder layer, T is the local temperature of the powder layer, T_a is the ambient temperature, U is the integrated heat transfer coefficient of radiation and convection, q is the density of laser heat flow.

On the underside of the powder layer, the density of heat flow is approximately zero:

$$\frac{\partial T}{\partial z}\bigg|_{z=+\infty} = 0 \tag{2.18}$$

The initial condition is that the start temperature of the powder layer is evenly distributed:

$$T|_{t=0} = T_0 \tag{2.19}$$

For the solution of heat transfer control Eq. (2.16), because the boundary conditions are complex, the finite difference method is used to perform numerical calculations on small time steps. In the calculation process, the process parameters are set unchanged, and the λ, C_p, ρ, etc. of the powder layer are functions of temperature and time, and these parameters are synchronously updated in the finite difference calculation process. The finite difference equation of the Crank-Nicholson form is unconditionally stable. The finite difference expression of the heat transfer control Eq. (2.16) is obtained through the Crank-Nicholson equation with the use of the uniform position step (Δz) and the time step (Δt):

$$\rho C_p \frac{T_i^{n+1} - T_i^n}{\Delta t} = \frac{1}{2}\left[\frac{\lambda_{i+1}-\lambda_i}{\Delta z} \cdot \frac{T_{i+1}-T_i}{2\times\Delta z} + \lambda_i \frac{T_{i+1}-2T_i+T_{i-1}}{\Delta z^2}\right]^{n+1}$$

$$+ \frac{1}{2}\left[\frac{\lambda_{i+1}-\lambda_i}{\Delta z} \cdot \frac{T_{i+1}-T_i}{2\times\Delta z} + \lambda_i \frac{T_{i+1}-2T_i+T_{i-1}}{\Delta z^2}\right]^{n} \tag{2.20}$$

where the superscript (n) represents the result at the nth moment, and the equation is as follows after summary:

$$[(\lambda_{i+1} - 3\lambda_i)T_{i-1} + 4(\lambda_i + m)T_i - \lambda' T_{i+1}]^{n+1}$$
$$= [(\lambda_i - \lambda_{i+1})T_{i-1} + 4(m - \lambda_1)T_i - \lambda' T_{i+1}]^n \quad (2.21)$$

$$i = 0, 1, 2, \ldots, N$$

$$m = \rho c_p \frac{(\Delta z)^2}{(\Delta t)}$$

$$\lambda' = \lambda_i + \lambda_{i+1}$$

The central difference equation is used for boundary condition (2.17), and the equation is as follows after summary:

$$[(m+B)T_0 - \lambda_0 T_1]^{n+1} = [(m-B)T_0 + \lambda_0 T_1]^n + A(q_1 + UT_n) \quad (2.22)$$

where $A = \Delta z(3 - \lambda_1/\lambda_0)$; $B = \lambda_0 + AU/2$.

The central difference equation is used for boundary condition (2.18), and the equation is as follows after summary:

$$[-\lambda_N T_{N-1} + (\lambda_N + m)T_N]^{n+1} = [\lambda_N T_{N-1} + (m - \lambda_N)T_N]^n \quad (2.23)$$

Combining Eqs. (2.21)–(2.23), the $(N+1)$ unknown temperatures $(T_0, T_1, \ldots, T_N)^{n+1}$ at the $(n+1)$th moment by $(N+1)$ known temperatures $(T_0, T_1, \ldots, T_N)^n$ at the nth moment. In this method, the $(N+1)$-order algebraic equations need to be solved for each time step (Δt). In the case where the start temperature $(T_0, T_1, \ldots, T_N)^0$ is known, the temperature at any time thereafter can be calculated.

2.1.1.5 Selective laser sintering mathematical model of crystalline polymer

The crystalline polymer has latent heat when sintered, and considering the heat is transferred in the transverse direction of the sintered part during the sintering process, the following two-dimensional model is adopted:

$$\rho c_p \frac{\partial T}{\partial t} = \lambda \left(\frac{\partial^2 T}{\partial y^2} + \frac{\partial^2 T}{\partial z^2} \right) + \frac{\partial \lambda}{\partial T} \left[\left(\frac{\partial T}{\partial y} \right)^2 + \left(\frac{\partial T}{\partial z} \right)^2 \right]$$
$$+ \left(\frac{\partial \lambda}{\partial y} \frac{\partial T}{\partial y} \right) + \left(\frac{\partial \lambda}{\partial z} \frac{\partial T}{\partial z} \right) - \rho c_p v \frac{\partial T}{\partial t} + \rho L \frac{\partial \alpha}{\partial t} \quad (2.24)$$

where α is the solid fraction of the polymer, v is the relative moving speed between the powder layer and the heat source, and L is the latent heat of fusion of the crystalline polymer.

The temperature recovery method can be used to solve heat transfer control Eq. (2.20) of crystalline polymer powder under unsteady state.

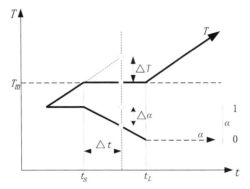

FIGURE 2.3 Solid−liquid transformation of crystalline polymer.

The calculation is carried out in two steps. The first step will not consider the latent heat. The finite element method is used to determine the temperature at a certain time. If the calculated temperature is ΔT higher than the melting point (T_m) of the polymer, then the latent heat is considered to correct the temperature calculated in the first step, the fusion of the crystalline polymer absorbs the heat, the solid fraction of the material decreases, and the temperature is restored to T_m, assuming that the solid fraction is decreased by $\Delta \alpha$, then:

$$\Delta \alpha = \left(\frac{c_p}{L}\right)\Delta T \qquad (2.25)$$

The schematic diagram of the calculation by the temperature recovery method is shown in Fig. 2.3.

2.1.2 Mechanism of selective laser sintering of polymer powder materials

The specific physical process of polymer material SLS-forming can be described as follows: when the high-intensity laser scans the powder layer under the control of the computer, the scanned area absorbs the energy of the laser, and the temperature of the powder particles in the area rises. When the temperature reaches to the softening point or melting point of the powder material, a sintering neck is formed between the particles due to the flow of the powder material, and then aggregation occurs. The process of formation of the sintering neck and the agglomeration of the powder particles is referred to as sintering. When the laser passes, the heat in the scanning area gradually disappears due to conduction under the powder layer and convection and radiation on the surface, the temperature is lowered, the powder particles are solidified, and the particles in the scanned area are bonded to each other to form a single-layer profile. Different from the general processing method of polymer materials, SLS is carried out under zero-shear stress. Ming-shen Martin Sun uses thermodynamic principles to prove that the driving force of sintering is the surface tension of powder particles.

2.1.2.1 Frenkel's two-liquid-drop model

Most of the polymer materials have low activation energy of viscous fluid flow, and the movement mode of the powder during the sintering process is mainly viscous flow. Therefore, the viscous flow is the main sintering mechanism of the polymer powder material. The mechanism of viscous flow sintering was first proposed by scholar Frenkel in 1945. The mechanism believes that the driving force of viscous flow sintering is the surface tension of powder particles, and the viscosity of powder particles hinders its sintering and acts on the work of the liquid drop surface tension (γ) in unit time and mutually balances with the energy dispersion rate caused by the viscous flow of the fluid, which is the theoretical basis of Frenkel's viscous flow sintering mechanism. Since the shape of the particles is extremely complicated, it is impossible to accurately calculate the "bonding" rate between the particles, so it is simplified to the spherical motion of the two spherical liquid drops to simulate the bonding process between the powder particles. As shown in Fig. 2.4, after two spherical liquid drops of equal radius begin point contact for a period of time (t), the liquid drops approach to form a circular contact surface while the remainder remains spherical.

Based on the "bonding" model of two spherical liquid drops, Frenkel derived the equation for calculating the length of Frenkel sintering neck by using the theoretical basis of the work of surface tension (γ) in unit time and the energy dispersion rate caused by the viscous flow of the fluid.

$$\left(\frac{x}{a}\right)^2 = \frac{3}{2\pi} \times \frac{\gamma}{a\eta} t \qquad (2.26)$$

In Eq. (2.26), x is the neck length of round contact surface within time t, that is, the radius of the sintering neck; γ is the surface tension of the material; η is the relative viscosity of the material, and a is the particle radius.

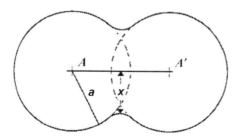

FIGURE 2.4 Frenkel's two-liquid-drop "bonding" model.

The Frenkel's viscous flow mechanism was first successfully applied to the sintering of glass and ceramic materials. Kuczynski et al. demonstrated that the polymer material is subject to zero-shear stress during sintering, and the melt is close to Newtonian fluid. The Frenkel's viscous flow mechanism is suitable for sintering of polymer materials, and it is concluded that the growth rate of sintering neck is proportional to the surface tension of the material and inversely proportional to the particle radius and melt viscosity.

2.1.2.2 "Sintered cube" model

Since the Frenkel's model only describes the sintering process of two spherical liquid drops, and SLS is the sintering of a powder bed formed by a large number of powder particles, so it is limited to describe the SLS-forming process by the Frenkel's model. Ming-shen Martin Sun proposed the "sintered cube" model based on the Frenkel hypothesis. This model considers that the powder packing in the SLS-forming system is similar to a cubic powder packing bed structure (shown in Fig. 2.5) and has the following assumptions:

1. The cubic packing powder bed consists of spheres of equal radius that initially contacts with each other (radius is a).
2. The particles are deformed in the densification process, but it always maintains a spherical shape with a radius of r. Thus, the contact between the particles is circular with a radius of $\sqrt{r^2 + x^2}$, where x represents the distance between the two particles.

The deformation process of a single powder particle is shown in Fig. 2.6. The "sintered cube" model is obtained through the principle of mutual balance of energy dispersion rate resulted from applying the surface tension (γ)

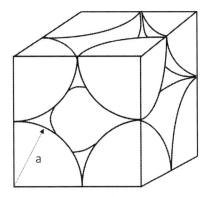

FIGURE 2.5 Structure of cube packing powder bed.

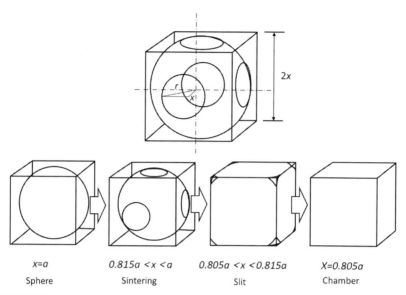

x=a	0.815a <x <a	0.805a <x <0.815a	X=0.805a
Sphere	Sintering	Slit	Chamber

FIGURE 2.6 Deformation process of a single powder particle during sintering.

of liquid drops work done in a unit of time and the viscous flow of the fluid. The energy balance equation is in the following form:

$$\gamma \dot{A} + \dot{e}_\varepsilon V = 0 \qquad (2.27)$$

In Eq. (2.27), "$\dot{A}$" refers to the change rate of surface area; refers to the change rate of volumetric strain energy; and "V" refers to the volume.

For a powder bed containing viscous materials, the change rate ($\dot{e}_\varepsilon$) of volumetric strain energy has the following relationship with the volumetric strain rate $\dot{\varepsilon}$:

$$\dot{e}_\varepsilon = \eta_b \dot{\varepsilon}^2 \qquad (2.28)$$

In Eq. (2.28), η_b refers to the apparent viscosity of the porous viscous structure, which is a function of the material viscosity and the void ratio. According to the Skorohod model:

$$\eta_b = \frac{4\eta\rho^3}{3(1-\rho)} \qquad (2.29)$$

Bring Eq. (2.28) into (2.27), and you can express the energy balance equation as:

$$\gamma \dot{A} + \eta_b \dot{\varepsilon}^2 V = 0 \qquad (2.30)$$

In the sintering neck stage, the equation of volume conservation is:

$$3x^3 - 9r^2 x + 4r^3 + 2a^3 = 0 \qquad (2.31)$$

The relative density in this stage is 0.502–0.965. If the powder particles are sintered in all six directions with other powder particles, the surface area retained by the particles is:

$$A_s = 12\pi rx - 8\pi r^2 \qquad (2.32)$$

In Eq. (2.31), r and x satisfy the volume conservation Eq. (2.30); A_s refers to the monotonically decreasing function of the relative density of the powder.

Many SLS or sintering tests in the oven have shown that the powder material stops densifying until its relative density reaches 0.96, indicating that for some reasons, some powder particles will not be sintered with other powder particles. The total surface area of the particles that do not sinter is:

$$A_u = 12\pi rx - 2\pi r^2 - 6\pi x^2 \qquad (2.33)$$

where A_u is the monotonically decreasing function of the relative density of the powder.

Now, assume that some of the powder particles in the powder bed are not sintered. Define the fraction of the sintered particles as ξ, and the sintering fraction as (ξ) which varies between 0 and 1, representing the probability of any two powder particles forming a sintering neck. If $\xi = 1$, it indicates that all powder particles are sintered; if $\xi = 0$, it indicates that no powder particles participate in sintering. From (2.31) and (2.32), we can conclude that the surface area of the partially sintered powder particles is:

$$A = \xi A_s + (1-\xi)A_u = 12\pi rx - (6\xi + 2)\pi r^2 - 6(1-\xi)\pi x^2 \qquad (2.34)$$

Thus, the change rate of surface area is:

$$\dot{A} = 12\pi(\dot{r}x + r\dot{x}) - 2(6\xi + 2)\pi r\dot{r} - 12(1-\xi)\pi x\dot{x} \qquad (2.35)$$

In Eq. (2.35), $\dot{r}$ and $\dot{x}$ satisfy the derivative of the volume conservation equation:

$$9x^2\dot{x} - 18r\dot{r}x - 9r^2\dot{x} + 12r^2\dot{r} = 0 \qquad (2.36)$$

The deformation of a volume unit containing a powder particle is considered, as shown in Fig. 2.6. Volume deformation can be expressed as:

$$\varepsilon = 3\left(1 - \frac{x}{a}\right) \qquad (2.37)$$

After the derivation of two sides of Eq. (2.35), the following result can be obtained:

$$\dot{\varepsilon} = \frac{3\dot{x}}{a} \qquad (2.38)$$

The volume is:

$$V = 8x^3 \qquad (2.39)$$

After bringing Eqs. (2.35)–(2.39) into Eq. (2.30), we can obtain the equation of sintering rate

$$\dot{x} = -\frac{3(1-\rho)\pi\gamma r^2}{24\eta\rho^3 x^3}\left\{r-(1-\xi)x+\left[x-\left(\xi+\frac{1}{3}\right)r\right]\frac{9(x^2-r^2)}{18rx-12r^2}\right\} \quad (2.40)$$

The sintering rate can also be expressed as a change in the relative density of the powder over time:

$$\dot{\rho} = -\frac{9\gamma}{4\eta a\rho}\left\{p-(1-\xi)+\left[1-\left(\xi+\frac{1}{3}\right)p\right]\frac{9(1-p^2)}{18p-12p^2}\right\} \quad (2.41)$$

In Eq. (2.41), $P = r/x$. The general sintering behavior can be seen from the sintering rate Eq. (2.41). It can be found that the densification rate is proportional to the surface tension of the material and inversely proportional to the viscosity (η) of the material and the radius (α) of the powder particles.

2.1.3 Influence of properties of polymer and its composite powder materials on selective laser sintering processing

Sintering materials are a key link in the development of SLS technology, which plays a decisive role in the forming speed and precision of sintering parts and their physical and mechanical properties. The polymer materials have various kinds of properties and different performances, which can meet the requirements of material properties for different occasions and applications. However, currently, very few polymer materials are widely used in SLS technology, mainly because the performance of SLS parts is strongly dependent on certain properties of the polymer. If these properties of the polymer do not meet the SLS-forming process requirements, then the accuracy or mechanical properties of its SLS parts are poor and cannot meet the requirements of actual use. Therefore, it is necessary to study the influence of the properties of polymer materials on the formation of SLS, so as to provide a theoretical basis for the selection and preparation of polymer materials for SLS.

2.1.3.1 Surface tension
2.1.3.1.1 Basic principle
The molecules on the surface of the material are only affected by the internal molecules, so the surface molecules increase the distance between the molecules along the parallel direction of the surface. The overall result is equivalent to a tension that expands the distance between the surface molecules, which is called surface tension, it makes the surface of the liquid always try to get the smallest area. The surface tension is related to the intermolecular forces, the surface tension is high in those with large intermolecular forces, and the surface tension is low in those with small intermolecular forces. If the Van der Waals force between the polymer melt molecules is small, the

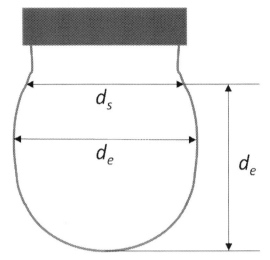

FIGURE 2.7 Shape of a hanging drop.

surface tension is low, ranging from 0.03–0.05 N/m; while the molten metal liquid has a very high surface tension due to the presence of strong metal bonds, usually at 0.1–3 N/m.

The hanging drop method is a common method for determining the surface tension and interfacial tension of a viscous polymer. The shape of the hanging drop (as shown in Fig. 2.7) is constant when the static pressure and the surface tension (or interfacial tension) is balanced. The relationship between the tension surface (or interfacial tension) and hanging drop shape is as follows:

$$\gamma = g\Delta\rho\left(\frac{d_e^2}{H}\right) \qquad (2.42)$$

where γ is the surface tension (or interfacial tension); g is the gravitational acceleration; $\Delta\rho$ is the density difference of two phases; d_e is the largest diameter of the hanging drop; H is the amount determined by the shape parameter factor $S(\frac{d_e}{d_s})$; and there is a functional relationship between $1/H$ and S, the corresponding value can be found in the relevant table; d_s is the diameter of the hanging drop from the end of the hanging drop to the length of the d_e, so the value of S can be measured after the d_s and d_e are measured in the experiment, and then the $1/H$ value can be obtained after checking the table, which is substituted into Eq. (2.42) to find γ.

2.1.3.1.2 Influence of surface tension on the selective laser sintering process

During the sintering process, the temperature of the polymer powder rises due to the absorption of laser energy. When the temperature of the polymer rises to its agglomeration temperature (melting temperature in terms of semicrystalline polymer and glass transition temperature in terms of amorphous polymer), the polymer molecular chain or segment begins to move freely. To reduce the surface energy of the powder material, the powder particles form sintering necks with each other driven by the surface tension, and even fuse together. Therefore, the surface tension is the driving force for the sintering formation thereof. In addition, the "sintered cube" model also shows that the sintering rate is proportional to the surface tension of the material. Therefore, surface tension is an important property that affects the formation of SLS of polymer materials. However, most polymers have relatively small surface tensions and they are very similar. So although surface tension is an important factor in determining the sintering rate of polymer, it is not a major factor causing a difference to sintering rate between polymers.

The balling effect is a problem that often frequently occurs in the selective laser melting forming process and seriously affects the surface precision of the sintered part. It is mainly due to the very large surface tension of the metal. After being heated and melted, the liquid phase sintering line is broken into a series of elliptical spheres after the action of the surface tension to reduce the surface area, thereby forming a surface shape of the sintered part composed of a series of semi-elliptical spherical protrusions. As can be seen from the above discussion, the surface tension of the polymer is much smaller than that of the metal, and the viscosity of the polymer melt is much higher than that of the metal during the sintering process. Therefore, during the SLS-forming process of the polymer, the balling part is not obvious, and its influence on the accuracy of the sintered part can often be neglected.

2.1.3.2 Particle size
2.1.3.2.1 Basic principle

When the physical property or physical behavior of the measured particle is closest to the homogenous sphere (or combination) of a certain diameter, the diameter (or combination) of the sphere is taken as the equivalent particle size (or particle size distribution) of the measured particle. When the particle size of the powder system is equal, the particle size of the powder can be expressed by a single particle size. In practice, the commonly used powder materials are composed of particles of different particle sizes, and the particle size refers to the average of the particle sizes of all the particles in the powder material. When there are n particles having a particle size d, there are four methods to solve the (weighted) average particle size.

Average particle size of count (arithmetic):

$$D_1 = \sum \left(\frac{n}{\sum n} d \right) = \frac{\sum (nd)}{\sum n} \qquad (2.43)$$

Average particle size of length:

$$D_2 = \sum \left(\frac{nd}{\sum (nd)} d \right) = \frac{\sum (nd^2)}{\sum (nd)} \qquad (2.44)$$

Average particle size of area:

$$D_3 = \sum \left(\frac{nd^2}{\sum (nd^2)} d \right) = \frac{\sum (nd^3)}{\sum (nd^2)} \qquad (2.45)$$

Average particle size of volume:

$$D_1 = \sum \left(\frac{nd^3}{\sum (nd^3)} d \right) = \frac{\sum (nd^4)}{\sum (nd^3)} \qquad (2.46)$$

At present, various particle size measuring methods have been developed, including sieving method, sedimentation method, laser method, and small-hole passage method. A brief overview of several common measurement methods is provided below.

1. Sieving method. The sieving machine can be divided into two types: electromagnetic vibration and acoustic vibration. Electromagnetic vibrating sieving machines are used for coarse particles (e.g., particles larger than 400 meshes), and sonic vibrating sieving machines are used for sieving of finer particles. The sieving method is an effective and simple method for analysis of powder particle size. It is widely used, but its precision is not high, so it is difficult to measure viscous and agglomerated materials such as clay.
2. Sedimentation method. When a beam of light passes through a measuring cell containing suspension liquid, part of the light is reflected or absorbed, and only a part of the light reaches the photoelectric sensor, which converts the light intensity into electrical signal. According to the Beer–Lambert law, it is related to the concentration of the suspension or the projected area of the particles through the light intensity. On the other hand, the particles settle in the force field, and the particle size can be calculated by Stokes' theorem to obtain cumulative particle size distribution.
3. Laser method. The basic principle is to measure the intensity of scattered light at different scattering angles by using concentric multiphotodetectors, and then calculate the average particle size and particle size distribution of the powder according to the Fraunhofer diffraction theory and the Mie scattering theory. Because of its high sensitivity, wide measurement range, and high reproducibility of measurement results, this method has become a widely used method to analyze the particle size of powder.

2.1.3.2.2 Influence of particle size on selective laser sintering processing

The particle size of the powder affects the surface finish, precision, sintering rate and powder layer density of the SLS parts. The particle size of the powder usually depends on the powdering method. The spray drying method and the solvent precipitation method usually can be used to obtain near-spherical powder with small particle size, and the low-temperature pulverization method can be used to obtain only irregular powder of large particle size.

In the SLS-forming process, the slice thickness of the powder and the surface finish of each layer are determined by the particle size of the powder. The thickness of the slice should not be smaller than the particle size of the powder. When the particle size of the powder is reduced, the SLS parts can be manufactured at a smaller slice thickness, which can reduce the staircase effect and improve the forming precision. At the same time, reducing the particle size of the powder can reduce the roughness of the single-layer powder after the powder spreading, so that the surface finish of the formed part can be improved. Therefore, the average particle size of the powder for SLS generally does not exceed 100 μm, otherwise, the formed part will have a very significant step part and the surface is very rough. However, the powder having an average particle diameter less than 10 μm is also unsuitable for the SLS process because such powder is adhesive on the rolls due to static electricity generated by friction during the spreading process, which makes it difficult to spread the powder.

The size of the particle also affects the rate of sintering of the polymer powder. From the "sintered cube" model, it is known that the sintering rate is inversely proportional to the radius of the powder particles. Therefore, the smaller the average particle size of the powder, the greater the sintering rate. Cutler and Henrichsen also inferred the same conclusion from the Frenkel model.

The density of the powder bed refers to the density of the powder in the working chamber after the powder spreading is completed, which can be approximated as the bulk density of the powder. It will affect the density, strength and dimensional accuracy of the SLS parts. Some studies have shown that the higher the density of the powder layer, the higher the density, strength and dimensional accuracy of the SLS parts. The particle size of the powder has a large influence on the bulk density. Next, a specific experiment will be used to explain the influence of the powder particle size on the bulk density.

The PS powder material is prepared by low-temperature pulverization method, and then the powder is divided into powders of three particle size ranges by sieving method: 30−45, 45−60, 60−75 μm, and the bulk density of three types of powders are measured respectively.

The device for measuring the density of powder bulk is shown in Fig. 2.8. First, accurately weigh the weight (W_0) of the measuring cylinder (the volume is 100 mL) to the 0.1 mg, then pour the sintered materials of powder into the measuring cylinder through the funnel, fill the measuring cylinder with powder through slight vibration, and scrape the powder sample along the receiver mouth with straight ruler. Accurately weigh the weight

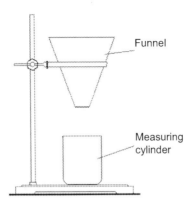

FIGURE 2.8 Device for measuring the powder density.

(W_1) of the cylinder full of the sample to 0.1 mg. The bulk density is calculated as follows:

$$\rho = \frac{W_1 - W_0}{V} \quad (2.47)$$

where ρ is the bulk density; W_0 and W_1 are the weight of the cylinders with and without the sample; and V is the cylinder volume.

Fig. 2.9 shows the bulk densities of powders of different particle sizes.

As can be seen from Fig. 2.9, the bulk density increases with the decrease of the particle size. When Ho et al. investigated the part of particle size of polycarbonate (PC) powder on bulk density, the same conclusion was reached, which may be due to the fact that small particles are more conducive to packing. However, both McGeary and Gray believed that when the particle size of the powder is too small (such as nanometer powder), the specific surface area of the material increases significantly, and the friction, adhesion and other surface forces between the powder particles become larger and larger. This affects the packing of the powder particle system, which in turn decreases as the particle size decreases.

2.1.3.3 Particle size distribution

2.1.3.3.1 Basic principle

The commonly used powder is not of a single particle size, but of different particle sizes. Particle size distribution, also known as size distribution, refers to the distribution of the particle size of the powder particles in a simple table, plot and function form. The particle size distribution can often be expressed in two forms, a frequency distribution and a cumulative distribution. The frequency distribution indicates the relative particle size (differential type) of each powder particle size, as shown in Fig. 2.10A; the cumulative distribution indicates that the relationship between the percentage

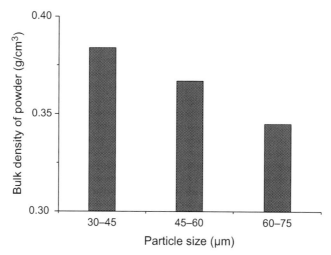

FIGURE 2.9 Bulk densities of powders of different powder particle sizes.

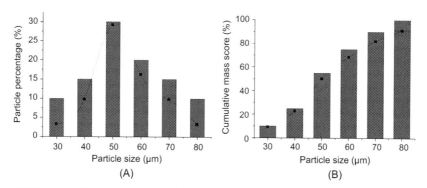

FIGURE 2.10 Frequency distribution (A) and cumulative distribution (B) of powder particle size.

content of a particle less than (or greater than) a particle size to the total particles and the particle size (integral type), as shown in Fig. 2.10B. The basis for the percentage content can be the number, volume and mass of particles.

2.1.3.3.2 Influence of particle size distribution on selective laser sintering processing

The particle size distribution affects the packing of solid particles, which affects the bulk density of the powder. An optimal relative bulk density is associated with a specific particle size distribution. For example, when the single-distributed spherical particles are orthogonally packed (Fig. 2.11), the relative density of the packing is 60.5% (i.e., the porosity is 39.5%).

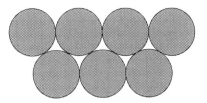

FIGURE 2.11 Orthogonal packing of single-distributed spherical powders.

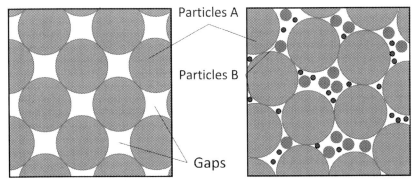

FIGURE 2.12 Powder packing: (A) single-distribution powder packing and (B) composite powder packing.

There is a certain gap between the single-distributed particles in orthogonal packing or other stacking modes. If smaller particles are placed in these voids, the porosity of the packing structure will decrease, the bulk density will increase. One method to increase the density of the powder layer is to combine powders of different particle size distributions. Fig. 2.12A and B are single powder packing pattern of the large particle size powder A and composite packing of the large particle size powder A and the small particle size powder B, respectively. It can be seen that there is a large pore in the single powder packing, and in the composite powder packing, since the small particle size powder occupies the pores in the large particle size powder packing, the bulk density thereof is increased.

2.1.3.4 Shape of powder particles
2.1.3.4.1 Basic principle

The particle shape of a polymer powder is related to the preparation method. Generally, the polymer powder prepared by the spray drying method is spherical, as shown in Fig. 2.13; the powder prepared by the solvent precipitation method is nearly spherical, as shown in Fig. 2.14; and the powder prepared by the low-temperature pulverization method is irregular, as shown in

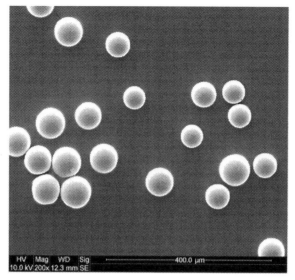

FIGURE 2.13 Microscopic shape of PS powder prepared by spray drying method.

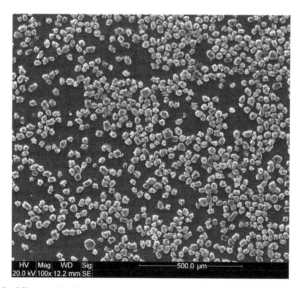

FIGURE 2.14 Microscopic shape of nylon powder prepared by solvent precipitation method.

Fig. 2.15. There is no quantitative test method to the shape of the powder particles, and qualitative analysis can only be performed by scanning electron microscope or the like.

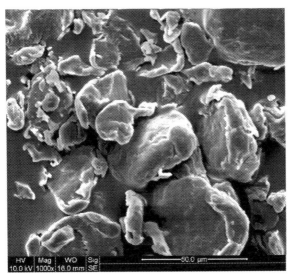

FIGURE 2.15 Microscopic shape of PS powder prepared by low-temperature pulverization method.

2.1.3.4.2 Influence of powder particle shape on selective laser sintering processing

The shape of powder particles has an influence on the shape accuracy, the powder spreading and the sintering rate of the SLS parts. The shape accuracy of SLS parts of the spherical powder is higher than that of the irregular powder; since the regular spherical powder has better fluidity than the irregular powder, the powder spreading part of the spherical powder is better, especially when the temperature rises and the flowability of powder decreases, this difference is more pronounced; Cutler and Henrichsen have found in the experiment that in the case of the same average particle size, the sintering rate of irregular powder particles is five times that of spherical powder, which may be because that the effective radius of the contact point between irregular particles is much smaller than the radius of the spherical particles, thus exhibiting a faster sintering rate.

2.1.3.5 Viscosity

2.1.3.5.1 Basic principle

Polymer melts are non-Newtonian fluids whose viscosity is dependent on shear rate. Since SLS is carried out at low shear stresses or even zero-shear stress, we are more concerned with the viscosity behavior of polymer melts at low shear rates. At low shear rates, non-Newtonian fluids can behave as Newtonian fluids, so the Newtonian viscosity can be obtained through the initial slope of the shear stress versus shear rate curve, also known as zero-shear viscosity, expressed as η_0, or a viscosity with shear rate tends to zero. The temperature and molecular weight of the polymer have a large part on the viscosity of the polymer. The influence of these two key factors on the viscosity of the polymer will be discussed below.

2.1.3.5.1.1 Temperature Above the viscous temperature of the polymer, the relationship between viscosity and temperature conform to the Arrhenius relational expression.

$$\eta = Ae^{\Delta E_\eta / RT} \tag{2.48}$$

where ΔE_η is activation energy of viscous fluid flow, and T is absolute temperature. As the temperature increases, the free volume of the melt increases and the mobility of the polymer segments increases, so that the fluidity of the polymer increases, and the melt viscosity decreases exponentially as the temperature increases.

When the temperature is lowered below the temperature of viscous flow, the apparent activation energy of viscous fluid flow (ΔE_η) is no longer a constant, but increases sharply with the decrease of temperature, and the Arrhenius equation is no longer applicable. The WLF equation is a good description of the relationship between the viscosity and temperature of polymer in the temperature range from T_g to $T_g + 100°C$.

$$\lg \left[\frac{\eta(T)}{\eta(T_g)} \right] = -\frac{17.44(T - T_g)}{51.6 + (T - T_g)} \tag{2.49}$$

For most amorphous polymers, the viscosity $\eta(T_g)$ at T_g is 10^{12} Pa s, and the viscosity of the polymer in the range from T_g to $T_g + 100°C$ can be calculated by substituting it to Eq. (2.49). The viscosity of the amorphous polymer decreases sharply above T_g with increasing temperature. The closer the temperature approaches T_g, the higher sensitivity the viscosity has to the temperature.

2.1.3.5.1.2 Molecular weight The molecular weight of the polymer has a great influence on its viscosity. The zero-shear viscosity (η_0) of the polymer melt and the weight-average molecular weight ($\overline{M}_W$) have the following empirical relationship:

$$\eta_0 = K_1 \overline{M}_W \left(\overline{M}_W < M_c \right) \tag{2.50}$$

$$\eta_0 = K_1 \overline{M}_W^{3.4} \left(\overline{M}_W < M_c \right) \tag{2.51}$$

where K_1 and K_2 are empirical constants. The various polymers have their own characteristic critical molecular weight (M_c). When the molecular weight is less than M_c, the zero-shear viscosity of the polymer melt is proportional to the weight-average molecular weight; and when the molecular weight is greater than M_c, the zero-shear viscosity increases sharply with the increase of molecular weight and is generally proportional to the 3.4 power of the weight-average molecular weight.

There are many methods for measuring the melt viscosity of the polymer, wherein the melt flow rate reflects the melt viscosity at a low shear rate, and thus the melt flow rate can better reflect the polymer flow properties of the polymer during the SLS process. Melt flow rate is defined as the weight (grams) of a molten polymer flowing from a standard capillary of a specified diameter and length within 10 minutes at a certain temperature under a certain load. The higher flow rate the melt has, the higher the fluidity is and the lower viscosity the melt has.

2.1.3.5.2 Influence of viscosity on selective laser sintering processing

From the Frenkel's two-liquid-drop model and the "sintered cube" model, the viscosity of the polymer has a great influence on the sintering rate. Next, some specific experiments will be used to study the influence of polymer viscosity on SLS-forming.

Three types of PS powders having different molecular weights were selected, and the relative viscosity of each sample in the benzene solution at 30°C was measured using an Ubbelohde viscometer. The test method is as follows: weigh a certain amount of sample to prepare a certain concentration of benzene solution, take 10 mL of this solution into the Ubbelohde viscometer with a pipette, measure the flow time of it, and then add 5, 5, 10 and 10 mL benzene for dilution and measure the flow time respectively time, then measure the flow time of pure benzene. Calculate the relative viscosity (η_r) and the specific viscosity (η_{sp}) of the solution using these measured values and the concentration of the known solution:

$$\eta_r = \frac{t}{t_0} \tag{2.52}$$

$$\eta_{sp} = \eta_r - 1 = \frac{t - t_0}{t_0} \tag{2.53}$$

According to Huggins equation:

$$\frac{\eta_{sp}}{c} = [\eta] + \kappa [\eta]^2 c \tag{2.54}$$

Or Kraemer equation:

$$\frac{\ln \eta_r}{c} = [\eta] - \beta [\eta]^2 c \tag{2.55}$$

where κ and β are both constants. When $\frac{\eta_{sp}}{c}$ or $\frac{\ln \eta_r}{c}$ is used to plot c or and extrapolated to $c \to 0$, the intercept of the straight line on the vertical coordinate is the intrinsic viscosity $[\eta]$, which can be expressed as:

$$[\eta] = \lim_{c \to 0} \frac{\eta_{sp}}{c} = \lim_{c \to 0} \ln \frac{\eta_r}{c} \tag{2.56}$$

When the chemical composition, solvent, and temperature of the polymer are determined, the $[\eta]$ value is related only to the molecular weight of the polymer, and can be expressed by using the Mark−Houwink empirical equation which contains two commonly used parameters:

$$[\eta] = KM^a \tag{2.57}$$

When PS at 30°C and the benzene is used as a solvent, K is 0.99×10^{-2} and a is 0.74.

The melt flow rate of the PS powder is measured by the melt flow meter. The sample is dried prior to testing. The sample is tested under the temperature of 200°C and the load of 5 kg. The melt flow rate is calculated as follows:

$$\text{MFI} = \frac{600 \times m}{t} \tag{2.58}$$

where MFI is the melt flow rate (the unit is g/10 minutes); m is the arithmetic mean of the mass of the cut sample band (the unit is g); t is the time interval of cutting sample band (the unit is s).

The viscosity-average molecular weight and melt flow rate of the three types of PS powders measured in the experiments are shown in Table 2.1.

The three PS powder materials were sintered by the HRPS-III laser sintering system developed by the Rapid Manufacturing Center of Huazhong University of Science and Technology.

Before the powder is sintered, in order to reduce the laser energy and the warping deformation of the sintered part, the powder bed is sufficiently preheated. PS is an amorphous polymer. Above the glass transition temperature (T_g), the movement of macromolecular segments becomes active. Due to the diffusion movement of molecular segments, PS powder particles will bond and agglomerate and lose fluidity, resulting in difficulty in powder spreading. Therefore, during the SLS process, its preheating temperature should be kept close to T_g.

The T_g of PS can be determined by differential scanning calorimetry (DSC). DSC analysis of PS under argon protection is carried out from room temperature to 250°C at a rate of 10°C/min, and the DSC curve of the temperature rising process is recorded. Fig. 2.16 shows the DSC curves of the three types of PS powders. The arrows indicate the glass transition of PS.

TABLE 2.1 Viscosity-average molecular weight and melt flow rate of three PS powders.

	PS-1	PS-2	PS-3
Viscosity-average molecular weight	0.75×10^4	1.21×10^4	1.82×10^4
Melt flow rate (g/10 min)	17.5	8.2	3.0

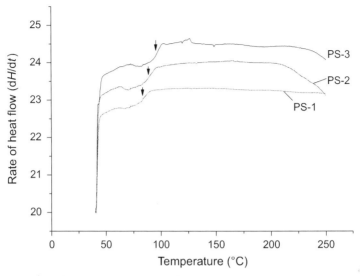

FIGURE 2.16 DSC curves of three types of PS powders.

TABLE 2.2 Glass transition temperature and preheating temperature of three types of PS powders.

	PS-1	PS-2	PS-3
Glass transition temperature (°C)	83.7	90.1	96.2
Preheating temperature (°C)	80	85	90

The corresponding T_g is listed in Table 2.2. The preheating temperature set according to T_g is also listed in Table 2.2.

The PS powder is sintered and formed by different laser energy densities, and the laser energy density is defined as the relative laser energy applied per unit area, which can be calculated by the following equation:

$$ED = \frac{P}{BS \times S} \qquad (2.59)$$

where ED is the laser energy density; P is the laser power; BS is the laser beam speed; S is the scan spacing. and BS is set to 2000 mm/s, S is set to 0.1 mm, and P is set from 8 to 16 W. Therefore, the variation range of ED is from 0.04 to 0.08 J/mm^2, and the slice thickness is set to 0.1 mm.

The PS powder is sintered into a sample of $4 \times 10 \times 80$ mm^3, and the length, width and height of the sample are measured by vernier caliper, indicated by l, w, h, respectively, and the weight (W) of the sample is accurately measured to the nearest 0.1 mg. The density (ρ) of the sample can be calculated by the following equation:

$$\rho = \frac{W}{l \times w \times h} \qquad (2.60)$$

Then the density of the sample can be calculated by the following equation:

$$\rho_r = \frac{\rho}{\rho_0} \qquad (2.61)$$

where ρ_r is the relative density of the sintered part; ρ_0 is the bulk density of PS (1.05 g/cm^3).

Fig. 2.17 shows the curves of the density of three PS powder sintered parts changing with the laser energy density. It can be seen from the figure that the densities of the three PS powder sintered parts increase with the increase of the laser energy density. This is because increasing the laser energy density can increase the absorption of laser energy by the powder, so that the temperature of the powder is increased, and the temperature has a greater influence on the viscosity of the PS. When the temperature is increased, the viscosity is decreased. The model of "sintered cube" shows that the viscosity of the material is lowered and the sintering rate is increased, so that the density of the sintered part is also improved.

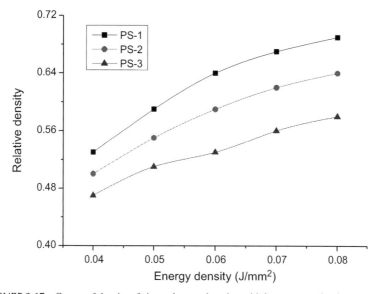

FIGURE 2.17 Curves of density of sintered parts changing with laser energy density.

Polymer materials for additive manufacturing—powder materials **Chapter | 2** 39

It can also be seen from Fig. 2.17 that the density of PS-1, PS-2, and PS-3 sintered parts decreases in sequence under the same laser energy density. It can be seen from Table 2.1 that the molecular weights of PS-1, PS-2, and PS-3 are sequentially increased, and the melt flow rates are sequentially decreased, that is, under the same conditions, the viscosities of PS-1, PS-2, and PS-3 are increased sequentially. Under the same laser energy density, the sintering rates of PS-1, PS-2, and PS-3 are successively decreased. Therefore, the densities of the PS-1, PS-2, and PS-3 sintered parts are successively decreased.

Fig. 2.18A−C show the microscopic shape of the cross section of PS-3, PS-2 and PS-1 powder sintered parts, respectively. It can be seen from Fig. 2.18A that in the PS-3 sintered part, the powder particles have sharp edges and corners, and are not rounded after sintering. The bonding between the particles is very weak, and there are a large number of pores in the sintered parts, so the relative density is the lowest; it can be seen from Fig. 2.18B that, in the PS-2 sintered part, the powder particles have been rounded due to sintering, and there are more sintering necks between the particles, which is denser than PS-3 sintered part; it can be seen from Fig. 2.18C that in the PS-1 sintered part, part of the powder particles have

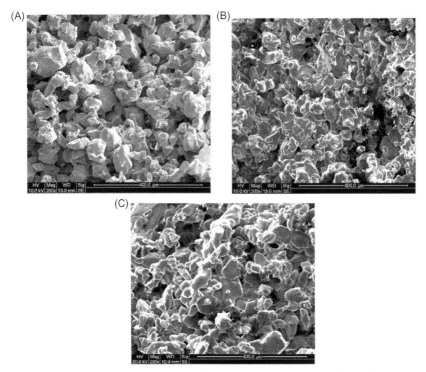

FIGURE 2.18 Microshape of the cross section of three types of PS material sintered parts: (A) PS-3; (B) PS-2; (C) PS-1. The laser energy density used is 0.06 J/mm^2.

been fused after sintering, and the density is the highest among the three sintered parts.

From the SLS-forming experiments of the above three types of PS powders, it can be concluded that the smaller the viscosity of the material, the larger the sintering rate and the higher the density of the sintered parts. In the SLS process, the temperature and molecular weight of the material are the main factors affecting the viscosity, which becomes the main factor affecting the sintering rate. The higher temperature the material is at, the smaller weight the molecular has, the smaller viscosity the molecular has, and thus the faster sintering rate the molecular has.

2.1.3.6 Bulk strength of materials

The strength of porous part varies with its relative density (ρ/ρ_0), and is subject to the following relationship:

$$\sigma = c\sigma_0 f\left(\frac{\rho}{\rho_0}\right) \tag{2.62}$$

where σ is the strength of the porous part of material; σ_0 is the bulk strength of the material; ρ is the density of the porous part; ρ_0 is the bulk density of the material; c is the empirical constant related to the material; and $f(\rho/\rho_0)$ is the function with the relative density as a variable. The researchers established the relationship between the strength of porous part and its relative density through different forms of $f(\rho/\rho_0)$ functions. The most commonly used relational expression is:

$$\frac{\sigma}{\sigma_0} = c\left(\frac{\rho}{\rho_0}\right)^m \tag{2.63}$$

Generally, the SLS sintered part of the polymer material is a porous part, and its relative density is defined as Eq. (2.61), and its porosity is defined as follows:

$$\varepsilon = 1 - \rho_r \tag{2.64}$$

where ρ_r is the density of the sintered part and ε is the porosity of the sintered part. From Eq. (2.63), it can be concluded that the relationship between the strength of the SLS sintered part and its bulk strength and density or porosity is:

$$\frac{\sigma}{\sigma_0} = c(\rho_r)^m \tag{2.65}$$

$$\frac{\sigma}{\sigma_0} = c(1-\varepsilon)^m \tag{2.66}$$

where σ is the strength of the SLS sintered part; σ_0 is the bulk strength of the polymer material; c and m are the constants related to the material. By plotting $\ln(\sigma/\sigma_0)$ with $\ln(\rho_r)$, the slope of the straight line obtained is a constant m, and the constant c can be found from the intercept.

It can be seen from Eq. (2.65) or (2.66) that the strength of the SLS part is closely related to the strength of the bulk material and the density of the sintered part, and increases as the strength of the bulk material and the density of the sintered part increase.

2.1.3.7 Aggregation structure

The polymers used in SLS are mainly thermoplastic polymers, and thermoplastic polymers can be classified into crystalline and amorphous depending on their aggregate structure. Since the thermal behavior of crystalline and amorphous polymers is absolutely different, there is a huge difference in the process parameter setting and the properties of the formed parts during the SLS-forming process. The amorphous polymer, PS, and the crystalline polymer, nylon 12, which are the most commonly used by SLS, are studied to investigate the influence of the aggregate structure of the polymer on the SLS-forming.

Sintering of the two types of polymer powders is carried out on HRPS-III laser sintering system. The sintering parameters are set as follows: BS is set to 1500 mm/s; SCSP is set to 0.1 mm; P is set to 6–20 W, and the slice thickness is set to 0.1 mm.

The method for determining the density of the sintered part is shown In Eq. (2.61), and the value of ρ_0 is given by the product performance table. The bulk density of the PS is 1.05 g/cm^3, and the bulk density of the PA12 is 1.01 g/cm^3.

Fig. 2.19 shows the design drawing of the dimensional accuracy test piece. The SLS test piece is manufactured from the design model and its dimensions are measured with vernier caliper. The dimensional accuracy is measured by the dimensional deviation A, and the dimensional deviation is calculated as follows:

$$A = \frac{D_1 - D_0}{D_0} \times 100\% \qquad (2.67)$$

where A is the dimensional deviation, D_0 is the design size, and D_1 is the actual size of the test piece.

2.1.3.7.1 Sintering temperature window

The sintering temperature window is to control the preheating temperature of the powder layer within a certain range before the laser sintering to prevent warping deformation during sintering, which can be expressed as [T_s, T_c], and only when the preheating temperature is controlled within the sintering temperature window can the warping deformation of the sintered layer can be avoided. Where T_s is the "softening point" of the powder

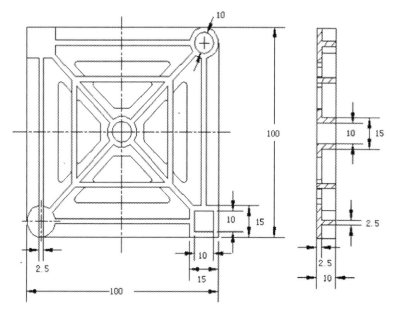

FIGURE 2.19 Dimensional accuracy test piece.

material. At T_s, the powder particles begin to adhere to each other and cannot flow freely. The storage modulus (G') of the powder material begins to drop sharply, because the storage modulus of the material is small when its temperature is greater than T_s, the stress relaxation is fast, so the shrinkage stress of the sintered layer is small without warping deformation; T_c is the "agglomeration temperature" of the powder material, and the powder layer will be completely agglomerated after the temperature reaches T_c, so that the powder cannot be cleaned after sintering, therefore, it is necessary to control the preheating temperature not to exceed T_c. The sintering temperature window is determined by the thermal properties of the material itself. The wider the sintering temperature window, the easier it is to control the sintering, and the sintered parts are less prone to warping deformation and vice versa.

For amorphous polymers, when the glass transition temperature is (T_g), the movement of the macromolecular segment becomes active. Due to the diffusion movement of the molecular segment, the powder particles will bond and the fluidity will decrease, and the storage modulus (G') begins to drop sharply, so T_s is T_g for amorphous polymers. Since the viscosity of the amorphous polymer is gradually decreased after T_g, its T_c cannot be determined by an amount having a certain physical meaning, and can only be determined by experimental observation. The DSC curve of PS-3 in Fig. 2.16 shows that the T_g is 96.2°C. It can be observed from the test that PS is completely agglomerated at 116°C and cannot flow, so its T_c is 116°C, and it can be concluded that the sintering temperature window of PS-3 is [96.2°C, 116°C].

For crystalline polymers, when the temperature reaches the start temperature of melting (T_{ms}), the viscosity drops sharply and the powder layer will completely agglomerate, so the T_{ms} of the crystalline polymer is T_c. The powder layer of crystalline polymer will be gradually cooled from the molten state after completion of sintering, when the temperature reaches the start temperature (T_{rs}) of recrystallization, the sintered layer begins to gradually transform from a liquid state to a solid state, since the polymer sintered layer is in a liquid state when it is greater than T_{rs}, it, the shrinkage stress is small, and the liquid does not carry stress, so warping deformation does not occur. Therefore, for a crystalline polymer, its T_{rs} is T_s. The sintering temperature window of the crystalline polymer can be obtained from the temperature rising DSC curve of the same sample and the subsequent temperature cooling DSC curve. The start temperature (T_{ms}) of melting on the temperature rising DSC curve is T_c, and the start temperature of recrystallization (T_{rs}) on the temperature cooling DSC curve is T_s, so its sintering temperature window is [T_{rs}, T_{ms}].

DSC analysis of PA12 was performed under argon protection using American Perkin Elmer DSC27 differential scanning calorimeter. First, the room temperature is increased to 200°C at a rate of 10°C/min, and then decreased to room temperature at a rate of 5°C/min, the DSC curve of the temperature rise and temperature reduction process is recorded, as shown in Fig. 2.20.

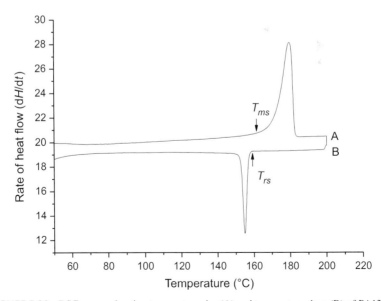

FIGURE 2.20 DSC curves showing temperature rise (A) and temperature drop (B) of PA12.

Curve A is the temperature rising DSC curve of PA12, and the temperature indicated by the arrow is the start temperature (T_{ms}) of PA12 melting, which is about 172.2°C. Curve B is the temperature cooling DSC curve of PA12, and the temperature indicated by the arrow is the start temperature of (T_{rs}) of PA12 recrystallization, which is about 156.9°C. Thus, the sintering temperature window of PA12 is [156.9°C, 172.2°C].

It can be seen from the above analysis that the sintering temperature window of PS is wide and the preheating temperature of the powder layer is low. Therefore, the sintering of PS is easier to control, and it is easier to sinter the qualified parts without warping; and the sintering temperature window of PA12 is narrower than PS and the preheating temperature of the powder layer is high. Therefore, the SLS-forming of the PA12 is very demanding on temperature control, and the sintered part is prone to warping deformation.

2.1.3.7.2 Density of sintered parts

Since the sintering temperature of the amorphous polymer powder is above T_g, the sintering of the crystalline polymer powder occurs above T_m, and generally, the viscosity difference between the two types of powders is very large when sintered, for example, the viscosity of the amorphous polymer at T_g is 10^{12} Pa s, and the viscosity of the crystalline polymer at T_m is about 10^3 Pa s, which results in a large difference in the sintering rate and the density of the sintered parts. Fig. 2.21 shows the curves of the density of PS and

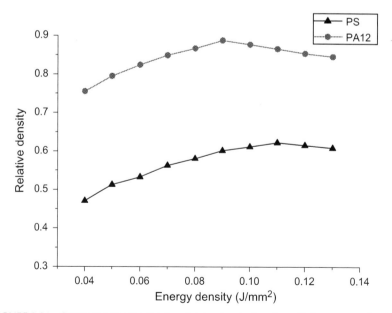

FIGURE 2.21 Curves showing the density of sintered parts changing with laser energy density.

PA12 sintered parts changing with the laser energy density. It can be seen that at the same laser energy density, the relative density of the PS sintered part is much smaller than that of the PA12 sintered part, which is because the viscosity of PS during sintering is much larger than the viscosity of PA12, resulting in the sintering rate of PS powder much lower than that of PA12. Thus, the relative density of the PS sintered part is much lower than that of the PA12 sintered part.

Theoretically, the viscosity of the amorphous polymer during sintering can be reduced by increasing the laser energy density, thereby increasing the density of the sintered part, and the sintered part having a density similar to that of the sintered part of crystalline polymer is obtained, but increasing the laser energy density will increase the secondary sintering (nonideal sintering of the powder outside the scanning area due to heat transfer), when the laser energy density is increased to a certain extent, it is difficult to remove the secondary sintered layer adhered outside the sintered part by postprocessing, so that the sintered part is scrapped. In addition, when the laser energy density is increased to a certain extent, the thermal degradation of the polymer is intensified due to the high temperature, so that the relative density of the sintered part is decreased. It can be seen from Fig. 2.21 that the densities of PS and PA12 sintered parts first increase with the increase of laser energy density. When the laser energy density increases to a certain value, the density reaches a maximum value, and then the laser energy density is increased, the density of the sintered part is decreased. This is because as the laser energy increases, the temperature of the sintered region increases, the viscosity of the polymer decreases, the sintering rate increases, and then the density of the sintered part increases, but when the laser energy increases to a certain value, the degradation of polymer material is intensified, resulting in a decrease in the relative density of the sintered part. In summary, it is difficult to obtain sintered parts with high density by SLS for amorphous polymers.

2.1.3.7.3 Mechanical properties of sintered parts

It can be seen from Eq. (2.65) that under the condition that the bulk strength of the material is constant, the strength of the sintered part is determined by its density, since the relative density of the crystalline polymer sintered part is higher and its strength closes to the bulk density of the polymer, so when its bulk density is large, the sintered part can be directly used as a functional part; while the amorphous polymer sintered part has a large number of pores, with very low relative density and strength, so it cannot be directly used as a functional part. Relative density is the main factor controlling the strength of amorphous polymer sintered parts. Only by proper postprocessing such as impregnating epoxy resin, the porosity of the sintered parts can be reduced, and the strength can be greatly improved while ensuring accuracy. However, the reinforcing methods commonly used in the plastics industry, such as the

46 Materials for Additive Manufacturing

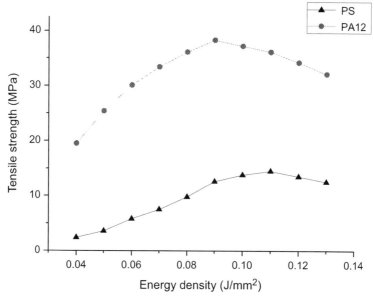

FIGURE 2.22 Curves showing tensile strength of sintered parts changing with laser energy density.

addition of inorganic fillers, generally fail to increase the density of the amorphous polymer sintered parts, and thus no obvious reinforcement can be achieved.

Fig. 2.22 shows the curves which show that tensile strength of PS and PA12 sintered parts changes with laser energy density. It can be seen that the tensile strength of PS sintered parts is much smaller than that of PA12, although the bulk strengths of PS and PA12 have no big difference (the bulk strength of PS is 42.5 MPa, and the tensile strength of PA12 is 46 MPa), since the density of PS sintered parts is much smaller than that of PA12, the tensile strength of PS sintered parts is much lower than that of PA12.

2.1.3.7.4 Fractured surface shape of sintered parts

The laser sintering behavior of amorphous polymers and crystalline polymers is quite different, which is more intuitively observed from the fractured surface shape of the sintered parts. After the impact fracture surfaces of the sintered parts of PS and PA12 powders are subject to metal spraying, the fractured surface shape is observed using Quanta 200 environmental scanning electron microscope from FEI company in the Netherlands, as shown in Fig. 2.23.

It can be seen from Fig. 2.23A that the powder particles in the amorphous polymer sintered part only form sintering necks at the contact site, and the

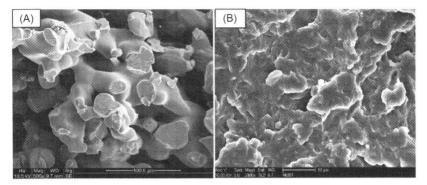

FIGURE 2.23 Microshape of PS (A) and PA12 (B) of the fractured surface of the sintered parts.

individual powder particles are still clear and distinguishable, and the relative position between the particles does not change to a large extent, there are a lot of pores inside the sintered part, and the relative density is very low. This is because the viscosity of the polymer during sintering is large, the sintering speed is slow, and the time of the laser action is extremely short, so that the sintering is not completely performed. It can be seen from Fig. 2.23B that the powder particles in the crystalline polymer sintered part are completely melted, the individual particles disappear, forming a dense whole with few pores and very high relative density, and thus the strength is close to the bulk strength of the polymer.

2.1.3.7.5 Dimensional accuracy of sintered parts

In the SLS-forming process of a polymer, volume shrinkage comes from two reasons: volume shrinkage of the polymer due to phase transformation and volume shrinkage due to sintering densification. Fig. 2.24 shows the specific volume−temperature curves of the amorphous polymer and the crystalline polymer. It can be seen from the figure that the crystalline polymer will produce large volume shrinkage due to crystal formation at the phase transition point T_m, which is about 4%−8%; on the contrary, the amorphous polymer has only a small volume change at the phase transition point T_g. For an amorphous polymer having a T_g of 110°C, a linear shrinkage of 0.8% will appear from 150°C to 30°C; for a crystalline polymer having a T_m at 150°C, a linear shrinkage of 3.9% will appear in the same temperature range. In the SLS process, since the powder particles in the sintered part of the amorphous polymer form a sintering neck at the contact site, the relative position between the particles does not change to a large extent, and a large amount of voids exist in the sintered parts, so that the volume shrinkage due to sintering densification is small; for crystalline polymers, since the loosely deposited powder becomes a dense whole after sintering, its sintering

48 Materials for Additive Manufacturing

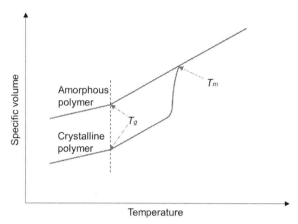

FIGURE 2.24 Specific volume–temperature curves of amorphous polymer and crystalline polymer.

densification produces a large volume shrinkage. In general, the relative density of the powder layer is between 0.4 and 0.6, and when the powder is completely densified, it will produce a linear shrinkage of 13%–20%. In summary, the volume shrinkage of the crystalline polymer due to phase change and sintering densification is much larger than that of the amorphous polymer, and thus the dimensional accuracy of a crystalline polymer sintered part is lower than that of an amorphous polymer.

In Table 2.3, the average dimension error of PS in directions X and Y is -1.37%, the shrinkage in Direction Z is small, and the dimensional error is -0.45%. The average dimension error of the PA12 sintered part in directions X and Y is -3.57%, which is 2.6 times that of PS; the dimensional error in Direction Z is -1.75%, which is 3.9 times that of PS.

Through the above discussion, the following conclusions can be drawn:

1. The sintering temperature window of amorphous polymer is wide, so it is easy to control the sintering process, and the sintered part is not easy to warp and deform; while the sintering temperature window of crystalline polymer is generally narrow, SLS-forming is very demanding on temperature control, and the sintered part is prone to warping and deformation
2. The amorphous polymer sintered part has a small relative density, so its strength is poor, which cannot be directly used as a functional part. The sufficient strength can be obtained by increasing its relative density through appropriate postprocessing; while the crystalline polymer sintered part has a high relative density and its strength is close to the bulk strength of the polymer, so that when the bulk strength is large, its sintered part can be directly used as a functional part.

TABLE 2.3 Dimensional deviation of PS and PA12 sintered parts.

Parameter		Design dimension (mm)	Actual dimension (mm)		Dimensional deviation (A) (%)	
			PS	PA12	PS	PA12
Side length	X_1	100	98.54	96.23	−1.46	−3.77
	Y_1	100	98.78	96.10	−1.22	−3.90
Height	Z_1	10	9.95	9.85	−0.5	−1.5
Inner diameter of the corner circle	R_1	10	9.82	9.62	−1.8	−3.8
Inner diameter of the central circle	R_2	10	9.85	9.63	−1.5	−3.7
Inner diameter of the corner square hole	X_2	10	9.88	9.63	−1.2	−3.7
	Y_2	10	9.89	9.64	−1.1	−3.6
Outer diameter of the corner square hole	X_3	15	14.80	14.55	−1.3	−3.0
	Y_3	15	14.79	14.54	−1.4	−3.1
Base plate thickness	Z_2	2.5	2.49	2.45	−0.4	−2.0

3. The powder particles in the amorphous polymer sintered part form sintering necks at the contact site, and the relative position between the particles does not change to a large extent, so the volume shrinkage is small and the dimensional precision is high; while the particles are completely melted when crystalline polymer powder is sintered, forming a dense whole, thus the volume shrinkage is large, and the dimensional accuracy of the sintered part is worse than that of the amorphous polymer.

2.2 Preparation, composition, and characterization of polymers and their composite powder materials

2.2.1 Preparation of polymer powder materials

The forming material used in the SLS technology is the powder material having a particle diameter of 100 μm or less, and the industrial product of the thermoplastic resin is generally a granule, and the granular resin must be made into powder to be used in the SLS process. Two methods are generally used for preparing the SLS polymer powder material: low-temperature pulverization and solvent precipitation.

2.2.1.1 Low-temperature pulverization

The polymer material has viscoelasticity. When pulverized at normal temperature, the pulverization heat generated increases the viscoelasticity, making the pulverization difficult, and the pulverized particles are re-bonded to reduce the pulverization efficiency, and even the phenomenon of drawing with melting may occur. Therefore, the powder required for the SLS process cannot be obtained by the conventional pulverization method.

It is difficult to prepare micron-sized polymer powder by mechanical pulverization at normal temperature, but at a low temperature, the polymer material has an embrittlement temperature T_b. When the temperature is lower than T_b, the material becomes brittle, which is favorable for pulverization by impact pulverization method. The low-temperature pulverization method utilizes the low-temperature brittleness of polymer material to prepare powder material. Common polymer materials such as PS, PC, polyethylene (PE), polypropylene (PP), polymethacrylate, nylon, acrylonitrile−butadiene−styrene (ABS), polyester, etc. can be used to prepare the powder by low-temperature pulverization method. Their embrittlement temperatures are shown in Table 2.4.

Low-temperature pulverization requires the use of refrigerant. The liquid nitrogen, featured by low boiling point, large latent heat of vaporization (199.4 kJ/kg at −190°C), inert liquefied gas performance, and abundant the source, is usually used as the refrigerant.

In the preparation of the polymer powder material, the raw material is first frozen with liquid nitrogen, the internal temperature of the pulverizer is maintained at a suitable low-temperature state, and then the frozen raw material is added for pulverization. The lower the pulverization temperature, the higher the pulverization efficiency, and the smaller the particle size of the obtained powder, but the larger the consumption of the refrigerant. The pulverization temperature can be determined according to the nature of the raw materials. For the brittle materials such as PS and polymethacrylate, the pulverization temperature can be higher, while for tougher materials such as PC, nylon, ABS, etc., the pulverization temperature should be kept low.

Low-temperature pulverization has simple process and can be continuously produced, but it requires special cryogenic equipment, which involves large investment and energy consumption, and the powder particles prepared

TABLE 2.4 Embrittlement temperatures of thermoplastic resins.

Resin	PS	PC	PE	PP	PA11	PA12
Embrittlement temperature (°C)	−30	−100	−60	−30 to −10	−60	−70

are irregular in shape, and the particle size distribution is wide. The powder is subjected to sieving treatment, and the coarse particles can be subjected to secondary pulverization and three times pulverization until the desired particle diameter is reached.

The polymer composite powder can be prepared by blending and granulating various additives and polymer materials through a twin-screw extruder to obtain granules, and then pulverizing the powder in a low temperature to obtain the powder material. The powder material prepared by the method has good dispersion uniformity, which is suitable for mass production, but not suitable for occasions where the sintering material equation needs to be changed frequently. Laboratory research usually uses the second method, that is, mechanical mixing of polymer powder with various additives in a 3D motion mixer, high-speed kneader or other mixing equipment. In order to improve the dispersion uniformity of the additives and the compatibility with the polymer material, some additives need to be pretreated before mixing. A small amount of additives such as antioxidants are directly mixed with the polymer powder and are difficult to disperse uniformly. The antioxidant can be dissolved in a suitable solvent such as acetone to prepare a solution of suitable concentration, mixed with the polymer powder, and then dried and sifted. For the convenience of preparation, antioxidants, lubricants and the like can be first mixed with a small amount of polymer powder to form a high concentration of the masterbatch, and then mixed with other raw materials [2].

2.2.1.2 Solvent precipitation

Solvent precipitation is to dissolve the polymer in a suitable solvent, and then uses a method of changing the temperature or adding a second nonsolvent (the solvent cannot dissolve the polymer, but can be mutually soluble with the former solvent) to precipitate the polymer in the form of powder. This method is particularly suitable for polymer materials with low-temperature flexibility like nylon. These materials are difficult to pulverize at low temperature and the yield of fine powder is very low.

Nylon is a kind of resin with excellent solvent resistance. It is difficult to dissolve in common solvents under normal temperature conditions, especially for nylon 11 and nylon 12, but it can be dissolved in a suitable solvent at high temperature. A solvent which dissolves nylon at a high temperature and has a very low solubility at a low temperature or a normal temperature is used, the nylon is dissolved at a high temperature, and the solution is cooled while stirring vigorously to precipitate the nylon in the form of a powder.

The process flow for preparing nylon 12 powder by solvent precipitation method is shown in Fig. 2.25.

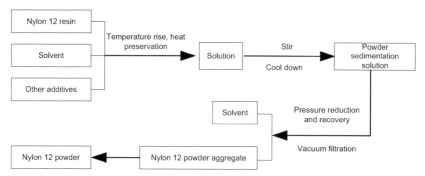

FIGURE 2.25 Process flow for preparing nylon 12 powder.

Ethanol can be used as the main solvent in the preparation of nylon 12 powder, supplemented by other co-solvents and additives. Put the nylon 12 granule, solvent and other additives into the jacketed stainless steel autoclave, and heat with the heating oil in the jacket, when the temperature inside slowly heats up to about 150°C, keep warm for 1−2 hours, stir vigorously, and cool at a certain speed to obtain the powder suspension. The cooled suspension was subjected to solid−liquid separation by vacuum filtration and pressure reduction and recovery. The obtained solid matter is an aggregate of nylon 12 powder, and after the aggregate is vacuum-dried, ground and sieved, the nylon 12 powder having a particle size of 100 μm or less and having a suitable particle size distribution is obtained.

The particle size and distribution of the nylon 12 powder prepared by the above method are affected by factors such as solvent dosage, dissolution temperature, heat preservation period, stirring rate, cooling rate, etc., and powder materials of different particle sizes can be prepared by changing these factors. In general, the larger the amount of the solvent dosage, the smaller the particle size of the powder. When the dissolution temperature is increased, the nylon 12 is completely dissolved, and the powder particle size is small, but since it is a closed container, when the temperature is increased, the system pressure is also increased, thus increasing the operation risk, and the excessive temperature causes the oxidative degradation of the nylon 12, affecting its performance, so the dissolution temperature should not be too high. Increasing the heat preservation time also reduces the particle size of the powder.

The particle shape of the powder prepared by the solvent precipitation method is close to the spherical shape, and the powder of the desired fineness can be produced by controlling the process conditions. To prevent the oxidative degradation of the nylon 12, a suitable antioxidant should be added.

In addition to the above two main preparation methods, some polymerization processes can be used to directly produce polymer powder. When a polymer such as polyacrylate, PS or ABS is synthesized by free radical initiated emulsion polymerization, the polymer powder can be obtained by spray drying to the polymer latex. The polymer powder prepared by this method has a spherical shape and good fluidity. When PC is produced by interfacial polycondensation, PC powder can also be obtained directly, but the powder obtained by this method is extremely irregular in shape and has a low bulk density.

2.2.2 Composition of selective laser sintering polymer materials

SLS polymer materials are composed of polymer powder and stabilizers, lubricants, dispersants, fillers and other additives. Among them, polymer matrix material is the main factor affecting the performance of sintered parts. Other additives make polymer materials suitable for SLS process requirements, thus improving the performance of sintered parts.

2.2.2.1 Selection of polymer matrix materials

2.2.2.1.1 Selection of amorphous polymers

There are many varieties of amorphous polymers, and the commonly used include PS, ABS, polymethyl-methacrylate (PMMA) and other acrylate polymers, polyvinyl chloride, PC, etc. In theory, these polymer powders can be used in the SLS process. However, these materials are difficult to form dense sintered parts due to high apparent viscosity during laser sintering. The high porosity results in the mechanical properties of the sintered parts being far lower than the mechanical properties of the materials. Therefore, the sintered part made of amorphous polymers cannot be used directly as a functional part. However, the porous sintered part can be greatly improved in strength after proper postprocessing, and the posttreated sintered part may be used as a functional part. A common postprocessing method is to impregnate the sintered part with liquid thermosetting resin, and a higher strength can be obtained after curing. Epoxy resin is easy to adjust curing temperature, has small curing shrinkage and excellent physical and mechanical properties, which is perfect for use as an impregnating agent.

Since the strength of the amorphous polymer laser-sintered part is mainly dependent on the postprocessing, and the degree of improvement of the strength of the sintered part depends on the compatibility of the sintered material with the impregnating resin. Therefore, the selection of the amorphous polymer used for the production of the functional part must take into account the compatibility with the impregnating resin in addition to the laser sintering properties of the material. PC is a kind of engineering plastic with excellent comprehensive performance. It has good thermal stability. Its apparent viscosity is dependent on temperature and it is easy to be formed after laser sintering. More importantly, PC and epoxy resin are polymer materials synthesized from the main raw material, bisphenol A, and have the same structural unit in their molecular structure, so that it is more compatible

TABLE 2.5 Basic performance of the PC.

Item	Value	Item	Value
Molecular weight ($\times 10^5$)	2.6–2.9	Tensile strength (MPa)	60
Density (g/cm^3)	1.18	Tensile modulus (MPa)	2130
Glass transition temperature T_g (°C)	145–150	Elongation at break (%)	85
Viscous flow temperature T_f (°C)	220–230	Bending strength (MPa)	95
Thermal decomposition temperature T_d (°C)	$>300 T_f$	Flexural modulus (MPa)	2100
Embrittlement temperature T_c (°C)	−100	Notched impact strength (kJ/m^2)	45
Specific heat capacity (J/g °C)	1.17	Heat distortion temperature (°C)	130
Thermal conductivity (W/(m K))	1.92×10^{-2}	Forming shrinkage (%)	0.5–0.8
Linear expansion coefficient (1/°C)	$(5-7) \times 10^{-5}$		

with epoxy resin than other amorphous polymers. Due to the high heat distortion temperature of PC, epoxy resin can be cured at a higher temperature, which is beneficial to improve performance, and its excellent toughness can make up for the shortcoming of large brittleness epoxy resin. The sintered part is expected to achieve excellent mechanical properties after being treated with epoxy resin. Therefore, PC is the most suitable sintering material for producing the functional part among the amorphous polymers. The basic performance of PC is shown in Table 2.5.

2.2.2.1.2 Selection of crystalline polymers

The crystalline polymer is laser-sintered to form a dense sintered part, and the performance of the sintered part is close to that of the molded part, so that it can be directly used as a plastic functional part. Since the properties of the polymer determine the properties that can be achieved by the sintered part, a high-performance polymer must be used as the sintered material in order to prepare a high-performance sintered part.

Commonly used crystalline polymers are PE, PP, nylon, thermoplastic polyester, polyoxymethylene (POM), etc. POM has poor thermal stability after processing, and is prone to thermal degradation and thermal oxidative degradation at sintering temperature, so it is not suitably used as the material for SLS-forming. PE has a high melt viscosity and mechanical properties which are not good, so it is also not suitably used as the plastic functional part for sintering. Nylon and thermoplastic polyester are general engineering plastics that have good comprehensive properties, good thermal stabilities, and low melt viscosity, which are favorable for laser sintering, so they are possible to become high-performance SLS-forming materials. Table 2.6 lists the properties of the major varieties of nylon and thermoplastic polyester.

As can be seen from Table 2.6, these resins have good mechanical properties and can meet the performance requirements of plastic functional parts. Nylon 6 and nylon 66 have high water absorption due to the high density of the amide group in the molecule. Water absorption destroys the hydrogen bond between the nylon molecules, and produces hydrolysis under high temperature that causes the molecular weight to decrease, thereby making the strength and modulus of the product significantly decreased, and the size changed a lot. The powder material has a large specific surface area and is more likely to absorb water, so the high water absorption rate is extremely disadvantageous for the powder sintering material, and the melting temperatures of the two nylons are high, and a high preheating temperature is required for sintering, which also brings big difficulty for sintering. Thermoplastic polyesters have very low water absorption rate, but their molding shrinkage ratio is large, forming precision is difficult to control, and their melting temperatures are also high. Among these resins, nylon 12 has the lowest melting temperature, low water absorption and mold shrinkage, and is perfect for use as a powder sintering material.

2.2.2.2 Stabilizer

The stabilizer used in SLS polymer materials is mainly antioxidant. The polymer powder material has a large specific surface area, and is susceptible to thermal oxidative degradation during the SLS-forming process, resulting in yellowing of the color and poor performance. The addition of antioxidant can solve the thermal oxidation problem of the polymer material, and at the same time prevent the thermal oxygen aging of the sintered part during use. Antioxidants are divided into two types: free radical trapping agents (also known as chain-stopping antioxidants) and hydroperoxide decomposers (also known as prophylactic antioxidants). The former function is to capture free radicals and make them not participate in the oxidative chain reaction cycle, and the latter function is to decompose the hydroperoxide so that it does not generate free radicals, which is usually used as auxiliary antioxidant. The chain-stopping antioxidants mainly include phenols and amines. The amine antioxidants have higher protective efficacy than phenols, but most amine antioxidants will undergo different degrees of discoloration after being exposed to light and oxygen, which are not suitable for light-colored

TABLE 2.6 Properties of nylon and thermoplastic polyester.

Property	Nylon 6	Nylon 66	Nylon 11	Nylon 12	Polyethylene terephthalate (PET)	Polybutylene terephthalate (PBT)
Density (g/cm^3)	1.14	1.14	1.04	1.02	1.4	1.31
Glass transition temperature (°C)	50	50	42	41	79	20
Melting point (°C)	220	260	186	178	265	225
Water absorption rate (%) (23°C, 24 h)	1.8	1.2	0.3	0.3	0.08	0.09
Forming shrinkage (%)	0.6–1.6	0.8–1.5	1.2	0.3–1.5	1.5–2	1.7–2.3
Tensile strength (MPa)	74	80	58	50	63	53–55
Elongation at break (%)	180	60	330	200	50–300	300–360
Bending strength (MPa)	125	130	69	74	83–115	85–96
Flexural modulus (GPa)	2.9	2.88	1.3	1.33	2.45–3.0	2.35–2.45
Notched impact strength of cantilever beam (J/m)	56	40	40	50	42–53	49–59
Heat distortion temperature (°C) (1.86 MPa)	63	70	55	55	80	58–60

products and less used in plastics. Preventive antioxidants mainly include phosphites, thioesters, etc. These antioxidants used together with the chain-stopping antioxidants often produce synergistic parts. Therefore, a composite antioxidant composed of a phenol and a phosphite or a thioester is used in the SLS polymer material.

In addition to antioxidants, light stabilizers are sometimes added to prevent photooxidation of polymer materials during processing and use.

2.2.2.3 Lubricant

Metal soap and metal salt such as calcium stearate and magnesium stearate may be used. Its main function is to reduce the mutual friction between the polymer powder particles, increase the fluidity of the processed materials, facilitate the powder spreading, and improve the thermal stability of the polymer powder.

2.2.2.4 Dispersant

Inorganic powders having a particle size of 10 μm or less, such as alumina, fumed silica, titanium dioxide, kaolinite, talcum powder, mica powder, and the like, are usually selected. Its main function is to reduce the agglomeration between the raw material particles, so that the polymer powder still has fluidity when the temperature is close to T_g (amorphous polymer) or T_m (crystalline polymer).

2.2.2.5 Selection and surface treatment of filler

There are many types of inorganic fillers, such as calcium carbonate, talcum powder, carbon black, kaolinite, wollastonite, mica, glass beads, aluminum hydroxide, and titanium dioxide. Different fillers have different geometries, particle sizes, and distribution, as well as different physicochemical properties that directly affect the properties of the filled polymer material. The strength of the filled polymer depends on the transfer of stress between the filler particles and the polymer matrix. If the applied stress can be effectively transferred from the matrix to the filler particles, and the filler particles can withstand a portion of the applied stress, the addition of the filler can increase the strength of the matrix, which would otherwise reduce the strength of the material. In general, fillers of large aspect ratio such as needles, fibers, and the like are advantageous to increase the strength of the matrix polymer, and the spherical filler is advantageous to improving the processability. Rigid particles of large particle size tend to form defects in the matrix, causing stress concentration, which reduces the mechanical properties of the material, the finer the particle size, the fewer surface defects, the better the interface bonding, and the better the mechanical properties of the filler material. However, the smaller the particle size, the larger the specific surface area, the higher the cohesive energy of the particles, and the easier it is to agglomerate, the more difficult it is to achieve uniform dispersion. When the specific surface area is constant, the larger the surface energy of the filler, the particles are more likely to agglomerate with each other, and the more difficult to disperse. The SLS-forming of polymer powder is completely different from other forming methods of polymer materials. Therefore, the influence of filler on the sintering processability of polymer powder and the performance of sintered parts should be comprehensively investigated when selecting the filler.

The molecular structure, physical form and surface properties of the inorganic filler are very different from those of the organic polymer. The two materials cannot be tightly bonded together, which directly affects the performance of the composite materials. Therefore, the surface treatment of the filler is required. A coupling agent is a kind of substance having an amphoteric structure, in which a part of the radical group in its molecule can react with the chemical radical group on the surface of the inorganic substance to form a strong chemical bond; and another part of the radical group has the organophilic property, and can react with or physically entangle with organic molecules, which combines two materials with very different properties. The most commonly used coupling agent is silane coupling agent.

The silane coupling agent has a structural equation of $RSiX_3$, and R refers to the reactive functional group having affinity or reaction ability with polymer molecule, such as amino group, mercapto group, vinyl group, epoxy group, cyano group, methacryloyloxy group or the like. X refers to the hydrolyzable alkoxy group or chlorine. The silane coupling agent is particularly suitable for fillers containing a large amount of silicic acid components such as glass microbeads, quartz powder, white carbon black, wollastonite or the like.

The type of the silane coupling agent should be selected according to the matrix polymer, and the silane coupling agent containing amino group such as γ-aminopropyltriethoxysilane (KH-550) can be selected for the nylon matrix material. When the filler is treated with KH-550, the ethoxy group is first hydrolyzed to form silanol, and then reacted with hydroxyl group on the surface of the filler, and the equation is as follows.

$$H_2N(CH_2)_3Si(OC_2H_5)_3 + 3H_2O \longrightarrow H_2N(CH_2)_3Si(OH)_3 + 3C_2H_5OH \quad (2.68)$$

$$H_2N(CH_2)_3Si(OH)_3 + HO-\text{Filler}-OH \longrightarrow H_2N(CH_2)_3Si(O-\text{Filler})(OH) + 2H_2O \quad (2.69)$$

The amino group in the KH-550 molecule reacts with the carboxyl group in the nylon molecule.

$$\text{filler}\diagup\!\!\diagdown\!\!\overset{OH}{\underset{|}{Si}}(CH_2)_3NH_2 + HOOC(CH_2)_{11}NH\sim\sim\sim \longrightarrow$$

$$\text{filler}\diagup\!\!\diagdown\!\!\overset{OH}{\underset{|}{Si}}(CH_2)_3NH\overset{O}{\overset{\|}{C}}(CH_2)_{11}NH\sim\sim\sim + H_2O \quad (2.70)$$

In this way, KH-550 is used as a bridge to firmly bond nylon and inorganic filler to form a composite material with excellent performance.

When the silane coupling agent is used to treat the filler, the theoretical amount can be calculated according to Eq. (2.71):

$$\text{Amount of silane (g)} = \frac{\text{Filler mass(g)} \times \text{filler specific surface area} (m^2)g}{\text{Wettable area of silane } (m^2g)} \quad (2.71)$$

The wettable area of KH-550 is 353 m²/g, and the specific surface area of the filler is substituted into Eq. (2.71) to determine the amount of KH-550. However, since the specific surface area of the filler is difficult to measure, it is difficult to accurately calculate the amount of KH-550. In practice, the filler is treated with silane coupling agent at a concentration of 1 wt.%. For most fillers, this amount is higher than the theoretical amount. The processing method is as follows:

The alcohol-water solution is prepared by using 95 wt.% ethanol and 5 wt.% water, and the silane coupling agent is added under stirring to obtain a mass fraction of 2%. After hydrolysis for 5 minutes, the hydrolyzate containing Si-OH is formed. The filler to be treated is added, stirred uniformly, naturally dried at room temperature for 1−2 days, and then dried in an oven at 60°C for 2 hours, and the dried filler is ground and sieved.

2.2.2.6 Other additives

Additives such as reinforcing agents, antistatic agents, and pigments may also be added as needed.

2.2.3 Characterization of selective laser sintering polymer materials

2.2.3.1 Size and shape of powder particles

The SLS process has high requirements for the particle size and shape of the powder material. The particle size affects the sintering speed and the dimensional accuracy and appearance quality of the sintered part, and the powder shape affects the fluidity of the powder. The polymer powders obtained by different methods have different shapes and sizes. Fig. 2.26 shows the microscopic shape of the polymer powder observed by scanning electron microscopy.

Fig. 2.26A shows the PC powder obtained by interfacial polycondensation, which has an extremely irregular particle shape and a rough surface, and has a wide particle size distribution. Fig. 2.26B shows nylon 12 powder prepared by solvent precipitation method, the particles of which are close to spherical type, and the particle size is mainly concentrated between 40 and 70 μm. Such powder has a high fluidity and is favorable for laser sintering.

2.2.3.2 Bulk density of powder

The bulk density of polymer powder is related to factors such as the shape of the powder, the size and distribution of the particles, and the type of the polymer material. In the SLS process, the powder bulk density affects the heat transfer during sintering and the densification of the sintered part, and the powder with a large bulk density is advantageous for increasing the density of the sintered part.

Table 2.7 shows the bulk density of several polymer powder sintered materials.

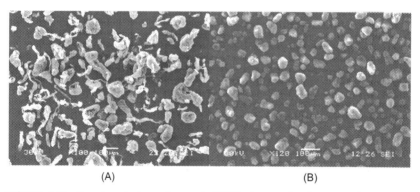

FIGURE 2.26 Scanning electron micrograph (SEM) pictures of polymer powders: (A) PC powder; (B) nylon 12 powder.

TABLE 2.7 Bulk density of polymer powder sintered materials.

Powder material	PC	Nylon 12	Nylon 12 added with 30 wt.% glass beads	Nylon 12 added with 30 wt.% talcum powder	Nylon 12 added with 30 wt.% wollastonite
Bulk density (g/cm³)	0.18	0.48	0.59	0.57	0.60

TABLE 2.8 Whiteness of polymer powder sintered materials.

Powder material	PC	Nylon 12	Nylon 12 added with 30 wt.% glass beads	Nylon 12 added with 30 wt.% talcum powder	Nylon 12 added with 30 wt.% wollastonite
Whiteness	93.5	96.8	90.9	93.9	91.8

2.2.3.3 Powder whiteness

The polymer material is prone to thermal aging under heating conditions, the color gradually turns yellow, the whiteness decreases, and the more severe the oxidation, the lower the whiteness value, so the whiteness can reflect the degree of thermal oxygen aging of the polymer material. The polymer material may appear thermal oxygen aging in the process of preparation and laser sintering. The oxidation rate is accelerated with the increase of temperature. When the nylon 12 powder is prepared by the solvent precipitation method, the nylon 12 has a risk of thermal oxygen aging due to high preparation temperature and long time. The whiteness of nylon 12 powder is an important indicator of control. The whiteness of the polymer powder sintered material is measured by ZBD whiteness meter according to GB2913-1982, as shown in Table 2.8.

In short, micron-sized polymer powder materials can be prepared by low-temperature pulverization method, solvent precipitation method, etc., and mixed with stabilizers, lubricants, dispersants, fillers and the like to obtain polymer powder materials that is suitable for SLS. Polymer powder is the main factor affecting the performance of sintered parts. PC and nylon 12 are the most suitable polymer materials for the production of functional parts, such as amorphous polymers and crystalline polymers. Stabilizers, lubricants, and dispersants are usually indispensable additives in polymer powder sintered materials. The filler can improve the sintering processability of the polymer powder sintered material and the performance of the sintered part, and reduce the material cost. To enhance the interface between the filler and the polymer material, the filler must be conducted with surface treatment.

2.3 Nylon 12 powder materials

Nylon 12 powder SLS part features high strength, good toughness, stable size and no need for postprocessing, which has become the material of first choice for plastic functional part of 3D printing and manufacturing, and has been widely used in the world. Both American 3D Systems Company and German EOS Company have successively launched special nylon 12 laser-sintered powder materials. Due to the strict requirements of laser sintering on nylon powder, it is required to have small particle size, narrow particle size distribution, and flat and smooth powder surface. The geometric shape is approximately spherical. Therefore, the price of nylon powder for SLS-forming is very expensive, which is difficult for domestic users to accept. Moreover, since the nylon powder for SLS is jointly developed by the nylon manufacturer and the SLS equipment manufacturing company, it is not sold outside. In order to realize the direct sintering of functional part by nylon powder, it is imperative to independently develop domestic nylon powder for SLS.

2.3.1 Preparation process of nylon 12 powder materials

At present, SLS-specific nylon 12 powder is mainly prepared by solvent precipitation method.

2.3.1.1 Selection of solvents

Alternative solvent systems include methanol, ethanol, ethylene glycol, dimethyl sulfoxide, nitroethanol, ε-caprolactam, etc. Cui Xiulan et al. studied the parts of different solvent systems on the preparation of nylon powder, including four solvent systems of diethylene glycol, diethylene glycol-water, ethanol-calcium chloride and ethanol-hydrochloric acid, comprehensively studied the properties of the powder, and found that the powder shape and particle size prepared by solvent precipitation method have a great relationship with the solvent. The average particle size of the powder prepared by using alcohol as solvent is 43 μm, the particle size of diethylene glycol-water system is the smallest under the same solvent dosage, only about 17 μm, and the particle size of ethanol-calcium chloride system is 37 μm. The average particle diameter of the ethanol-hydrochloric acid system is about 66 μm. In addition, it was found that the powder particles prepared by the ethanol-calcium chloride system have a porous structure, and the powder prepared by the ethanol-hydrochloric acid system has better thermal stability at 230°C or higher.

Among the above solvent systems, methanol, dimethyl sulfoxide, nitroethanol, and the like are relatively toxic, and are not suitable for manual operation; solvents such as diethylene glycol have a high boiling point, which is unfavorable for solvent recovery and high in price. The powder prepared by the ethanol-calcium chloride system has uniform particle size and moderate particle diameter, but the prepared powder has a porous structure, which is disadvantageous for SLS-forming and has certain corrosion to the equipment. The powder prepared by the ethanol-hydrochloric acid system has good stability, but the corrosion of the equipment is more serious. In addition, hydrochloric acid easily volatilizes hydrogen chloride gas, which is toxic and irritating, so the method is poor. In order to avoid these shortcomings, Ding Shuzhen et al. studied another alcohol solution system. The particle diameter of the nylon powder produced by this method can be controlled at 53−75 μm, but the powder color is yellow and the melting point is above 200°C.

TABLE 2.9 Influence of different moisture contents on powder particle size [dissolution temperature: 145°C; m (nylon): m (solvent) = 1:5].

Moisture content of solvent (%)	0.3	0.5	1	2	5
Average particle size (μm)	53.5	56.7	78.8	125	>500

Ethanol is an excellent solvent having a low toxicity and low irritation, which is low in price and easy to recycle, so it is the solvent system of first choice. Powder with a particle size distribution of 30–50 μm is most advantageous for SLS-forming. Because during the SLS-forming process, the small particle size will make powder fluffy, and decrease the bulk density, and it is easy to adhere to the power spreading roller, which is disadvantageous for the flattening of the powder. When the particle size is too coarse, the forming property is deteriorated, and the surface of the part is rough.

Studies have shown that the nylon powder prepared by the ethanol system is difficult to meet such requirements. Generally, the average particle size is about 75 μm. Therefore, the fine nylon (average particle size of 40 μm) introduced by 3D Company is mainly obtained by sieving. This method is featured by low yield and high cost. Although the particle size of the powder can be lowered by decreasing the solute-solvent ratio, the reported solute-solvent ratio is very low, ranging from 1:10 to 1:20, so it is uneconomical to reduce the ratio. Solvent polarity has a significant part on the particle size of the powder. For example, the presence of moisture in the solvent increases the particle size of the powder. Table 2.9 shows the influence of different moisture contents (mass fraction) on the particle size of the powder.

It can be seen from Table 2.9 that as the moisture content in the solvent increases, the particle size of the prepared nylon 12 powder increases rapidly. In the experiment, the moisture content in the solvent should be controlled below 0.5%, and the maximum should not exceed 1%. It has been found that the addition of a weakly polar solvent such as 2-Butanone or diethylene glycol to the solvent is advantageous for reducing the particle size of the powder. Therefore, the solvent used has not more than 10% butanone and diethylene glycol in addition to ethanol, and the purpose of preparing powders of different particle diameters can be achieved by adjusting the solvent ratio.

2.3.1.2 Dissolution temperature

Solvent precipitation is used to prepare nylon 12 powder, which is essentially the process of dissolving nylon 12 at high temperature and precipitating at low temperature. Therefore, temperature control plays an important role in the preparation of nylon 12 powder. Solvent precipitation method must ensure complete dissolution of nylon 12, the higher the temperature, the longer the dissolution

TABLE 2.10 Influence of dissolution temperature and time on powder.

Dissolution temperature (°C)	Dissolution time (h)	Powder fineness	Color
130	8	Coarse, larger than 500 μm	White
135	4	Relatively coarse, larger than 200 μm	White
135	8	Relatively coarse, larger than 100 μm	White
140	1	Relatively coarse, larger than 80 μm	White
140	2	Fine	White
145	1	Fine	White
150	0.5	Fine	White
150	4	Fine	Light yellow
170	1	Fine	Light yellow

time, the more favorable the dissolution of nylon 12, the finer the powder produced; in contrast, the lower the dissolution temperature, the shorter the dissolution time is, the more incomplete dissolution of nylon 12, the coarser the particle size of the produced powder. However, nylon 12 will oxidize and degrade at high temperature, which has an adverse part on the performance of nylon 12. Therefore, a lower dissolution temperature and time should be used on the premise of ensuring dissolution. Table 2.10 shows the influence of dissolution time and temperature on powder particle size and the color.

According to the above experiment, the dissolution temperature is selected from 140°C to 145°C, and the dissolution time is 2 hours.

2.3.1.3 Cooling method and speed

The cooling method and speed have significant part on the precipitation of the powder. The influences of several cooling methods are as follows [3].

2.3.1.3.1 Natural cooling

The cooling rate of natural cooling is related to the ambient temperature. When the temperature is low, a faster cooling rate can be obtained, and when the temperature is high, the cooling rate is slow. When the nylon 12 precipitates and crystallizes, the crystallization enthalpy will be released to increase the temperature of the system. Therefore, the precipitation temperature of the nylon 12 can be judged according to the temperature transition. Fig. 2.27 shows the temperature drop curve at an ambient temperature of 13°C, as shown in the figure, when the precipitation and crystallization temperature is 106°C, the cooling rate is 26°C/h. The enthalpy released during crystallization raises the temperature of the entire system by more than 1°C, it can conclude that the crystallization enthalpy is very large.

The powder produced by the natural cooling method is not uniform in size and irregular in shape (as shown in Fig. 2.28), and the SLS-forming property of the powder is not good.

Further research found that the particle size and distribution of the nylon 12 powder produced by the natural cooling method are greatly affected by the ambient temperature. The higher the temperature, the slower the cooling rate, the finer the nylon 12 powder produced, but the particle size distribution is wider, and the content of fine powder (less than 10 μm) with irregular geometry is increased. Fig. 2.29 shows the cooling curve with an ambient temperature of 31°C, the cooling curve with a cooling rate of 19°C/h and the micro photo of prepared nylon 12 powder.

It can be seen from Fig. 2.29B that the powder prepared at a high temperature has a wide particle size, the shape of the powder becomes more

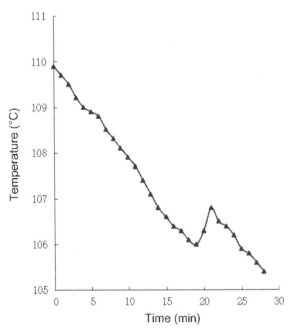

FIGURE 2.27 Curve showing temperature in the kettle changing with time (natural cooling, room temperature: 13°C).

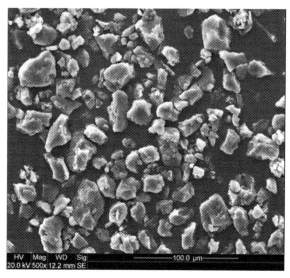

FIGURE 2.28 Picture of nylon 12 powder (natural cooling, room temperature: 13°C).

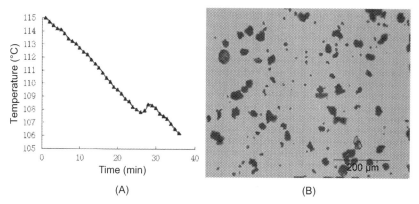

FIGURE 2.29 Curve of temperature in the kettle changing with time and picture of nylon 12 powder (natural cooling, room temperature: 31°C): (A) cooling curve; (B) picture of powder.

irregular, and a large part of the powder has a particle size of 10 μm or less, and the fluidity of the powder is not good, which is easy to agglomerate, and is difficult to disperse after drying. SLS has large shrinkage during forming and is easy to warp and deform.

2.3.1.3.2 Cooling directly through the cooling water

Natural cooling cannot control the cooling rate, especially when the room temperature is high, due to the slow cooling rate, the crystallization enthalpy during precipitation cannot be taken away in time, the temperature of the solution system rises obviously, which is unfavorable to the powder geometrical shape and particle size. In order to obtain a faster cooling rate, the cooling coil in the kettle is used to cool.

However, when supplying water in the cooling coil, it is found that the nylon 12 is completely precipitated surrounding the cooling coil, and the innermost layer of the cooling coil is a layer of nylon 12 film, followed by obviously coarse powder, and then the powder becomes finer gradually powder from the inside to the outside. From this, it is understood that as the cooling rate increases, the particle size of the powder increases.

2.3.1.3.3 Cooling oil temperature by supplying cooling water in the jacket

When supplying water in the cooling coil in the kettle directly, it produces excessive temperature difference and is not conducive to the formation of powder, so the method of cooling the oil temperature in the jacket is used.

However, after supplying cooling water in the jacket, it is found that the oil temperature becomes uneven, and a layer of nylon 12 film also appears in the center of the autoclave. Moreover, it is difficult to accurately control the temperature in the kettle through the oil temperature. Although the geometric shape of the powder is approximately spherical, the particle size is relatively coarse, most of which is above 70 μm, and even more than 100 μm. The fluidity of the powder is good, but the SLS-forming performance is poor (Fig. 2.30).

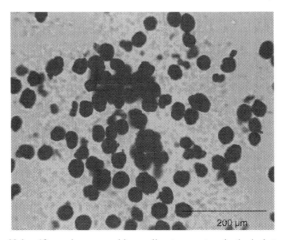

FIGURE 2.30 Nylon 12 powder prepared by cooling temperature in the jacket.

2.3.1.3.4 Cooling outside the kettle and distillation cooling

In order to obtain nearly spherical nylon 12 powder having a narrow particle size distribution, it is necessary to strictly control the cooling rate at the time of precipitation and crystallization, in particular, to quickly remove the crystallization enthalpy, so as to prevent the temperature from rising during precipitation and crystallization, and ensure the uniformity of the temperature of the system. To this end, the fan is turned on when cooled to near the precipitation temperature, and the heat is taken away by air convection. After adding the air blow, the temperature rising range decreases during the cooling process (as shown in Fig. 2.31A), the particle size distribution of the powder is narrowed, and the geometrical shape of most powder is nearly spherical, but there are still some irregular powders, as shown in Fig. 2.31B.

Although the amount of powder of irregular particle size is reduced after cooling with a fan, the particle size is not uniform, and there are still some fine-grained powders with irregular geometrical shapes. It has been found through a large number of experiments that the nylon 12 with temperature rise during the precipitation process has a very adverse part on the regularity of the geometrical shape of powder. Obviously, it is impossible to take away the crystallization enthalpy that is suddenly released by air convection. Therefore, the cooling water is directly sprayed outside of the autoclave. The specific method is: when the temperature is cooled to near the precipitation temperature, the water is cool sprayed to the outer surface of the kettle for cooling until the precipitation is completed. After using this method, the temperature rise range during the precipitation and crystallization process is effectively reduced to less than 0.5°C. After the completion of the precipitation, the reaction vessel is opened to find that a large amount of powder is adhered to the inner surface of the lid and the particle diameter is coarse. For this reason, the filling amount of the kettle is reduced. When the filling

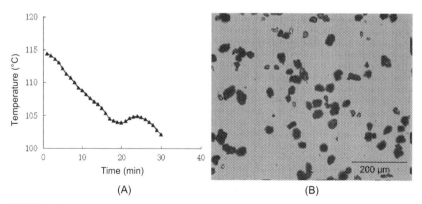

FIGURE 2.31 Cooling curve of forced convection cooling outside the kettle and prepared nylon 12 powder (room temperature: 25°C): (A) cooling curve; (B) picture of powder.

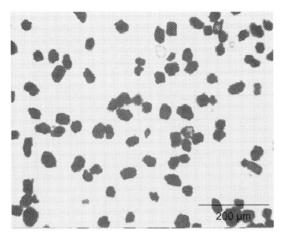

FIGURE 2.32 Nylon 12 powder prepared when the cooling rate is strictly controlled.

amount is less than 70% of the volume in the kettle, no powder appears on the inner surface of the kettle lid, which may be due to the fact that the liquid does not come into contact with the kettle lid during stirring. The prepared powder has a relatively uniform particle size, and the fine powder basically disappears, but the powder has a coarse particle diameter and an average particle diameter of 55 μm, as shown in Fig. 2.32.

The above method can well solve the cooling problem of the small kettle, but for the large reaction kettle, because of the large heat capacity, it is impossible to achieve heat transfer by forced convection through the outside air. Previous experiments have shown that heat transfer through direct contact of the cryogenic liquid with the kettle body is detrimental to the powdering process. The latent heat of evaporation of the solvent is very large, and a large amount of heat is absorbed during evaporation, so that the evaporated solvent can be used to cool down. This method does not cause local low-temperature phenomenon, and can maintain relative stable temperature of the kettle body, so as to control the temperature drop rate by controlling the evaporation amount of the solvent. The specific method is: opening the distillation valve when cooling, adjusting the evaporation amount of the solvent to achieve the appropriate cooling rate, when the nylon 12 begins to precipitate, the temperature rises due to the heat released from crystallization, at this time, increase the solvent evaporation to keep the temperature rise does not exceed 0.5°C until the end of precipitation.

2.3.1.4 Stirring

The particle size distribution and its range are related to the stirring speed. Table 2.11 shows the particle size and distribution of nylon 12 powder at different stirring speeds.

TABLE 2.11 Particle size and distribution of powder at different stirring speeds.

Stirring speed (rpm)	500	600	700
$D_{0.5}$ (μm)	75	66	53
$D_{0.1-0.9}$ (μm)	83	67	40

As can be seen from Table 2.11, as the stirring speed increases, the particle size of the powder decreases and the particle size distribution becomes narrow. Therefore, a higher stirring speed should be selected as much as possible.

2.3.1.5 Nucleation during powder precipitation

The nylon 12 powder is prepared by solvent precipitation method. The movement is random when the macromolecular chain is in a dissolved state. During the cooling process, as the temperature decreases, the movement of the molecular chain is gradually limited. When saturated, an infinite number of ordered crystals aggregated by several segments are formed in the solution, but the crystal nucleus is too small and the solution is still in a supersaturated state, the solution is transparent, so that the nylon 12 does not precipitate. As the temperature continues to decrease, the size of the crystal nucleus becomes larger and larger, and once the crystallite size reaches a critical value, it is stably present, thereby forming a crystal nucleus. At this point, nylon 12 begins to precipitate in quantity around these crystal nuclei.

Above is the mechanism for homogeneous nucleation. During the precipitation process, homogeneous nucleation and heterogeneous nucleation are simultaneously present. For example, nylon 12 precipitated centering on the condensation coil and the inner wall of the kettle is called heterogeneous nucleation. When directly cooled by water, the nylon 12 is deposited around the walls due to the lower temperature of the coil and the inner wall of the kettle, that is, a film of nylon 12 is formed. When directly cooled by water, the temperature is lower as closer to the wall, that is, a temperature gradient is formed, and thus the powder also exhibits a gradient precipitation. Therefore, in order to obtain a uniform powder, the jacket temperature should be kept slightly higher than the temperature inside the kettle, and the cooling coil in the kettle should be removed.

In the absence of an additional nucleating agent, the precipitation and crystallization of nylon 12 is dominated by homogeneous nucleation. Therefore, the formation of crystal nucleus becomes the key to control the particle size and distribution of the nylon 12 powder.

To prepare the powder of uniform particle size, the nucleus before precipitation must be uniform. The temperature is directly cooled from the dissolution temperature to the precipitation temperature. After the temperature drops to the saturation temperature of the solution, the nucleus begins to appear until the end of the precipitation. During this period, as the time increases and the temperature decreases, the number of crystal nucleus continuously increases and grows. Therefore, the size of the crystal nucleus appearing at different stages is different. The crystal nucleus that appears first has coarse particles due to sufficient growth time. Due to the complete growth, the surface is smooth and the geometrical shape is regular. The crystal nucleus that appears afterwards is incomplete, so the size of powder particles is small and the shape is irregular.

In order to obtain a uniform crystal nucleus, a nucleation stage of 0.5−1 hour can be maintained at a certain temperature before actual precipitation. Table 2.12 shows the influence of different nucleation temperatures on the preparation of powder.

Through further experiments, it is found that the nucleation temperature is controlled at 120°C−122°C, and the nucleation stage of half an hour can achieve good results. The prepared powder has uniform particle size and regular geometrical shape, and most of the particle size can be controlled at 30−50 μm.

The formation mechanism of the nucleus can also explain the appearance of fine powder during precipitation. Because at a higher temperature, the thermal motion of the molecule is too intense, the crystal nucleus is not easy to form, or the generated crystal nucleus is unstable, and is easily destroyed by molecular thermal motion. As the temperature decreases, the rate of homogeneous nucleation gradually increases, so the slower the cooling rate, the more crystal nucleus formed, and the finer the powder. When precipitated and crystallized, the crystal nucleus formed in different periods is different in degree of perfection. Excessive crystal nuclei interact with each

TABLE 2.12 Influence of different nucleation temperatures on nylon 12 powder.

Nucleation temperature (°C)	Results
130	No change
125	Finer particle size, regular geometrical shape
120	Finer particle size, most of them are regular and, but there are still some fine powder
115	Fine particle size, but there are some fine powder
110	Fine particle size, almost all are fine powder with irregular geometrical shape

other, and many imperfect crystal nuclei may aggregate with each other. Therefore, the geometrical shape of the powder is irregular and the particle size distribution is widened. If the heat released during precipitation and crystallization is not taken away in time, the temperature will rise, during which a large number of crystal nucleus will be produced, resulting in a large number of irregular fine powder.

The above studies show that the particle size and distribution of the powder can be improved by adding a nucleation stage during the precipitation process, but the particle size distribution of the powder prepared by this method is still wide, especially the powder of fine particle size and regular geometrical shape cannot be obtained at the same time. The powder cannot meet the needs of SLS for nylon powder. It must be sieved before use. For this purpose, an additional nucleating agent is used to adjust the particle size and distribution of the powder.

There are many nucleating agents for nylon 12. Commonly used nucleating agents include silica (SiO_2), colloidal graphite, lithium fluoride, boron nitride, aluminum borate, and some polymers. The ordinary inorganic material has coarse particle size and is almost equivalent to the particle size of the prepared powder, which is not suitable as a nucleating agent. The fumed silica has small particle size and rapidly swells and disperses after being dissolved in ethanol. Therefore, fumed silica is selected as a nucleating agent. 0.1 wt.% fumed silica is added during the precipitation process. Subsequent experiments showed that the natural cooling and precipitation occur after adding silica, but the particle size and geometrical shape of the powder are not improved. When the nucleation stage is maintained for a period of time, the particle size distribution and geometrical shape of the powder are improved. It may be that the particle size of the fumed silica is too small to reach the size of the crystal nucleus when directly precipitated, but the fumed silica promotes the formation of the nucleus, that is, the nylon 12 may form a crystal nucleus around the fine fumed silica. So the crystal nucleus formed is more uniform and stable. However, if there is no nucleation stage, nylon 12 nucleates at low temperature. Because of the lower temperature, the homogeneous nucleation speed is faster, so the part of fumed silica is small. Fig. 2.33 shows the nylon 12 powder prepared after the addition of fumed silica and undergoing a nucleation stage.

The mass fraction of the nucleating agent has a significant influence on the particle size of the powder. As the mass fraction of the nucleating agent increases, the particle size of the powder becomes finer, but the regularity of the geometric shape deteriorates. When fumed silica is used as a nucleating agent, the viscosity of the solution increases remarkably after the mass fraction exceeds 1%, and the bulk density of the prepared powder rapidly decreases. The powder absorbs a large amount of solvent due to large specific surface area, resulting in failure to discharge. Therefore, the mass fraction of silica should be controlled within 1%.

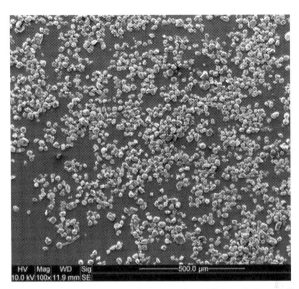

FIGURE 2.33 Nylon 12 powder prepared after nucleation of 0.1 wt.% fumed silica.

2.3.1.6 Influence of thermal history on the preparation of powder

The thermal history has a great influence on the powder. The prepared powder is added to the reaction kettle and powdered several times. The prepared powder is shown in Fig. 2.34. It can be seen that the particle size distribution of the powder is widened, and not only the powder of irregular shape appears, but also the cracks appear in some powder particles.

2.3.1.7 Postprocessing of nylon powder

The prepared nylon 12 powder slurry is separated into a solvent in a centrifuge, and then dried in a double cone vacuum dryer, and sieved through a ball mill to obtain the desired nylon 12 powder. The nylon 12 powder need to be vacuum-dried at 70°C for 4 hours before use.

2.3.2 Selective laser sintering process characteristics of the nylon 12 powder

2.3.2.1 Influence of thermal oxygen stability of nylon 12 powder on selective laser sintering process

In the laser sintering process, the nylon 12 powder has high preheating temperature and large specific surface area, the thermooxidative aging is serious. After sintering and using the nylon 12 powder without antiaging treatment one time, the sintered part and the powder in the intermediate working

74 Materials for Additive Manufacturing

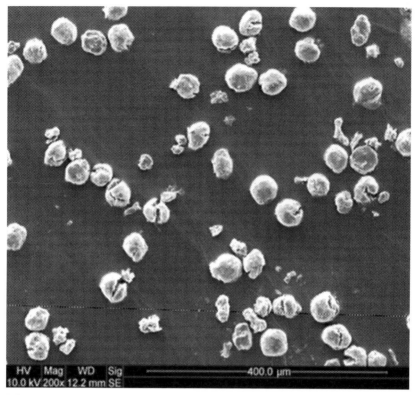

FIGURE 2.34 Nylon 12 powder prepared by repeated heating.

cylinder become yellow, which not only affects the appearance quality of the sintered part, but also has a great influence on its physical and mechanical properties, and the yellow powder cannot be reused due to the decrease in formability, which greatly increases the material cost. Therefore, it is necessary to conduct an in-depth study on the thermooxidative stability of nylon 12 to reveal its thermooxidative aging mechanism and its influencing factors, and then study its stabilization method to improve the recycling times of nylon 12 powder.

The thermogravimetric (TG) curve measured by the temperature rise of nylon 12 at different speeds in N_2 is shown in Fig. 2.35. As can be seen from Fig. 2.35, in the N_2 atmosphere, nylon 12 has high stability and almost no mass loss at 350°C. The thermal degradation residue when heated to 550°C is only about 1%, indicating that the thermal degradation of nylon 12 mainly produces volatiles, and rarely produces crosslinked structure, which is quite different from the thermal degradation of nylon 6.

For the thermooxidative aging mechanism of nylon, a lot of research has been done by the predecessors. Although the mechanism is still not very

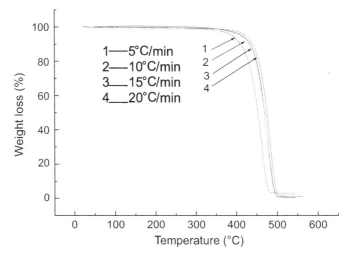

FIGURE 2.35 TG diagram of nylon 12.

clear, a variety of nylon antiaging equations have been developed and can be used as a reference. Since the antiaging research on nylon is mostly for molding, the antiaging agent can be well mixed with the melt in the molding to achieve the purpose of antiaging, but for the nylon powder prepared by solvent precipitation method, the antiaging agent will remain in the solution, so its antiaging properties are greatly reduced.

The simplest way to antiaging of nylon 12 powder is to add an antioxidant to the nylon 12 powder. In order to uniformly disperse the antioxidant in the nylon 12, the antioxidant may be dissolved in a solvent, and then blended with the nylon 12 powder, followed by vacuum drying. The aging test is carried out in an oven, and the nylon 12 powder and the nylon 12 powder treated by the antiaging are aged in an oven at 150°C for 4 hours.

The most important characteristic of nylon 12 aging is the yellowing of the color and the decrease of mechanical properties. Therefore, the part of the antiaging agent is determined by color change and mechanical property test. The antiaging parts of different antioxidants on nylon 12 powder are shown in Table 2.13.

It can be seen from Table 2.13 that although a variety of nylon antiaging schemes, which are considered to be ideal, have been tested, the results are not satisfactory, and the antiaging properties are hardly changed (the color of samples containing KI and nantokite is red due to the parts of iodine and copper). Subsequent SLS-forming experiments also confirmed this. The prepared SLS samples are yellow to red in color, and the powder used in the secondary cycle is warped severely during SLS-forming.

Poor antiaging properties of the powder may be related to the dispersion of the antioxidant. Although the antioxidant has been dissolved in the solvent, it

TABLE 2.13 Influence of antioxidants on antiaging properties of nylon 12 powder.

No.	Antioxidant	Color before aging	Color after aging
1	None	White	Reddish yellow
2	1098	White	Yellow
3	1010	White	Yellow
4	DNP	Slight dark green	Yellow
5	m(1098): m(168) (1:1)	White	Yellow
6	m(1098): m(168) (3:1)	White	Yellow
7	m(1010): m(168) (1:1)	White	Yellow
8	$CuCl_2$	Light green	Reddish yellow
9	KI	White	Yellow
10	CuI	White	Red
11	m($CuCl_2$): m(KI) (1:10)	Red	Red
12	$CuCl_2$/KI/$K_3P_2O_6$	Red	Red
13	1098/168/$CuCl_2$/KI/$K_3P_2O_6$	Red	Red
14	1098/168/KI/$K_3P_2O_6$	White	Yellow

can only be in contact with the surface of the powder, and a good degree of antiaging part must be achieved at a molecular level. Therefore, the addition of antioxidants during process of powder preparation is shown in Table 2.14.

It can be seen from Table 2.14 that after the addition of the antioxidant, the powder containing nantokite is red, indicating that the nantokite has decomposed. While the surface of the powder containing KI is yellow when naturally dried, and disappears after heating and drying, indicating that iodine is formed. The antiaging property of the antioxidant 1098 and the powder containing KI/$K_3P_2O_6$ is improved, but it is still unsatisfactory. Although the powder after SLS-forming does not turn yellow, the sample is yellow. This is widely different from the reported part, probably because most of the antioxidants do not work in the solvent. The powder containing antioxidant 1098 has certain antiaging properties, probably because the amide structure of the antioxidant 1098 is similar to that of nylon and can be precipitated together with the nylon 12. Since the color of the powder containing nantokite is red, the antioxidant properties cannot be judged by its color, but red is not conducive to the absorption of heat of the infrared heating tube, which is disadvantageous for SLS-forming.

It can be seen from Table 2.15 that the four-component antioxidant system consisting of 1098, 168, KI, and $K_3P_2O_6$ has good antiaging effect, and the

TABLE 2.14 Antiaging properties of nylon 12 powder with antioxidant added during the powdering process.

No.	Antioxidant	Powder color	Color of sintered sample
1	None	White	Light yellow on upper surface, red on lower surface
2	1098	White	White on upper surface, light yellow on lower surface
3	DNP	Dark green	Dark black
4	m(1010): m(168) (1:1)	White	Light yellow on upper surface, red on lower surface
5	m(1098): m(168) (1:1)	White	White on upper surface, light yellow on lower surface
6	$CuCl_2$	Red	Red
7	CuI	Red	Red
8	$CuCl_2$/KI/ $K_3P_2O_6$	Red	Red
9	KI/$K_3P_2O_6$	Yellow on upper surface, and disappear after heating	White on upper surface, light yellow on lower surface
10	1098/168/ KI/$K_3P_2O_6$	Yellow on upper surface, and disappear after heating	White on upper surface, light yellow on lower surface

mechanical properties of the sample are obviously improved compared with no addition of antioxidant. The system in which the antioxidant is not added cannot be carried out in the secondary cycle, and the system in which the antioxidant is added can perform the secondary cycle. It can be seen from Table 2.15 that the influence of aging on tensile strength is small and impact strength is large, so aging mainly makes the material brittle.

The aging of nylon 12 not only affects the mechanical properties and color of the part, but also has a significant part on the laser sintering performance. The aging of nylon is mainly caused by thermal oxidation crosslinking and degradation. Crosslinking will increase the melting point and the viscosity of the polymer. For example, when nylon 66 is heated in air at 260°C for 5–10 minutes, nylon 66 becomes insoluble and infusible. Crosslinking causes a significant increase in the melt viscosity of the nylon during laser sintering and an increase in the temperature required for sintering. Oxidative degradation produces part of oligomers, when the melting point of the oligomers

TABLE 2.15 Influence of antioxidants on mechanical properties of nylon 12 SLS samples.

Antioxidants	Mechanical properties					
	Primary laser sintering		Secondary laser sintering		Tertiary laser sintering	
	Tensile strength (MPa)	Impact strength (kJ/m^2)	Tensile strength (MPa)	Impact strength (kJ/m^2)	Tensile strength (MPa)	Impact strength (kJ/m^2)
None	41.5	23.6	Forming failure			
1098/168	42.2	36.2	41.7	28.5	41.3	20.1
KI/K$_3$P$_2$O$_6$	43.1	35.3	42.4	29.6	40.5	21.3
1098/168/KI/K$_3$P$_2$O$_6$	44.5	37.2	42.3	33.6	40.8	26.9

decreases, the crystallization rate increases, and a large amount of spherulites are formed during crystallization, which increases the shrinkage of the polymer and lowers the strength of the polymer.

The SLS of aged nylon 12 powder is formed to be easy to agglomerate, difficult to melt, poor in fluidity, and easy to warp. The nylon 12 powder used for multiple cycles requires higher laser energy to completely melt it. Even when laser scanning is performed while the nylon 12 is agglomerated, the sintered body still warps. Therefore, aging is very unfavorable for the formation of nylon 12. Nitrogen protection is required when carrying out laser sintering with nylon 12 powder at home and abroad, and at least 30 wt. % of new powder is added to the old powder before use.

2.3.2.2 Influence of melting and crystallization characteristics of nylon 12 on selective laser sintering process

The SLS process of nylon 12 powder is a melting and solidification process, so the melting and crystallization characteristics of nylon 12 powder play a decisive role in the sintering processability and the quality of the sintered part.

The melting process of the crystalline polymer is different from the melting process of the low molecular crystal, and the melting of the low molecular crystal is carried out in a narrow temperature range of about 0.2°C, and the melting process is maintained at a certain temperature at which the two phases are balanced until the crystal is completely melted. While the melting of the crystalline polymer occurs over a wide temperature range, which is called the melting range, and in the melting range, the crystalline polymer is heated while melting. This is because the crystalline polymer contains crystals of different degrees of perfection, the less perfect crystals melt at lower temperatures, while the more perfect crystals melt at higher temperatures. The melting point and melting range of the crystalline polymer have little to do with the molecular weight size and distribution, but are related to the crystallization history, the degree of crystallinity and the size of the spherulites. The lower the crystallization temperature, the lower the melting point and the wider the melting range; the higher the crystallinity, the larger the spherulite and the higher the melting point. Fig. 2.36 shows the DSC melting curve of nylon 12 powder material measured at a heating rate of 10°C/min by using Perkin Elmer DSC-7 differential scanning calorimeter.

As can be seen from Fig. 2.36, the start temperature, summit temperature, and end temperature of the melting peak of the nylon 12 powder material are 174.2°C, 184.9°C, and 186.7°C, respectively, and the latent heat of melting measured by DSC is 93.9 J/g. The melting peak is steep, the start temperature of melting is high, the melting range is narrow, and the latent heat of melting is large, and these characteristics are beneficial to the sintering process. Since the start temperature of melting is high, the preheating temperature of the powder can be increased, and the temperature gradient of the sintered layer and the surrounding powder can be reduced. The high latent heat of melting prevents the

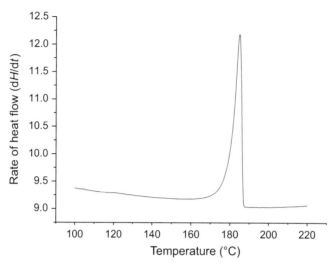

FIGURE 2.36 DSC temperature rising curve of nylon 12 powder materials.

powder particles adjacent to the laser scanning region from melting due to heat conduction, which is advantageous for controlling the dimensional accuracy of the sintered part.

Crystallization occurs when nylon 12 is cooled from a molten state, and the crystallization rate is highly dependent on temperature. Since the crystallization rate is the sum of the nucleation rate and the grain growth rate, the temperature dependence of the nylon 12 crystallization rate is a result of the combined part of the nucleation rate and the grain growth rate on temperature dependence. At a temperature close to the melting point, the nylon 12 molecular segment moves vigorously, the crystal nucleus is not easy to form or the formed nucleus is unstable, the number of nucleation is small, and the total crystallization rate is small; as the temperature decreases, the nucleation forming speed is greatly increased. At the same time, since the polymer chain has sufficient mobility, it is easy to diffuse into the crystal nucleus and discharge into the crystal lattice, so the grain growth rate also increases, and the total crystallization rate increases and reaches the maximum until a certain temperature; when the temperature continues to decrease, although the nucleation rate continues to increase, as the melt viscosity increases, the diffusion capacity of polymer segment decreases, the formation rate of grain slows down, so that the total crystallization rate decreases; when the temperature is below T_g, the segment motion is "frozen," the nucleation and grain growth rate are very low, and the crystallization process cannot actually proceed. Fig. 2.37 shows a DSC curve of the nylon 12 powder materials from a molten state at 220°C to a room temperature at a rate of 10°C/min.

As can be seen from Fig. 2.37, the start temperature of crystallization of the nylon 12 powder material is 154.8°C, the summit temperature of the

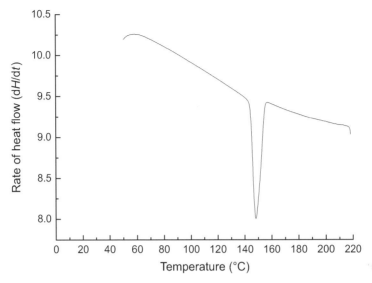

FIGURE 2.37 DSC temperature drop curve of nylon 12 powder materials.

crystallization peak is 148.2°C, and the end temperature of crystallization is 144.3°C. From this, it is understood that the crystallization mainly occurs between 144.3°C and 154.8°C. At a temperature of 154.8°C or higher, the crystallization nucleus is difficult to form, the crystallization rate is slow, and the crystallization process is difficult to proceed. The crystallization rate can be adjusted by controlling the operating temperature during the sintering process, and the tendency of the sintered part to warp due to the shrinkage stress generated by the crystallization is reduced.

Theoretically, the SLS preheating temperature of the crystalline polymer is the temperature range at which the powder begins to melt and the melt begins to crystallize, so the theoretical preheating temperature window can be calculated using the following equation:

$$\Delta T_0 = T_{im} - T_{ic} \qquad (2.72)$$

where T_{im} is the start temperature of melting and T_{ic} is the start temperature of crystallization.

But in fact, because nylon 12 is a semicrystalline polymer, before T_{im}, the powder begins to agglomerate due to the activity of the amorphous part of the molecular chain, so the actual maximum preheating temperature is lower than T_{im}. The actual preheating temperature window is much narrower than the theoretical calculation, and is related to various factors such as the characteristics, composition and preparation method of nylon 12 powder.

Table 2.16 shows the influence of nucleating agent addition in the stage of nylon 12 powdering on the preheating temperature.

TABLE 2.16 Influence of nucleating agent on the preheating temperature.

Nucleating agent	None	Fumed silica	Wollastonite	Wollastonite	Talcum powder
Mass fraction (%)	–	0.1	0.1	0.5	0.5
Preheating temperature (°C)	167–169	167–169	168–169	169–170	–170
Preheating temperature window (°C)	2	2	1	1	<1

It can be seen from Table 2.16 that a nucleating agent is added during the powdering, and other nucleating agents, in addition to the fumed silica, narrow the preheating temperature window and deteriorate the forming properties.

The nylon 12 powder prepared by the solvent precipitation method tends to agglomerate after drying, and is easily compacted by the grinding ball during ball milling without being easily dispersed. The extremely fine inorganic powder can be used as dispersing agent to break the bonding force between the powder and improve the efficiency of the ball milling.

The fluidity of the powder increases and the agglomerates disappears when 0.1% mass fraction of fumed silica is added during the ball milling. After the SLS-forming experiment is carried out using this powder, the preheating temperature of the first few layers is significantly increased to 169°C–170°C, similar to the parts of adding other nucleating agents. It can be seen that fumed silica plays a role as a nucleating agent in the laser sintering process. After the multilayer sintering, it is found that the sintered body and the surrounding powder are cracked, and the sintered body is also transparent. After the removal, it is found that the transparent sintered body is solidified. This phenomenon indicates that the addition of silica accelerates the crystallization rate and refines the spherulites, so the addition of the inorganic dispersant narrows the preheating temperature window, which is disadvantageous for SLS-forming.

The filler also functions as a nucleating agent, and the difference from the nucleating agent is mainly in the amount of the content and the particle size. On the one hand, the addition of the filler accelerates the crystallization of the melt and narrows the temperature of the preheating window; on the other hand, the filling reduces the shrinkage of the melt. At the same time, the filler acts as a release agent for the polymer powder, thereby preventing the mutual adhesion of the nylon 12 powder particles and increasing the agglomeration temperature of the nylon powder. Table 2.17 shows the

TABLE 2.17 Influence of different fillers on sintering.

Type of filler	Glass beads (200–250 mesh)	Glass beads (400 mesh)	Talcum powder (325 mesh)	Wollastonite (600 mesh)
Preheating temperature (°C)	167–170	168–170	Failed	Failed

preheating window temperature of nylon 12 powder material added with 30% of mass fraction of different fillers.

As can be seen from Table 2.17, the glass beads have less influence on the agglomeration temperature of the nylon 12 powder, but enlarge the preheating temperature window, which may be because the glass beads are large in size, smooth in surface, and spherical in shape, so the part on the crystallization of nylon 12 is small. The addition of nonspherical talc and wollastonite deteriorates the formability.

2.3.2.3 Shrinkage and warping deformation of nylon 12 during selective laser sintering

Shrinkage and warping deformation during SLS-forming are the main causes of forming failure. The shrinkage of nylon 12 powder during SLS-forming process mainly includes: (1) densification shrinkage, (2) melt solidification shrinkage, (3) temperature shrinkage, and (4) crystal shrinkage. The shrinkage is large and it is prone to warping and deformation.

The volume shrinkage of nylon 12 due to densification in laser sintering mainly occurs in the height direction, that is, the powder is reduced in height after laser sintering, which has little part on the warpage of the sintered body on the horizontal surface. When the temperature of the melt continues to decrease, the viscosity of the melt rises and cannot even flow, and the contraction stress cannot be released by the microscopic material flow, thereby causing the macroscopic displacement of the sintered body, that is, warping deformation occurs. This is an important reason why the preheating temperature must be higher than the crystallization temperature of nylon 12 when SLS is formed. Nylon 12 is prone to warpage during SLS-forming, especially in the first few layers. The reasons are manifold: First, due to the lower temperature of the first layer of powder layer, the laser-scanned sintered body and surrounding powder have large temperature difference, the periphery of the sintered body cools rapidly, causing shrinkage and warping of the edge of the sintered body. Second, the shrinkage of the first layer of sintered body occurs on the surface of the loose powder, and only a small stress is required to cause the sintered layer to warp, so the formation of the first layer is the most critical. In the subsequent forming, the tendency of warping is gradually reduced due to the anchoring action of the underlayer.

Strict control of the powder bed temperature is an important means to solve the warpage problem in the SLS-forming of nylon 12. When the temperature of the powder bed is close to the melting point of nylon 12, the energy input by the laser just melts the nylon 12, that is, the laser only provides the heat required for the melting of the nylon 12. Due to the small temperature difference between the melt and the surrounding powder, and the nylon 12 is in a completely molten state in the single-layer scanning process, and after the sintering, the melt is cooled, and the stress is gradually released, so that the occurrence of warping deformation can be avoided.

Although the warping and deformation of nylon 12 powder during laser sintering are mainly due to the solidification shrinkage and temperature shrinkage after powder melting, a large number of studies have shown that the geometrical shape of powder has a significant influence on the warping and deformation of laser sintering.

As shown in Fig. 2.38, the nylon 12 powder prepared by the low-temperature pulverization method has irregular geometrical shape. Although the particle size is very fine, the SLS-forming performance is still not good. The preheating temperature exceeds 170°C, and the powder has agglomerated. The edge of the sintered body is still severely warped so that the warpage occurs, as shown in Fig. 2.39. Since the powder is excessively fine, the powder spreading performance is not good. A large amount of powder adheres to the roller without adding the glass beads, and a large amount of dust is accompanied during the powder spreading.

It can be seen from Fig. 2.39 that very serious warpage occurs to the nylon 12 powder prepared by low-temperature pulverization during SLS, especially in the middle of the scanning edge, and the shape of the warpage is a half-moon

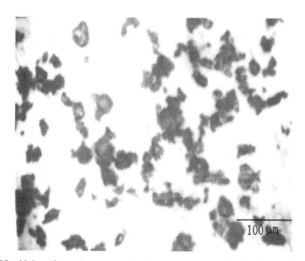

FIGURE 2.38 Nylon 12 powder prepared by low-temperature pulverization method.

FIGURE 2.39 Single-layer scanning photo of nylon 12 powder prepared by low-temperature pulverization.

shape, indicating that the stress at the center position is large. After careful observation, the warpage occurs almost simultaneously with the laser scanning, that is, the warpage occurs during the melting of the nylon 12. This phenomenon can be explained by several stages of powder sintering:

1. Free packing stage between particles: the powder particles are completely freely stacked and independent of each other.
2. Forming a sintering neck for mutual bonding: the surface in contact with the powder particles is melted, and the particles are bonded to each other, but volume shrinkage has not yet occurred.
3. Powder spheroidization: As the temperature increases further, the crystal melts, but at this time the viscosity is high and the melt cannot flow freely, however, driven by the surface tension, the surface area of powder tends to reduce and shrink into a spherical shape, that is, the so-called spheroidization.
4. Complete fusion and densification: The melt viscosity is further reduced, the powder is completely melted into liquid, the air in the powder is extruded, and the powder is completely melted into one.

Figs. 2.40 and 2.41 show laser sintering of nonspherical powder particles and spherical powder particles respectively. When the nonspherical powder particles are sintered, the powder particles first stick to each other to form a sintering neck, and then spheroidized to remelt. Since the powder particles

FIGURE 2.40 Schematic diagram of laser sintering of nonspherical powder particles.

FIGURE 2.41 Schematic diagram of laser sintering of spherical powder particles.

have adhered to each other before spheroidization, the stress of the spheroidization of the powder causes the shrinkage to occur not only in the height direction but also in the horizontal direction, resulting in edge warpage during laser sintering.

In the sintering process of spherical powder, only the sintering neck grows and the powder is completely melted and densified, and there is no spheroidization process, so the shrinkage in the horizontal direction is small. Moreover, the bulk density of the spherical powder is higher than that of the nonspherical powder, and the volume shrinkage of the densification is small. For the above reasons, the shrinkage of the spherical powder during laser sintering is lower than that of the nonspherical powder.

The particle size of the powder has a significant influence on the SLS-forming. To study the influence of powder particle size on the preheating temperature, nylon 12 powder of narrow particle size distribution are prepared. The particle size distribution range of all powder particles is less than 10 μm. The preheating temperature of the powder of different particle sizes is measured as shown in Table 2.18.

As shown in Table 2.18, as the particle size increases, the preheating temperature increases, and at the same time, the agglomeration temperature also increases, but the preheating temperature window narrows. When the particle size is larger than 65.9 μm, the preheating temperature of the powder exceeds 170°C, and the SLS-forming process cannot proceed.

In order to determine the part of particle size distribution on the preheating temperature, powders of different particle sizes are mixed for SLS-forming experiments. The results are shown in Table 2.19.

It can be seen from Table 2.19 that the agglomeration temperature of the powder is mainly affected by the small particle size powder, and the lower limit of the preheating temperature is limited by the coarse powder, so the preheating temperature window of powder with narrow particle size

TABLE 2.18 Influence of powder particle size on preheating temperature of SLS-forming.

Average particle size (μm)	28.5	40.8	45.2	57.6	65.9
Preheating temperature (°C)	166–168	167–169	167–169	168–169	–170

TABLE 2.19 Influence of mixture of powders of different particle sizes on preheating temperature.

Particle size (μm)	28.5/65.9	28.5/65.9	28.5/65.9	28.5/40.8/65.9	28.5/40.8/65.9
Proportion	1:2	1:1	2:1	1:1:1	2:1:1
Preheating temperature (°C)	–	–	167–168	–168	167–168
Agglomeration temperature (°C)	169	168	168	168	168

distribution is wide, while the preheating temperature window of powder with wide particle size distribution is narrow.

The finer the powder is, the larger the surface area is, the larger the corresponding surface energy is; the larger the surface energy is, the lower the sintering temperature is, therefore, the sintering temperature decreases as the particle size of the powder decreases. When the laser power is constant, the penetration depth of the laser increases with the increase of the particle size of the powder. When the first layer is scanned, the sintered body is most likely to warp and deform. The increase of the penetration depth reduces the energy obtained by the surface, and the temperature of the melt is decreased. The deeper the penetration depth, the higher the sintering depth and the larger the shrinkage stress. Therefore, the coarser the particle size, the easier the warping deformation is when the first layer is sintered. Since the heat is transferred from the outside to the inside during sintering, the melting of coarse powder is slower than the fine powder during sintering. If the powder is excessively coarse, some of the powder may not be completely melted during sintering, and acts as crystal nucleus during cooling, thereby accelerating the crystallization rate of the powder. In summary, powder of coarse particle size is very detrimental to SLS-forming.

88 Materials for Additive Manufacturing

For the laser scanning formed in multilayer, after the powder is completely melted, the shrinkage crystallization is completely independent of the particle size of the powder, and the sintering temperature of the fine powder is low, which is favorable for the sintering of the first layer, but in order to prevent the agglomeration of the powder, it is required to maintain low preheating temperature in forming, which may cause deformation of the entire sintered body. Therefore, in order to obtain good laser sintering performance, the particle size of the nylon 12 powder needs to be maintained within a certain range. According to the experiment, the nylon 12 powder can obtain a good effect with a particle size ranging from 40 to 50 μm.

The powder prepared by the solvent precipitation method is easy to agglomerate after drying. This agglomeration belongs to soft agglomeration, from which the powder can be dispersed after ball milling. However, when the particle size of the powder is small, the effect of ball milling is not good, and the powder is even compacted by the grinding ball. During the laser sintering process, if the temperature is too high, the powder will agglomerate. If the agglomerated powder is only sieved and not ball-milled, the particles will agglomerate each other. The agglomerated powder has large voids, which will not cause low density, but also produce significant influence on laser sintering.

Fig. 2.42 shows the results of single-layer scanning of the agglomerated powder, in which the warpage of four corners can be seen, which is similar to that of the nonspherical powder.

FIGURE 2.42 Single-layer scanning photo of agglomerated powder.

Even if the preheating temperature is raised, this phenomenon cannot disappear, so the agglomerated powder has a poor SLS-forming property.

2.3.3 Performance of nylon 12 powder selective laser sintering parts

2.3.3.1 Physical and mechanical properties

Table 2.20 shows the comparison of performance between nylon 12 SLS part and nylon 12 molded part. As can be seen from Table 2.20, the density of the SLS part is 0.96 g/cm^3, which is 94% of the density of the molded part, indicating good sintering performance (96% is the upper limit of powder sintering), which has big difference from the SLS-forming of the amorphous polymer. The performance indicators such as tensile strength, flexural modulus and heat distortion temperature of SLS parts are similar to those of molded parts. However, the fracture behavior is quite different from that of the molded part, and the elongation at break of the molded part reaches 200%, while the SLS part has no necking during the tension process, and breaks at the yield point, and the elongation at break is only one-tenth of that of the molded part, because a small amount of pores in the SLS part concentrate the stress, causing the material to change from ductile fracture to brittle fracture, and the impact strength is much lower than that of the molded part.

2.3.3.2 Precision of parts

The low precision is an important problem that restricts the application of laser-sintered nylon 12 powder, and the main factors affecting the precision of the parts include deformation of the part, shrinkage of the dimension and beyond the sintering boundary. As mentioned earlier, warping deformation may cause complete failure of the SLS-forming process. The effect of dimensional shrinkage on the part is also very significant. Nylon 12 is a crystalline polymer, which is accompanied by large shrinkage during crystallization. The shrinkage causes the deformation of the part frequently. To reduce the deformation caused by the shrinkage stress, it is sometimes necessary to tilt the part during forming to avoid scanning a large plane.

The preheating temperature of nylon 12 powder during SLS-forming is very high, which is close to the melting point. The heat conduction after laser scanning often causes the powder around the sintered body to start to melt. Although the latent heat of melting of nylon 12 powder is large, it is beneficial to prevent this phenomenon, but for large areas of sintering, this phenomenon is still difficult to avoid due to heat concentration. The heat conduction in the melting zone melts and sinters the surrounding powder, making the boundary of the parts unclear, the size becomes larger, and the pores shrink and even disappear. When the nylon 12 crystallizes, a large

TABLE 2.20 Comparison of performance between SLS parts of nylon 12 powder and molded parts of nylon 12 powder.

Performance	Density (g/cm^3)	Tensile strength (MPa)	Elongation at break (%)	Bending strength (MPa)	Flexural modulus (GPa)	Impact strength (kJ/m^2)	Heat distortion temperature (°C) (1.85 MPa)
SLS parts	0.96	41	21.2	47.8	1.30	39.2	51
Molded parts	1.02	50	200	74	1.4	No fracture	55

amount of latent heat is released. If the size of the part is large, the latent heat will also melt the surrounding powder. Measures such as reducing the laser power and preheating temperature can alleviate this phenomenon.

The volume shrinkage produced in the SLS-forming will cause the actual size of the part to be smaller than the design size, that is, the dimensional error is negative. The average shrinkage ratio in the horizontal direction of nylon 12 during SLS-forming is about 2.5%, and the shrinkage ratio in the height direction is from 1% to 1.5%. This dimensional error caused by material shrinkage can be compensated by setting the dimensional correction factor of the parts on the computer; for the melting of powder in the unscanned area due to heat conduction, the preheating temperature should be lowered in addition to the newly added cross section; for large-area scanning, the laser power should be appropriately reduced. However, if a new cross section is added while scanning in a large-area, it is difficult to make balance. This problem is very difficult to solve, so laser-sintered nylon 12 powder are not suitable for manufacturing parts of large size.

2.4 Composite powder material of nylon 12

Nylon 12 powder has been proved to be the best material for the direct preparation of plastic functional parts by SLS technology, but the strength, modulus and heat distortion temperature of pure nylon 12 are not ideal, and cannot meet the higher requirements for plastic functional parts. Besides, it is easy to cause warping deformation during laser sintering due to its high shrinkage rate and relatively low precision. To this end, research institutes and companies engaged in SLS at home and abroad have focused on the research of reinforced composite materials of nylon 12.

The formed parts of nylon composite materials sintered from the nylon 12 composite powder prepared previously have some more outstanding performance than the formed parts of pure nylon 12, so that they can meet the requirements for the performance of the plastic functional parts in different occasions and applications. Unlike amorphous polymers, sintered parts of crystalline polymers are nearly completely dense, so density is no longer a major factor affecting the performance. Adding inorganic fillers can greatly improve the performance in some aspects, such as mechanical properties and heat resistance. At present, the inorganic fillers and reinforcing materials commonly used to reinforce nylon 12 SLS parts are glass beads, silicon carbide, wollastonite, talcum powder, titanium dioxide, hydroxyapatite, layered silicate, carbon fiber powder, and metal powder [4].

In addition, very active researches have been conducted on the preparation of nanonylon composites by using nanoparticles in recent years. Due to the surface effect, volume effect, and macroscopic quantum tunneling part of nanomaterials, the performance of nanocomposites is better than the physical and mechanical properties of conventional composites with the same components. Therefore, the preparation of nanocomposites is one of the important methods to obtain high-performance composite materials. To this end, many institutions and scholars at home and abroad have conducted a lot of research on this and developed some high-performance nanonylon composite materials. However, the dispersion of nanomaterials is difficult. How to prepare polymer-based nanocomposites with uniform dispersion of nanoparticles is still a difficult task. At present, the in situ polymerization and intercalation polymerization of monomers are relatively successful cases, and the nanomaterials used are mainly layered montmorillonite and silica.

Ordinary inorganic fillers significantly reduce the impact strength of the nylon 12 sintered parts, and cannot be used for functional parts requiring high impact strength. Therefore, it is necessary to adopt other reinforced modification methods to improve the performance of the sintered parts. Since the particle diameter of material used for SLS is less than 100 μm, which cannot be reinforced by the reinforcing method commonly used for polymer materials such as glass fiber, and even the powdery filler having an aspect ratio of 15 or more has an adverse part on the SLS process. Although the nano inorganic particles have good reinforcing part on the polymer material, the conventional mixing method cannot achieve the dispersion on the nanometer scale, and thus cannot exert the enhancement of nanoparticles. The polymer/layered silicate nanocomposite (PLS) which has appeared in recent years not only has excellent physical and mechanical properties, but also has practical and economical preparation process, especially the polymer melting intercalation method, featured by simple process, low in cost and flexible and strong application, provides a good way to prepare high-performance composite sintered materials. When the layered silicate is added to the laser-sintered powder material, if the intercalation of the polymer with the layered silicate can be achieved during the sintering process, a high-performance sintered part can be prepared.

2.4.1 Composite powder material of nylon 12/rectorite

Rectorite is a natural mineral material that is easily dispersed into nanoscale microchips and is named after its discoverer, E.W Rector. In 1981, the New Mineral and Mineral Nomenclature Committee of International Minerals Association defined it as "a 1:1 regular interlayer mineral composed of dioctahedral mica and dioctahedral montmorillonite." There are more than ten rectorite producing areas in China, among which the Zhongxiang Yangzha Rectorite Deposit in Hubei is a large-scale industrial deposit, and its deposit reserves and grades are rare at home and abroad.

Rectorite is a layered silicate mineral that is hydrophilic and has poor dispersibility in the polymer matrix. However, hydrated cations such as Ca^{2+}, Mg^{2+}, K^+, and Na^+ are contained between the montmorillonite layers of the rectorite. These metal cations are adsorbed on the surface of the sheet by weak electric field force, so they are easily exchanged out by the organic cationic surfactant. The cation exchange reaction of the organic cation with the rectorite mineral causes the organic matter to enter the montmorillonite layers of the rectorite to form organic complex of rectorite. Since the organic matter enters the mineral layers and covers the surface thereof, the rectorite changes from the original hydrophilicity to the lipophilicity, which enhances the affinity between the rectorite and the polymer, which is not only beneficial for the uniform dispersion of rectorite in the polymer matrix, but also makes the polymer molecular chain easier to be inserted in the layers of the rectorite.

As layered silicate clay, rectorite is very similar to montmorillonite but has its unique structural characteristics. It has the same cation exchange property as montmorillonite. After entering the organic cations between layers, it can be expanded and peeled off. Since the layer charge of the montmorillonite layer in the rectorite mineral structure is lower than that of the montmorillonite layer, it is easier to disperse, intercalate and exfoliate than the montmorillonite. Moreover, in the unit structure of the rectorite, one crystal layer has a thickness of 2.4−2.5 nm, a width of 300−1000 nm, and a length of 1−40 μm, and its aspect ratio is much larger than that of the montmorillonite, and the crystal layer is also 1 nm thicker than that of the montmorillonite, which is incomparable by montmorillonite with small aspect ratio in terms of the reinforcement and barrier properties of the polymer. In addition, since the rectorite contains nonexpanded mica layer, its thermal stability and high temperature resistance are superior to those of montmorillonite. Therefore, rectorite has greater advantages in the preparation of high-performance PLS nanocomposites.

2.4.1.1 Preparation of nylon 12/rectorite composite powder sintered materials

2.4.1.1.1 Preparation of organic rectorite

The rectorite produced in Zhongxiang, Hubei, is silver-gray and shows silky oily gloss. In the experiment, fine sodium-based rectorite is used, and organic rectorite (OREC) is prepared by using trimethyloctadecyl ammonium as an organic treatment agent. OREC is prepared as follows: place a certain amount of rectorite in an appropriate amount of distilled water, and stir the rectorite at a high speed so that they are sufficiently dispersed and heated to 40°C−50°C, add the required amount of organic treatment agent of quaternary ammonium salt dropwise, and stir the rectorite for 2 hours until it cools down to room temperature naturally, and then carry out suction filtration and rinsing for several times until the OREC filter mass is obtained. Finally, dry the filter mass at a temperature of 80°C, and mill and sieve it for future use. For the microscopic shape of OREC, see Fig. 2.43 [5].

Fig. 2.43A shows the overall appearance of the OREC powder. The particle shape is irregular and the particle size distribution is wide. Most of the

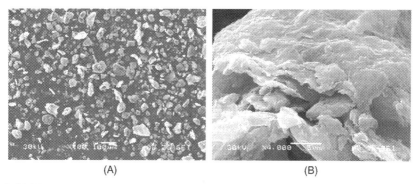

FIGURE 2.43 The scanning electron micrograph (SEM) pictures of organic rectorite under the magnification of (A) 100x and (B) 4000x.

particles have a particle size of 10–80 μm. Fig. 2.43B shows the shape of the OREC particles which have been magnified by 4000 times, from which we can see its topographical structure.

2.4.1.1.2 Preparation of composite powder sintered materials

The composite powder sintered material is composed of nylon 12, OREC and other auxiliaries, and the mass fraction of OREC is 3%–10%. The composite powder sintered material is prepared as follows: Mix the vacuum-dried nylon 12 powder, OREC and nylon 12 masterbatch added with stabilizer, dispersant, lubricant and other additives in a high-speed mixer for 5 minutes, sieve the mixed powder through a 200-mesh sieve, further mix the powder in a high-speed mixer for 3 minutes until the composite powder sintered material is obtained.

2.4.1.2 Properties of selective laser sintering parts of nylon 12/ rectorite composite powder

Standard test samples such as tensile, impact and heat distortion temperature of nylon 12/OREC composites are prepared on HRPS-III SLS-forming machine. The preparation parameters of the samples are as follows: laser power: 8–10 W; scanning speed: 1500 mm/s; sintering pitch: 0.1 mm; sintered layer thickness: 0.15 mm; preheating temperature: 168°C–170°C.

2.4.1.2.1 Mechanical properties

Table 2.21 shows the mechanical properties of nylon 12 and nylon 12/OREC composites formed by SLS.

It can be seen from Table 2.21 that the mechanical properties of composite material SLS parts are better than those of pure nylon 12 parts in terms of tensile strength, bending strength, flexural modulus and impact strength. With the increase of the mass fraction of OREC, the mechanical strength of the composite

TABLE 2.21 Mechanical properties of SLS parts.

Rectorite content (mass fraction) (%)	0	3	5	10
Tensile strength (MPa)	44.0	48.8	50.3	48.5
Elongation at break (%)	20.1	22.8	19.6	18.2
Bending strength (MPa)	50.8	57.8	62.4	58.9
Flexural modulus (GPa)	1.36	1.44	1.57	1.58
Impact strength (kJ/m^2)	37.2	40.4	52.2	50.9

material first increases and then decreases. When the mass fraction of OREC is 5%, the parts show the best mechanical properties. Compared with the pure nylon 12, the tensile strength increases by 14.3%, the bending strength and flexural modulus increase by 22.8% and 15.4%, respectively, and the impact strength increases by 40.3%. The structural characterization of the composite material has proved that the composite powder of nylon 12 and OREC is laser-sintered to realize the intercalation of nylon 12 to OREC, forming a nanocomposite. Since the rectorite is dispersed in the nylon 12 matrix by the nanometer-scale layer, the specific surface area is extremely large, and the interface bonding with the nylon 12 is strong. When the composite material is broken, in addition to the fracture of the matrix material, the rectorite layer is also required to be pulled out from the matrix material or the rectorite layer is required to be broken, thus significantly improving the mechanical properties of the composite material. In particular, the impact strength of the sintered part is greatly improved, which is unmatched by ordinary inorganic fillers. Therefore, nylon 12/OREC is of great significance for laser sintering of high-performance plastic functional parts [6].

2.4.1.2.2 Thermal properties

TG analysis of nylon 12 and sintered nylon 12/OREC composite material is carried out by a comprehensive thermal analyzer manufactured by Netzsch, Germany. The temperature is raised from room temperature to 450°C at a rate of 10°C/min under the protection of N_2. The TG curve of the heating process is record, as shown in Fig. 2.44.

Curves a and b in Fig. 2.44 are the TG curves of nylon 12 and nylon 12/OREC composite materials (OREC mass fraction is 10%). Comparing the two curves, it can be seen that initiation temperature of thermal decomposition of nylon 12 is 358°C, and the thermal weight loss at 450°C is 55.77%; while the initiation temperature of thermal decomposition of the composite material is 385°C, and the thermal weight loss at 450°C is only 15.84%, and the thermal stability of the composite material is significantly better than nylon 12. It may be that the rectorite layer dispersed on the nanometer scale has the function of

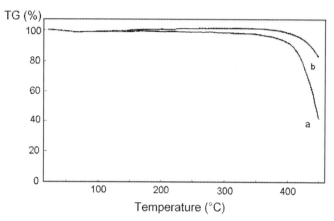

FIGURE 2.44 TG curves of nylon 12 and nylon 12/OREC composite materials.

TABLE 2.22 Heat distortion temperature of nylon 12/OREC composite material SLS parts.

Rectorite content (%)	0	3	5	10
Heat distortion temperature (°C) (1.85 MPa)	52	101	>120	>120

blocking the diffusion of volatile thermal decomposition products, so the thermal decomposition temperature of the composite material is greatly improved.

Table 2.22 shows the heat distortion temperature of nylon 12 and nylon 12/OREC composite material SLS parts under a load of 1.85 MPa.

It can be seen from Table 2.22 that when the OREC content is only 3%, the heat distortion temperature of the composite powder SLS parts is 101°C, which is 46°C higher than that of the pure nylon 12 SLS parts, and the heat deformation temperature is further increased with the increase of the OREC content. Because the nylon 12 molecular chain has a strong interfacial interaction with the rectorite layer, the rectorite layer can effectively help the matrix material maintain good mechanical stability at high temperatures. At the same time, the restriction effect of the rectorite layer on the nylon 12 molecular chain can reduce the deformation of parts due to the rearrangement of the molecular chain and improve the dimensional stability of the composite material.

2.4.1.3 Intercalation mechanism in selective laser sintering process

In the SLS process, nylon 12 of the composites powder absorbs the energy of the laser and melts, and solidifies into a solid material after cooling. At the same time, the nylon 12 molecules are intercalated into the rectorite layer. The intercalation method is called polymer melt intercalation and it is static. Fig. 2.45 is the schematic diagram of the melt intercalation of nylon 12.

Analyzed from thermodynamic perspective, whether the intercalation process of polymer macromolecular chain to OREC can be carried out depends on the free energy change (ΔG) of the system in the corresponding process. Only when $\Delta G < 0$, the process can proceed spontaneously. For the isothermal process, the following relationship applies:

$$\Delta G = \Delta H - T\Delta S \tag{2.73}$$

In the equation, ΔG, ΔH, and ΔS represent free energy change, enthalpy change and entropy change, respectively, and T represents absolute temperature.

According to the mean field theory of Vaia et al., the entropy change of a polymer melt intercalation system consists of two parts:

$$\Delta S = \Delta S_{\text{polymer}} + \Delta S_{\text{intercalator}} \tag{2.74}$$

In the process of melt intercalation, on the one hand, the polymer molecular chain changes from random coil conformation under free state to the confined chain conformation constrained by the quasi-two-dimensional space between the clay layers, and the entropy value is decreased; on the other hand, the intercalator distributed between the organic clay layers obtains a larger degree of conformational freedom due to the enlargement of the interlayer spacing, and the entropy value is increased. When the layer spacing of the layered silicate is not changed significantly, the total entropy value of the system changes very little, so the enthalpy value will play a decisive role on the change of the free energy of the system, that is, the degree of the mutual relationship between the polymer molecular chain and the organic clay is a key factor in determining whether the intercalation can proceed. Nylon 12 is a kind of polar polymer that forms a strong polar interaction with the polar surface of the OREC. Thus, this system facilitates the formation of intercalated composites.

Polymer melt intercalation is usually carried out under external forced mechanical forces, but such forced mechanical forces are not absolutely necessary. Some systems can form PLS nanocomposites in good dispersion condition in quasistatic state. It is mainly driven by enthalpy that the polymer

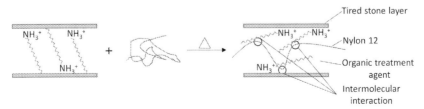

FIGURE 2.45 Schematic diagram of nylon 12 melt intercalation.

molecular chain enters the interlayer of the layered silicate. As long as the polymer molecular chain enters the interlayer of the organic clay, it is not easy to escape from the constraint of the layer to restore the relatively free state. Because it takes some energy to force the molecular chain from the coiled state of the random coil to the straight chain before entering the clay layer, the layer structure of the organic clay restricts the movement of the polymer molecular chain in space, and more importantly, the action of the polymer molecular chain and the surface of the clay layer prevent it from escaping from the surface of the clay layer.

Melt intercalation studies of PS/clay systems by Vaia et al. have shown that the rate of diffusion of the polymer melt between the clay layers is similar to its self-diffusion rate, which means that the melt intercalation does not require additional processing time. In other researches, it has been found that the diffusion rate of polymer molecules to organic clay is much faster than its self-diffusion rate in a uniform polymer body. Therefore, analyzed from the kinetic perspective, it is possible to make nylon 12 intercalate between the OREC layers during laser sintering.

In the laser sintering process, the nylon 12 melt intercalation is carried out in two steps: the nylon 12 diffuses into the primary particle aggregates of rectorite and the silicate layer. Nylon 12 melt has a very low viscosity and can flow within the sintered layer or between the sintered layers. It can quickly wet the surface of the rectorite particles and infiltrate into the voids of the rectorite agglomerates, that is, the rectorite primary particle aggregates; under the polar interaction of nylon 12 and OREC, nylon 12 macromolecules further diffuse into the layers of rectorite to form intercalated composites. Although the nylon 12 is in a molten state for a short period of time during the laser sintering process, the sintered part is always at a temperature close to the melting point of the nylon 12, and at such a temperature, the crystallization rate of the nylon 12 is very slow, and the nylon 12 has sufficient time to diffuse into the layers of the rectorite before crystallization to form the intercalated composite.

2.4.2 Nanosilica/nylon 12 composite powder materials

At present, there are generally two methods for preparing nanofiller/nylon composite powder for SLS: the one is mechanical mixing, which is to mechanically mix the nanofiller and nylon powder to obtain composite powder. Since the nanoparticles are easily agglomerated, it is difficult to uniformly disperse them in the polymer matrix in the nanometer size through the mechanical mixing method. The other is the low-temperature pulverization method, which is to obtain the polymer nanocomposite by the conventional method first, and then impact it to pulverize into the powder suitable for SLS-forming under extremely low-temperature conditions. Although this method can uniformly disperse the nanoparticles, the prepared powder usually has a large particle size, a wide particle size distribution and an extremely irregular powder shape, which are very disadvantageous for the precision of the SLS parts. On the basis of previous work, this chapter proposes a new method for preparing inorganic nanoparticles/nylon composite powder, namely solvent precipitation method. The basic process is as follows: first, the nanoparticle is surface-organized, and then the ultrasonic vibration, intense stirring and other means are used to uniformly disperse in the nylon alcohol solution. When the above mixing system is slowly cooled, the nylon is gradually crystallized and precipitated by using the nanoparticles as a heterogeneous nucleating agent to obtain composite powder, so that the nanoparticles are dispersed evenly in the nylon matrix in nanometer size. The nanosilica/nylon 12 composite powder is prepared by the above method. The dispersion of nanosilica in the sintered parts and the impact fracture surface are observed by field emission scanning electron microscopy. The influence of nanosilica on the thermal properties of nylon 12 and the mechanical properties of its SLS parts are studied, and the properties of nanosilica/nylon 12 composite powder prepared by mechanical mixing method and solvent precipitation method are compared and researched.

2.4.2.1 Preparation of nanosilica/nylon 12 composite powder
2.4.2.1.1 Surface modification of nanosilica

The method for surface modification of nanosilica is as follows: (1) preheat nanosilica and fully disperse it in the solvent by ultrasonic vibration to form nanoparticle suspension; (2) use ethanol and water to formulate into alcohol-water solution at a mass ratio of 95:5, after stirring, add the silane coupling agent KH-550 to achieve a mass fraction of 2%; after 1 hour of standing, fully hydrolyze the coupling agent; (3) add the hydrolyzed KH-550 to the above nanoparticle suspension, and stir the mixture at room temperature for 2 hours, and then condense and reflux at 75°C for 4 hours; (4) centrifuge the mixture to recover the solvent, and wash the precipitate with ethanol to remove excess KH-550 adsorbed on the surface of the nanosilica; (5) Vacuum dry the obtained precipitate at 110°C for 1 hour, and then conduct vacuum drying at 50°C for 12 hours.

2.4.2.1.2 Preparation process of powder

The steps of preparing nanosilica/nylon 12 composite powder (D-nanosilica/PA12) by solvent precipitation method are as follows: (1) Add the surface-modified nanosilica to a certain amount of solvent, and ultrasonically shake at 30°C for 2 hours to form nanosilica suspension; (2) Pour the nylon 12 particles, solvent and nanosilica suspension into a jacketed stainless steel autoclave at a certain ratio, sealed, and protected by nitrogen gas; (3) slowly heat up to 150°C to make nylon 12 completely soluble in ethanol. At the same time, vigorously stir to make the nanosilica uniformly dispersed in the alcohol solution of nylon 12; (4) slowly cool to room temperature at a certain speed, so that nylon 12 slowly forms powder with nanosilica as crystallization nucleus; (5) vacuum dry, ball mill and sieve the obtained powder aggregate to obtain the D-nanosilica/PA12, wherein the nanosilica has a mass fraction of 3%.

Following the same procedure as above, pure nylon 12 powder (NPA12) can be prepared without the addition of nanosilica.

The process of preparing nanosilica/nylon 12 composite powder (M-nanosilica/PA12) using mechanical mixing method is as follows: a certain mass ratio of surface-modified nanosilica and NPA12 are mixed, and the mixture is placed in a planetary ball mill for 5 hours, after that, the M-nanosilica/PA12 is obtained, wherein the nanosilica had a mass fraction of 3%.

2.4.2.2 Interface bonding between nanosilica and nylon 12

In order to improve the interface bonding between the nanosilica and the nylon 12 matrix, the surface of the nanosilica particles is organically treated with silane coupling agent KH-550. The reaction process of KH-550 with nanosilica and nylon 12 is shown in Fig. 2.46.

First, KH-550 is hydrolyzed to form a hydrolyzate containing silanol groups (Si-OH). Secondly, the surface of the nanosilica contains a large amount of Si-OH, which can undergo polycondensation reaction with the hydrolyzate of KH-550 to form siloxane, so that the amino group ($-NH_2$) is grafted onto the surface of the nanosilica particles. Finally, the amino energy grafted onto the surface of the nanosilica reacts with the carboxyl group in the nylon 12 to form amide bond, such that the interface bonding of nanosilica and nylon 12 matrix is improved.

Fourier infrared spectroscopy (FTIR) is used to qualitatively analyze the structural changes of nanosilica before and after surface modification.

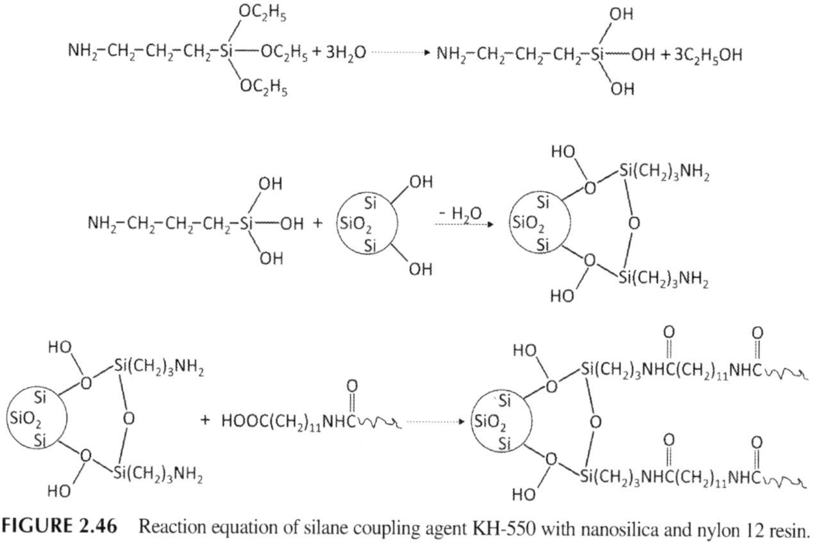

FIGURE 2.46 Reaction equation of silane coupling agent KH-550 with nanosilica and nylon 12 resin.

The instrument used is a German VERTEX 70 Fourier transform micro infrared/Raman spectrometer. Fig. 2.47 is the FTIR spectrum of nanosilica before (a) surface modification and (b) after surface modification.

From the infrared spectrum before surface modification of nanosilica (Fig. 2.47a), it can be seen that there is a broad and strong peak at 3387 cm^{-1}, which is the OH stretching vibration peak of Si-OH on the surface of nanosilica. It has a strong Si-O-Si absorption peak at 1100 and 467 cm^{-1}, and a weak Si-OH absorption peak at 963 cm^{-1}. From the infrared spectrum after the surface modification of nanosilica (Fig. 2.47b), it can be seen that new absorption peaks appear at 2920 and 2895 cm^{-1} in the infrared spectrum after surface treatment in comparison with the infrared spectrum before surface treatment, of which 2920 cm^{-1} is the absorption peak of $-CH_3$, and 2895 cm^{-1} is the absorption peak of $-CH_2$, because the organic carbon chain is grafted to the surface of silicon oxide; the absorption peaks at 3387 and 963 cm^{-1} are weakened because the reaction equation in Fig. 2.46(2) consumes a certain amount of Si-OH on the surface of the nanosilica; and the absorption peaks at 1100 and 467 cm^{-1} become stronger because the reaction equation in Fig. 2.46(2) generates Si-O-Si, so Si-O-Si is increased and the absorption peak is strengthened. From the above Fourier

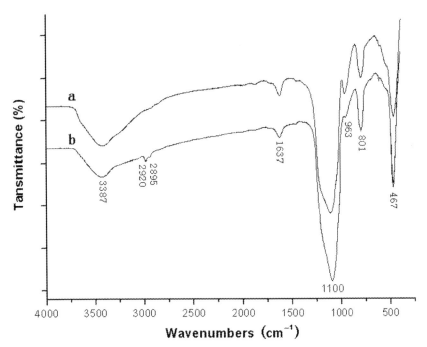

FIGURE 2.47 FTIR spectrum of nanosilica before surface modification (a) and after surface modification (b).

transform infrared spectroscopy, it shows that the coupling agent KH-550 is successfully grafted onto the surface of the nanosilica.

2.4.2.3 Analysis of powder characteristics

The particle size and particle size distribution of NPA12 and D-nanosilica/PA12 are analyzed using MAN5004 laser diffraction particle size analyzer manufactured by Nalvern Instruments of the United Kingdom. After the powder sample is subjected to gold spray treatment, the microscopic shape is observed by the Sirion 200 field scanning electron microscope of FEI company, the Netherlands.

Fig. 2.48A and B are scanning electron micrograph (SEM) microscopic pictures and particle size distribution of D-nanosilica/PA12, respectively. As can be seen from the figure, D-nanosilica/PA12 has irregular shape and rough surface, with a particle size distribution of 6–89 μm, and the main particle size distribution of 14–36 μm. Analyzed from the laser particle size, the average particle size is 25.08 μm. Fig. 2.49A and B are SEM microscopic pictures and particle size distribution of NPA12, respectively. As can be seen from the figure, NPA12 also has irregular shape and rough surface with a particle size distribution of 10–90 μm, and the main particle size distribution of 31–56 μm. Analyzed from the laser particle size, the average particle size is 37.42 μm. From the above experimental results, it can be found that although both powders are prepared by solvent precipitation method, the particle size of D-nanosilica/PA12 is much smaller than that of NPA12, mainly because the nanosilica plays the role of nucleating agent in the crystallization process of nylon 12. Thus the nucleation center increases, so that the number of powder particles increases and the particle size of the powder particles decreases. The small particle size of D-nanosilica/PA12 allows for a faster sintering rate, and the SLS parts have clearer details and sharper outline.

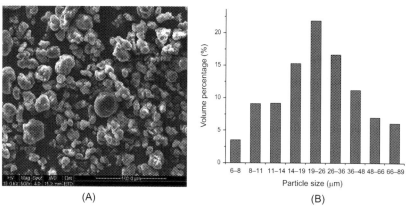

FIGURE 2.48 SEM picture (A) and particle size distribution (B) of D-nanosilica/PA12.

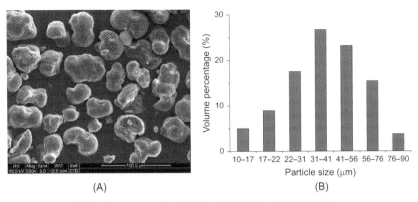

FIGURE 2.49 SEM picture (A) and particle size distribution (B) of NPA12.

2.4.2.4 Dispersion of nanosilica in nylon 12 matrix

The degree of dispersion of the nanoparticles in the matrix is critical to the properties of the composite material. If the nanoparticles are not dispersed uniformly in the matrix due to agglomeration of the nanoparticles, the composite material will exhibit the same or worse properties than the ordinary microparticle reinforcement material. Figs. 2.50 and 2.51 show the microscopic shape of the low-temperature brittle section of the sample observed by Sirion 200 field scanning electron microscope of FEI company, the Netherlands.

Fig. 2.50 shows the SEM microscopic shape of the low-temperature brittle section of the SLS parts of D-nanosilica/PA12. As can be seen from the figure, in the low-temperature brittle section of the SLS part of D-nanosilica/PA12, a large amount of white particles are very uniformly dispersed in the nylon 12 matrix material, and by measurement, the size of these particles ranges from 30 to 100 nm, indicating that the nanosilica is uniformly dispersed in the nylon 12 matrix at a nanometer scale. There are two main reasons for this: first, the surface treatment of the nanosilica by silane coupling agent is conducted to increase the compatibility of nanosilica with the nylon 12 matrix, thereby facilitating the dispersion of the nanosilica; more importantly, in the process of preparing the composite powder by the solvent precipitation method, the nanosilica is first uniformly dispersed in the alcohol solution of the nylon 12, and when the mixture is cooled, the nylon 12 is formed into powder by using the nanosilica as crystallization nucleus, then the nanosilica is uniformly dispersed in the nylon 12 matrix.

Fig. 2.51 shows the SEM microscopic shape of the low-temperature brittle section of the SLS parts of M-nanosilica/PA12. As can be seen from the figure, in the low-temperature brittle section of the SLS parts of M-nanosilica/PA12, there are a large number of aggregates of nanosilica, and the dispersion of these aggregates is uneven, and these aggregates are measured at $2-10\,\mu m$. This indicates that the mechanical mixing method cannot

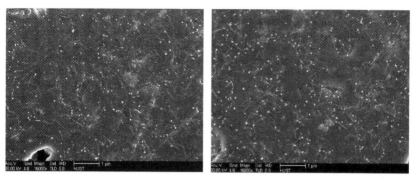

FIGURE 2.50 SEM microscopic shape of the low-temperature brittle section of SLS parts of D-nanosilica/PA12.

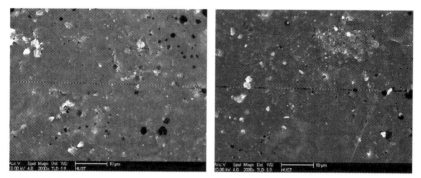

FIGURE 2.51 SEM microscopic shape of the low-temperature brittle section of SLS parts of M-nanosilica/PA12.

uniformly disperse the highly agglomerated nanomaterials in the nylon 12 matrix. In the SLS parts of M-nanosilica/PA12, the nanosilica exists in the form of micron-sized agglomerates.

2.4.2.5 Effects of nanosilica on the melting and crystallization of nylon 12

DSC is used to study the effect of nanosilica on the melting and crystallization of nylon 12. DSC test conditions are as follows: under argon protection, the temperature is first increased from room temperature to 200°C at a rate of 10°C/min, and then drop to room temperature at 5°C/min after maintained for 5 minutes, the DSC curves of temperature rise and temperature drop are recorded.

Fig. 2.52 shows the DSC curves of temperature rise and temperature drop of D-nanosilica/PA12, M-nanosilica/PA12 and NPA12. The initial

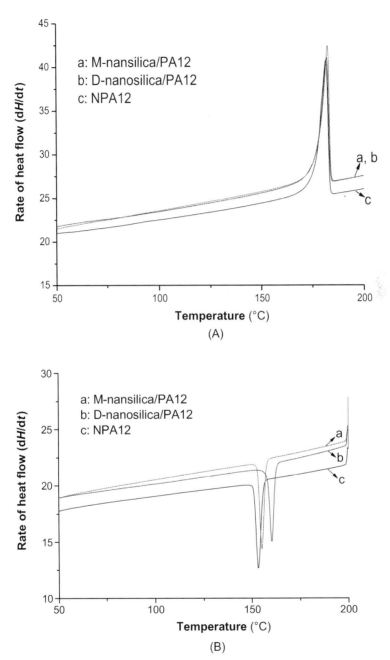

FIGURE 2.52 DSC curves of (A) temperature rise and (B) temperature drop of D-nanosilica/PA12, M-nanosilica/PA12, and NPA12.

temperature (T_{im}) of melting, initial temperature of crystallization (T_{ic}), melting enthalpy (ΔH_m) and crystallization enthalpy (ΔH_c) obtained from the DSC curve are listed in Table 2.23, and the relative crystallinity (*CI*) in Table 2.23 is calculated by Eq. (2.75).

$$CI = \frac{\Delta H_m - \Delta H_c}{\Delta H_f^0 \times (1-f)} \times 100\% \qquad (2.75)$$

In Eq. (2.75), H_f^0 is the melting enthalpy of nylon 12 of 100% crystal, which is 209.2 J/g as known from the literature, and f is the mass fraction of nanosilica.

It can be seen from Fig. 2.52A and Table 2.23 that the melting temperatures of the three types of powders of D-nanosilica/PA12, M-nanosilica/PA12 and NPA12 are not much different, indicating that nanosilica has little effect on the melting temperature of nylon 12. Since the preheating temperature of the crystalline polymer in the SLS process should be close to but not higher than the initial temperature of melting, so the three types of powders can be set to the same preheating temperature. The preheating temperature of the powders is set to 170°C in this experiment.

Comparing the DSC curve of temperature drop in Fig. 2.52B, it is known that D-nanosilica/PA12 has the highest crystallization temperature, while NPA12 has the lowest crystallization temperature, which indicates that nanosilica has a heterogeneous nucleation effect. However, when the nanosilica content (mass fraction) is same, D-nanosilica/PA12 has higher crystallization temperature than M-nanosilica/PA12, which may be because nanosilica in D-nanosilica/PA12 is uniformly dispersed in the nylon 12 matrix in nanometer size, while the nanosilica in M-nanosilica/PA12 is present in micron-sized agglomerates, so D-nanosilica/PA12 has more nucleation centers at the same level. As can be seen from Table 2.23, D-nanosilica/PA12 has the highest relative crystallinity, while NPA12 has the lowest relative crystallinity, indicating that the nanosilica makes the crystal content of nylon 12 improved. Similar to the crystallization temperature, D-nanosilica/PA12 has a higher relative crystallinity than M-nanosilica/PA12 at the same nanosilica content.

TABLE 2.23 Thermal properties data obtained from the DSC curves.

Samples	T_{im} (°C)	T_{ic} (°C)	ΔH_m (J/g)	ΔH_c (J/g)	*CI* (%)
D-nanosilica/PA12	178.00	162.33	76.70	40.03	18.0
M-nanosilica/PA12	178.22	157.38	77.16	48.17	14.2
NPA12	177.80	155.83	85.47	58.30	13.0

2.4.2.6 Effects of nanosilica on the thermal stability of nylon 12

Fig. 2.53 shows the TG curves of D-nanosilica/PA12, M-nanosilica/PA12 and NPA12, and Table 2.24 shows the *TG* temperatures of the three powders at 5% and 10% weight loss (respectively recorded as Td-5% and Td-10%). It can be found that the Td-5% and Td-10% of M-nanosilica/PA12 and NPA12 have no great difference, indicating that the nanosilica in M-nanosilica/PA12 has no effect on the thermal stability of the nylon 12 matrix. However, the Td-5% of D-nanosilica/PA12 is 33.6°C higher than that of NPA12, and the Td-10% is also 37.52°C higher than that of NPA12, indicating that the thermal stability of D-nanosilica/PA12 is significantly better than that of

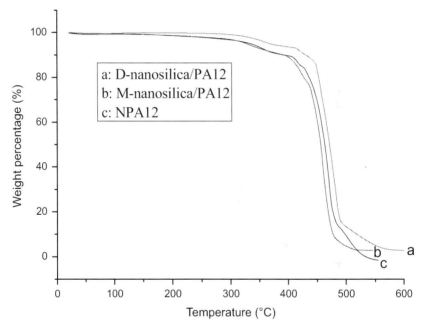

FIGURE 2.53 TG curves of D-nanosilica/PA12, M-nanosilica/PA12, and NPA12.

TABLE 2.24 Thermogravimetric temperature of D-nanosilica/PA12, M-nanosilica/PA12 and NPA12.

Samples	$T_d - 5\%$ (°C)	$T_d - 10\%$ (°C)
D-nanosilica/PA12	368.49	430.19
M-nanosilica/PA12	328.32	393.31
NPA12	334.89	392.67

NPA12, that is, the nanosilica in D-nanosilica/PA12 can increase the thermal stability of nylon 12 matrix. This may be because that the strong interfacial interaction between the nanosized uniformly dispersed nanosilica particles and the nylon 12 matrix limits the thermal decomposition of the nylon 12 molecular chain.

2.4.2.7 Effects of nanosilica on the mechanical properties of nylon 12 selective laser sintering parts

The three types of polymer powders are sintered by HRPS-III laser sintering system. Sintering parameters are set as follows: BS is set to 2000 mm/s; SCSP is set to 0.1 mm; P is set to 8 to 24 W; the range of ED varies from 0.04 to 0.12 J/mm^2; the slice thickness is set to 0.1 mm.

Fig. 2.54 shows the curves of tensile strength of SLS parts of D-nanosilica/PA12, M-nanosilica/PA12 and NPA12 changing with energy density. It can be seen from the figure that the tensile strength of SLS parts of three types of powders varies with laser energy density. The tensile strength of SLS parts increases with the increase of laser energy density until it reaches a maximum value, after that, the continuous increase of the laser energy density will decrease the tensile strength. This is because increasing the energy density will increase the sintering rate, so that the tensile strength of the

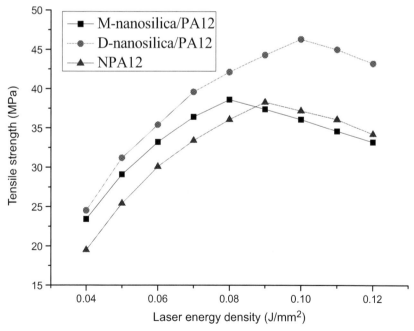

FIGURE 2.54 Curves of tensile strength of SLS parts of D-nanosilica/PA12, M-nanosilica/PA12, and NPA12 changing with energy density.

sintered part increases, but when the energy density increases to a certain value, the material is decomposed violently, so that the mechanical properties of the sintered parts are degraded.

The laser energy density corresponding to the maximum tensile strength is called the optimum laser energy density, and the optimum laser energy density of each powder is used to manufacture the respective mechanical property test pieces. Table 2.25 lists the mechanical properties of sintered parts of D-nanosilica/PA12, M-nanosilica/PA12, and NPA12 at optimum laser energy density. From the data in the table, the tensile strength, tensile modulus and impact strength of SLS parts of D-nanosilica/PA12 are 20.9%, 39.4%, and 9.54% higher than the SLS parts of NPA12, respectively, while the elongation at break is 3.65% lower than that of NPA12. The tensile strength and tensile modulus of SLS parts of M-nanosilica/PA12 are 0.78% and 22.5% higher than NPA12, respectively, and the elongation at break and impact strength decreased by 17.4% and 17.2%, respectively, compared with NPA12. These mechanical properties show that the nanosilica in D-nanosilica/PA12 has a good reinforcing effect, so that the tensile strength, modulus and impact strength of nylon 12 SLS parts are simultaneously improved; while the nanosilica in M-nanosilica/PA12 has a very limited reinforcing effect. While the tensile strength and modulus of the nylon 12 SLS part are slightly improved, the elongation at break is greatly reduced, and the reinforcing effect is similar to that of the conventional micron-sized filler. These results are mainly caused by the following two reasons: First, the nanosilica dispersed uniformly in nanosize in D-nanosilica/PA12 has a strong interfacial interaction with the nylon 12 matrix, while the nanosilica present aggregated in micron-size in M-nanosilica/PA12 has a weak interfacial interaction with the nylon 12 matrix, and the loose agglomerates form stress concentration points, which destroy the mechanical properties of the SLS parts, and thus the nanosilica in D-nanosilica/PA12 has better reinforcing role than nanosilica in M-nanosilica/PA12. Secondly, D-nanosilica/PA12 has better thermal

TABLE 2.25 Optimum energy density of D-nanosilica/PA12, M-nanosilica/PA12, and NPA12 and mechanical properties of their sintered parts.

Sample	Optimum energy density (J/mm^2)	Tensile strength (MPa)	Elongation (%)	Tensile modulus (GPa)	Impact strength (kJ/m^2)
D-nanosilica/PA12	0.1	46.3	20.07	1.98	40.2
M-nanosilica/PA12	0.08	38.6	17.21	1.74	30.4
NPA12	0.09	38.3	20.83	1.42	36.7

stability than M-nanosilica/PA12 and NPA12, so in the SLS process, the material decomposition has smaller influence on the mechanical properties of SLS parts than those of M-nanosilica/PA12 and NPA12. Therefore, the mechanical properties of D-nanosilica/PA12 SLS parts are higher than those of M-nanosilica/PA12 and NPA12.

Fig. 2.55A and B show the SEM microscopic shape of the impact sections of SLS parts of NPS12 and M-nanosilica/PA12, respectively. From the figure, large smooth and banded areas on the section of the SLS parts of NPA12 and M-nanosilica/PA12 can be seen, which are presented as brittle fractures, indicating that the fractures are easily extended and the fractured samples require less energy. On the SLS parts section of M-nanosilica/PA12, the presence of aggregates of micron-sized nanosilica is found. The adhesion of these microscale aggregates to the nylon 12 matrix is very poor, which leads to the brittle fracture of impact samples, thus reducing the impact strength.

Fig. 2.56 shows the SEM microscopic shape of the impact section of the SLS part of D-nanosilica/PA12 SLS. It can be seen from the figure that the

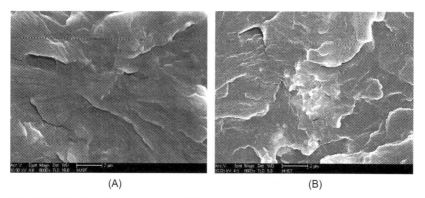

FIGURE 2.55 (A) Microscopic shape of the impact section of the SLS parts of NPA12 and (B) M-nanosilica/PA12.

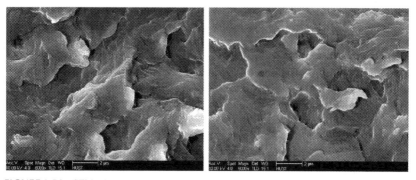

FIGURE 2.56 SEM microscopic shape of the impact section of the D-nanosilica/PA12 SLS parts.

section of SLS part of D-nanosilica is rougher compared with the section of SLS part of NPA12 and M-nanosilica/PA12, and there are a large number of shear yield area and split pins. Due to the need to consume more energy to form these fracture features, the SLS parts of nanosilica/PA12 have higher impact strength.

2.4.3 Composite powder material of nylon 12 coated aluminum

Filling nylon SLS parts with metal powder can not only improve the mechanical properties such as strength, modulus and hardness of nylon SLS parts, but also give the metal appearance, high thermal conductivity and heat resistance to the parts, therefore, it is widely concerned. 3D System and EOS, which are world leaders in SLS material research and development, have introduced commercial aluminum powder-filled nylon composite powder materials.

At present, the filler/nylon composite powder for SLS is mainly prepared by mechanical mixing method, however, when the particle size of the filler powder is very small (such as powder particle size is less than 10 μm), or when the filler (such as metal powder) has a specific gravity density much greater than nylon, the mechanical mixing method is difficult to uniformly disperse the inorganic filler particles in the nylon matrix, and for the mechanically mixed metal/nylon composite powder, the metal powder particles tend to be segregated due to powder transportation and SLS powder spreading process when there is large difference in the density of the metal and nylon, which causes a certain nonuniform distribution of metal particle agglomerates in the SLS parts of the composite material, and these agglomerates often form mechanical defects, so that the mechanical properties of formed parts are degraded.

Another method for preparing the polymer/metal composite powder is film coating method in which the polymer is coated on the outer surface of metal particle by a certain process to form a polymer-coated metal composite powder. In the polymer-coated metal composite powder, since the polymer is uniformly coated on the outer surface of the metal particles, the metal and the nylon powder are mixed uniformly, and segregation does not occur during transportation and powder leveling, so that the above disadvantages of the mechanically mixed metal/nylon composite powder are completely overcome. At present, polymer-coated metal composite powder is mainly used to prepare metal parts of SLS through indirect method, but few work is done to use polymer-coated metal composite powder for metal powder reinforced polymer SLS parts, and the current polymer-coated metal powder is mainly prepared by spray drying method, and the polymer used is emulsion PMMA and its derivatives.

This section proposes a new film coating process: the solvent precipitation method is used to prepare nylon 12 coated aluminum composite powder for aluminum powder reinforced nylon SLS parts, and the microscopic shape, particle size and particle size distribution of nylon 12 coated aluminum composite powder, thermal properties and sintering properties of aluminum powder to nylon 12, as well as the influence of aluminum powder content (mass fraction) and particle size on the mechanical properties and precision of nylon 12 SLS parts.

2.4.3.1 Preparation and characterization of nylon 12 coated aluminum composite powder

2.4.3.1.1 Preparation process of nylon 12 coated aluminum composite powder

The nylon 12, the solvent, the aluminum powder and the antioxidant are put into the jacketed stainless steel reaction kettle in proportion, the reaction kettle is sealed, vacuumed, and protected by N_2 gas, and gradually increased to 150°C at a rate of 2°C/min gradually, so that nylon 12 is completely dissolved in the solvent, holding temperature and pressure for 2 hours. Under vigorous stirring, the temperature is gradually cooled to room temperature at a rate of 2°C/min, and the nylon 12 takes the aluminum powder particles as nucleus gradually, and crystallizes on the outer surface of the aluminum powder particles to form a nylon-coated metal powder suspension. The coated metal powder suspension is taken out from the reaction kettle and distilled under reduced pressure to obtain the powder aggregate. The recycled solvent can be reused. The obtained powder aggregate is vacuum-dried at 80°C for 24 hours, and then ball-milled in a ball mill at a speed of 350 rpm for 15 minutes and sieved, and the powder having a particle diameter of 100 μm or less is selected to obtain the nylon 12 coated aluminum composite powder material. The mass fractions of aluminum powder are 10%, 20%, 30%, 40%, and 50%, respectively. Five types of nylon 12 coated aluminum composite powders are respectively recorded as Al/PA (10/90), Al/PA (20/80), Al/PA (30/70), Al/PA (40/60), and Al/PA (50/50).

2.4.3.1.2 Characterization of powder materials

2.4.3.1.2.1 Particle size and particle size distribution The particle size distributions of Al powder, Al/PA (50/50) and Al/PA (20/80) powders measured by laser diffraction particle size analyzer are shown in Fig. 2.58. The average particle sizes measured have the following meanings: volume average particle size, which refers to the data representing the average particle size calculated by volume distribution; median diameter, this value accurately divides the total amount into two equal parts, that is, 50% of the particles are larger than this value, and 50% of the particles are smaller than this value. The average particle size of the powders is listed in Table 2.26.

It can be seen from Fig. 2.57A that the particle size distribution of Al powder is 6−89 μm, and the particle size is mainly concentrated in 19−36 μm. It

TABLE 2.26 Average particle size of Al powder, Al/PA (50/50) and Al/PA (20/80).

Powder type	Volume mean diameter (μm)	Median diameter (μm)
Al powder	27.99	23.08
Al/PA(50/50)	38.90	35.65
Al/PA(20/80)	40.93	39.42

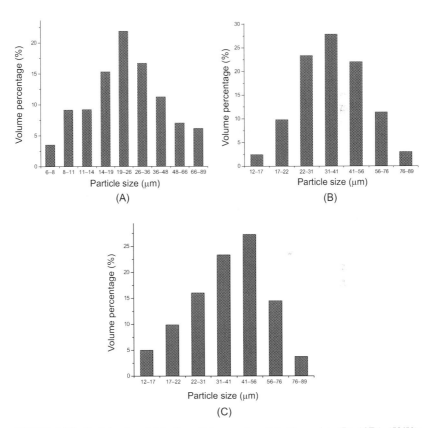

FIGURE 2.57 Particle size distribution of the powder: (A) Al powder; (B) Al/PA (50/50); (C) Al/PA (20/80).

can be seen from Fig. 2.57B that the particle size distribution of Al/PA (50/50) is 12−89 μm, and the particle size distribution is mainly concentrated in 22−41 μm, indicating the particle size distribution of Al/PA(50/50) is narrower than Al powder. Al/PA (50/50) does not contain 6−11 μm particles in

Al powder, and the larger particle size is greater than Al powder in quantity, which is mainly due to the coating of nylon 12 on the outer surface of Al particles, so that the particle size of the powder is increased. It can be seen from Fig. 2.57C that the particle size distribution of Al/PA (20/80) is 12–89 μm, and the particle size distribution is mainly concentrated in 22–41 μm, indicating that although the particle size distribution of Al/PA (20/80) is the same as Al/PA (50/50), but the main particle size is larger than Al/PA (50/50). This is because the content of Al/PA (20/80) aluminum powder is reduced compared to Al/PA (50/50), so that the nylon 12 is further increased, and the nylon 12 coating layer outside the aluminum powder particles is further increased.

As can be seen from Table 2.26, as the mass fraction of aluminum powder decreases, the content of nylon (mass fraction) increases correspondingly, leading to an increase in the thickness of the coating layer, and the average particle size of the powder also increases. In general, the particle size of two types of nylon 12 coated aluminum powders 10–100 μm, which is suitable for the SLS-forming process.

2.4.3.1.2.2 Microscopic shape of the powder The SEM pictures of Al powder are shown in Fig. 2.58. It can be seen from the figure that the Al powder particles used in the experiment have smooth surface and nearly spherical shape, thus ensuring that the composite powder particles obtained by coating the nylon 12 on its surface can also be close to spherical shape. This is very advantageous for the powder spreading in the SLS process.

Fig. 2.59 shows the SEM pictures of Al/PA (50/50). It can be seen from the figure that the particle shape of Al/PA (50/50) is similar to that of Al and is also nearly spherical. When the nylon 12 is cooled and crystallized, the Al powder particle is used as the nucleus to be gradually coated on the outer surface of the Al powder particles, and thus the obtained composite powder has a shape similar to that of the Al powder. Moreover, the surface of the

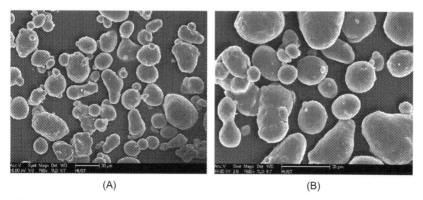

(A)   (B)

FIGURE 2.58 SEM pictures of Al powder. (A) 750×; (B) 1500×.

Polymer materials for additive manufacturing—powder materials Chapter | 2 **115**

FIGURE 2.59 SEM pictures of Al/PA (50/50). (A) 600×; (B) 1000×.

FIGURE 2.60 SEM pictures of Al/PA (40/60). (A) 600×; (B) 1000×.

particles in the composite powder is very rough, and no particles having a smooth surface are found, indicating that the Al powder particles are coated with the nylon 12, and no exposed Al particles are present.

Fig. 2.60 is the SEM pictures of Al/PA (40/60). It can be seen from the figure that the shape of Al/PA (40/60) is similar to that of Al/PA (50/50), but the irregularity is increased, and the powder particles are porous. Since the nylon content of Al/PA (40/60) is higher than that of Al/PA (50/50), the nylon coating layer of Al/PA (40/60) is thicker, which makes the particles tend to be irregular. The pores on the surface of the Al/PA (40/60) particles may be due to the excessive rate of solvent distillation.

Figs. 2.61, 2.62, and 2.63 are SEM pictures of Al/PA (30/80), Al/PA (20/80), and Al/PA (10/90), respectively. It can be seen that as the content of

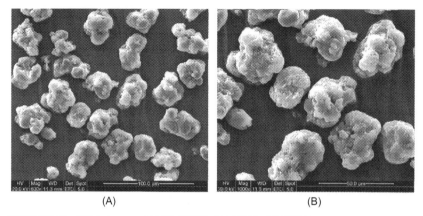

FIGURE 2.61 SEM pictures of Al/PA (30/80). (A) 600×; (B) 1000×.

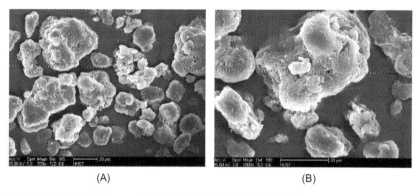

FIGURE 2.62 SEM pictures of Al/PA (20/80). (A) 750×; (B) 1500×.

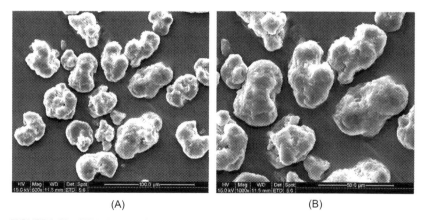

FIGURE 2.63 SEM pictures of Al/PA (10/90). (A) 600×; (B) 1000×.

nylon 12 increases, the coating layer of the particles gradually thickens and the composite powder becomes more irregular. Moreover, there are a plurality of Al powder particles being coated together by nylon, which is related to the thickening of the coating layer and the process control of ball milling, and there are many variable factors worth considering in the process of ball milling, such as the amount of ball loading, ball milling speed, ball milling time, etc. In general, in the nylon 12 coated aluminum composite powder prepared by the solvent precipitation method, the aluminum powder particles can be completely coated without the presence of exposed aluminum powder particles; the shape of the composite powder is similar to aluminum powder, and the less the nylon 12 content, the shape of the composite powder is closer to the aluminum powder; some composite powder particles have a large amount of pores on the surface, which may be caused by the solvent distillation rate being too fast; in the composite powder of thick nylon 12 coating layer, there are a plurality of aluminum powder particles being coated together, which can be solved by controlling the ball milling process.

2.4.3.1.2.3 Energy spectrum analysis of powder Fig. 2.64 shows the SEM picture and EDX analysis chart of Al/PA (50/50). The EDX analysis window is shown by the cross on the SEM picture. It can be seen from the EDX analysis chart of Fig. 2.64B that the particles mainly contain C, N, O and Al elements, indicating that the nylon 12 has crystallized and coated on the outer surface of the Al powder particles. It can be seen from Fig. 2.64B that the particles also contain a very small amount of Si element, which may be due to a small amount of impurities being introduced

2.4.3.1.2.4 Analysis on thermal weight loss of powder Fig. 2.5 shows the TG curves of NPA12, Al/PA (50/50), and Al/PA (10/90) powders. Table 2.27 shows the thermal weight loss temperature of three types powders

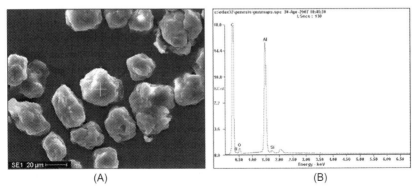

(A)   (B)

FIGURE 2.64 (A) SEM picture of Al/PA (50/50) and (B) EDX analysis chart (EDX analysis window is shown by cross on SEM picture).

TABLE 2.27 Thermogravimetric temperature of NPA12, Al/PA (50/50), and Al/PA (10/90).

Powder	$T_d - 5\%$ (°C)	$T_d - 10\%$ (°C)
NPA12	334.9	392.7
Al/PA(10/90)	376.6	418.2
Al/PA(50/50)	382.0	434.5

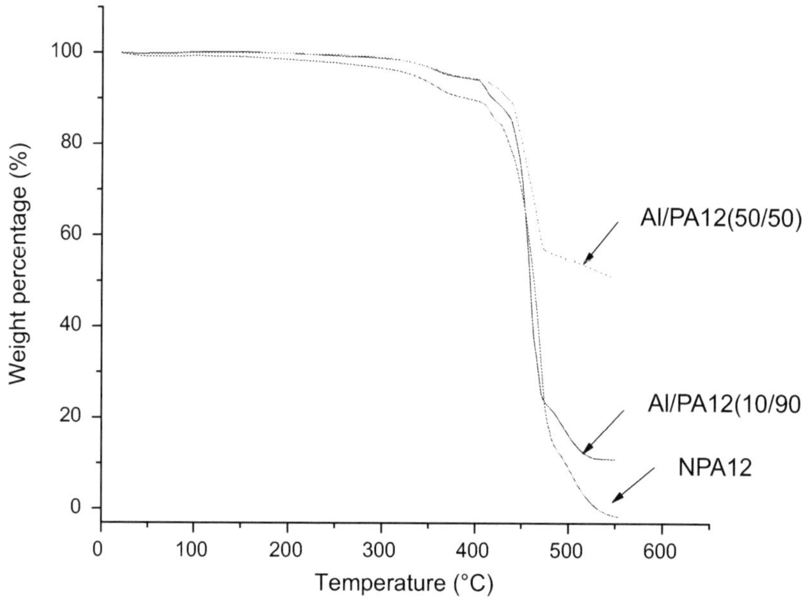

FIGURE 2.65 TG curves of NPA12, Al/PA (50/50), and Al/PA (10/90).

at the weight loss of 5% and 10% (recorded as $T_d - 5\%$ and $T_d - 10\%$, respectively).

It can be seen from Fig. 2.65 and Table 2.27 that the thermal stability of the composite powder is significantly better than that of pure nylon 12. The $T_d - 5\%$ and $T_d - 10\%$ of Al/PA (10/90) are 41.7°C and 25.5°C higher than that of pure nylon 12, respectively. The $T_d - 5\%$ and $T_d - 10\%$ of the 50% aluminum powder content are 47.1°C and 41.8°C higher than that of pure nylon 12, respectively. This may be caused by the barrier effect of the aluminum powder particles. When the active end of the nylon 12 main chain degrades, it loses its activity when it touches the aluminum powder particles, and no longer causes degradation of the remaining part of the main chain.

Polymer materials for additive manufacturing—powder materials Chapter | 2 119

2.4.3.1.2.5 Differential scanning calorimetry analysis of powder

Fig. 2.66 shows the DSC curves for the melting and crystallization processes of NPA, Al/PA (50/50) and Al/PA (10/90). The melting peak temperature (T_{mp}), melting start temperature (T_{ms}), melting end temperature (T_{me}), crystallization peak temperature (T_{cp}), crystallization start temperature (T_{cs}), and crystallization end temperature of these three types of powders can be obtained from Fig. 2.66 and listed in Table 2.28.

The crystallization time is obtained by Eq. (2.76):

$$t_c = \frac{(T_{cs} - T_{ce})}{r} \qquad (2.76)$$

where r is the rate of cooling, in this case, it is 5°C/min.

The width of sintering temperature window (W) is calculated according to Eq. (2.77):

$$W = T_{ms} - T_{cs} \qquad (2.77)$$

It can be seen from Fig. 2.66A that as the content of aluminum powder increases, the melting temperature of nylon 12 increases, which may be because that the amide group forms hydrogen bonding with the active hydrogen on the surface of the aluminum powder when the aluminum powder is coated on the nylon 12 during crystallizing, so that the nylon 12 molecular chains are arranged more closely, resulting in an increase in the melting temperature thereof.

It can be seen from Fig. 2.66B that the crystallization peak temperature, crystallization start temperature and crystallization end temperature of Al/PA (50/50) are higher than that of NPA, and the crystallization time is also greatly shortened, indicating that the heterogeneous nucleation of aluminum powder is exerted during the crystallization of nylon 12, so that the crystallization temperature of nylon 12 is increased and the crystallization rate is increased.

From Table 2.28, it can be seen that as the aluminum powder content increases, the sintering temperature window becomes wider. Generally, the wider the sintering temperature window of the powder material, the easier the temperature of the powder layer is controlled in the sintering temperature window, and the less the sintered member is warped and deformed. It is proved that the nylon-coated aluminum composite powder is more suitable for sintering than the pure nylon powder.

2.4.3.2 Selective laser sintering process characteristics of nylon 12 coated aluminum composite powder

2.4.3.2.1 Powder spreading performance

In order to obtain a better sintering effect, the powder is first required to have good powder spreading properties. The quality of the powder spreading is

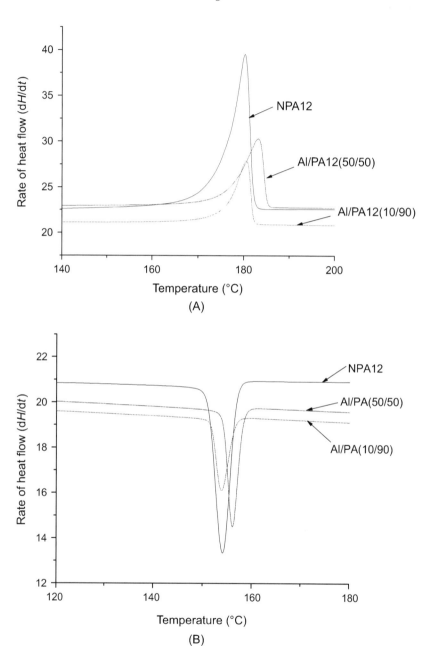

FIGURE 2.66 DSC curves of NPA, Al/PA (50/50), and Al/PA (10/90): (A) temperature rise; (B) temperature drop.

TABLE 2.28 Thermal transition temperature and sintering temperature window width of NPA, Al/PA (50/50), and Al/PA (10/90).

Powder type	T_{mp} (°C)	T_{ms} (°C)	T_{me} (°C)	T_{cp} (°C)	T_{cs} (°C)	T_{ce} (°C)	t_c (min)	W (°C)
NPA	180.2	175.2	182.0	154.1	156.9	151.2	1.14	18.3
Al/PA(10/90)	180.5	175.8	182.1	153.9	156.6	151.8	0.96	19.2
Al/PA(50/50)	183.2	177.3	185.1	156.1	157.2	154.2	0.60	20.1

directly related to the precision and mechanical properties of the formed parts. If the powder spreading is unevenly distributed, defects will be generated in the formed parts, resulting in a decrease in the precision and mechanical properties of the SLS parts. When there is good fluidity between the powder particles, the powder is evenly dispersed on the desired powder spreading surface during the powder spreading process. In general, spherical powder particles have better fluidity than irregular powder particles. It can be seen from the above analysis that the shape of the nylon 12 coated aluminum composite powder particles prepared by the solvent precipitation method is determined by the shape of the aluminum powder particles, and since the spherical or near-spherical aluminum powder is selected, the shape of the composite powder is also nearly spherical, thereby having good fluidity. In addition, the powder spreading performance of the powder is also related to the particle size. The larger the particle size, the better the powder spreading performance of the powder; the smaller the particle size, the greater the friction between the powder particles and the poorer the powder spreading performance of the powder. Generally, the average particle size of the powder should not be less than 10 μm, otherwise it would cause difficulty in powder spreading. From the analysis of the particle size of composite powder, it is found that the average particle size of the composite powder is 10 μm or above.

Based on the above factors affecting powder spreading performance, it can be seen that the nylon 12 coated aluminum composite powder prepared by the solvent precipitation method has good powder leveling performance.

2.4.3.2.2 Preheating temperature

The preheating temperature of the crystalline polymer is generally controlled to be close to T_{ms} but lower than T_{ms}, and T_{ms} can be obtained from the DSC curve of the powder. In practical applications, the preheating temperature is obtained by T_{ms} and gradual approximation experiment. The specific method is as follows: the preheating temperature experiment starts from 160°C, and the warping deformation and the purification of the formed parts are carefully observed. If the formed part is observed to be warped and deformed, the preheating temperature is raised by 3°C−4°C until the formed part does not occur warping deformation, and the formed part is easier to be purified, thus the preheating temperature of the material is found. The experimental results of the gradual approximation of the preheating temperature of NPA, Al/PA (50/50) and Al/PA (10/90) powders are listed in Table 2.29. From this experiment, it can be concluded that the preheating temperatures of NPA, Al/PA (50/50) and Al/PA (10/90) are 167°C, 168°C, and 170°C, respectively. It can be seen that the higher the aluminum powder content, the higher the preheating temperature of the composite powder. From the DSC curve of Fig. 2.66A, the higher the aluminum powder content, the larger the T_m of the composite powder, and the higher the preheating temperature.

TABLE 2.29 Experimental results of gradual approximation to preheating temperature.

Preheating temperature (°C)			
NPA	Al/PA (10/90)	Al/PA (50/50)	Observation results
160	160	160	The formed part warps severely, and the sintered layer appears to move when the powder is spread. The formed part is easy to be purified and the outline is clear.
162	162	163	The formed part warps, and the sintered layer appears to move when the powder is spread. The formed part is easy to be purified and the outline is clear.
165	166	168	The formed part warps slightly, and the sintered layer does not move when the powder is spread. The formed part is easy to be purified and the outline is clear.
167	168	170	The formed part does not warp, and the sintered layer does not move when the powder is spread. The formed part has clear outline.
169	170	172	The formed part does not warp, and the sintered layer does not move when the powder is spread. The formed part is not easy to be purified and the outline is not clear.

2.4.3.2.3 Effects of aluminum powder content on mechanical properties of selective laser sintering parts

The test pieces of different aluminum powder contents are sintered to carry out test for mechanical properties. The sintering process parameters of the test pieces are as follows: initial preheating temperature is 168°C–172°C, laser power is 20 W, scanning speed is 2000 mm/s, and slice thickness is 0.08 mm. Fig. 2.67 is the physical pictures of some sintered tensile samples, bent samples, and impact samples before and after testing.

Fig. 2.68 shows the curve of tensile properties of SLS parts changing with the content of aluminum powder. It can be seen from the figure that the tensile strength of sintered parts increases with the increase of aluminum powder content, while the elongation at break decreases with the increase in the powder content, indicating that the addition of the rigid aluminum powder particles increases the tensile strength of the sintered part, but reduces the flexibility of the nylon 12 matrix.

Fig. 2.69 shows the curve of bending properties of the sintered part changing with the aluminum powder content. It can be seen from the

FIGURE 2.67 (A) Tensile samples and (B) bent samples and impact samples.

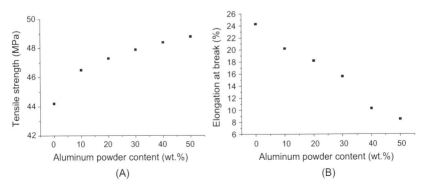

FIGURE 2.68 Effects of aluminum powder content on tensile properties: (A) tensile strength; (B) elongation at break.

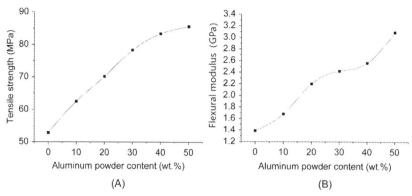

FIGURE 2.69 Effects of aluminum powder content on bending properties: (A) bending strength; (B) flexural modulus.

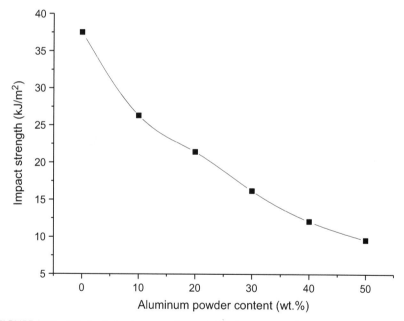

FIGURE 2.70 Effects of aluminum powder content on impact strength.

figure that the bending strength and flexural modulus of the sintered part increase with the increase of the aluminum powder content, indicating that the addition of rigid aluminum powder particles increases the bending strength and also increases the rigidity of the composite material.

Fig. 2.70 shows the curve of the impact strength of the sintered part changing with the aluminum powder content. It can be seen from the figure that as the aluminum powder content increases, the impact strength of the sintered part is significantly reduced, which is also due to the limiting effect of rigid aluminum powder particles to the molecular chain thermal motion of nylon 12. This also indicates that the aluminum powder cannot improve the strength and toughness of the sintered part at the same time. In actual use, the content of the aluminum powder can be adjusted according to the requirements of strength and toughness, thereby achieving a balance between the strength and toughness.

2.4.3.2.4 Effects of aluminum powder content on dimensional accuracy of selective laser sintering parts

Fig. 2.71 shows the curves of dimensional accuracy of the sintered samples in the X, Y, and Z directions changing with the aluminum powder content under the same sintering parameters. It can be seen from the figure that the dimensional accuracy of various composite powder sintered samples is a negative deviation due to volume shrinkage during powder sintering. The negative deviation of the sintered samples in the X, Y, and Z directions decreases with the increase of the aluminum powder content. For example, in the X direction, the negative deviation at 50% aluminum powder content is 1.53%, and the negative deviation of pure nylon is 3.2%, which is a reduction of 50%, indicating

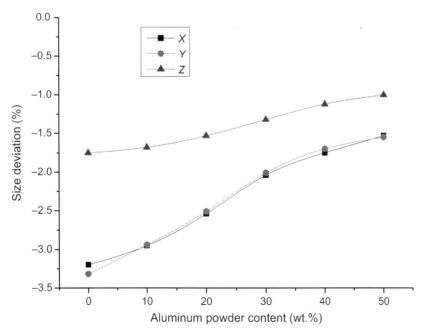

FIGURE 2.71 Effects of aluminum powder content on dimensional deviation in the X-, Y-, and Z-axis directions of the formed part.

that the dimensional accuracy of the sintered sample increases as the aluminum powder content increases. This is because as the content of aluminum powder increases, the content of the sinterable molten powder, that is, nylon 12, in the powder material is gradually reduced, thereby reducing the volume shrinkage during sintering and forming, and reducing the negative deviation of the size of the sintered part. The accuracy is improved.

It can also be seen from Fig. 2.71 that the dimensional deviations in the X and Y directions are not much different, and the deviation in the Z direction differs greatly from the dimensional deviation in the X and Y directions. This is because the scanning strategy used in the experiment is grouping turning, that is, the deviation along X-axis and the Y-axis are alternate layer by layer, so that the volume shrinkage and the secondary sintering degree of the formed parts in the X and Y directions are the same. Therefore, the dimensional deviations of the formed parts in the X and Y directions are not much different. Since there is a Z-axis "surplus" phenomenon in which the size is increased in the Z direction, the negative deviation in the Z direction is smaller than the negative deviation in the X and Y directions.

2.4.3.3 Dispersion state of aluminum powder particles and interface bonding with nylon 12

The dispersion state of the inorganic filler particles and the interface bonding with the polymer matrix has great influence on the properties of the composite material. In general, if the filler particles can be uniformly dispersed in the polymer matrix, and has good interface bonding with the polymer matrix, then the resulting composite material has higher performance.

Fig. 2.72 shows the sectional microscopic shape of Al/PA (50/50) bent sample. As can be seen from Fig. 2.72A, the aluminum powder particles are

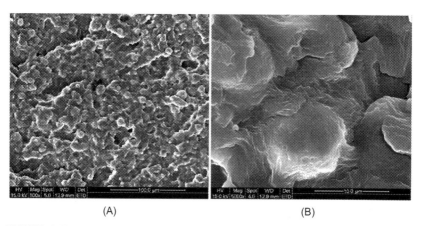

FIGURE 2.72 Sectional microscopic shape of Al/PA (50/50) bent sample: (A) 500×; (B) 5000×.

uniformly dispersed in the nylon 12 matrix without aggregates of aluminum powder particles. From the previous analysis, it is known that in the nylon 12 coated aluminum composite powder, the nylon 12 is preferably coated on the outer surface of the aluminum powder particles, so that the nylon 12 and the aluminum powder are mixed very uniformly, and the segregation occurs in the transportation and powder spreading can be effectively avoided. Therefore, in the SLS part of the nylon 12 coated aluminum composite powder, the aluminum powder particles are uniformly dispersed in the nylon 12 matrix without aggregation.

It can be seen from Fig. 2.72B that the outer surface of the aluminum powder particles on the cross section of the sample is very rough, and a layer of nylon 12 resin is attached, and the broken part is in the body of the nylon 12, and the results indicate aluminum powder and nylon 12 matrix have good interfacial bonding. The surface of aluminum powder with high polarity generally adsorbs many small polar molecular substances such as H_2O. The amide group in the nylon 12 has relatively high polarity, and the N, O elements in the amide group have lone pair electron, and it is easy to form a hydrogen bond with a polar small molecule adsorbed on the surface of the aluminum powder. Therefore, the aluminum powder forms a good interfacial bond with the nylon 12 matrix.

2.4.3.4 Effects of aluminum powder particle size on the properties of selective laser sintering parts

Nylon 12 coated aluminum composite powder with aluminum powder content of 50% mass fraction is prepared from three types of aluminum powders with average particle diameters of 9.36, 18.37 and 27.99 μm, respectively, and recorded as Al-9.36/PA (50/50), Al-18.37/PA (50/50), and Al-27.99/PA (50/50). Tensile samples and impact samples are formed using HRPS-III laser sintering system to study the effect of aluminum powder particle size on tensile properties and impact strength of SLS parts.

Fig. 2.73 shows the curves of tensile strength and elongation at break of the formed part changing with the particle size of the aluminum powder. Fig. 2.74 shows the curve of the impact strength of the formed part changing with the particle size of the aluminum powder. It can be seen from Figs. 2.73 and 2.74 that the tensile strength, elongation at break, and impact strength of the formed part increase with the decrease of the average particle size of the aluminum powder. The tensile strength, elongation at break and impact strength of the SLS parts of Al-9.36/PA (50/50) are greatly improved compared with the SLS parts of Al-27.99/PA (50/50), and the tensile strength is increased by 15.6%, the elongation at break is increased by 94.1% and the impact strength is increased by 103.1%. The use of aluminum powder with smaller particle size can increase the interface. Because the aluminum powder in the formed part is evenly distributed and the interface is well bonded,

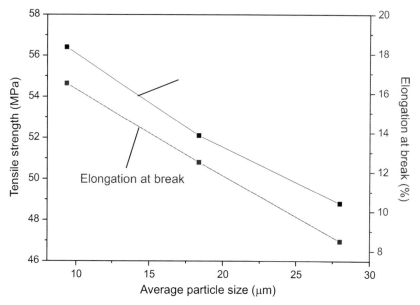

FIGURE 2.73 Effects of aluminum powder particle size on tensile properties of formed parts.

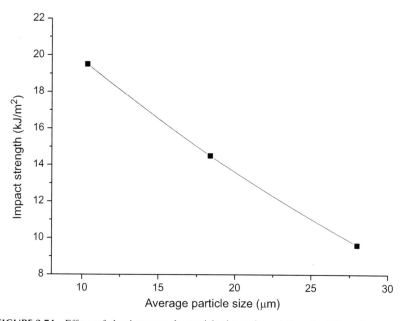

FIGURE 2.74 Effects of aluminum powder particle size on impact strength of formed parts.

the stress distribution is more uniform under stress, and the randomness of the crack propagation path is greatly increased, thus making the strength and toughness of the formed part improved.

Fig. 2.75 shows the sectional microscopic shape of the impact samples of Al-9.36/PA (50/50), Al-18.37/PA (50/50), and Al-27.99/PA (50/50). It can be seen from Fig. 2.75A−C that the smaller the particle size of the aluminum powder, the finer the cross-sectional crack of the sintered parts. When the particle size of the aluminum powder is large, the fracture holes appear obviously on the cross section. The holes are left when the aluminum powder particles fracture at the interface and separate from the section. The smaller the particle size of aluminum powder, the larger the interface between aluminum powder and nylon 12 resin, the randomness of crack propagation path is greatly increased, so there are more and smaller cracks, and the stress distribution is more uniform, thus improving the mechanical properties of the formed parts.

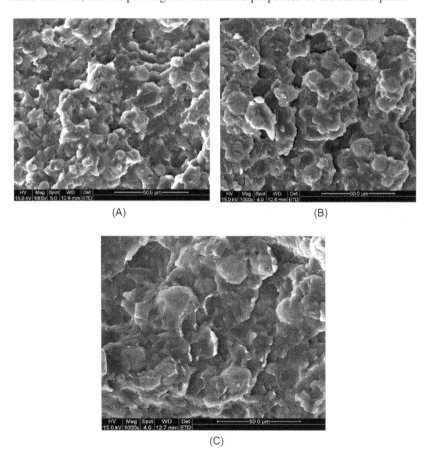

FIGURE 2.75 Sectional microscopic shape of the impact samples: (A) Al-9.36/PA (50/50); (B) Al-18.37/PA (50/50); (C) Al-27.99/PA (50/50).

2.4.4 Nylon 12/copper composite powder material

The manufacturing cycle can be greatly shortened and the cost can be greatly reduced by directly manufacturing the composite powder material of polymer and metal to injection mold through SLS technology, thus bringing benefits to the enterprise. Moreover, this forming method is not limited by the complicated shape of the mold, and the postprocessing process is simple, thereby eliminating many complicated processes of the machining mold. In 1998, DTM developed a nylon 12/metal composite for the rapid manufacturing of injection molds through the SLS process. In the composite material, the copper accounts for 70%, nylon 12 accounts for 30%, the density of the formed part is 3.5 times that of pure nylon 12, about 3.45 g/cm^3, the thermal conductivity is good, and the tensile strength is up to 34 MPa, the stretching modulus can reach 3.4 GPa. The mold prepared by the SLS-forming of the composite material does not need to be degreased, copper infiltrated, etc., as long as it is surface-sealed, polished, and finally packaged in a metal alloy for injection molding. Because of its low price and simple process, it is widely used in the trial production of new products and the production of small batches of plastic products.

2.4.4.1 Preparation and characterization of nylon 12/copper composite powder material

Nylon 12/copper composite powder materials are prepared by two methods: mechanical mixing and solvent precipitation coating.

The procedure of preparing the nylon 12/copper composite powder material by mechanical mixing method is as follows: first, the nylon 12 powder as a matrix material is sieved to within 50 μm; the copper powder is added in proportion and mixed in a ball mill for 2 hours. The copper powder is selected from electrolytic copper powder of 200−400 mesh. The mass ratio of nylon 12 to copper powder is 1:5−1:2, and the volume ratio of the mixing ball to the powder raw material is about 1:3.

The preparation process of nylon 12/copper composite powder material by solvent precipitation coating method is the same as that of nylon 12 coated aluminum composite powder.

Figs. 2.76, 2.77, 2.78, and 2.79 are SEM pictures of pure nylon 12 powder, electrolytic copper powder, nylon 12 coated copper composite powder, and nylon 12/copper mechanical mixed powder, respectively.

It can be seen from Fig. 2.76 that the nylon 12 powder has fewer spherical particles, most of the particles have irregular shapes, sharp edges and corners, and the particle size is not uniform. Most of the particles have a diameter of 40 μm and a maximum particle diameter of 50 μm.

It can be seen from Fig. 2.77 that the electrolytic copper powder has a small particle size but agglomerates to form a spike-like morphology, and the maximum size of the agglomerates is also about 50 μm.

Polymer materials for additive manufacturing—powder materials Chapter | 2 **131**

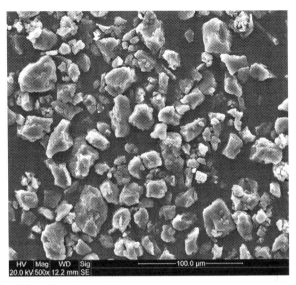

FIGURE 2.76 Nylon 12 powder particles.

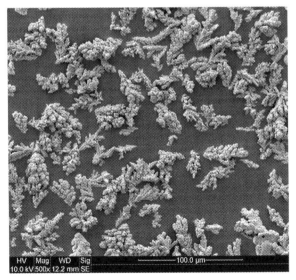

FIGURE 2.77 Electrolytic copper powder particles.

No bare copper powder was observed in Fig. 2.78, indicating that the copper powder was all coated with nylon 12, and the powder shape was relatively uniform. The particle size was mostly 30–50 μm, and the maximum particle diameter was not more than 100 μm.

FIGURE 2.78 Nylon 12 coated copper composite powder particles.

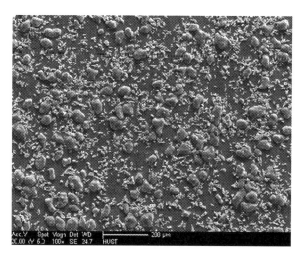

FIGURE 2.79 Topography of nylon 12/copper mechanically mixed powder particles.

The large powder particles in Fig. 2.79 are nylon powder. The small powder particles are mainly formed by the dispersion of copper powder agglomerates by ball milling. There is also a small amount of nylon 12 fine powder. The dispersion state of the composite powder is not very uniform.

2.4.4.2 Selective laser sintering-forming process of nylon 12/copper composite powder and its parts properties

2.4.4.2.1 Selective laser sintering-forming process

In the SLS process of nylon 12/copper composite powder, nylon 12 absorbs laser energy and melts. The liquid nylon 12 produces viscous flow filled

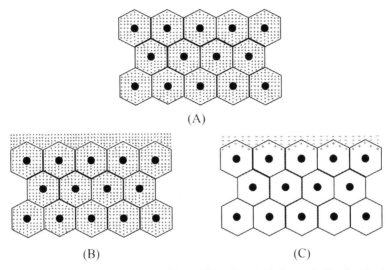

FIGURE 2.80 Changes in powder particle configuration (A) before and (B) after the laser scanning process. The white is polymer and the black is metal particles.

void of copper powder particles while wetting the surface of the copper particles, and the copper particles close to each other under the action of Laplace force due to the surface tension of the liquid resulting in position reconfiguration. When the particles are reconfigured, the friction between the irregularly shaped powder particles constitutes resistance thereto. The entire process can be represented by Fig. 2.80.

$$\sigma = \gamma \left(\frac{1}{r_1} + \frac{1}{r_2} \right) \tag{2.78}$$

In Eq. (2.78), γ represents the surface tension of the binder liquid, and r_1 and r_2 represent two radii of curvature of the curved liquid surface mutually perpendicular, the equation indicates that the bending liquid level of the viscous flow polymer between the two particles generates pressure σ that can affect the metal particles, especially when the curvature of the liquid surface is small, and the surface tension of the liquid binder is large, the pulling force is also large. Under the dual actions of Laplace force and polymer solidification shrinkage traction, the metal particles produce position reconfiguration. As the temperature is lowered after the scanning, the nylon 12 is solidified, and the copper particles are wrapped therein to form a SLS part having relatively high strength and good thermal conductivity, which is based on nylon 12.

2.4.4.2.2 Interface formation between nylon 12 and copper

Due to the lack of 3D translational symmetry of the metal surface atoms, the electron wave function near the surface changes, causing the surface to form an electronic state different from the inside of the crystal, which inevitably affects the arrangement of surface atoms. The change of the atomic arrangement is fed back to the surface electron wave function, which leads to the establishment of a self-consistent potential in the surface region which is different from that in the crystal. The main feature is to form a surface electrical double layer, as shown in Fig. 2.81. In the figure, the hexagon is the primitive cell of each atom. The small black dot at the center of the hexagon indicates the center position of the atom, and the dense black dots around it represent the free electron cloud. If there is no surface effect, the density of the electron cloud is uniform (Fig. 2.81A). Due to the existence of the surface, electrons with larger kinetic energy can pass through the surface barrier through tunneling effect and sparse electron cloud is outside the metal adjacent to the surface (Fig. 2.81B), while the electron cloud density in the metal adjacent to the surface is reduced, thus forming an electrical double layer (Fig. 2.81C).

Nylon 12 has strong polarity throughout the molecular chain due to the amide group in the structure, that is, the positive charge center and the negative charge center do not coincide with each other in the molecule. When an external electrical field exists, the positive and negative charge centers in the polar molecule will change under the influence of the electric field, as shown in Fig. 2.82.

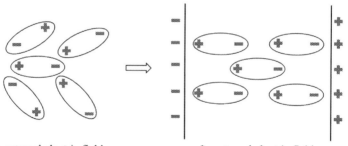

No external electric field   In external electric field

FIGURE 2.81 Schematic diagram of forming and structure of the electrical layer on the surface of metal: (A) uniform electron cloud without surface influence; (B) sparse electron cloud adjacent to the outside of the surface metal; (C) surface electrical double layer structure.

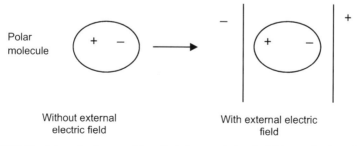

FIGURE 2.82 Schematic diagram of the effect of external electric field on molecular polarity.

Metallic copper and nylon 12 have different electronic band structures, and when they are in contact, electron transfer occurs to form an electric double layer. The electrostatic force generated by such an electrical double layer is a major contributor to the bond strength.

2.4.4.2.3 Crystallization and melting characteristics of nylon 12/copper composite powder

Fig. 2.83 shows the DSC melting curve of nylon 12/copper mechanically mixed composite powder measured at a heating rate of 10°C/min. The initial temperature, peak top temperature, and end temperature of the melting peak of nylon 12 in the mixed powder are 176.1°C, 183.0°C, and 185.9°C, respectively, and the latent heat of fusion is 28.1 J/g. Compared with the melting peak of pure nylon 12 (see Fig. 2.36), the initial temperature is 1.9°C higher, the end temperature is 0.9°C lower, that is, the melting range is shorter.

Fig. 2.84 is the DSC curve of nylon 12/copper mechanically mixed composite powder dropping from a molten state at a rate of 10°C/min to room temperature. The crystallization start temperature is 161.3°C, the peak top temperature is 154.8°C, and the end temperature is 151.8°C. Compared with the crystallization peak of pure nylon 12 (see Fig. 2.37), the crystallization start temperature of the composite powder is increased by 6.5°C. It indicates

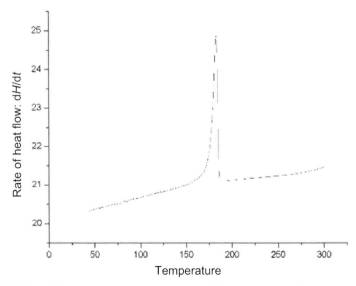

FIGURE 2.83 DSC temperature rise curve of nylon 12/copper mechanically mixed composite powder.

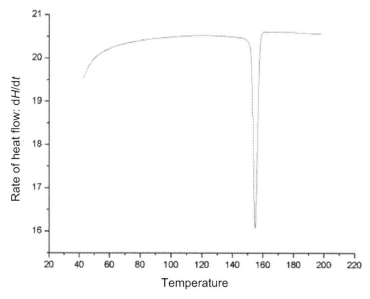

FIGURE 2.84 DSC temperature drop curve of nylon 12/copper mechanically mixed composite powder.

that the copper powder plays a heterogeneous nucleation role in the crystallization process of nylon 12.

2.4.4.2.4 Properties of sintered parts

The nylon 12/copper composite powder is sintered by HRPS-III laser sintering system. Sintering parameters are set as follows: preheating temperature is 165°C, single-layer thickness is 0.15 mm, laser power is 15 W, and scanning speed is 2000 mm/s. The formed test pieces are shown in Figs. 2.85 and 2.86.

In the case where the weight ratio of the nylon 12 to the copper powder is the same, the strength of the samples of the film-coated powder and the mechanically mixed powder after sintering is different. Fig. 2.87 reflects the above situation.

It can be seen from Fig. 2.87 that the bending strength of the two composite powder sintered parts increases with the increase of the laser energy density, and at the same laser energy density, the strength of the coated powder sintered parts can be twice of the mechanically mixed powder sintered part, which can be attributed to the following reasons: (1) Because the density difference between nylon 12 and copper is large, it is difficult to mix uniformly through the mechanical mixing, and it is easy to cause segregation. Therefore, insufficient adhesion occurs in the position where the amount of nylon 12 particles is small during laser scanning, so that the

FIGURE 2.85 Sintered tensile samples of nylon 12/copper mechanical composite powder.

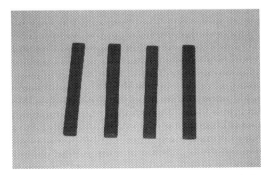

FIGURE 2.86 Sintered impact test samples of nylon 12/copper mechanical composite powder.

strength of the sintered part is poor. The coated copper powder eliminates the above weakness from the powder preparation, so the sintered part obtains a relatively high strength. (2) The absorption ratio of nylon 12 and copper powder to CO_2 laser with a wavelength of 10.6 μm is 0.75 and 0.26, respectively, which has a big difference. The film powder when subjected to laser scanning is basically equivalent to scanning of nylon 12 itself, while mechanically mixed powder, due to the obvious low absorption rate component, has the absorption rate much lower than 0.75, so that the temperature rise of nylon 12 is small in the mechanically mixed powder under the same laser energy density, which affects the adhesion. (3) The film-coated powder is basically bonded to the surface of the same material (surface of nylon 12) under the action of laser heat. Similar to Frenkel's flow bonding theory, it does not need to infiltrate and spread the heterogeneous surface, but nylon 12 particles in the mechanical mixed powder should not only adhere to the surface of the copper powder under the action of laser, but also complete the bonding between them. The effect is not as good as that of the film powder in a very short scanning time, so the strength is relatively poor.

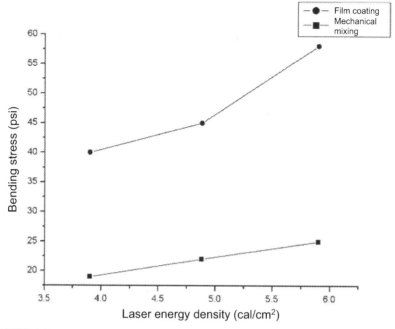

FIGURE 2.87 Comparison of bending strength between copper powder and nylon mechanically mixed and coated materials after SLS sintering.

2.4.5 Composite powder material of nylon 12/potassium titanate whisker

The blending modification method has been widely used in the reinforced modification of polymers because of its low cost and ease of use. However, since the particle size of the powder required for SLS-forming is below 100 μm, in order to ensure good laser sintering performance of the powder, the geometric shape of the powder requires a spherical shape or a near-spherical shape. For the above reasons, the existing commercial nylon composite powder materials are mainly glass beads-reinforced nylon powder, and the use of fiber reinforcement materials with good reinforcing effects in SLS materials is limited.

Nanoreinforced materials and whisker-reinforced materials are new types of reinforcing materials developed in recent years. They not only have small particle size, and a small amount of addition in the polymer can greatly improve the mechanical properties of the material, and the effect on the processing performance is also small. However, due to the particle size being too small, or having a large aspect ratio structure, the dispersion is difficult, and the conventional blending method is not suitable for preparing the SLS composite material. Therefore, this section will explore the preparation of composite nylon powder suitable for SLS by solvent precipitation method and study its properties.

2.4.5.1 Preparation of nylon 12/potassium titanate whisker composite powder

The nylon 12, potassium titanate whisker (PTW), solvent, and auxiliary agent are added into the reaction kettle. The ratio of nylon 12 granules to solvent is 1:7. The mixture is heated up to 150°C under nitrogen protection, after stirring, it is kept under a constant temperature for 1–2 hours, and discharged after natural cooling, then the nylon-coated PTW powder can be obtained by distilling of the ethanol under vacuum.

2.4.5.2 Properties of nylon 12/potassium titanate whisker composite powder

The microscopic shape of PTW and nylon 12/PTW composite powder is shown in Figs. 2.88–2.91. It can be seen from Fig. 2.88 that the aspect ratio of the PTW is large, which is beneficial to the reinforcement of the material. Fig. 2.89 and Table 2.30 shows the nylon 12 powder of PTW with a mass fraction of 10%. The particle size is uniform, the average particle size is 36.7 μm, and the geometrical shape is regular, which is close to the spherical shape. No exposed PTW can be observed in the figure, indicating that the PTW is completely covered by nylon 12. PTW plays a role of heterogeneous nucleation during the precipitation of nylon 12. Nylon 12 grows almost simultaneously on the surface of PTW, so the particle size is basically the same, and there is a nucleation when pure nylon 12 precipitates. In the process, the powder particles which nucleate early grow for a longer period of time than the powder particles which nucleate later, so the particle size is also larger, and thus the particle size distribution is not uniform. After

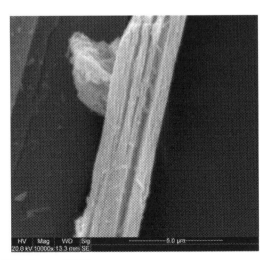

FIGURE 2.88 SEM picture of PTW.

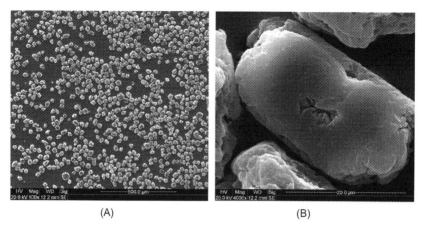

FIGURE 2.89 Nylon 12 composite powder material with 10 wt.% PTW under the magnification of (A) 100x and (B) 4000x.

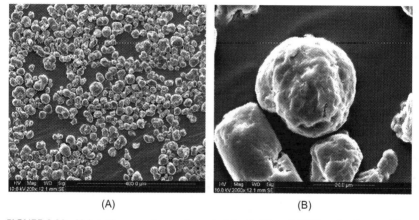

FIGURE 2.90 Nylon 12 composite powder material with 20 wt.% PTW under the magnification of (A) 200x and (B) 2000x.

continuing to magnify the powder particles and observe Fig. 2.89B, it is found that the powder surface is smooth but has holes.

Fig. 2.91 and Fig. 2.92 shows the nylon 12 powder of PTW with a mass fraction of 30%. The particle size of the powder is not uniform. Two peaks are visible from the particle size distribution map (Fig. 2.93D). One peak is 40.69 μm, the other peak is 19.80 μm. After further magnification and observation, it is found that the surface of the powder is extremely not smooth, and a large number of pores exist, unlike a single particle and more like an aggregate of many fine particles, as shown in Fig. 2.91B.

Comparing Figs. 2.89−2.91, it can be found that when a small amount of PTW is present in the solution, the powder has a more uniform particle size and

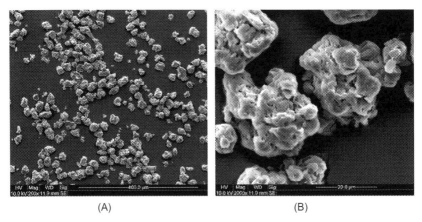

FIGURE 2.91 Nylon 12 composite powder material with 30 wt.% PTW under the magnification of (A) 200x and (B) 2000x.

TABLE 2.30 Particle size of pure nylon 12 and its composite powder.

Type of powder	Average diameter (μm)	Standard deviation
Pure nylon	34.90	11.03
Nylon 12 powder containing 10 wt.% PTW	36.66	7.20
Nylon 12 powder containing 20 wt.% PTW	44.49	10.04
Nylon 12 powder containing 30 wt.% PTW	38.01	10.10

a smoother surface due to the heterogeneous nucleation of PTW (Fig. 2.89). However, as the PTW content (mass fraction) increases, the geometrical shape of the powder begins to become more irregular, and a large number of pores appear in the particles. When the mass fraction of PTW reaches 30%, the particles are more like the aggregates of multiple particles, indicating that the growth of powder particles is not a single process, but a result of growth in multiple directions in space. This may be because PTW is easy to bridge due to its special high aspect ratio structure, and it is difficult to disperse. Especially when the mass fraction of PTW in the solution is large, it cannot be dispersed by mechanical stirring. Many PTWs reunite together due to the special form of PTW. The agglomerates are elongated in all directions of the space, which makes it have a plurality of growth points in the same particle, and the particles exhibit conditions of disorderly pile and porous hole due to the interaction of the plurality of growth points. Due to the excessive number of crystal nuclei,

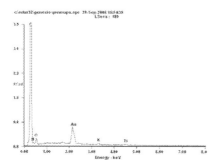

Element	Mass fraction %	Atomic fraction %
CK	51.78	59.99
NK	19.57	19.44
OK	20.77	18.06
KK	03.41	01.21
TiK	04.47	01.30
Matrix	Correction	ZAF

FIGURE 2.92 Energy spectrum of nylon 12 powder with 30 wt.% PTW.

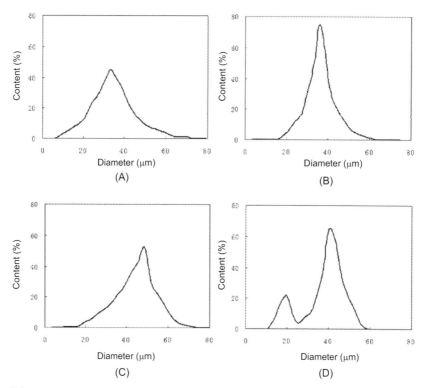

FIGURE 2.93 Particle size curve of powder (A) pure nylon 12 powder; (B) nylon 12 powder containing 10 wt.% PTW; (C) Nylon 12 powder containing 20 wt.% PTW; (D) Nylon 12 powder containing 30 wt.% PTW.

the particles with less growth points have not grown up, and the nylon 12 has settled, so that some fine powder appears (Table 2.30).

The bulk density of pure nylon 12 powder and nylon 12/PTW composite powder is shown in Table 2.31. The nylon 12 powder containing 10 wt.% PTW

TABLE 2.31 Bulk density of pure nylon 12 powder and nylon 12/PTW.

Type	Pure nylon 12	PTW containing 10 wt.%	PTW containing 20 wt.%	PTW containing 30 wt.%
Bulk density (g/cm^3)	0.41	0.44	0.40	0.35

has the highest bulk density, while the density decreases with the increase of PTW content; the density of nylon 12 powder containing 30 wt.% PTW is not only 79.5% of nylon 12 powder containing 10 wt.% PTW, but also 85.3% of pure nylon 12 powder, which is closely related to the shape of the powder. When the particle geometry is more regular, the bulk density will be higher. Among the above several types of powders, the shape of the nylon 12 powder containing 10 wt.% PTW is the most regular, and as the PTW content increases, the surface of the powder becomes rougher and coarser, and there are a large number of pores in the powder particles, so the bulk density is reduced.

2.4.5.3 Laser sintering properties of nylon 12 potassium titanate whisker composite

2.4.5.3.1 Powder spreading performance

Good powder spreading is the premise of SLS-forming. The shape, size and aggregation state of the filler have different effects on the powder spreading. Spherical fillers are advantageous for powder spreading, and fibrous, crystalline and easily agglomerated fine powders are not conducive to powder spreading. The composite powders containing 10 and 20 wt.% PTW show good powder spreading performance, while the composite powder containing 30 wt.% PTW is also able to spread the powdering, but due to the fluffy powder of low density and partial adsorption on the powder roller, this part of the adsorbed powder falls to the surface of the powder layer after accumulating to a certain amount, so it is necessary to clean the powder after every period of time to ensure the normal progress of laser sintering.

2.4.5.3.2 Crystallization performance

The DSC temperature rise and temperature drop curves of pure nylon 12 and its composite powder materials are shown in Fig. 2.94. It can be seen from Fig. 2.94 that both pure nylon 12 and nylon 12/PTW composite powder materials have only one melting peak and the melting peaks are similar, indicating that there is only one crystal form, and the addition of PTW does not change the crystal structure of nylon 12.

Table 2.32 is the specific data obtained from Fig. 2.94. The composite nylon 12 powder containing 10 wt.% PTW has the highest melting point and

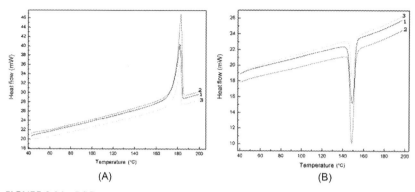

FIGURE 2.94 DSC temperature rise and temperature drop curves of pure nylon 12 and PTW composite powder materials: (A) DSC temperature rise curves; (B) DSC temperature drop curve. 1: nylon 12; 2: composite nylon 12 powder containing 10 wt.% PTW; 3: composite nylon powder containing 20 wt.% PTW.

TABLE 2.32 Basic thermal properties of nylon 12/PTW composite nylon 12 powder.

	T_{im} (°C)	T_{mp} (°C)	T_{em} (°C)	ΔT_m (°C)	T_{ic} (°C)	T_{ec} (°C)	ΔH_m (J/g)	ΔH_c (J/g)
Nylon 12	176.5	181.8	184.1	7.6	152.9	144.7	81.9	−51.9
10 wt.% PTW	178.0	182.6	184.5	6.5	152.5	145.3	83.3	−50.0
20 wt.% PTW	176.7	181.7	183.2	6.5	153.2	146.0	74.1	−43.7

the largest melting enthalpy. When the content of PTW is increased, the melting point and melting enthalpy decrease, which may be that PTW acts as a nucleating agent in the powder, but as the whisker content increases, excessive whiskers may become defects in the nylon 12 crystal, resulting in a decrease in the melting point of the nylon 12 and a decrease in melting enthalpy.

The crystallinity (CI) of nylon 12 and its composite powder materials can be calculated from the data in Table 2.32:

$$CI = \left(\Delta H_m / \Delta H_m^0\right) \times 100\% \tag{2.79}$$

where ΔH_m is the melting enthalpy; ΔH_m^0 is the melting enthalpy of the completely crystallized nylon 12, which is a constant. For the composite material, the crystallinity should be deducted from the filler portion, so Eq. (2.79) can be corrected to:

$$CI = \left(\Delta H_m / \Delta H_m^0\right) / (1 - f) \times 100\% \tag{2.80}$$

where f is the content of the filler, the crystallinity of the pure nylon 12 and the composite material can be calculated from Eq. (2.80), wherein the crystallinity of the nylon 12 powder containing 10 wt.% PTW is 12% higher than that of the pure nylon 12 powder, while the crystallinity of nylon 12 powder containing 20 wt.% PTW is 8% higher than that of pure nylon 12 powder, which further proves that the nucleation of whiskers promotes the crystallization of nylon 12, while the excessive whiskers cause the defect of lattice, on the contrary, reduces the crystallinity.

The crystallization rate of the nylon 12 powder can also be calculated by the DSC temperature drop curve:

$$t_c = \frac{(T_{ic} - T_{ec})}{r} \qquad (2.81)$$

where T_{ic}, T_{ec}, and r are the crystallization start temperature, the crystallization end temperature, and the temperature drop rate, respectively.

From this, the crystallization rates of pure nylon 12 powder, composite nylon 12 powder containing 10 wt.% PTW and 20 wt.% PTW are calculated to be 0.82, 0.72, and 0.72 minutes, respectively, indicating the crystallization initial temperature of the three types of powders is similar, but the crystallization rate of nylon 12 powder containing PTW is higher than that of pure nylon 12.

2.4.5.3.3 Laser sintering performance

The nylon 12 composite powder containing 10 and 20 wt.% PTW has good laser sinterability (Fig. 2.95), which is basically the same as pure nylon 12, and the surface is flat and smooth, and the nylon 12 composite powder containing 30 wt.% PTW has flat scanning surface, but there is curl at the corners, so the boundary is jagged, as shown in Fig. 2.96, but it can still be formed. The side of the laser-sintered body is not smooth, as shown in Fig. 2.97. A single-layer laser scanning image of PTW and nylon 12 powder

FIGURE 2.95 Single-layer laser scanning photo of nylon composite powder with 20 wt.% PTW.

FIGURE 2.96 Single-layer laser scanning photo of nylon composite powder with 30 wt.% PTW.

FIGURE 2.97 Picture of laser-sintered body of nylon composite powder containing 30 wt.% PTW.

FIGURE 2.98 Single-layer laser scanning photo of 20 wt.% PTW blended with nylon powder.

TABLE 2.33 Preheating temperature of composite powder with different PTW contents.

Type	10 wt.% PTW composite	20 wt.% PTW composite	30 wt.% PTW composite
Preheating temperature range (°C)	167–169	168–169	169

directly blended is shown in Fig. 2.98. The color of the laser-sintered body is lighter in Fig. 2.98, indicating that the dispersion of PTW is not good (PTW is yellow) and the surface is not smooth. There are a large number of shrinkage cavities, the boundaries are not neat, severely curled and shrunk to the center, and the laser sintering process cannot be carried out at all.

It can be seen from the above experimental results that the influence of the geometrical shape of the powder on the laser sintering process is very significant. The addition of PTW is not conducive to the laser sintering of the nylon 12 powder, but if it is coated by the nylon 12, the laser sintering at this time is changed to the laser sintering of nylon 12 powder, so that the effect of PTW on laser sintering can be minimized. However, if the mass fraction of the PTW is large, the dispersion is not good and the geometrical shape of the nylon powder is affected, which is not suitable for laser sintering. The large number of shrinkage cavities appearing on the surface of the single-layer laser scanning shown in Fig. 2.98 may be due to the poor dispersion of the PTW, which are aggregates that bridge each other. When the laser scans, the wettability of the melt to the PTW is poor, and it cannot be leveled due to the surface tension, so a large number of shrinkage cavities occur. Table 2.33 shows the preheating temperatures and laser sintering conditions of several types of powders.

2.4.5.3.4 Mechanical properties of selective laser sintering parts

Table 2.34 shows the mechanical properties of laser-sintered samples of nylon 12/glass beads and nylon 12/PTW composite powder. Glass beads have poor reinforcing effect on nylon 12, compared with pure nylon sintered samples, the tensile strength is almost constant, and only the bending strength and flexural modulus are improved. Even for the 40 wt.% glass beads-reinforced nylon with the best effects, it has a bending strength and flexural modulus of only 60.7 MPa and 1.84 GPa, more importantly, it is achieved at the expense of impact performance. As the content (mass fraction) of glass beads increases, the impact strength decreases drastically. When the content of glass beads increases from 30 to 50 wt.%, the impact strength is 56.2%, 50.3%, and 41.1% of pure nylon 12, respectively. This is because the modulus of the glass beads is much larger than the modulus of the nylon 12, so the modulus of the filling system is significantly increased. The glass beads are rigid, it is not subject to deformation under stress, and cannot terminate cracks or generate crazes to absorb impact energy, so the brittleness increases and the impact strength decreases.

The tensile strength, bending strength and flexural modulus of the nylon 12 composite powder material containing PTW are greatly improved. When the mass fraction of PTW reaches 20 wt.%, the tensile strength, bending

TABLE 2.34 Mechanical properties of laser-sintered samples of nylon 12 and its composite powder.

Properties	Tensile strength (MPa)	Impact strength (kJ/m^2)	Bending strength (MPa)	Flexural modulus (GPa)
Pure nylon 12	44.0	37.2	50.8	1.14
Nylon 12/glass beads (30 wt.%)	44.5	20.9	59.8	1.68
Nylon 12/glass beads (40 wt.%)	45	18.7	60.7	1.84
Nylon 12/glass beads (50 wt.%)	45.3	15.3	59.4	1.81
Nylon 12/PTW (10 wt.%)	52.5	34.3	72.18	1.518
Nylon 12/PTW (20 wt.%)	68.3	31.2	110.90	2.833
Nylon 12/PTW (30 wt.%)	52.7	20.3	85.29	2.682

strength and flexural modulus reach the maximum, which are 1.55 times, 2.18 times and 2.48 times, respectively, of pure nylon laser-sintered samples, and 1.52 times, 1.82 times, and 1.54 times of 40 wt.% glass beads filled samples. The impact strength does not decrease much, which is 83.9% of pure nylon 12 and 1.69 times of 40 wt.% glass beads filled samples. This indicates that PTW has a significant reinforcing effect on SLS formation of nylon 12 powder and is an more excellent reinforcing material than glass beads. However, the filling amount is low, when the filling amount exceeds 20 wt.%, the mechanical properties are significantly decreased; while the mass fraction of the glass beads can reach 40 to 50 wt.%. In theory, when injection molding is used, the mass fraction of the maximum value of PTW enhancement is between 30% and 35%, which is closely related to the degree of dispersion of PTW in nylon 12 and the geometrical shape of nylon 12 powder.

2.4.5.3.5 Analysis of impact section morphology

Figs. 2.99 and 2.100 are SEM graphs of the impact section of nylon 12/glass beads (40 wt.%) and nylon 12/PTW (20 wt.%) powder laser-sintered samples, respectively. It can be seen from Fig. 2.99 that a large number of glass beads are exposed on the cross section, and a large number of smooth round holes due to the removal of the glass beads are visible on the cross section, which may be due to the fact that the glass beads are too smooth, it is not well bonded with nylon 12 even after being treated with coupling agent. When a crack is generated by an external force, the glass beads and the nylon 12 body are first detached, and the crack cannot prevented, and the crack is more easily spread along the

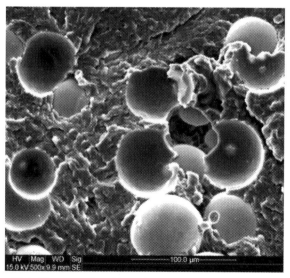

FIGURE 2.99 SEM of impact section of laser-sintered samples of nylon 12/glass beads (40 wt.%) composite powder.

Polymer materials for additive manufacturing—powder materials Chapter | 2 149

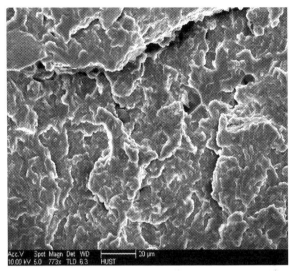

FIGURE 2.100 SEM of impact section of laser-sintered samples of nylon 12/glass beads (20 wt.%) composite powder.

joint of the glass beads and the nylon. Therefore, the impact strength of the glass bead-filled nylon powder laser-sintered sample is greatly reduced. It can be seen from Fig. 2.100 that the impact section of the nylon 12/PTW powder laser-sintered sample has no exposed PTW, and no void left after the PTW is pulled out, indicating that the PTW and the nylon 12 matrix are well bonded. The surface of the section is uneven, and there are a large number of filaments and cracks due to stretching, indicating that the nylon undergoes a toughness deformation before being fractured by an external force. The appearance of the crack indicates that the crack has changed direction due to being blocked during the growth process. This is why nylon 12/PTW exhibits good mechanical properties.

According to the mechanical property test results of Table 2.34, when the mass fraction of PTW is more than 20 wt.%, the mechanical properties of the laser-sintered sample begin to decrease, which can be explained from the SEM image of the impact section (as shown in Fig. 2.101), part of the exposed PTW is obviously agglomerated, and there are cavities in the agglomeration part. The cavity is not caused by the impact, it is the original cavity in the sample. It is these defects that decrease the density and the mechanical properties. When forming SLS, in order to ensure the forming precision and prevent the melting of the unsintered portion, the temperature of the sintered part of the melt can only be slightly higher than the melting point of the polymer, or even the outside of the powder is melted and the center is not melted, so the viscosity of the melt is large and liquidity is poor. Therefore, the nylon 12/PTW composite powder must be well dispersed before SLS-forming, and its geometrical shape is also highly demanded.

FIGURE 2.101 SEM of impact section of laser-sintered samples of nylon 12/glass beads (30 wt.%) composite powder under the magnification of (A) 2000x and (B) 4000x.

2.4.5.3.6 Thermal stability of composite powder

Curves a and b of Fig. 2.102 show the TG curves for pure nylon and nylon 12/PTW (20 wt.%) composite powders. The initial degradation temperature of pure nylon is 323°C, while the nylon 12/PTW (20 wt.%) composite powder is 360°C. At 450°C, pure nylon has degraded with weight loss of 50%, while nylon 12/PTW (20 wt.%) composite powder is only degraded by 31%, indicating that the addition of PTW is beneficial to improve the thermal stability of nylon 12.

2.4.6 Carbon fiber/nylon 12 composite powder material

The development of new formed powder materials is one of the keys to the continuous development, improvement and expansion of SLS technology. The carbon fiber reinforced nylon composite powder material is expected to provide sintered parts with better mechanical properties, so that it can satisfy some applications with high mechanical properties.

By preparing the carbon fiber composite powder coated with nylon 12, the fibers can be uniformly dispersed in the matrix of the sintered parts, thereby ensuring the reinforcing effect of the fiber on the matrix. The study of the prepared powder can further understand the various aspects of the properties of the composite powder and the effect of addition of fiber on some properties of the powder.

The addition of carbon fiber not only improves the mechanical properties of the sintered part, but also improves the thermal conductivity, electrical conductivity, wear resistance, and heat distortion temperature of the sintered parts. The forming process of composite materials has always been the research focus of composite materials. The forming method of composite materials is the bridge between the composition and practical application of

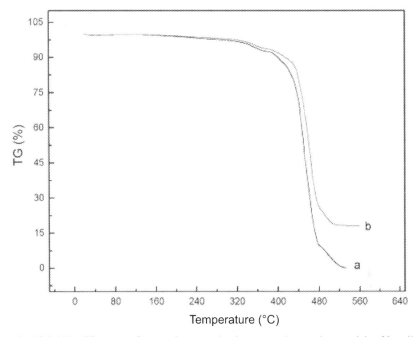

FIGURE 2.102 TG curves of pure nylon (a) and nylon composite powder containing 20 wt.% whisker (b).

composite materials, and how to obtain composite parts with complex shapes and structures has always been the problem in the development of composite materials. More importantly, the forming process has a very large impact on the internal structure and dispersion of the composite, so choosing the right forming method also has a decisive influence on the properties of final parts.

2.4.6.1 Preparation and characterization of carbon fiber/nylon 12 composite powder

2.4.6.1.1 Surface treatment of fiber powder

The carbon fiber without surface treatment has a smooth surface and lacks active groups bonded to the resin, and the interface bonding between the fiber and the matrix resin material is weak, which is not conducive to the effective transfer of stress when the composite material is under load, thereby reducing the mechanical properties of composite material. At present, there are a large number of research literatures on the surface treatment of carbon fibers. The treatment methods for surface modification of carbon fibers at home and abroad mainly include liquid phase oxidation, gas phase oxidation, anodic electrolytic oxidation, plasma oxidation treatment, coupling agent coating, and the like.

Combining the treatment effects of various oxidation methods and their requirements for equipment, the nitric acid oxidation treatment method which is easy to operate and has been widely recognized is finally selected.

Nitric acid is one of the most studied oxidants in liquid phase oxidation. Oxidation of carbon fibers with nitric acid produces carboxyl groups, hydroxyl groups, and acidic groups on the surface, and the amount of these groups increases as the oxidation time increases and the temperature increases. The various oxygen-containing polar groups and gullies contained on the surface of the oxidized carbon fiber are increased, which is advantageous for improving the interfacial bonding force between the fiber and the resin. A strong oxidizing agent and an aqueous solution of a high concentration of oxyacid are considered to be the most effective of various oxidizing agents. The increase of the carboxyl group increases the polarity of the fiber surface, thereby improving the wettability of the fiber and the resin, and facilitating the interface bonding. Moreover, the oxidizing agent has controllability on the surface of the fiber, and does not cause damage to the fiber, and the etching depth on the surface of the fiber is not large, which is beneficial to improve the bonding of the fiber and the resin.

The carbon fiber powder is treated with concentrated nitric acid at a concentration of 67%. The carbon fiber powder is placed in concentrated nitric acid, and treated by ultrasound at 60°C for 2.5 hours, then diluted with distilled water, and the diluted solution is subjected to vacuum filtration, and thus repeated until the pH of the filtrate is 7, and the filtered powder is placed in an oven and dried at 100°C for 12 hours.

2.4.6.1.2 Preparation process of composite powder

The specific process for preparing carbon fiber (CF)/nylon 12 (PA12) composite powder is as follows:

The PA12 granules, the surface-treated CF powder and the antioxidant are proportionally put into a jacketed stainless steel reaction kettle, adding a sufficient amount of ethanol solvent, and the reaction kettle is sealed, vacuumed, and protected by N_2 gas. The kettle is gradually increased from room temperature to 150°C−160°C at a rate of 1−2°C/min, so that nylon 12 is completely dissolved in the solvent, holding pressure and temperature for 2−3 hours. Under vigorous stirring, the temperature is gradually cooled to room temperature at a rate of 2°C−4°C, and the nylon 12, taking the CF powder as the nucleus, is crystallized and coated on the surface to form PA12 coated CF powder suspension. The coating powder suspension taken out from the reaction kettle is distilled under reduced pressure to obtain powder aggregate. The recovered ethanol solvent can be reused. The obtained aggregate is vacuum-dried at 80°C for 24 hours, and then ball-milled in a ball mill at a speed of 350 rpm for 20 minutes, and sieved to obtain the powder having a particle diameter of 100 μm or less.

Three types of CF/PA12 composite powders with different carbon fiber contents (mass fraction) are prepared by the above method, wherein the carbon fiber content is 30, 40, and 50 wt.%, respectively. The obtained CF/PA12 is grayish black powder and has no greasy feel.

2.4.6.1.3 Characteristics of composite powder

2.4.6.1.3.1 Particle size and particle size distribution Fig. 2.103 shows the particle size distribution diagram of three types of CF/PA12 composite powders. It can be seen from the figure that the particle size distribution of the prepared composite powder is wide, and as the carbon fiber content increases, the particle size distribution of the powder gradually widens.

Table 2.35 shows the particle size related parameter values calculated by the software of the laser particle size analyzer.

It can be seen from Table 2.35 that the average particle size of CF/PA12 composite powder increases with the increase of carbon fiber content, and the average particle size of the three types of composite powders is all suitable for SLS process. However, the parameter D_{90} indicating the powder coarse-end particle size index exceeded 100 μm, indicating that some powder particles with large long diameters in the composite powder could not be removed in the sieve, which has a certain influence on the fluidity of the

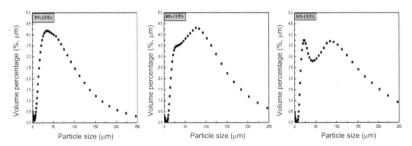

FIGURE 2.103 Particle size distribution diagram of three types of CF/PA12 composite powders.

TABLE 2.35 Measurement values of particle size related parameters of three types of powders.

	30 wt.% CF/PA12	40 wt.% CF/PA12	50 wt.% CF/PA12
$D[3,4]$ (μm)	51	67.38	68.54
D_{50} (μm)	37.59	52.20	46.86
D_{90} (μm)	111.62	143.71	157.35

powder, in SLS process, it may result in a nonsmooth surface of the powder bed and a relatively rough surface of the sintered part.

2.4.6.1.3.2 Microscopic shape of the powder
Fig. 2.104 shows SEM picture of surface-treated carbon fiber powder. The overall composition and distribution of the carbon fiber powder can be seen in Fig. 2.104A. The length of the fiber varies from 100 μm to as short as a few μm. From Fig. 2.104B, the surface shape of the carbon fiber can be observed, and the surface has several axial gullies. These gullies are very advantageous for increasing the surface roughness of the fiber, thereby increasing the bonding between the fiber and the resin.

Fig. 2.105 shows the microscopic shape of the CF/PA12 composite powder. As can be seen from Fig. 2.105A, the composite powder appears to consist of carbon fiber coated with nylon 12 and nearly isometric nylon 12 powder particles. It can be found from Fig. 2.105B that the surface of the carbon fiber is coated with a layer of nylon 12, and the relatively smooth surface of the exposed carbon fiber is not seen. A typical nylon 12 polymer surface appears, and the shape remains the original fiber shape. The nearly equiaxial powder particles in the figure may be formed by homogeneous nucleation of nylon 12, or may be formed by nucleation of carbon fiber powder slag as shown in Fig. 2.104.

2.4.6.1.3.3 Melting/crystallization of composite powder
Fig. 2.106 is a comparison of DSC curves for the melting and crystallization processes of three types of CF/PA12 composite powders. By comparing the melting curves of the three types of composite powders, it can be seen from Fig. 2.106A that as the carbon fiber content increases, a new small melting peak is added to the

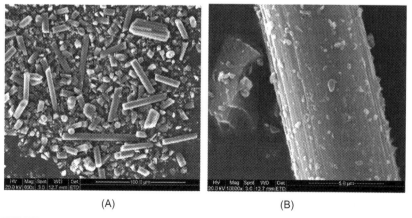

(A) (B)

FIGURE 2.104 SEM picture of surface-treated carbon fiber powder under the magnification of (A) 600x and (B) 10000x.

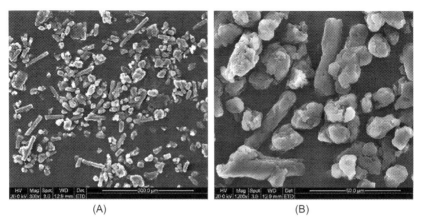

FIGURE 2.105 Microscopic shape of the CF/PA12 composite powder under the magnification of (A) 300x and (B) 1200x.

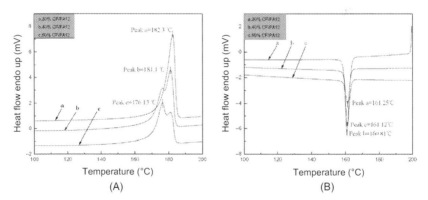

FIGURE 2.106 DSC curves of three types CF/PA12 composite powders: (A) melting process; (B) crystallization process.

left side of the original melting peak, and when the carbon fiber content is 50 wt.%, the height of this peak even exceeds the original peak. This peak at around 176°C is most likely formed by the nylon 12 coated on the carbon fiber. Because of the high thermal conductivity of the carbon fiber, the nylon 12 attached to the fiber has only a thin layer. During the heating process, due to good thermal conductivity, the nylon 12 coated on the carbon fibers preferentially melts and absorbs heat. As the carbon fiber content increases, the content of nylon 12 coated on the carbon fiber also increases, and the absorbed heat increases, and this peak also gradually increases.

In Fig. 2.106B, the crystallization peak positions of the three types of composite powders hardly change. Careful observation shows that as the carbon fiber content increases, the width of the crystallization peak becomes narrower,

TABLE 2.36 Comparison table of powder melting/crystallization parameter value.

	Pure nylon 12	30 wt.% CF/PA12	40 wt.% CF/PA12	50 wt.% CF/PA12
T_{mp} (°C)	180.2	182.3	181.1	176.13
T_{cp} (°C)	154.1	161.25	160.81	161.12

and the crystallization exotherm process is more rapid and concentrated, which can also be attributed to the good thermal conductivity of the carbon fiber.

Table 2.36 compares the melting/crystallization peak temperatures of the three types of CF/PA12 composite powders with pure nylon 12. It can be seen from the table that the crystallization peak temperature of the composite powder is about 6°C higher than that of the pure nylon 12 powder, which indicates that the carbon fiber powder plays a role of heterogeneous nucleation and contributes to the crystallization process of the nylon 12. For the SLS process, an increase in the crystallization peak temperature will result in a narrowing of the sintering temperature window. This means that the composite powder is more prone to warping deformation during sintering, which imposes stricter requirements on the temperature control of the equipment.

2.4.6.1.3.4 Thermal weight loss analysis of composite powder

Fig. 2.107A is a comparison of TG curves for three types of CF/PA12 composite powders and nylon 12 powder. It can be seen from the figure that the initial weight loss temperatures of the three types of composite powders are significantly higher than that of pure nylon 12, indicating that the addition of carbon fibers improves the thermal stability of the material. Fig. 2.107B shows the degradation differential curve. It can be seen that the weight loss peak of the 30 wt.% CF/PA12 composite powder is significantly shifted to the right compared with the pure nylon 12, indicating that the thermal stability of the composite powder is significantly improved.

Since the carbon fiber does not degrade in the test temperature range and the residual weight of the powder is taken into account, the weight loss rate of the composite powder at a certain temperature cannot be directly compared with the pure nylon 12. In order to eliminate the influence of carbon fiber on the calculation of weight loss rate, the weight loss rate of the composite powder with a carbon fiber content of 30 wt.% is calculated according to Eq. (2.82).

$$MR_1 = \frac{(MR_0 - 0.3)}{0.7} \tag{2.82}$$

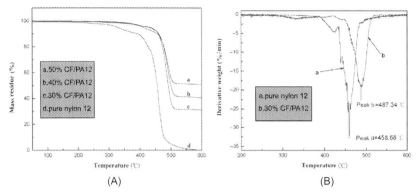

FIGURE 2.107 (A) TG curves and their (B) differential curves of CF/PA12 composite powders with different CF contents.

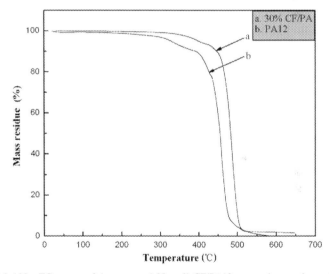

FIGURE 2.108 TG curves of the corrected 30 wt.% CF/PA12 composite powder and nylon 12 powder.

where MR_0 is the residual mass percentage of the composite powder and MR_1 is the residual mass percentage of nylon 12 in the composite powder.

The data is processed according to Eq. (2.82), and the plot is obtained, as shown in Fig. 2.108. It can be seen that the weight loss curve of the composite powder is significantly shifted to the right as compared with the pure nylon 12, further indicating that the thermal stability of the composite powder is improved.

2.4.6.2 Selective laser sintering properties of carbon fiber/nylon 12 composite powder

2.4.6.2.1 Powder spreading performance

Since there are some relatively large-diameter particles in the CF/PA12 composite powder, the powder has poor fluidity, the powder spreading performance is not so good, and the surface of the powder layer is rough. When the flow modifier such as calcium stearate is added, the powder spreading performance of the composite powder is improved, and SLS-forming can be performed.

2.4.6.2.2 Mechanical properties of sintered parts

It can be seen from Fig. 2.109 that the addition of carbon fiber makes the bending strength and flexural modulus of the sintered part of the composite powder material greatly improved compared with the sintered part of the pure nylon 12 powder material, and as the carbon fiber content increases, the bending strength and flexural modulus also increase. The bending strength of the three types of CF/PA12 composite powders sintered parts increased by 44.5%, 83.3%, and 114%, respectively; the flexural modulus increased by 93.4%, 129.4%, and 243.4%, respectively.

Fig. 2.110 shows the effects of different filler contents on the impact strength of sintered parts when the fillers are carbon fiber and aluminum powder, respectively. It can be seen that the impact strength of the sintered part of CF/PA12 composite powder material is gradually reduced as the fiber content increases compared with the pure nylon 12 sintered part, but the degree of decline is much smaller than that of the Al/PA12 composite powder. When the filler mass fraction is 50%, the impact strength of the Al/PA12 composite powder sintered part is only 15.9% of the pure nylon

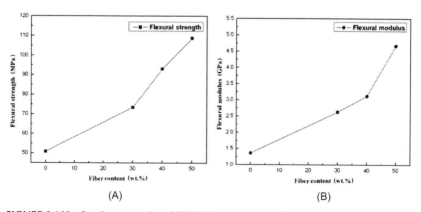

FIGURE 2.109 Bending properties of CF/PA12 composite powder sintered parts: (A) bending strength; (B) flexural modulus.

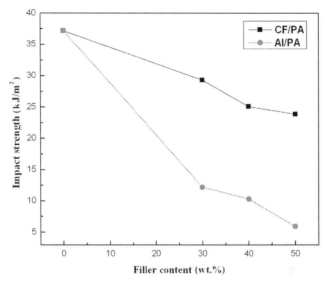

FIGURE 2.110 Effects of filler content on the impact strength of sintered parts.

12, and the CF/PA12 composite powder sintered part is 64.2% of the pure nylon 12. The composite powder formed by spherical or near-spherical fillers such as glass beads and aluminum powder and nylon 12 has improved strength and rigidity, but its impact resistance also decreases sharply with the increase of filler content. However, since the carbon fiber is fibrous filler, the fiber not only delays the crack propagation during the fracture process of the material, but also absorbs additional energy due to the fiber pull-out during the fracture, so that the decline in impact resistance is lowered relative to other fillers, still maintaining a certain degree of impact resistance.

2.4.6.2.3 Sectional shape of the sintered part

Fig. 2.111 shows a low-magnification SEM photograph of the bending test section of 40 wt.% CF/PA12 composite powder sintered test piece. It can be seen from the figure that the section is very rough, in which the carbon fiber dispersion is relatively uniform, and it can be seen that there is a nylon matrix between the carbon fibers, and no carbon fibers are overlapped. The orientation of the carbon fibers is randomly distributed, and it can be seen that the carbon fibers of various angles are exposed on the cross section. As shown in the position "a" of Fig. 2.111, the holes left after pulling out the fibers can also be seen. The nylon 12 matrix is distributed around the carbon fiber and both are cloud-like, which is the shape left by the nylon 12 matrix after undergoing large plastic deformation, indicating that the toughness of the nylon 12 matrix has been fully exerted. It can be considered that the uniform distribution of carbon fibers in the matrix is the

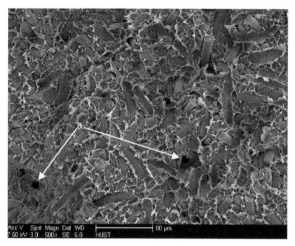

FIGURE 2.111 Overall shape of the section of the sintered part.

result of the fiber coated, because in the preparation stage of the composite powder, the surface of the carbon fiber is coated with a layer of nylon 12, and in the process of sintering the composite powder, the coated nylon 12 is melted and recrystallized, and the carbon fibers are still wrapped by the surrounding nylon 12, so that such uniform dispersion can be achieved. The uniform dispersion of carbon fibers in the matrix is a guarantee for the good mechanical properties of the sintered part, because once two or more carbon fibers are directly joined together, the weak interface between these fibers will become tiny sources of cracks around it. A large concentration of stress will occur, which will accelerate and promote the fracture of the entire substrate. The random orientation of the fibers causes the entire sintered part to exhibit approximately isotropic mechanical properties. From the shape of the nylon 12 around the carbon fibers, it can be said that the addition of the carbon fibers does not affect the nylon 12 to exhibit good plasticity in the fracture of the materials, so that the composite material also has certain toughness. The effect of uniform distribution of carbon fibers on the matrix can also be explained by "dispersion strengthening." The addition of carbon fibers limits the free movement of the nylon 12 molecular chains during deformation, thereby increasing the resistance of the plastic deformation of the matrix and improving the strength of the composite.

Two SEM pictures of high multiples of the cross section of two sintered parts were shown in Fig. 2.112. From the figure, the holes left after pulling out the fibers, the shape of the fibers, and the plastic deformation of the nylon 12 matrix around the fibers are observed. It can also be observed from the figure that the original smooth surface of the fiber is still maintained on the side wall of the fiber, indicating that the bonding strength between the fiber and the nylon 12 matrix is not as strong as that of the nylon 12 substrate; however, the nylon matrix left at the end of the fiber can still be seen, exhibiting

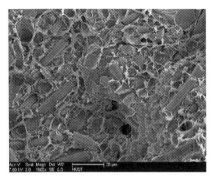 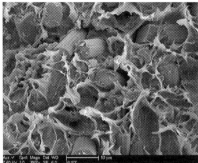

FIGURE 2.112 SEM pictures of enlarged sintered parts section details.

plastic deformation, which may be due to the presence of more unsaturated chemical bonds at the ends of the carbon fibers, resulting in a large number of chemical bonds between the active sites and the nylon 12 matrix.

Destruction of short-fiber composites typically begins with microscopic voids and mesoscopic cracks that are present in the reinforcing phase, matrix, and interphase. In the preparation of the composite material, defects are also generated, especially for the SLS process, it is difficult to avoid the presence of a small amount of minute pores inside the sintered part. The ultimate destruction of short-fiber composites is the result of several mesomechanical processes, and the macroscopic appearance of the fracture depends on which of these processes controls the entire destruction process.

As shown in Fig. 2.113, the main destruction mechanisms of short-fiber composites include: A is fiber fracture; B is fiber pull-out; C is fiber/matrix debonding; D is plastic deformation and destruction of resin matrix.

From the previous cross-sectional SEM photograph, it can be found that the fracture mode of the CF/PA12 composite powder sintered parts mainly consists of the latter three modes, namely, fiber pull-out, fiber/matrix debonding, plastic deformation and destruction of the resin matrix. For the improvement of the bending strength and flexural modulus of the composite material, the following explanation can be made: on the one hand, the content of the plastically deformable matrix is correspondingly reduced due to the addition of the carbon fiber; on the other hand, since the crack propagation is required to bypass the fiber when the destruction mechanism is B and C, thereby increasing the path of crack propagation, and in the destruction mechanism C, the bridging effect of the fiber can be decreased to a certain extent, weakening the further propagation of the crack, thereby hindering the fracture of the entire matrix; due to the rigidity limitation of the fiber, the plastic deformation of the nylon 12 matrix is hindered and the deformation resistance is increased. The combination of the above points has led to a significant increase in the bending strength and flexural modulus of the composite material.

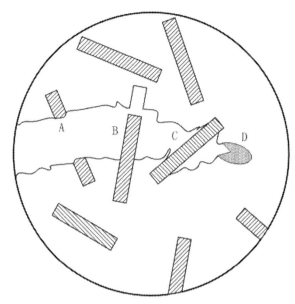

FIGURE 2.113 Schematic diagram of the path of a crack passing through a short-fiber reinforced resin.

The impact resistance of the composite material can be explained as follows: Since the nylon 12 matrix has good toughness, the energy absorption by the matrix is higher than the energy absorption caused by the carbon fiber during the fracture process, and thus, compared with the nylon 12 matrix, the impact resistance of the composite material gradually decreases with the increase of the carbon fiber content; compared with other filler materials, it maintains better impact resistance due to the additional energy absorption caused by addition of the fiber.

If the fiber ends are not well bonded to the matrix and there are voids at the ends, there is a high concentration of stress at the fiber/matrix interface, thereby promoting crack propagation. From the SEM pictures, the nylon 12 matrix after plastic deformation remains at the end of the carbon fiber, indicating that the end and the matrix are well bonded, which is one of the advantages of the carbon fiber surface coated nylon 12, further ensuring the improvement of the mechanical properties of the composite material.

2.5 Styrene-based amorphous polymer powder materials

The amorphous polymer has high melt viscosity and slow sintering rate during SLS, so its sintered part density is low, which determines its low mechanical properties. But its sintered parts have higher dimensional accuracy due to small forming shrinkage. SLS parts of amorphous polymers are often used for investment casting wax molds and functional parts with low performance requirements by impregnating wax and resin, respectively.

2.5.1 Polystyrene powder materials

PS is easy to pulverize at low temperature, and the powdering cost is low. PS powder has a low sintering temperature and high precision of SLS parts. It is a widely used SLS material and is mainly used for investment casting.

The SLS process characteristics of PS powder and its parts properties are given in Section 2.1.3.5.

2.5.2 Styrene–acrylonitrile copolymer powder material

The amorphous polymer has high melt viscosity and slow sintering rate during sintering, so its sintered part density is low, which determines its low mechanical properties, but its sintered part has higher dimensional accuracy due to small forming shrinkage. SLS parts of amorphous polymers are often used for investment casting wax molds and functional parts with low performance requirements by impregnating wax and resin, respectively.

2.5.2.1 Preparation and characterization of styrene–acrylonitrile copolymer powder

The styrene–acrylonitrile (SAN) granules produced by Zhenjiang Chimei Industrial Co., Ltd. are selected. Its melt index is 3 g/10 minutes and density is 1.06 g/cm^3. The SAN powder is prepared by low-temperature pulverization method, and then the powder having a particle diameter of 10−100 μm is selected by using gas stream-screen classifier.

Fig. 2.114 is the SEM photograph of SAN powder. As can be seen from Fig. 2.114, the shape of the SAN powder particles is very irregular and the surface is extremely rough.

Fig. 2.115 shows the particle size distribution of the SAN powder, which is obtained by laser particle size analyzer. From Fig. 2.115, it can be concluded that the particle size distribution of the SAN powder is 10−85 μm, and the powder particles having a particle size in the range of 31−65 μm account for the majority, and the measured average particle size of the powder is 59.08 μm.

2.5.2.2 Selective laser sintering-forming and postprocessing of styrene–acrylonitrile powder material

The SAN powder is sintered and formed by HRPS-III laser sintering system. The sintering parameters are set as follows: scanning rate is 2000 mm/s; scanning pitch is 0.1 mm; laser power range is 8−28 W, thus, the laser energy density ranges from 0.04 to 0.14 J/mm^2, and the slice thickness is 0.1 mm.

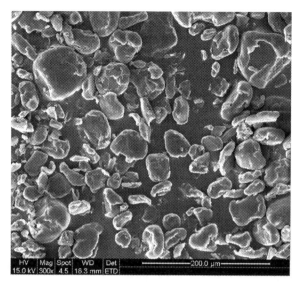

FIGURE 2.114 Microscopic shape of SAN powder.

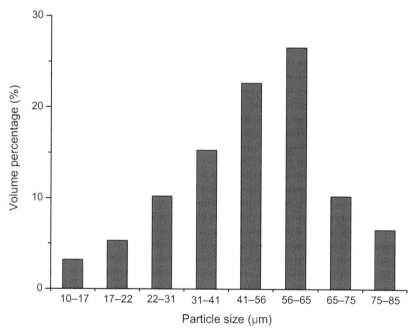

FIGURE 2.115 Particle size distribution of SAN powder.

Since SAN is an amorphous polymer, its SLS parts are mainly porous, and the relative density and strength are very low. In order to improve the strength of the parts, the postprocessing of coating epoxy resin is used for reinforcement.

The steps of coating the epoxy resin on the SLS parts are as follows: accurately weigh the epoxy resin component A and component B according to the mass ratio 1:1, and mix the two components uniformly; heat the mixed resin liquid and the SLS parts to 30°C; use a brush to apply epoxy resin twice: (1) coat SLS parts with the first epoxy resin until it cannot be penetrated, place the SLS parts in the vacuum oven at a temperature of 40°C and vacuum for 10−20 minutes. (2) Take the SLS parts out of the vacuum oven, apply a second epoxy resin coating until it cannot be penetrated, and clean the resin remaining on the surface of the formed parts; finally, put the SLS parts coated with epoxy resin in an oven for curing, and the stepwise curing conditions are as follows: cure at 40°C for 10 hours, and further cure at 75°C for 6 hours.

2.5.2.3 Properties of selective laser sintering parts styrene−acrylonitrile powder material

2.5.2.3.1 Mechanical properties

The bending strength, flexural modulus and impact strength of SLS parts and resin-coated parts of SAN and PS can be seen in Table 2.37.

As can be seen from Table 2.37, the mechanical properties of SAN SLS parts are low and cannot be directly used as plastic functional parts. Fig. 2.116 shows the microscopic shape of the fracture surface of SAN SLS parts. There are a large number of pores in the SLS parts, and the powder particles are not melted together but are formed by the bonding of sintering neck, which results in low mechanical properties of the SLS parts. After the SLS parts are coated with epoxy resin, the mechanical properties are greatly improved. The bending strength, flexural modulus and impact strength of the coated-resin parts are 2.3, 44.9, and 3.5 times of the SLS parts, respectively. Fig. 2.117 shows the microscopic shape of the fracture surface of the SAN

TABLE 2.37 Mechanical properties of SLS parts and resin-coated parts of SAN and PS.

Parts	SAN		PS	
	SLS parts	Resin-coated parts	SLS parts	Resin-coated parts
Bending strength (MPa)	15.3	35.6	9.8	29.7
Flexural modulus (MPa)	60.1	2701	33.5	2263
Impact strength (kJ/m^2)	3.5	12.4	2.4	9.6

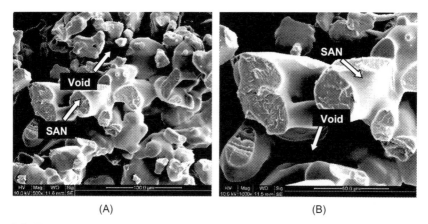

FIGURE 2.116 Microscopic shape of the fracture surface of the SLS part of the SAN: (A) 500×; (B) 1000×.

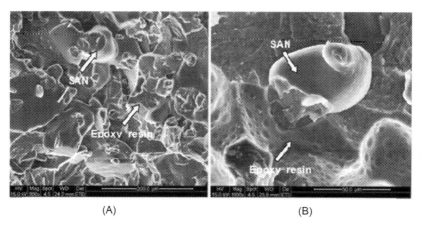

FIGURE 2.117 Microscopic shape of the fracture surface of the SAN resin-coated parts: (A) 300×; (B) 1000×.

coated-resin parts. As can be seen from Fig. 2.117, the pores in the part are completely filled with epoxy resin, and the interface between the SAN resin and the epoxy resin is well bonded. Therefore, after the SLS parts are coated with epoxy resin, the mechanical properties have been greatly improved and can be used as plastic functional parts.

It can be seen from Table 2.37 that the mechanical properties of the PS coated-resin parts are lower than that of the SAN coated-resin parts. The bending strength, flexural modulus and impact strength of the SAN coated-resin parts are 1.2, 1.2, and 1.3 times of PS coated-resin parts, respectively. There are two main reasons: First, the mechanical properties of PS

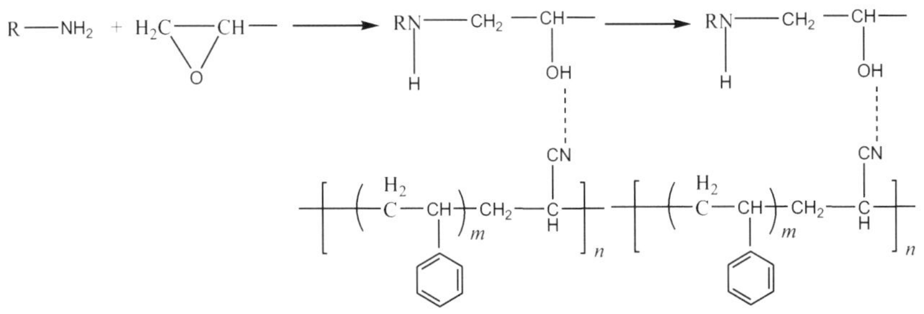

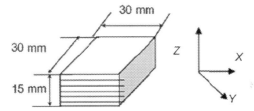

FIGURE 2.118 Reaction equation of the hydrogen bond formed between the SAN and the hydroxyl group of the epoxy resin.

FIGURE 2.119 Test piece of dimensional deviation.

SLS parts are lower than that of SAN SLS parts; second, due to the presence of a large number of hydroxyl groups in the cured epoxy resin of aliphatic diamine, the cyano nitrogen atom in the SAN structure has a lone pair of electrons, which can form a hydrogen bond with the hydroxyl group in the epoxy resin. Therefore, the SAN and the epoxy resin can form a good interfacial bond. However, since there is no group capable of forming a hydrogen bond with hydroxyl group in the molecular structure of the PS, and the polarity of the PS and the epoxy resin are largely different, the interface bonding between the PS and the epoxy resin is relatively poor (Fig. 2.118).

2.5.2.3.2 Dimensional accuracy

The dimensional accuracy is characterized by the dimensional deviation. The design dimension of the test piece is 30 mm × 30 mm × 15 mm, as shown in Fig. 2.119, and the manufacturing direction is the Z-axis direction.

Table 2.38 lists the dimensional deviations of the SLS parts and the resin-coated parts of the SAN in the X, Y, and Z directions. It can be seen that the SLS parts of the SAN and the resin-coated parts have high dimensional accuracy in the X, Y, and Z directions, the negative deviation in all three directions is below -1%, and the negative deviations of the SLS parts in the three directions after the epoxy resin is applied are increased due to curing shrinkage of epoxy resin.

TABLE 2.38 Dimensional deviation of SLS parts and coated permeability resin parts in the X, Y, and Z directions.

Parts	Dimensional deviation (%)		
	X direction	Y direction	Z direction
SLS parts	−0.64	−0.60	−0.32
Coated-resin parts	−0.75	−0.73	−0.45

2.5.3 High impact polystyrene powder material

High impact PS (HIPS) is a modified version of PS with components of PS and rubber. The HIPS produced by the graft copolymerization method can overcome the disadvantages of uneven dispersion of the rubber phase produced by the blending method. The HIPS appears as white opaque pearl globular or granular particles, which has excellent impact properties and has most of the advantages of PS.

2.5.3.1 Selective laser sintering process characteristics of high impact polystyrene powder material

The particle size distribution of the HIPS powder used is 30–80 μm, and the shape of the powder particles is irregular. The linear expansion coefficient of HIPS is small, and the thickness of the spreading layer is 0.10 mm. The power leveling effect is very good.

Fig. 2.120 shows the DSC curves of PS and HIPS. The curves show that the T_g of PS and HIPS are 102°C and 97°C, respectively. From the experiments in Table 2.39, the sintering window temperatures of PS and HIPS are 92°C–102°C and 88°C–98°C, although the glass transition temperatures of the two polymers are different, the sintering window temperature is not the same, but the sintering window is 10°C, indicating that the sintering performance of the two polymers is similar, both have good sintering properties. The HIPS preheating temperature is set to 95°C.

The sintering scanning speed is set to 2000 mm/s, the single-layer thickness is 0.10 mm, and the laser effective energy (P) is adjusted to the range of 6.5–8.0 W. Fig. 2.121 shows the relationship between the laser sintering power (P) and the powder bed temperature in the scanning area. The powder bed temperature is measured by an infrared thermometer.

As can be seen from Fig. 2.121, the change of the laser power (P) has a great influence on the sintering temperature. Fig. 2.122 shows the microstructure of the surface of the sintered part when the laser power P is 7.5 W. Fig. 2.122A is a SEM picture of the upper surface of the sintered part (50 mm × 50 mm × 5 mm),

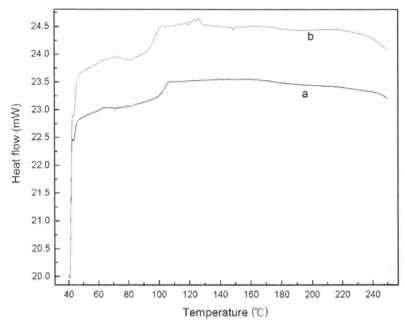

FIGURE 2.120 DSC curves for PS and HIPS (a: PS; b: HIPS).

in which the particles in the sintered layer are uniformly melted and bonded, there is no difference between the sintered lines, and the pores between the particles are very small. Fig. 2.122B is a SEM picture of the side surface of the sintered part, and the slice level in sintering is inseparable [7].

2.5.3.2 Properties of selective laser sintering parts of high impact polystyrene powder material

The mechanical properties of PS and HIPS sintered parts are shown in Table 2.40. It can be seen that HIPS sintered parts have better mechanical properties than PS. This may be because the addition of rubber components in HIPS is beneficial to the adhesion between powder particles. Fig. 2.123 verifies this speculation.

Although HIPS is superior to PS in terms of mechanical properties of sintered parts, they are equivalent in terms of sintering properties, but the viscoelasticity of the rubber component in HIPS makes it relatively difficult to purify the powder after sintering. During the sintering process, the rubber is easily decomposed and emits unpleasant odor, therefore, the application of HIPS powder in SLS is not as popular as PS, and it is more suitable for the case where the mechanical properties of sintered parts are highly demanded, such as producing large thin-walled parts.

TABLE 2.39 Sintering performance of PS and HIPS (scanning pitch: 0.10 mm; scanning rate: 2000 mm/s; layer thickness: 0.1 mm; laser power: 14 W).

Preheating temperature (°C)	86	88	90	92	96	98	100	102
PS	–	–	Warpage	Successful sintering				Agglomeration
HIPS	Warpage	Successful sintering				Agglomeration	–	–

Polymer materials for additive manufacturing—powder materials Chapter | 2 **171**

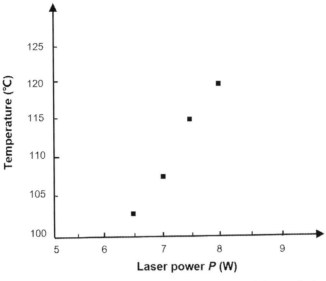

FIGURE 2.121 Relationship between laser sintering power (P) and the powder bed temperature of HIPS.

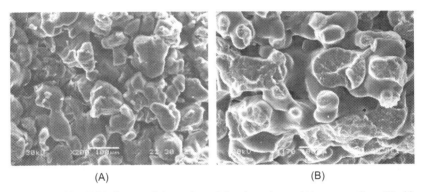

(A)   (B)

FIGURE 2.122 SEM pictures of the surface of the sintered part: (A) upper surface; (B) side surface.

TABLE 2.40 Mechanical properties of PS and HIPS sintered samples.

Parts	Tensile strength (MPa)	Elongation at break (%)	Young modulus (MPa)	Bending strength (MPa)	Impact strength (kJ/m^2)
PS	1.57	5.03	9.42	9.8	2.4
HIPS	4.49	5.79	62.25	18.93	3.30

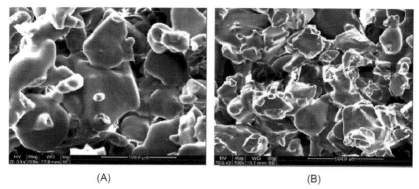

FIGURE 2.123 SEM pictures of the sintered part: (A) PS; (B) HIPS.

2.5.3.3 Influence of postprocessing on the properties of selective laser sintering parts of high impact polystyrene powder material

2.5.3.3.1 Selection and use of reinforced resins

The reinforced resin liquid is infiltrated into the SLS part to fill the gap between the powder particles, thereby enhancing the SLS part. In theory, in order to make the parts have higher mechanical properties, it is hoped that the reinforced resin and the SLS bulk material can be well bonded, that is, they have good compatibility, and only they diffuse and penetrate each other can achieve the best enhancing results.

The compatibility of materials must first satisfy the thermodynamic possibilities. A process can only be carried out if the free energy is reduced. It is expressed as:

$$\Delta G = \Delta H - T \Delta S \tag{2.83}$$

The process is only possible when the free energy (ΔG) is less than zero, and in general the mixing entropy (ΔS) is increasing, so the key to determining compatibility is ΔH, and

$$\Delta H = V \varphi_1 \varphi_2 (\delta_1 - \delta_2)^2 \tag{2.84}$$

It can be seen from Eq. (2.84) that the compatibility is determined by the size of solubility parameter($|\delta_1 - \delta_2|$), and the closer the solubility parameters δ_1 and δ_2 are, the better the compatibility is, the better the enhancement effect is, and the solubility parameter principle combined with the polarity principle can accurately determine the compatibility of the polymer. The solubility parameter (δ) of the PS powder used in SLS is 8.7–9.1, which is close to that of the polyester, and the solubility parameter (δ) of the epoxy resin is 9.7–10.9, the solubility parameters are different when combined with different curing agents and diluents. By proper adjustment, it can achieve better compatibility with powder or complete incompatibility [8].

However, for the reinforced resin used for the postprocessing of SLS parts, it is necessary to consider the influence on the precision of the parts. For example, although the 502 glue has good compatibility with PS, the good compatibility makes the liquid reinforced resin completely dissolves the powder material in the postprocessing, and the AB glue and the polyester also have good compatibility with the powder, even if they cannot dissolve the powder material, the part is softened.

According to the principle of solubility parameter, the solubility parameters of epoxy resin and PS and HIPS materials are not very different; from the principle of polarity, the polarity of the them is not much different, moderate compatibility and adjustability is an important reason for choosing epoxy resin as a postprocessing material. In order to make the post-processed parts stronger, the compatibility can be improved by adjusting the curing agent and the diluent; to reduce the deformation during the postprocessing, the compatibility should be lowered. The SLS parts of PS and HIPS are only bonded by weak combination of powders, so that the strength is very low. When the liquid reinforced resin is infiltrated, the bond between the powders is easily destroyed, so that the part is deformed due to its own gravity during the period from the penetration of the resin to the curing of the resin. Table 2.41 shows the effect of the glycidyl ether (5748) containing 12–14 carbon long chains and butyl glycidyl ether (660 A) containing 4 carbons as diluents on the deformability of the part.

As can be seen from the above experiments, since the diluent 5748 has a long chain, the compatibility with the PS is increased, and the bonding between the powder particles is broken, resulting in deformation of the parts during the postprocessing. Therefore, in order to ensure the precision of the parts, the compatibility between the reinforced resin and the powder should neither too good nor too bad so that it cannot be wetted, which is disadvantageous to the improvement of the appearance and the reinforcement.

The solubility parameter of the reinforced resin is determined by the epoxy resin, the diluent and the curing agent. Since the solubility parameter of the epoxy resin does not change significantly, the compatibility of the reinforced resin with the powder is mainly determined by the curing agent and the diluent. However, there are many varieties of curing agents and diluents for epoxy resin, and there are many means for modification. In

TABLE 2.41 Effects of different diluents on the deformability of SLS parts.

Diluents	Effects
5748	SLS parts are slightly softened after infiltrated by the resin, and have certain bending deformation.
660A	SLS parts are not softened after infiltrated by the resin, and have no bending deformation.

particular, the curing agents are mostly mixtures, and their solubility parameter values cannot be found from the manual, it is impossible and not necessary to measure the solubility parameters of each curing agent. Therefore, estimate is needed when making the selection. Combined with the polarity principle, the compatibility of the reinforced resin with the powder can be estimated initially. The equation for estimating the solubility parameter is:

$$\delta = \frac{\sum F}{M} \times \rho \qquad (2.85)$$

where $\sum F$ is the molar attraction constant of each group in the repeating unit; M is the molecular weight of the repeating unit; and ρ is the density.

The solubility parameter (δ) of epoxy resin is generally between 9.7 and 10.9, and has a certain polarity. The solubility parameter (δ) of PS is between 8.7 and 9.1, and it is a nonpolar material, so the group having a large polarity and molar attraction constant is introduced into the curing agent, such as cyano group, hydroxyl group, etc.; reducing the length of the nonpolar chain in the diluent can reduce the compatibility and improve the dimensional stability of the part during the operation process; The introduction of the group having a small polarity and a low molar attraction constant in the curing agent can increase the compatibility, and the introduction of a long chain in the diluent can simultaneously increase flexibility and compatibility, but reduce the dilution effect.

2.5.3.3.2 Effects of postprocessing on the properties of selective laser sintering parts

Table 2.42 shows the dimensional test data of the HIPS powder sintered parts. A_0 is the design size of the sintered part, A is the size of the HIPS sintered part, and B and C are the dimensions of the HIPS sintered part treated with the epoxy resin HC2 and HC1, respectively. The size of the HIPS sintered part after the resin treatment tends to increase, which is because the density of the sintered HIPS powder material is relatively high, resulting in a thick resin layer on the surface of the HIPS sintered part after the resin treatment.

TABLE 2.42 Dimensional accuracy of HIPS sintered parts (50 mm × 50 mm).

Test points	A_0 (mm)	Series 1		Series 2	
		A (mm)	B (mm)	A (mm)	C (mm)
1	50	50.601	51.082	50.842	51.300
2	50	50.804	51.330	50.726	51.422
3	50	50.662	51.264	50.602	51.140

Table 2.43 shows the mechanical properties of HIPS powder sintered parts before and after resin treatment. Among them, A is the sintered part of HIPS powder, and B and C are the sintered parts treated by epoxy resin HC2 and HC1, respectively. The tensile strength and bending strength of the HIPS powder sintered parts are 4.49 and 21.80 MPa, respectively. The tensile strength of the B and C parts of the HIPS powder sintered parts treated with epoxy resin are increased by 3.7 times and 4.2 times, respectively, and the impact strength is increased by 1.8 times.

Fig. 2.124 is the SEM pictures of the tensile section of the HIPS powder sintered parts. Fig. 2.124A shows that the section of HIPS powder sintered part is not smooth when subjected to tensile stress, indicating that the HIPS material has certain flexibility after rubber modification. At the same time, SEM also shows that the pores between the particles in the HIPS powder sintered parts are small, indicating that the sintered parts were relatively dense. In Fig. 2.124B and C, the pores in the resin-treated sintered parts disappear, and the density is further increased.

Fig. 2.125 is the SEM pictures of the impact section of HIPS powder sintered parts. It can be clearly seen from Fig. 2.125A that the impact section of the HIPS sintered part is quite uniform, and the fusion between the particles is good, so that the bending resistance is good. It can be seen from Fig. 2.125B and C that although the epoxy resin filled in the voids of the sintered parts increases the density of the sintered parts, the proportion of the resin is small, and it becomes an isolated "island." This weakens the reinforcement of the filled resin. When subjected to an external force, the epoxy resin is broken first, and the main body subjected to stress is HIPS.

The HIPS powder sintered parts manufactured by the SLS process can be directly used as prototype test pieces, and can be assembled as functional parts after being reinforced by epoxy resin. Fig. 2.126 shows the impeller sintered with HIPS, and Fig. 2.127 shows the reinforced part of HIPS sintered part treated with epoxy resin. Compared with the PS powder, the HIPS powder can be sintered into a thin-walled member having a wall thickness of

TABLE 2.43 Mechanical properties of HIPS sintered parts.

	Tensile strength (MPa)	Elongation at break (%)	Tensile modulus (MPa)	Bending strength (MPa)	Flexural modulus (MPa)	Impact strength (kJ/m^2)
A	4.49	5.79	62.25	21.80	1056.31	2.62
B	16.61	6.68	254.89	28.55	1791.84	4.71
C	18.79	6.25	346.95	31.59	1838.31	4.78

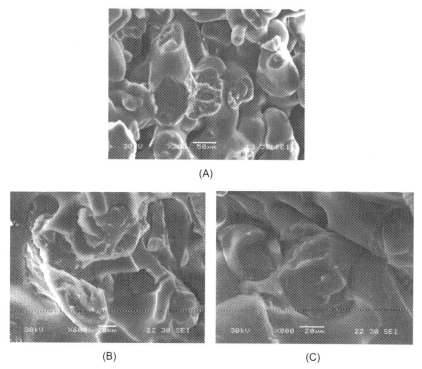

FIGURE 2.124 SEM pictures of the tensile section of HIPS powder sintered part: (A) sintered parts without resin treatment, × 300; (B) HC2 treatment parts, × 600; (C) HC1 treatment parts, × 800.

0.20 mm, and the integrity of the sintered part can be maintained after the floating powder is removed.

2.6 Polycarbonate powder material

PC is an excellent engineering plastic with outstanding impact strength and creep resistance, high tensile and bending strength, high elongation and rigidity, and low water absorption. PC powder can be directly obtained by interfacial polycondensation. It is the first commercial SLS material and has applications in both prototype and functional parts. At present, due to the better sintering performance of PS materials and the emergence of direct-sintered nylon with better mechanical properties, the use of PCs in prototypes and functional parts has been replaced, making the importance of PCs in the SLS field decline, but PC is still a good SLS material, occupying a certain position in the SLS material family, because the strength of the PC prototype is much higher than PS, and the performance of the prototype is very good after the resin is infiltrated, and the sintering process is not as strict as nylon [9].

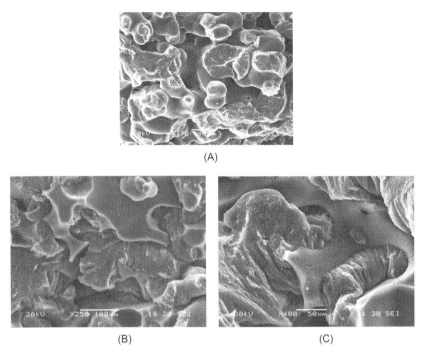

FIGURE 2.125 SEM pictures of the impact section of HIPS powder sintered parts: (A) sintered parts without resin treatment, ×170; (B) HC2 treatment parts, ×250; (C) HC1 treatment parts, ×400.

FIGURE 2.126 HIPS powder sintered impeller.

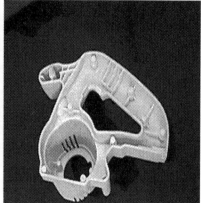

FIGURE 2.127 Enhanced parts after postprocessing.

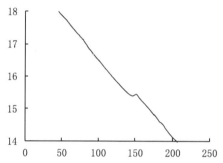

FIGURE 2.128 DSC curve of PC.

2.6.1 Selective laser sintering process and properties of polycarbonate powder material

Fig. 2.128 shows the DSC curve of PC, from which the T_g of PC is 150°C.

Laser sintering of PC powder is carried out on HRSP-III rapid prototyping machine. The preheating temperature of the powder bed is controlled at 138°C–143°C, when the preheating temperature exceeds 143°C, the powder in the intermediate working cylinder agglomerates severely, making powder spreading difficult.

Fig. 2.129 is a cross-sectional SEM of PC sintered parts prepared at different laser powers.

When the laser power is very low, the powder particles in Fig. 2.129A are slightly sintered together at the locations where they are in contact with each other, and the powder particles still keep their original shape. As the laser power increases, the powder particles in Fig. 2.129B change significantly from the original irregular shape to be close to the spherical shape, and the surface

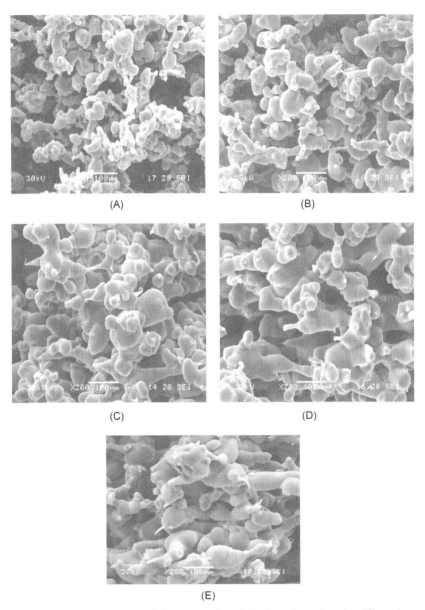

FIGURE 2.129 SEM pictures of the cross section of the sintered sample under different laser powers: (A) 6 W; (B) 7.5 W; (C) 9 W; (D) 10.5 W; (E) 12 W.

becomes smooth. Because with the increase of laser power, the energy absorbed by the powder increases, and the temperature rises significantly. Above the T_g temperature, the apparent viscosity of PC decreases rapidly with the increase of temperature, and the activity of macromolecular segments increases, under the action of surface tension, the particles tend to be spherical and the surface becomes smooth. When the laser power continues to increase, the sintered neck in Fig. 2.129C–E increases significantly, small particles merge into large particles, the pores become smaller, and the relative density increases.

The density and mechanical properties of PC sintered parts vary with laser power as shown in Table 2.44.

It can be seen from Table 2.44 that the density, tensile strength, tensile modulus and impact strength of PC sintered parts increase with the increase of laser power, on the contrary, the elongation at break decreases with the increase of laser power. When the laser power increases from 6 w to 13.5 w, the density of the PC sintered part increases from 0.257 to 0.463 g/cm^3, and the tensile strength increases from 0.39–2.29 MPa, which are increased by 80% and 487%, respectively. Nevertheless, the density and tensile strength of the PC sintered parts are much lower than those of the PC molded parts with density of 1.18 g/cm^3 and the tensile strength of 60 MPa, which are only 39% and 3.8% of the molded parts, respectively. Although it is possible to further increase the density of the sintered part after continuing to increase the laser power, when the laser power is 13.5 W, the color of the sintered part has become yellow, indicating that the PC has partially degraded, and it is not suitable to continue to increase the laser power.

It can be seen that the strength of the PC sintered part is mainly affected by the porosity of the sintered part, but has little relationship with the strength of the PC. When the density of the sintered part is higher, the strength will be higher. Increasing the laser power can make the PC powder well sintered, which can increase the density of the sintered part. However, excessive energy input will cause the powder under direct

TABLE 2.44 Density and mechanical properties of PC sintered parts.

Laser power (W)	Density (g/cm^3)	Tensile strength (MPa)	Elongation at break (%)	Tensile modulus (MPa)	Impact strength (kJ/m^2)
6	0.257	0.39	52.1	2.19	0.92
7.5	0.343	1.32	35.6	7.42	1.37
9	0.384	1.89	32.8	10.62	2.14
10.5	0.416	2.04	31.4	13.24	2.81
12	0.445	2.18	30.7	15.97	2.98
13.5	0.463	2.29	30.1	17.13	3.13

irradiation of the laser beam to produce overheat, which brings the following problems:

1. The thermal oxidation of PC is aggravated, causing discoloration and deteriorated properties of the sintered part. When the local temperature exceeds the decomposition temperature of PC, the PC will be strongly decomposed, and the properties of the sintered part will be further deteriorated.
2. The temperature gradient between the powder under laser irradiation and the surrounding powder is increased, and the PC sintered part is prone to warping deformation.
3. Since the PC has no latent heat of fusion, the heat transfer causes the powder outside the scanning area to adhere to the sintered part, so that the sintered part loses clear outline and affects the forming precision.

Therefore, optimizing the sintering process parameters can only improve the density and mechanical properties of the PC sintered parts to a certain extent, and cannot fundamentally eliminate the porosity of the sintered parts, so the functional parts cannot directly sintered with the PC powder.

The PC powder is subjected to laser sintering to prepare a 50 mm × 50 mm × 4 mm square. The dimensional errors of the sintered part in the X direction and the Y direction vary with the laser powers as shown in Fig. 2.130.

As can be seen from Fig. 2.130, the dimensional error of the PC sintered part is a negative value. When the laser power is small, the error of the sintered part is large because the excessively low laser power is insufficient to make the powder particles well bonded, especially at the edge site of the sample where the size of the sintered part is smaller than the range of the laser scanning. As the laser power increases, the sintering at the edge of the sample is improved and the dimensional error is reduced. The negative dimensional error is caused by shrinkage of the PC powder during sintering. The forming shrinkage of the PC material is not large, and the sintered part produces a large shrinkage, which is related to the

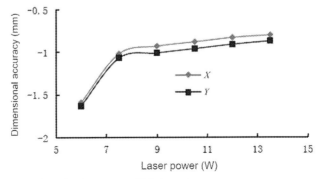

FIGURE 2.130 Dimensional accuracy of PC sintered parts.

low bulk density of the powder used. Since the initial density of the powder is very low, relatively large densification occurs during sintering, so that a large shrinkage is produced. The dimensional error in the Y direction is slightly larger than the X direction, which may be related to the orientation of the nonspherical powder under the action of the powder roller moving in the X direction. The powder is tightly arranged in the X direction relatively and the sintering shrinkage is small.

The dimensional error caused by material shrinkage can be compensated by adjusting the scaling factor in the X and Y directions on the SLS-forming equipment [10].

2.6.2 Effects of postprocessing on the properties of polycarbonate sintered parts

2.6.2.1 Postprocessing of polycarbonate sintered parts

In the postprocessing of PC sintered parts, the porous PC sintered parts are impregnated with liquid epoxy resin system. The epoxy resin system is immersed in the inside of the sintered parts by capillary action, filling the voids therein, and then the epoxy resin is cured at a certain temperature to form dense parts.

The epoxy resin system consists of liquid epoxy resin, curing agent and diluent. Epoxy resin should be selected from varieties of low molecular weight and low viscosity such as CYD-128 to facilitate the impregnation of the sintered parts. It is critical to select the curing agent. In order to avoid deformation of the sintered parts during curing, the curing temperature should be lower than the heat distortion temperature of PC, which is preferably not more than 120°C. Therefore, only the medium and low-temperature curing agent can be selected. However, it is not suitable to use a curing agent which has a large activity at room temperature, because such curing system is possible to start curing during the impregnation process due to fast curing speed and short pot life, resulting in the inability to immerse the sintered parts completely, which will seriously affect the postprocessing effects. The functions of diluent is to adjust the viscosity of the epoxy resin, it is preferably to choose the reactive diluent containing monoepoxy group or biepoxy group. Because the reactive diluent can participate in the curing reaction of the epoxy resin, it has less damage on the properties of the epoxy resin cured product, and the amount thereof is such that the epoxy resin system saturates the sintered parts completely, and it is not suitable to add excessively.

2.6.2.2 Effects of postprocessing on the properties of polycarbonate sintered parts

Table 2.45 shows the density and mechanical properties of sintered parts treated with epoxy resin system.

TABLE 2.45 Density and mechanical properties of treated PC sintered parts.

Laser power (W)	Density (g/cm³)	Tensile strength (MPa)	Elongation at break (%)	Tensile modulus (MPa)	Impact strength (kJ/m²)
6.0	1.02	38.87	10.31	385.60	6.47
7.5	1.09	42.19	14.50	581.50	7.93
9.0	1.12	44.70	15.10	600.60	8.83
10.5	1.08	42.04	15.70	547.20	7.52
12.0	1.06	41.18	16.20	515.97	7.08
13.5	1.03	39.24	15.90	475.13	6.93

Comparing Tables 2.44 and 2.45, it can be seen that the density and mechanical properties of PC sintered parts are greatly improved after being treated by epoxy resin system, and the density is increased by 2.22−3.97 times, tensile strength and modulus are increased by 17.1−99.7 times and 26.7−176.1 times respectively, showing the maximum increase ratio, the impact strength is increased by 2.2−7.03 times, and the elongation at break is decreased by 47%−80%. The mechanical properties of the treated sintered parts are still related to the density; the greater the density, the greater the tensile strength, tensile modulus and impact strength. However, the density of the treated sintered parts does not increase with the increase in density before the treatment, and the density of the sintered parts having the medium density is the largest after treatment. This is related to the impregnation of the epoxy resin system. The epoxy resin shows slow penetration rate in the sintered part with high density and small porosity, and does not easily penetrate into all the voids, which affects the increase of density. The density and mechanical properties of PC sintered parts differ greatly at different laser power before treatment, but the difference after treatment is greatly reduced, indicating that postprocessing plays a decisive role in the properties of sintered parts. The sintered parts prepared with 9 W laser power have the best mechanical properties after epoxy resin treatment, and their performance indexes can meet the needs of plastic functional parts with low performance requirements such as impact strength. Fig. 2.131 is the SEM picture of the impact section of a post-processed PC sintered part.

As can be seen from Fig. 2.131, the pores in the PC sintered part are filled with epoxy resin to form dense material. When the sample is subjected to an external force, the epoxy resin becomes a main body that withstands the external force, which greatly reduces the damage of the external force to the bond between the PC particles, thereby greatly improving the mechanical strength of the sintered part.

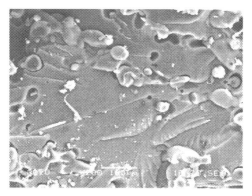

FIGURE 2.131 SEM picture of the impact section of the post-processed PC sintered part.

The 50 mm × 50 mm × 4 mm squares sintered at different laser powers are post-processed with epoxy resin, and the dimensions of the treated samples in the X direction and the Y direction are slightly increased, but the added values are all below 0.1 mm. It can be seen that the post-treatment has little effect on the dimensional accuracy of the PC sintered part, and the dimensional accuracy of the final sample depends on the precision of the sintered part before the treatment.

2.7 Acrylonitrile–butadiene–styrene powder material

2.7.1 Basic characteristics of acrylonitrile–butadiene–styrene materials

ABS is a polymer obtained by copolymerization of three monomers of acrylonitrile (A, 23%–41%), butadiene (B, 10%–30%), and styrene (S, 29%–60%). The macromolecular main chain of ABS is repeatedly connected by three structural units. Different structural units give ABS different properties: acrylonitrile has good chemical resistance and high surface hardness; butadiene has good toughness; and styrene has good transparency and processing performance. When the three monomers are combined, the tough, hard, and rigid ABS resin is formed.

The ABS shows opaque ivory granules in appearance, which combines well with other materials. The heat distortion temperature of ABS is 93°C–118°C. ABS can be used in a temperature range of $-40°C-100°C$. The specific properties are shown in Table 2.46.

ABS has good thermal stability, but it has high water absorption and should be dried at 80°C–85°C for 2–4 hours before sintering.

The ABS powder used for sintering in the experiment is pulverized from ABS at a low temperature. The largest particle is screened through a 200-mesh screen, the shape of which is irregular. After adding antioxidant and antistatic agent, the ABS powder sintered material is formed.

TABLE 2.46 Properties of ABS materials (plastic injection parts).

Relative density	Water absorption rate (%)	Molding shrinkage (%)	Tensile strength (MPa)	Bending strength (MPa)	Rockwell hardness	Heat distortion temperature (°C)	Linear expansion coefficient ($\times 10^{-5} K^{-1}$)
1.02–1.05	0.2–0.45	0.3–0.8	35–44	51–81	R65–109	93–103	9.5–10.5

2.7.2 Sintering performance of acrylonitrile–butadiene–styrene powder

2.7.2.1 Dimensional accuracy

The dimensional accuracy of ABS sintered part is shown in Table 2.47, where A_0 is the design size of the sintered part, A is the actual size of the ABS sintered part, and B and C are the sizes of ABS sintered parts reinforced by the epoxy resin HC2 and HC1, respectively. The data in the table indicates that the sintered part has a negative deviation, which is 4.20% smaller than the design size on average. After the epoxy resin treatment, the size of the sintered part shrinks again, the average shrinkage of the B part is 1.01%, and the average shrinkage of the C part is 0.91%.

2.7.2.2 Mechanical properties

Table 2.48 compares the mechanical properties of ABS sintered parts before and after epoxy resin treatment. The tensile strength of ABS sintered part is only 1/13 of ABS plastic injection part, and the tensile strengths of ABS plastic injection parts reach 2/7 and 3/7 of ABS plastic injection parts respectively. The bending strength of ABS sintered parts is only 1/10 of that of

TABLE 2.47 Dimensional accuracy of ABS sintered parts (50 mm × 50 mm).

Test points	A_0 (mm)	Series 1		Series 2	
		A (mm)	B (mm)	A (mm)	C (mm)
1	50	48.024	47.540	48.265	47.860
2	50	48.110	47.612	48.304	47.892
3	50	48.268	47.720	48.055	47.506

TABLE 2.48 Mechanical properties of ABS sintered parts.

	Tensile strength (MPa)	Elongation at break (%)	Tensile modulus (MPa)	Bending strength (MPa)	Flexural modulus (MPa)	Impact strength (kJ/m^2)
A	2.61	4.80	54.99	5.66	364.23	1.05
B	9.65	5.86	196.91	20.72	1362.51	4.26
C	16.16	6.23	257.55	25.33	1665.80	3.77

ABS plastic injection parts, and the bending strength of ABS sintered parts treated with epoxy resin is increased to 2/5 and 1/2 of ABS plastic injection parts. The impact strength of the ABS sintered part is only 1.05 kJ/m^2, and the impact strengths of the B and C parts after the epoxy treatment are 4.1 times and 3.6 times that of the prototype, respectively.

Fig. 2.132 is SEM pictures of the tensile section of the ABS sintered part. Fig. 2.132A shows the existence of a large number of pores in the ABS sintered part. These pores reduce the compactness of the sintered part and directly affect the mechanical properties of the sintered part. Compared with the PC sintered part, the thickness of the powder layer is different due to the different powder characteristics, and the density of the ABS sintered part is significantly larger than that of the PC sintered part. Fig. 2.132B and C show that the B and C parts are broken in a different way. The filling resin in the part B has a smooth cross section, which indicates that the resin HC2 is brittle fracture, and the fracture surface of the

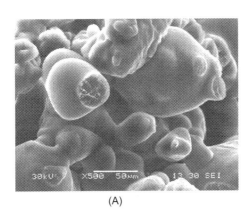
(A)

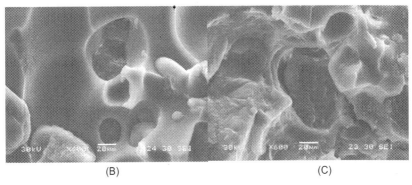
(B) (C)

FIGURE 2.132 SEM pictures of the tensile section of ABS powder sintered parts: (A) sintered part without resin treatment, × 500; (B) HC2 treated part, × 600; (C) HC1 treated part, × 600.

filling resin HC1 in the part C is rough, which indicates that the part C has good toughness.

Fig. 2.133 show SEM pictures of the impact section of the ABS sintered part. Fig. 2.133A shows that the ABS matrix resin has a rough cross section when subjected to impact stress, and is not brittle fracture. At the same time, due to the presence of a large number of pores in the sintered body, the overall impact resistance of the sintered part is still poor. After treated with the epoxy resin, the density of the part B and the part C is increased, and the impact resistance is also enhanced. Fig. 2.133B shows that the part C has a smooth cross section when subjected to the impact stress, indicating that the HC2 resin is broken earlier than the ABS material, and Fig. 2.133C shows that when subjected to the impact stress, the port shape of two components of the part C is consistent, indicating that the HC1 resin and the ABS material are cooperative bearing, and thus the strength of the part C is higher than that of the part B.

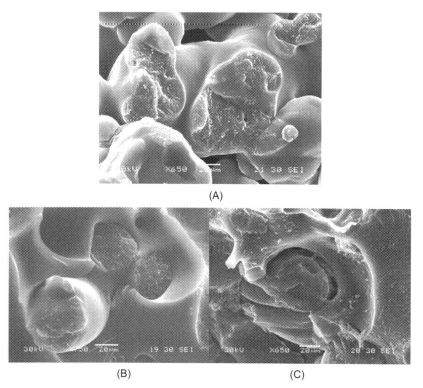

FIGURE 2.133 SEM pictures of the impact section of ABS powder sintered parts: (A) sintered part without resin treatment, × 650; (B) HC2 treated part, × 650; (C) HC1 treated part, × 650.

References

[1] J. Liu, Research on indirect manufacturing of metal parts by selective laser sintering (Ph.D. thesis), Huazhong University of Science and Technology, Wuhan, 2006.
[2] W. Yan, Research on selective laser sintering polymer materials and their parts (Ph.D. thesis), Huazhong University of Science and Technology, Wuhan, 2005.
[3] Y. Jinsong, Research on selective laser sintering materials for plastic functional parts and complex castings (Ph.D. thesis), Huazhong University of Science and Technology, Wuhan, 2008.
[4] Y. Zhang, L. Hao, M.M. Savalani, et al., Characterization and dynamic mechanical analysis of selective laser sintered hydroxyapatite-filled polymeric composites, J. Biomed. Mater. Res. A 86A (2008) 607−616.
[5] Y. Wang, Y. Shi, S. Huang, Structure and properties of laser sintered nylon 12/rectorite composites, Acta Materiae Compositae Sinica 22 (2) (2005) 52−56.
[6] Y. Wang, Y. Shi, S. Huang, Preparation of nylon 12/rectorite nanocomposites by laser sintering, Acta Polymerica Sinica (5) (2005) 683−686.
[7] J.S. Yang, Y.S. Shi, Q.W. Shen, et al., Selective laser sintering of HIPS and investment casting technology, J. Mater. Process. Tech. (2008). Available from: https://doi.org/10.1016/j.jmatprotec.2008.04.056.
[8] Y.S. Shi, Y. Wang, J.B. Chen, et al., Experimental investigation into the selective laser sintering of high-impact polystyrene, J. Appl. Polym. Sci. 108 (1) (2008) 535−540.
[9] L. Lin, Y. Shi, F. Zeng, et al., Research on enhanced post-treatment of polymer powder sintered parts, J. Funct. Mater. 34 (1) (2003) 67−72.
[10] C. Wang, X. Li, S. Huang, Analysis on precision of SLS formed parts, J. Huazhong U. Sci-Med. 29 (6) (2001) 77−79.

Chapter 3

Polymer materials for additive manufacturing: liquid materials

3.1 Overview of stereolithography apparatus formed photopolymer

Since Inmont Corporation in the United States first published the patent for unsaturated polyester/styrene ultraviolet (UV) curing inks in 1946, and Germany first commercialized the particle plate coated UV curing in the 1960s, UV curing materials have attracted more and more attention due to its fast curing rate, low energy consumption, low environmental pollution, high efficiency, and good film forming performance, and are widely used in the coatings industry, adhesive industry, printing industry, microelectronics industry, and other optical imaging fields. UV curing technology and materials have developed rapidly in recent decades, with a growing speed at almost 10%–15% per year.

UV curing material is a general term for materials such as UV curing coatings, inks, and adhesives. This material, based on oligomers (also known as prepolymers), is formulated with specific active diluent monomers (also known as reactive diluents), photoinitiators, and various additives. The proportion and function of each component are shown in Table 3.1.

Stereolithography apparatus (SLA) is the first additive manufacturing (AM) technology that is used in industrial applications, which usually has two kinds of working modes (see Fig. 3.1). Fig. 3.1A shows that the UV light is irradiated onto the surface of the photopolymer through a light-shielding mask to expose and cure the reception surface of the resin. The method shown in Fig. 3.1B is to use a scanning head to scan the beam onto the surface of the resin to expose it. At present, the most common working method is to irradiate the UV laser beam onto the surface of the liquid photopolymer through the galvanometer scanning system to cure and form the resin.

The SLA working process is shown in Fig. 3.2. First, the 3D solid model of the product is constructed on the computer using a 3D CAD system (Fig. 3.2A), and then the model in the STL file format is generated and output (Fig. 3.2B). Then, using the slicing software, the model is sliced in layers in the height direction to obtain the 2D data group S_n ($n = 1, 2,...,N$) of each layer of a cross section of the model, as shown in Fig. 3.2C. Based on these data, the computer extracts the data in order from the lower layer S_1, controls

TABLE 3.1 Basic components and functions of ultraviolet curing materials.

Names	Functions	Common content (mass fraction) (%)	Types
Photoinitiator	Absorbs ultraviolet light and initiates polymerization	<10	Radical type, cationic type
Oligomer (prepolymer)	The main part of the material, it determines the main properties of the cured material	>40	Epoxy acrylatePolyester acrylatePolyurethane acrylate, etc.
Diluting monomer (reactive diluent)	Adjusts viscosity and participates in curing reaction, affecting the performance of cured films	20–50	MonofunctionalityBifunctionalityPolyfunctionality
Others [pigment (dye), stabilizer, surfactant, wax, etc.]	Depends on different purposes	0–30	

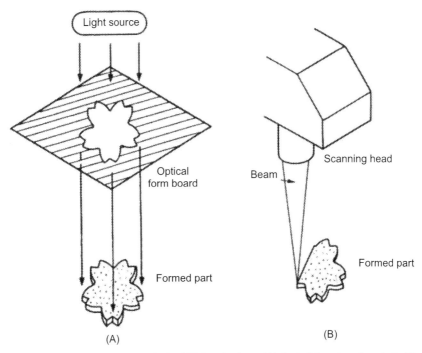

FIGURE 3.1 Two exposure methods of SLA technology. (A) Optical form board method. (B) Light scanning method.

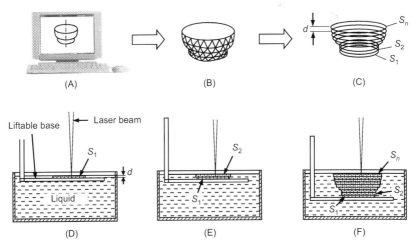

FIGURE 3.2 Stereolithography forming process. (A) CAD 3D model. (B) STL format model. (C) Model slicing. (D) Curing of the first layer S_1. (E) Curing of the second layer S_2. (F) Curing of the last layer S_n.

the UV laser beam through a scanning head, and scans the cross-sectional shape of the first layer model on the surface of the liquid photopolymer. The portion irradiated by the UV laser beam is cured by the polymerization of the prepolymer and the reactive monomer due to the action of the photoinitiator, resulting in a thin layer of the cured layer (Fig. 3.2D). After the cured layer of the first layer of cross section is formed, the base is lowered by a set height D, and a layer of liquid resin is further applied to the surface of the cured layer. Then repeat the aforementioned process, the data of the second layer S_2 cross section is scanned, exposed, and cured (Fig. 3.2E). When the height D of the slice layer is less than the thickness at which the resin can be cured, the upper layer of the cured resin can be bonded to the lower layer of cured resin. Then the third layer S_3, the fourth layer S_4, are cured, bonded, and stacked step by step until the S_n layer, finally forming a solid prototype (Fig. 3.2F).

The light source used in SLA includes UV and visible light, mostly using UV light, so this book mainly introduces UV light.

The material used for SLA forming is liquid photopolymer, such as acrylate system, epoxy resin system, etc., when UV light is irradiated onto the liquid, the exposed portion is cured after photoinduced polymerization, and a chemical reaction occurs mainly during forming. SLA-forming material is an extension of the application of UV curing resin, and the development of UV curing coating has also promoted the development of SLA-forming material to a certain extent.

Photopolymers used in SLA technology can be roughly divided into three generations.

The early commercialized SLA materials are all radical type photopolymers using acrylate or urethane acrylate as prepolymer, and the reaction mechanism is to convert the double bond into a single covalent bond by an addition reaction, such as 5081, 5131, 5149 developed by Ciba-Geigy Cibatool and 2100, 2110, 3100, 3110 developed by Du Pont Company, and the like. Such a resin has the advantages of low price and fast curing speed, but the surface layer has aerobic inhibition and large curing shrinkage, the part formed by the resin is easy to warp and deform, especially for the workpiece having large planar structure, the manufacturing precision is not very high.

To improve the shrinkage of the acrylate resin, researchers have added various fillers to reduce the shrinkage in the acrylate resin. Zak et al. treated the SLA resin with glass fiber, which not only improved the shrinkage of the resin, but also enhanced the mechanical properties of the material, but the disadvantage was that the viscosity of the resin is too high, resulting in difficulty in operation and an increase in brittleness of the material. Karrer et al. modified the SLA resin with porous polystyrene and quartz powder. When the filler reached 40%, the resin shrinkage decreased from 8% to about 2%, but the disadvantage was that the viscosity of the resin was too high, which was extremely unfavorable to the operation. Xi'an Jiaotong University has also done a lot of experiments to modify with a fine powder of resin polymer. The amount of polymer powder should be added under the premise that the viscosity of the SLA resin will not increase significantly, which can also make the resin shrinkage greatly improved.

The second generation of commercialized SLA materials is mostly epoxy resin (or vinyl ether) based photopolymer. Compared with the first-generation resin, it is featured by lower viscosity, smaller curing shrinkage, and a lower degree of warpage of parts, high precision, and good timeliness.

The third generation of SLA materials was born with the development of SLA technology. The parts made from this kind of resin have special properties, such as better mechanical properties, optical properties, etc. Parts made on SLA-forming equipment can be directly used as functional parts [1].

3.1.1 Stereolithography apparatus material

3.1.1.1 Composition of stereolithography apparatus material

1. Photopolymer

The photopolymer material refers to the resin material which exhibits special functions under the action of light. It has a wide range. Table 3.2

TABLE 3.2 Application and classification of photopolymer.

Classification of photoreactions	Classification of chemical reactions	Types of resins	Application
Photodegradation type	Addition reaction	Azide system	Plate making material photoresist
	Rearrangement reaction	Naphtha-benzoquinone system	
Photocrosslinking type	Chelate crosslinking	Dichromic acid system	Plate making material photoresist
	Dimerization	Polyvinyl cinnamate system	
Photoinitiated polymerization type	Free-radical polymerization	Unsaturated polyester	Woodworking painting
		Acrylate	Printing ink, adhesive, plastic, paper coating, metal coating, solder resist
	Free-radical addition reaction	Thiol/ene	
	Cationic polymerization	Epoxy resin	

lists the application of photopolymers in various aspects. Among them, those converted from liquid state to solid state under the action of photochemical reaction are named as SLA resins. It is a mixture composed of main components of photopolymerization prepolymers or oligomers, monomers and photoinitiators. Oligomers such as acrylates and epoxy resins are the main components of the photopolymer, which determine the physical properties of the SLA product.

Since the viscosity of the oligomer is generally high, the monomer should be added as a diluent to improve the fluidity of the resin as a whole, and the monomer also participates in the molecular chain reaction upon curing. The photoinitiator is an active species which can be decomposed under irradiation of light to generate polymerization reaction. In order to increase the sensitivity of the resin reaction, the sensitizer is added sometimes. The sensitizer does not directly react after light absorption to generate an active center capable of initiating polymerization, but acts on the initiator by energy transfer to expand the absorption wavelength band and absorption coefficient, so as to increase light efficiency. In addition, additives such as defoamers and leveling agents are added to the system.

With the continuous development of SLA technology, SLA resins have evolved from the first generation of unsaturated polyester system to the second generation of acrylate system and cationic curing system. Different curing mechanisms have different requirements for the resin and the type of initiator used is also different. The resins and photoinitiators currently used for free-radical and cationic polymerization are shown in Table 3.3 [2].

TABLE 3.3 Classification of photopolymer.

Type		Prepolymer	Initiator
Free-radical polymerization type	Unsaturated polyester	Unsaturated polyester resin	Benzoin alkyl ether
	Acrylic ester	Polyester acrylate Polyether acrylate Acrylic polyurethane Epoxy acrylate, etc.	Benzoin alkyl ether Benzophenone Acetylbenzene Michler's ketone
Cationic polymerization type		Epoxy resin	Diazonium salt of Lewis acid Super acid diphenyl iodide Triphenyl sulfide Ferrocenium salt
		Polycyclic monomer	
		Vinyl ether	

2. Reactive diluent

The reactive diluent is a functional monomer, which functions to adjust the proper viscosity of the photopolymer, control the curing crosslink density of the resin, and improve the physical and mechanical properties of the cured parts. Most of the reactive diluents currently used are acrylate monomers, which can be classified into monofunctional, difunctional and polyfunctional reactive diluents depending on the number of double bonds contained in each molecule. Due to their different structures and activities, therefore, factors such as the solubility, volatility, flash point, odor, toxicity, reactivity, functionality, homopolymer, glass transition temperature, polymerization shrinkage and surface tension should be considered in the selection of reactive diluents.

Commonly used monofunctional, difunctional, and polyfunctional diluents are shown in Table 3.4.

From the dilution effect of the reactive diluent: monofunctional > difunctional > polyfunctional; but from the SLA speed: polyfunctional > difunctional > monofunctional, modified ethoxylation > unmodified ethoxylation > propoxylation; from irritation to the skin: unmodified ethoxylation > modified ethoxylation (propoxylation).

In the actual formulation, depending on the property requirements of the cured parts, two or more reactive diluents are often used in

TABLE 3.4 Types of reactive diluents.

Type	Monomer name	Code	Viscosity/MPa s (25°C)
Monofunctional	Styrene	St	2.07
	N-vinylpyrrolidone	N-VP	1.70
	Isooctyl acrylate	EHA	5.34
	Hydroxyethyl acrylate	HEA	7.5
	Isobornyl acrylate	IBOA	
Bifunctional	Triethylene glycol diacrylate	TEGDA	18
	Tripropylene glycol diacrylate	TPGDA	13–15
	Ethylene glycol diacrylate	PEGDA	8
	Polyethylene glycol diacrylate	NPGDA	20
	Neopentyl glycol diacrylate	PONPGDA	10
	Propoxy neopentyl glycol diacrylate		5
Polyfunctional	Trimethylolpropane triacrylate	TMPTA	70–100
	Ethoxylated trimethylolpropane triacrylate	EO-TMPTA	25
	Pentaerythritol triacrylate	PETA	600–800
	Propoxylated pentaerythritol triacrylate	PO-PETA	225

combination. Generally, the polyfunctional reactive diluents such as TMPTA, EO-TMPTA, PETA, and PO-PETA are selected to increase the SLA speed, the crosslink density and the hardness.; to improve the flexibility of the cured film, choose 2-EHA, TEGDA, TPGDA and HDDA; a-EHA, HEA, NPGDA, and N-VP are useful to reduce the viscosity of the system, especially N-VP, it is particularly good at reducing viscosity due to its ability to destroy intermolecular hydrogen bonds in UV systems. It has relatively low volatility in monofunctional diluents, and the highest rate of SLA speed, and can increase the tensile strength of the cured film, which has little effect on the elasticity of the system similar to the crosslinking monomer.

3. Photoinitiator

Photoinitiator is a major factor in determining the degree and speed of curing. According to the difference of active radicals, photoinitiators can be classified into three types: cationic type, radical type, and radical-cationic combined type. According to the different mechanism of excitation and decomposition of photoinitiators, they are mainly divided into the following categories:

Cationic initiator

The earliest commercialized cationic photoinitiator system is diazonium salt, such as aryl diazonium fluoroborate, which produces BF_3, N_2, and fluoro aryl ArF after photodecomposition:

$$ArN_2 + BF_4^- \rightarrow ArF + BF_3 + N_2$$

Since the diazonium salt precipitates N_2 during photolysis, bubbles or pinholes are formed when the polymer is formed into a film, which limits its practical application. Another disadvantage is that it is unstable and cannot be stored for a long period of time.

The most common cationic photoinitiators currently available are diaryliodonium salts and triarylsulfonium salts. The cationic photoinitiator is cleaved under the action of light to produce protonic acid or Lewis acid which initiates ring-opening polymerization of the epoxy compound, and also produce free radicals, so that they can also act as initiators of free-radical photopolymerization.

It has been reported that the free radical–cation hybrid photoinitiator synthesized by combining the structural characteristics of benzophenone with diphenyl iodonium salt has higher initiation efficiency than the mixed system of benzophenone/diphenyl iodonium salt, and its molecular formula is as shown in Fig. 3.3.

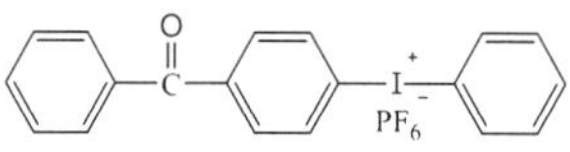

FIGURE 3.3 Free radical–cation hybrid photoinitiator.

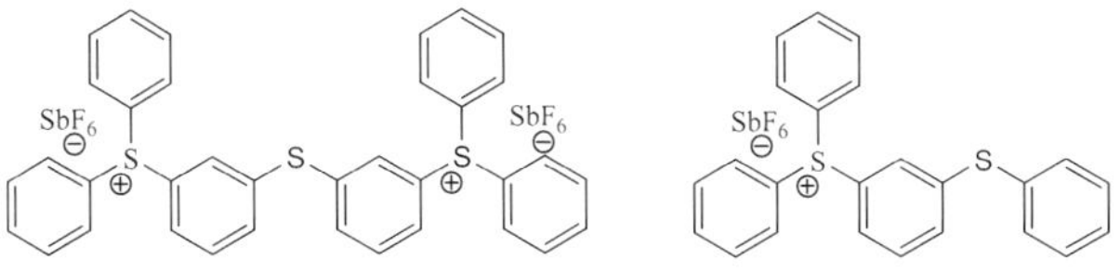

FIGURE 3.4 TPS-SbF$_6$.

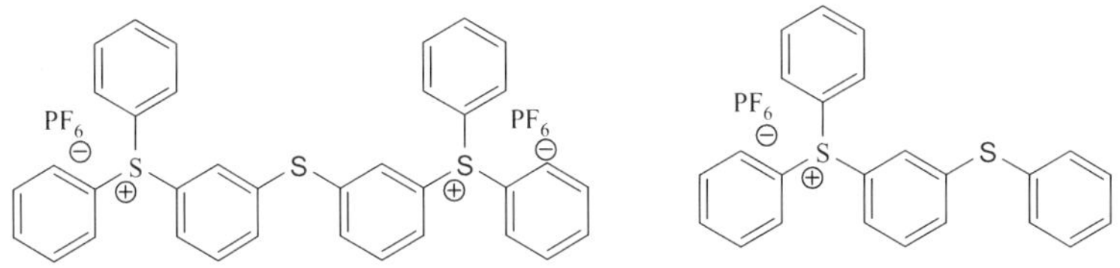

FIGURE 3.5 TPS-PF$_6$.

In addition, there are reports on the use of sulfonium salts as cationic photoinitiators.

There are few commercial cationic initiators. There are mainly four kinds of cationic photoinitiators: triaryl hexafluoroantimonate sulfonium salt (TPS-SbF$_6$) as shown in Fig. 3.4, triaryl hexafluorophosphate sulfate salt (TPS-PF$_6$) as shown in Fig. 3.5, aryl hexafluoroantimonate iodonium salt (DPI-SbF$_6$) and diaryl hexafluorophosphate iodonium salt (DPI-PF$_6$). If the SLA device uses a semiconductor diode laser, its wavelength is 355 nm. According to reports, the absorption wavelength of DPI-PF$_6$ is about 250 nm, which is not well matched with the laser; DPI-SbF$_6$ has strong absorption at 300 nm, which can make full use of the light source; and two kinds of sulfonium salts TPS-SbF$_6$ and TPS-PF$_6$ both have maximum absorption wavelength above 300 nm, which are greater than the maximum absorption wavelength of DPI-PF$_6$ and DPI-SbF$_6$, and well matched with the light source, showing good initiation efficiency.

Since the cationic initiation is essentially the photoreaction initiated by the super acid formed after the decomposition of the initiator, the rate of initiation is faster as the acidity increases. The acidity of the corresponding super acid is HPF$_6$ < HSbF$_6$. That is to say, the initiation efficiency of TPS-PF$_6$ is inferior to that of TPS-SbF$_6$. It has been reported that the hexafluoroantimonate-containing initiator system has a greater depth of cure than the hexafluorophosphate-containing initiator system.

In addition, there is ferrocenium salt. Although it has strong absorption in the near UV region, it also has absorption in the visible region. Considering the particularity of the SLA-forming process, it is required that the SLA photopolymer can be stored in the presence of air and visible light. Therefore, such initiators are not considered. There are two forms of ionic polymerization, cations and anions. The industry is of particular interest in photoinitiated cationic polymerization. The cationic photoinitiator is not sensitive to aerobic conditions and therefore does not require inert encapsulation, which can be used to cure compounds such as epoxy, cyclic ether, sulfide, acetal, lactone saturated monomer, and olefins, and has advantages such as postcuring; however, it also has disadvantages, for example, the Lewis acid generated by the cationic photoinitiator during polymerization causes corrosion to the equipment, and polymerization is inhibited due to the presence of moisture in the system.

Free-radical initiator

The free-radical photoinitiators are relatively mature and have a wide variety of commercial products. However, according to the generation mechanism of free radicals, they can be divided into two categories: one is the split photoinitiator, also known as the first-type photoinitiator; the other is the hydrogen-extraction photoinitiator, also known as the second-type photoinitiator.

A common feature of the split initiator is that, under the Norrish I mechanism, the carbon−carbon σ bond adjacent to the carbonyl group in the molecule is broken after absorption of UV, including some aromatic carbonyl compounds capable of Norrish type I cleavage, most of them are benzoin and its derivatives, benzyl ketal, acetophenone derivatives and some sulfur photoinitiators; 1-hydroxycyclohexylacetophenone is taken as an example, as shown in Fig. 3.6.

The hydrogen-extraction initiator is generally an aromatic ketone compound, for example, benzophenone and its derivatives, thioxanthone and most of sulfur initiators. Such initiators absorb UV light and photolyze according to the Norrish II mechanism, and form active radicals which can initiate polymerization after extracting hydrogen. The initiation mechanism of the common hydrogen-extraction initiator, benzophenone, is shown in Fig. 3.7.

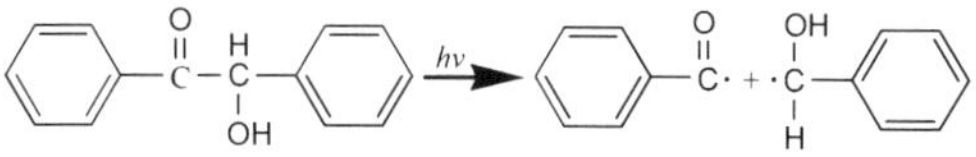

FIGURE 3.6 Hydroxycyclohexylacetophenone.

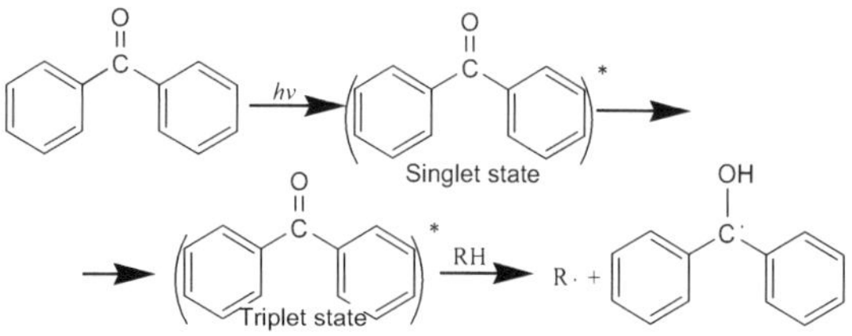

FIGURE 3.7 Initiation mechanism of benzophenone.

Polymer photoinitiation system

Photoinitiators are often not completely exhausted in SLA, and the unphotosolved parts will migrate to the surface of the coating and cause yellowing and aging of the coating, affecting product quality; on the other hand, some initiators are not compatible with the system, its application is limited. Polymer photoinitiators can overcome these shortcomings. The macro molecularization of the photoinitiator allows the initiator to be directly attached to the polymer or oligomer. For example, benzophenone, thioxanthone, acylphosphine oxide or the like are introduced into the polymer chain.

Compound initiator

The introduction of radical-initiated group and cation-initiated group into the same molecule results in a free radical−cation hybrid SLA system. He et al. combined the structural characteristics of benzophenone and diaryliodonium salt to design and synthesize a suitable photoinitiator.

The compatibility of various photoinitiators is also a research direction in recent years, which can reduce the cost, expand the region of absorption wavelength, and increase the absorption of UV radiation. The compatibility of various photoinitiators can be between the same type as is the free-radical type, for example, Ciba-Grtiey's newly introduced Irgacure-1700 that is composed by 25% bis(2,6-dimethoxybenzyl)-2,4,4-trimethylpentylphosphonium compound and 75% Darocure-1173; or a combination of different types, such as the combined use of free-radical type and cationic photoinitiator. For example, when triaryl sulfide salt and benzophenone are mixed as an initiation system, the initiation of the epoxy compound can improve the curing speed.

Photosensitizer additive

The photosensitizer additive itself has no photosensitizing effect, but it can produce antioxidation effect, increase the sensitivity, and promote the change of light energy as an auxiliary agent. For example, tertiary amine compound belongs to such an additive, and the commonly used compounds include dimethylethanolamine, triethylamine, N,N-dimethylbenzylamine, etc.

The SLA resin itself is also heat sensitive. When some photoinitiators are heated, they can also thermally decompose and initiate polymerization. Therefore, it is necessary to add additive to prevent the resin from being polymerized when heated. SLA resin mainly uses thermal polymerization inhibitor. The terpenoid itself has thermal inhibition function and is also a photoinitiator. The initiation mechanism is shown in Fig. 3.8.

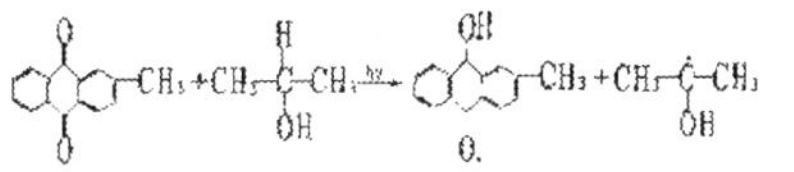

FIGURE 3.8 Initiation mechanism of terpenoids.

3.1.1.2 Requirements of stereolithography apparatus for resin materials

Although the SLA resin is similar to general UV resin in terms of composition, similar to paint before curing and general plastic after curing, it is different from ordinary UV resin due to the uniqueness of SLA-forming process. It is required to be fast and accurate in manufacturing the prototypes with SLA technology, which is demanding on the precision and properties of the parts, and it is required to be easy to operate during the forming process. Its property requirements are special, which should generally meet the following requirements: the viscosity and photosensitive properties of the resin before curing, the precision and mechanical properties of the material after curing. So this material must have the following characteristics:

1. Stable properties before curing

    It is easy to transport and store, and basically has no dark reaction.

    The photopolymer used for SLA is usually injected into the resin tank but not taken out, and is added later with the continuous use and consumption. Therefore, the resin is generally used for a long time, that is, the resin is required not to appear thermal polymerization under normal conditions, and also has high stability to visible light to ensure stable resin properties during long-term forming process.

2. Low viscosity

    The manufacturing process of the SLA method is superimposed layer by layer at a time. When one layer is completed, it is difficult for the resin to automatically cover the surface of the cured layer due to the surface tension of the liquid. Therefore, after the layer is completed, the liquid surface of the resin is flattened once by using an automatic scraper, and is scanned when the liquid surface is flattened and stabilized, otherwise the parts will be defective. Therefore, the viscosity of the resin becomes an important performance index. The viscosity of the resin should be as small as possible in the case that the other performance kept unchanged, which not only shortens the production time, but also facilitates the feeding of the resin and the cleaning of the waste liquid.

3. Small curing shrinkage

    The main problem of the SLA method lies in manufacturing precision. The shrinkage during forming not only reduces the precision of the part, but more importantly, the curing shrinkage also causes warpage, deformation, cracking, etc. of the part. In severe cases, the part will be moved by the scraping plate during the forming process, causing the complete failure of forming. Therefore, the resin used for SLA should be made of material with low shrinkage.

4. High degree of cure at one time

    Some SLA materials cannot be directly applied after the parts are made, and need to be post-cured in the UV exposure chamber. However,

it is impossible to ensure that the light intensity received in all directions and on each surface is exactly the same in the postcuring process, thus causing overall deformation of the parts, which seriously affects the precision of the parts.
5. Small swelling

    During the forming process, the cured product is infiltrated into the liquid resin. If the cured product swells, the parts will not only lose strength, but also the cured part will swell and overflow, which will seriously affect the precision. The surface of the formed parts has a large amount of uncured resin which needs to be cleaned by solvent. It is desirable to remove only the uncured portion during cleaning, so that it has no effect on the surface of the parts, therefore, it is desirable that the cured product has good solvent resistance.
6. Fast curing speed and high absorption and response speed for light with a wavelength of 355 nm

    SLA forming generally uses UV lasers. The concentrated energy of the laser can ensure high precision of the forming, but the scanning speed of the laser is very fast, generally more than 1 m/s, so the time of light acting on the resin is extremely short, and the resin only has large absorption and high response speed on the light of this band, and then cures quickly.
7. High strength of semifinished products

    This is to ensure that the parts do not deform, swell, bubble, and have separated layer during the postcuring process.
8. The cured product has good mechanical properties

    For example, the cured product has high breaking strength, impact strength, hardness and toughness, chemical resistance, and is easy to wash and dry, as well as having good thermal stability. Among them, precision and strength are the two most important indicators of rapid prototyping. The strength of rapid prototyping parts is generally not high, especially for SLA materials, which used to be generally brittle, and are difficult to meet the requirements of functional parts. In recent years, some companies have launched materials with better toughness.
9. Low toxicity

    Toxic oligomers, monomers, and initiators should be avoided to ensure the health of the operators while avoiding environmental pollution. Future rapid prototyping can be done in the office, so it is important to consider this when designing a recipe.

3.1.2 Stereolithography apparatus reaction mechanism

The SLA AM technology is developed by utilizing the liquid photopolymer to undergo chemical reaction and rapid curing under UV light irradiation. According to the different active centers generated by the initiation, it can be classified into free-radical SLA system, cationic SLA system, and free radical—cation hybrid SLA system.

The free-radical SLA system has the advantages of fast curing speed, rich raw materials, and adjustable performance. The main disadvantages include severe oxygen resistance, severe shrinkage, and low-forming precision and poor adhesion between layers. The main advantages of the cationic SLA system include wear resistance, high hardness, good mechanical properties, small volume shrinkage, and strong adhesion between layers. Therefore, it is especially suitable for SLA AM technology that requires high precision. The main disadvantages of the cationic SLA system are that the curing is affected by moisture, the curing speed is slow, the types of raw materials are small, the price is high, and the performance is difficult to adjust. The biggest disadvantage is that the curing speed is slow. The free radical—cation hybrid SLA system combines the advantages of above two systems, while avoiding the shortcomings of them, broadening the use scope of the SLA system to a certain extent. The free radical—cation hybrid polymerization system exhibits a good synergistic effect. Unlike the copolymerization process of several free-radical monomers, the free radical—cation hybrid polymerization produces not copolymer but polymer alloy. During the reaction polymerization process, Free-radical polymerization and cationic polymerization are carried out separately to obtain an interpenetrating network (IPN) structure product, which makes the SLA product have better comprehensive properties; in addition, the radical initiating system and the cationic initiating system are synergistic and promote each other to make reaction.

Since the curing process of the SLA system is a polymerization reaction caused by light irradiation, there are some unavoidable defects in the SLA method, such as the curing depth is limited; it is difficult to be applied well in the colored system; the shadow portion cannot be cured; and the shape of the cured object is limited by the SLA device. In response to these shortcomings, people have developed multiple curing systems such as photothermal dual curing system, photothermal moisture triple curing system.

The following sections describe the chemical reaction processes and characteristics of free-radical SLA systems, cationic SLA systems, and free radical—cation hybrid SLA systems.

3.1.2.1 Free-radical photoinitiated polymerization mechanism

The photopolymer of free-radical system is typically characterized by the occurrence of free-radical polymerization. The main reaction mechanism can be divided into three stages: chain initiation, chain growth, and chain termination reactions.

First, the photoinitiator absorbs the UV light energy under the irradiation of a certain wavelength of light; the initiator molecule changes from the ground state to the excited state; the covalent bond in the molecular structure undergoes singlet state or triplet state or hydrogen extraction to generate the primary free radical of active fragments; and the primary free radical, with the addition of monomer, forms monomer free radical. The activity of the monomer free radical generated during the chain initiation phase is not reduced, and generates new free radical through reaction with the second monomer or oligomer molecule, and the activity of the new radical is still not reduced and continues to combine with other reactive molecules to form chain radicals with more repeating units, eventually forming macromolecules. With the rapid increase of the degree of reaction, the photopolymer cures. Except that a very small number of free-radical active centers cannot be terminated, the active radicals eventually terminate by interaction in a coupling mode and a disproportionation mode.

1. Free-radical photoinitiated polymerization mechanism

   The split photoinitiator is decomposed according to the Norrish I mechanism, and after absorption of UV light, the C-C single bond adjacent to the carbonyl group in the molecule is broken, as shown in Fig. 3.9.

   Depending on the chemical structure, most of these photoinitiators are benzoin and its derivatives. For example, the most important photoinitiator in the UV curing industry, 2,2-Dimethoxy-2-phenylacetophenone (commonly known as benzoin dimethyl ether, i.e. Irgacure-651) is an aromatic carbonyl compound. After it absorbs UV light, the splitting mechanism is shown in Fig. 3.10.

   The two free radicals produced by the photolysis of benzoin dimethyl ether can be added to the unsaturated double bond of the oligomer or monomer to initiate the polymerization reaction. The reaction equation of benzyl dimethyl ether free radical is shown in Fig. 3.11.

   The reaction equation of another type of free radical, benzophenone series free radical, is shown in Fig. 3.12.

   Most of the hydrogen-extraction photoinitiators are aromatic ketone compounds such as benzophenone series, thioxanthen, anthraquinone, 3-ketocoumarin, and aromatic ketone/mercaptans photoinitiators. Taking

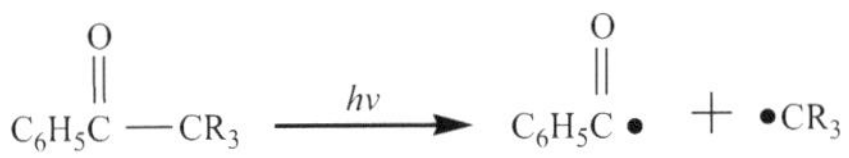

$$C_6H_5C(=O)-CR_3 \xrightarrow{h\nu} C_6H_5C(=O)\bullet + \bullet CR_3$$

FIGURE 3.9 Photolysis equation of aromatic carbonyl compounds.

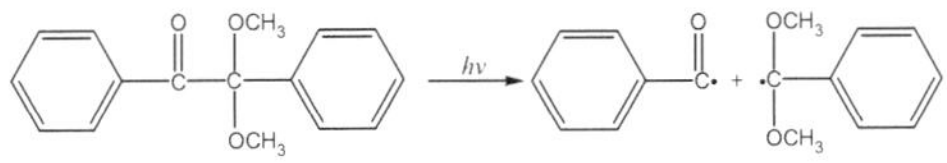

FIGURE 3.10 Photolysis equation of benzyl dimethyl ketal.

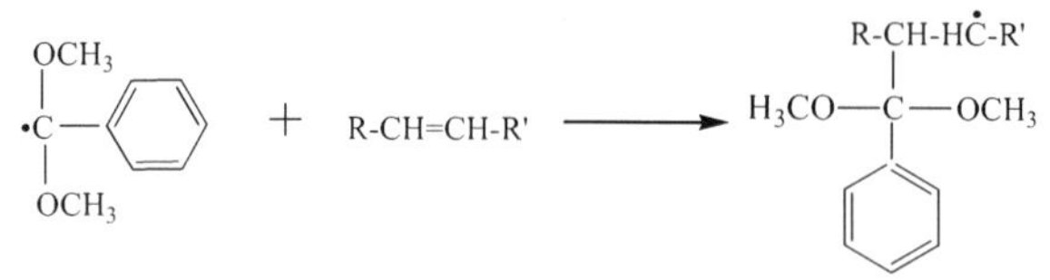

FIGURE 3.11 Reaction equation of benzyl dimethyl ether free-radical initiated monomer polymerization.

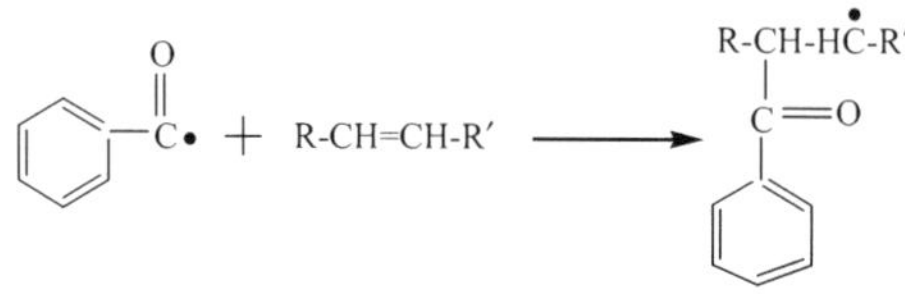

FIGURE 3.12 Reaction equation of benzophenone series free-radical initiated monomer polymerization.

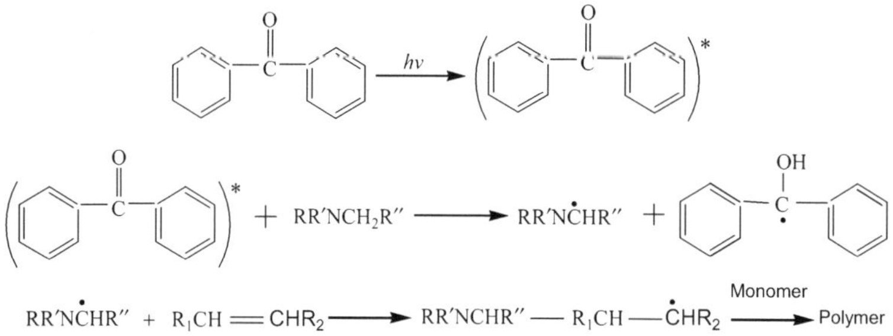

FIGURE 3.13 Mechanism of photopolymerization initiated by benzophenone series.

a common benzophenone series photoinitiator as an example, after the benzophenone series is irradiated with UV light, it does not undergo a splitting reaction when transitioning to an excited state, but extracts hydrogen from a hydrogen donor molecule (mostly triallylamine) to produce a benzhydrol free radical and an aminealkyl free radical. The initiation of polymerization is generally achieved by an aminealkyl free radical. The benzhydrol free radical finally forms benzophenone series and benzhydrol generated by two same radicals after disproportionation, or tetraphenyl pinacol ether generated by biradical coupling. Triallylamine is the most commonly used hydrogen donor. Once exposed to UV light, one of the electrons in the lone pair of electrons on the nitrogen is transferred to the benzophenone series carbonyl oxygen and forms exciplex. The mechanism of photopolymerization initiated by benzophenone series is shown in Fig. 3.13.

2. Chemical reaction mechanism of shrinkage of free-radical system

In terms of the chemical reaction process, the curing process of the photopolymer is that the photopolymer transforms from small molecule to long-chain or macromolecule, and its molecular structure changes significantly, so shrinkage is inevitable during the curing process. The shrinkage of the photopolymer is mainly composed by two parts, one is the shrinkage caused by chemical reaction curing; the other is thermal expansion and cold shrinkage caused by temperature change of the photopolymer. Generally, the thermal expansion coefficient of photopolymer is about 10^{-4}, so that the amount of shrinkage caused by the temperature change is extremely small and can be neglected. The volume shrinkage caused by the curing of free-radical chemical reactions cannot be ignored.

From the perspective of polymer physics, the main reason for the volumetric shrinkage of chemical reactions is that before the photopolymer is uncured, the force between the monomers and oligomers is van der Waals force, the distance between them is van der Waals distance, and the distance between the molecules becomes the length of covalent bond when the photopolymer is polymerized and cured, the length of covalent bond is much shorter than the operating distance of van der Waals force, and the large change in the intermolecular distance will definitely cause volume shrinkage during polymerization and curing. The severe volume shrinkage means that there is shrinkage stress, so it inevitably produces warping deformation of the formed part. The reason why the radical system resin shrinks during the polymerization and curing reaction is shown in Fig. 3.14.

3.1.2.2 Cationic photoinitiated polymerization mechanism

The photopolymerization curing process of the cationic system is carried out by cationic mechanism. The initiator generates cationic active center under UV light irradiation in a certain wavelength range, and the cation further initiates polymerization of epoxy or vinyl ether. The cationic photoinitiator includes diazonium salt, diaryliodonium salt, triarylsulfonium salt, alkylsulfonium salt, iron arene salt, sulfonyloxyketone, and triarylsiloxysiloxane. The monomers suitable for cationic photopolymerization are mainly epoxy compounds, vinyl ethers, lactones, acetals, cyclic ethers, and so on.

In the cationic photoinitiation system, the light is activated to the excited state, and the initiator molecules undergo a series of decomposition reactions

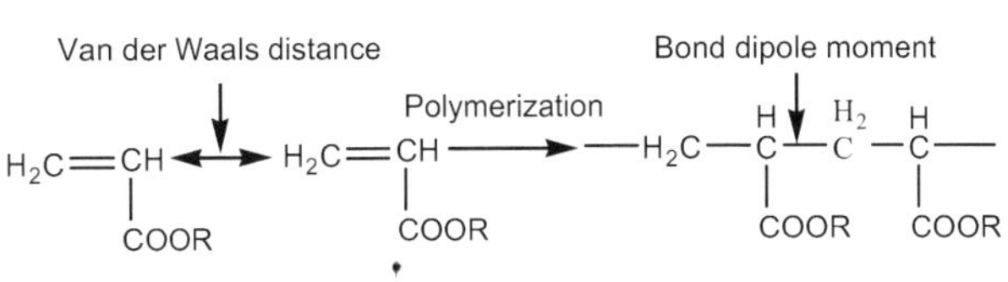

FIGURE 3.14 Change in structural unit distance during polymerization of free-radical monomers.

to produce super protonic acid or Lewis acid, which initiates cationic photopolymerization. The strength of the acid is the key to whether the cationic polymerization can be initiated and proceeded. If the acidity is not strong, the paired anion has strong nucleophilicity and is easily combined with the carbon cation center to prevent polymerization. Compared with the free-radical photoinitiation system, the cationic system has the characteristics of dark reaction after initiation of polymerization, no oxygen resistance, relatively slow curing, and curing affected by moisture.

1. Diazonium salt

    The earliest studied cationic photoinitiator system was diazonium salt, such as aryl diazonium tetrafluoroborate, which was irradiated with UV light to produce BF_3, N_2, and fluoro aryl group ArF upon photolysis:

    $$ArN_2^+ + BF_4^- \rightarrow ArF + BF_3 + N_2$$

    The generated BF_3 is a Lewis acid which can directly initiate cationic polymerization, or can react with H_2O or other compounds to form protons, and then initiate cationic polymerization:

    $$BF_3 + H_2O \rightarrow H^+ + BF_3(OH)^-$$

    The cationic polymerization process of the epoxy compound initiated by the aryl diazonium tetrafluoroborate is shown in Fig. 3.15.

    In the aforementioned reaction, the anion of the diazonium salt must be an anion having a very weak nucleophilicity, and besides BF_4^-, it may be PF_6^-, AsF_6^-, and SbF_6^-. Cationic polymerization is easy to cause chain transfer, while substances with strong nucleophilicity tend to cause chain termination reactions, such as amines, thiols, and the like. It

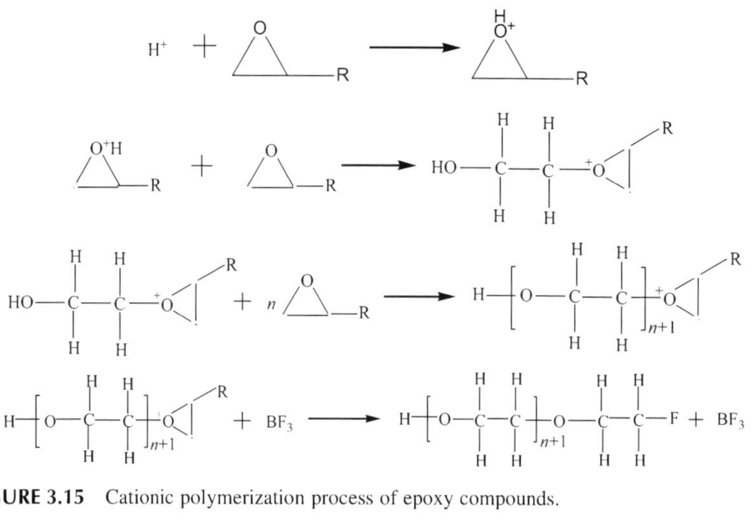

FIGURE 3.15 Cationic polymerization process of epoxy compounds.

can also be seen from the aforementioned polymerization process that a Lewis acid is generated during the chain transfer, and the polymerization reaction can still be initiated. Therefore, the life of the cationic active center is relatively long, and after the light source is removed, the dark reaction can still be performed. At the same time, there is also a problem that after the photopolymerization reaction, protonic acid may still remain in the curing system, which may cause long-term damage to the formed part.

The biggest disadvantage of the diazonium salt as a cationic photoinitiator is that nitrogen is released during photolysis, so that the cured polymer may have bubbles or pinholes, which limits its practical application, and another disadvantage is its poor long-term storage.

2. Iodonium salt and sulfonium salt

In 1977, Crivello and Lan of GE Company first conducted researches on metal halide complexes of diaryliodonium salt and triarylsulfonium, and developed a second generation of cationic photoinitiator. These two types of photoinitiators have good thermal stability and generate no nitrogen.

Diphenyl iodonium salt and triphenyl sulfonium salt are the most important types of cationic photoinitiators, which are commercially available. Both onium salts also release super strong proton acids when exposed to UV light and generate free radicals, and the reaction mechanisms are shown in Figs. 3.16 and 3.17.

In the photolysis reaction of Figs. 3.16 and 3.17, the strength of acidity of corresponding super protonic acids of $X^- = PF_6^-, A_sF_6^-, S_bF_6^-$ is: $HSbF_6 > HAsF_6 > HPF_6$. As the acidity increases, the rate at which it initiates cationic curing is also faster, and the RH in the reaction equation may be a solvent, a monomer or an oligomer.

The disadvantage of diphenyl iodonium salt and triphenyl sulfonium salt is that the maximum absorption wavelength of UV light is about 220–280 nm, the monomer will strongly absorb this part of UV light,

$$Ph_2I^+X^- \longrightarrow P_h\cdot + P_hI^+X^-$$

$$P_hI^+X^- + RH \longrightarrow P_hI + R\cdot + H^+X^-$$

FIGURE 3.16 Photolysis of diphenyl iodonium salt.

$$Ph_3S^+X \xrightarrow{h\nu} [Ph_2S\bullet\bullet\bullet Ph]^+X^-$$

$$[Ph_2S\bullet\bullet\bullet Ph]^+X \longrightarrow Ph-C_6H_4-S-Ph + H^+X$$

$$[Ph_2S\bullet\bullet\bullet Ph]^+X \longrightarrow Ph_2S^+X + Ph\bullet$$

$$Ph_2S^+X + RH \longrightarrow Ph_2S + R^\bullet + H^+X$$

FIGURE 3.17 Photolysis of triphenyl sulfonium salt.

and the UV light in this band will also attenuate seriously. Therefore, the maximum absorption wavelength of onium salt is generally shifted to the long-wave direction by enlarging its degree of conjugation. For example, the structure shown in Fig. 3.18 can be obtained by introducing thiophenol group into the triphenyl sulfonium salt.

The absorption region of the thiophenol group triphenyl sulfonium salt can be extended to 300−360 nm, and it is also a kind of triarylsulfonium salt, which is commercially available. Taking UVI6976 and UVI6992 of DOW Chemical Company of the United States as examples, UVI6976 is obtained by dissolving the two aryl sulfonium salts in 50% propylene carbonate, as shown in Fig. 3.18, wherein the counter ion is fluoroantimonic acid salt anion; UVI6992 is obtained by dissolving the two aryl sulfonium salts in 50% propylene carbonate, as shown in Fig. 3.18, wherein the counter ion is hexafluorophosphate anion.

3. Ferrocenium salt

   The aromatic ferrocene salt is a new cationic photoinitiator developed after the diaryliodonium salt and the triarylsulfonium salt. The products that have been commercialized include Irgacure-261 from Cibay, which has strong absorption in the far UV region (240−250 nm) and the near UV region (390−400 nm), and also absorbs in the visible region (530−540 nm). Therefore, it is a dual photoinitiator for both UV and visible light. The photoinitiated polymerization process of aromatic ferrocene salt is shown in Fig. 3.19.

   In the reaction formula as shown in Fig. 3.19, the benzene ring of $X^- = PF_6^-, A_sF_6^-, S_bF_6^-$ aromatic ferrocene salts may also be cumene, and when the counter ion is hexafluorophosphate anion, it is Irgacure-261.

4. Chemical reaction mechanism of shrinkage of cationic system

   Taking hexafluorophosphate as cationic initiator to initiate photopolymerization of dimethyl adipate (3,4-epoxycyclohexylmethyl) as an example, the polymerization process is shown in Fig. 3.20.

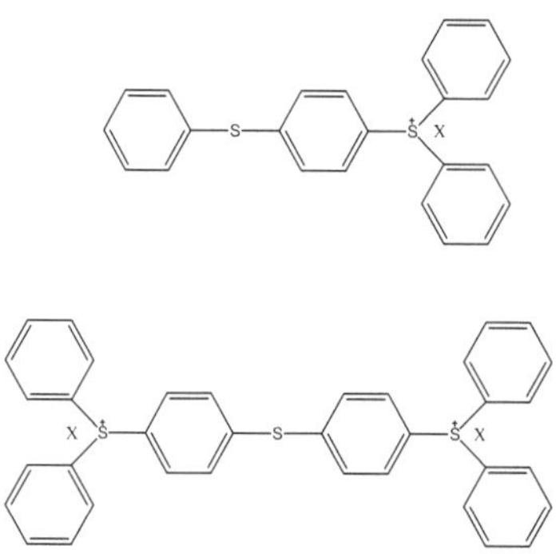

FIGURE 3.18 Thiophenol group triphenyl sulfonium salt.

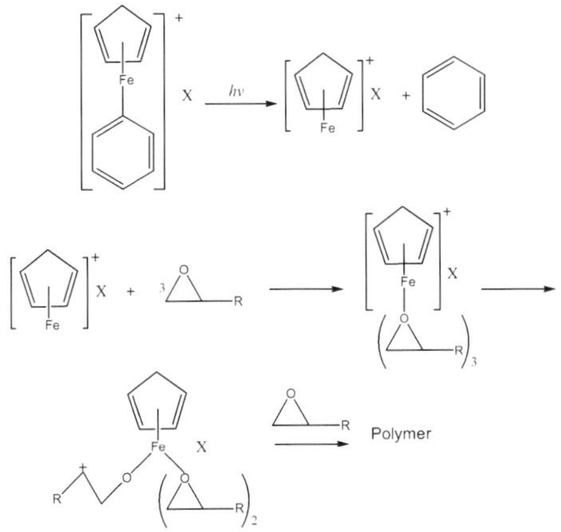

FIGURE 3.19 Photoinitiated polymerization process of ferrocenium salt.

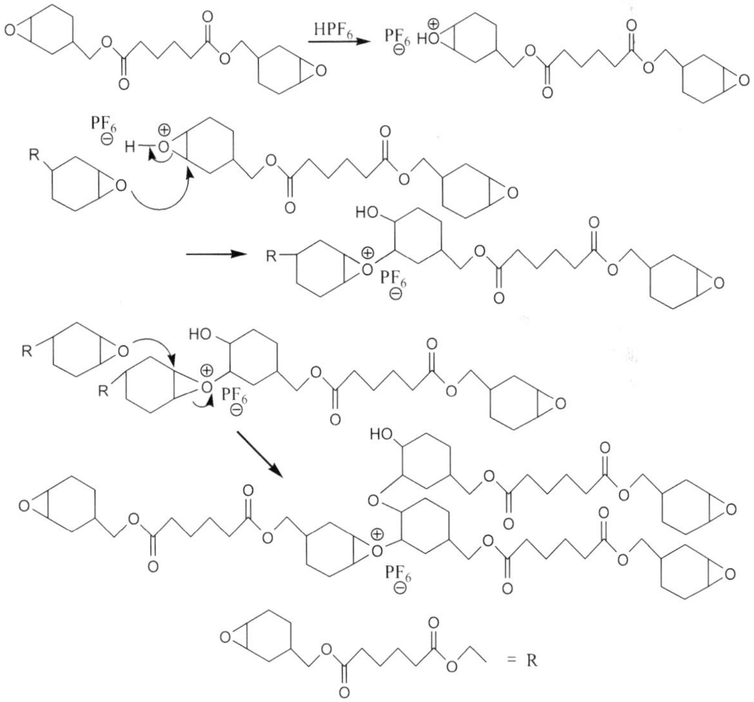

FIGURE 3.20 Cationic photopolymerization of dimethyl adipate (3,4-epoxycyclohexylmethyl).

In the cationic photopolymerization as shown in Fig. 3.20, there is also shrinkage: the interaction force of the monomer or prepolymer before the reaction is van der Waals force, the distance between them is van der Waals

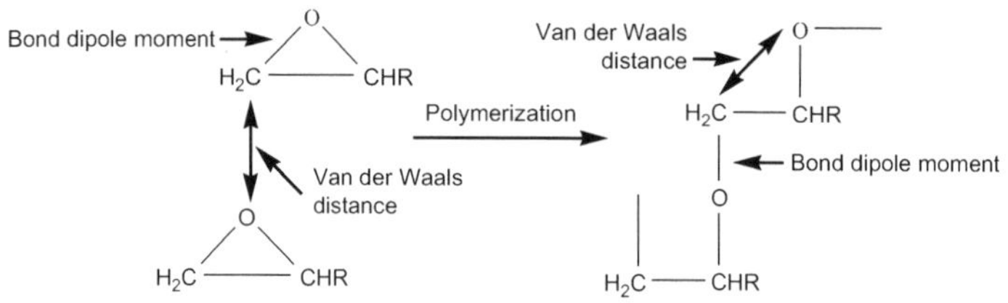

FIGURE 3.21 Change in structural unit distance during ring-opening polymerization of epoxy monomer.

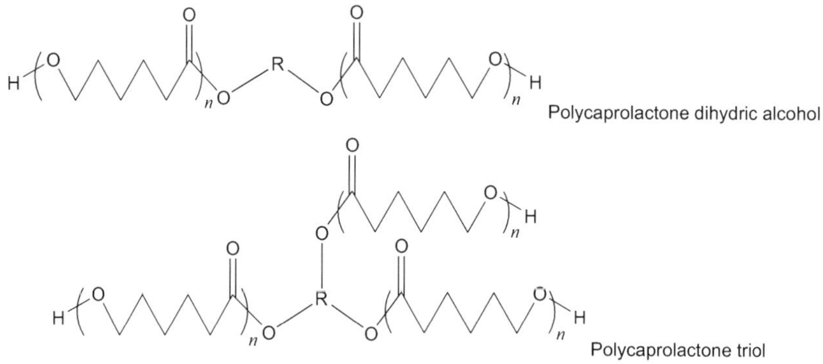

FIGURE 3.22 Structural formulas of typical polycaprolactone polyol.

distance, and after the cationic active center initiates the polymerization reaction, the monomer or oligomer is linked to each other in the form of covalent bond, and the distance between them is covalent bond having a very short distance.

However, in the ring-opening reaction of the epoxy group, after the two C-O single bonds on the alicyclic ring are broken and become van der Waals force distance. That is, the covalent bond distance becomes the van der Waals force distance, which is a factor of the volume expansion of the cation system. The combined result of the aforementioned two factors is that the epoxy compound exhibits small volume shrinkage in the ring-opening polymerization.

The change in structural unit distance during ring-opening polymerization of epoxy monomer or oligomer is shown in Fig. 3.21.

5. Reaction mechanism when polyol participates in the cationic polymerization

In cationic SLA system, an appropriate amount of polyol is often added to increase the SLA rate of the cationic SLA system and to improve the performance of the SLA product. Polycaprolactone polyols are commonly used because of their narrow molecular weight distribution, good toughness, and difficulty in hydrolysis. Typical polycaprolactone polyols have the following two structural formulas, as shown in Fig. 3.22.

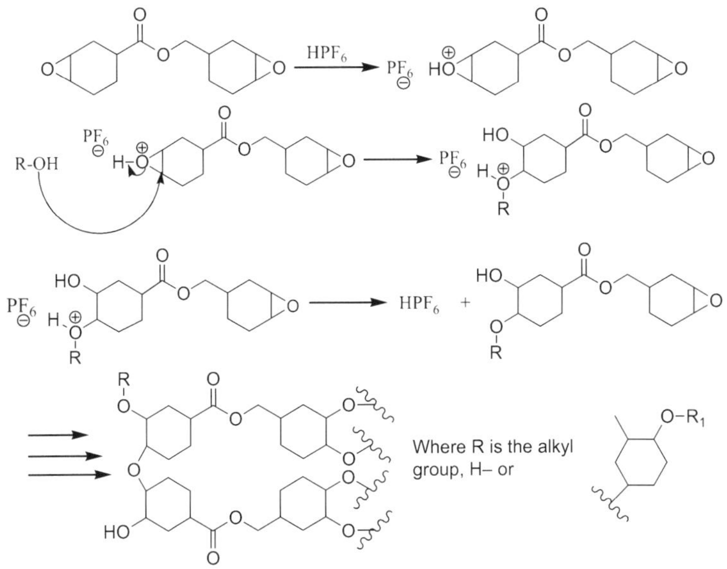

FIGURE 3.23 Reaction mechanism when polyol participates in the cationic polymerization.

The reaction mechanism when polyol participates in the cationic polymerization is shown in Fig. 3.23.

Since the cationic photopolymerization reaction causes chain transfer easily, the crosslinking point density of the polymer is very large. When the polyol is added to the cationic system, the reaction formula as shown in Fig. 3.23 shows that the polyol having long-chain structure makes the molecular chain of the product longer and reduce the crosslinking point density, thus increasing the toughness of the final product.

3.1.2.3 Hybrid polymerization system

The free radical−cation hybrid SLA system generally comprises two types: hybrid system composed by acrylate and epoxy compound, and hybrid system composed by acrylate and vinyl ether. Hybrid systems have good synergistic effects in terms of SLA speed, volume change complementation, and performance adjustment. SLA resins are mainly based on hybrid SLA systems. The photopolymer of the hybrid polymerization system can also be applied to AM solid materials and exhibits high performance.

In the hybrid polymerization system, sulfonium salt or iodonium salt cationic photoinitiator is generally used in conjunction with a free-radical photoinitiator to initiate polymerization of the resin system. Under the irradiation of UV light, the system can simultaneously generate cations and free radicals, thereby respectively initiating polymerization of free-radical monomers and cationic monomers in the system, and obtaining SLA product having an IPN structure.

Due to the short absorption wavelength of the cationic photoinitiators of diphenyl iodonium salt and triphenyl sulfonium salt, in order to match the UV light emitted by the metal halide lamp, some sensitizers are usually added to make full use of the long-wave UV light. The free-radical photoinitiator can effectively sensitize the onium salt cationic photoinitiator by electron transfer. The sensitization mechanism is as follows: the radical fragments produced from photolysis of free-radical photoinitiators further reduce the cerium salt, generate cations and free radicals, and initiate cationic and free-radical polymerization, respectively. The radical photoinitiator herein is required to have an appropriate oxidation—reduction potential to reduce the corresponding cationic photoinitiator.

The α,α-diethoxy acetophenone (DEAP) is free-radical photoinitiator, and the diaryliodonium salt is cationic photoinitiator. The mechanism of synergistic initiation is shown in Fig. 3.24.

As can be seen from Fig. 3.24, the radical initiator and the cerium salt constitute a spontaneous oxidation—reduction reaction, and electron transfer occurs to achieve indirect photolysis. In addition to electron transfer, some hydrogen abstraction photoinitiators can directly sensitize sulfonium salts such as benzophenone series and acetophenone. The hydrogen abstraction photoinitiator has a higher absorption wavelength, and after absorbing the UV light energy to the excited state, the energy is transferred to the strontium salt by physical action (collision, electromagnetic field, etc.) to promote the photolysis of the strontium salt. In this process, there is no electron transfer and any chemical reaction between the free-radical initiator and the cerium salt, which is a pure energy transfer process. In the direct photosensitization process, the photosensitizer (i.e., the hydrogen abstraction photoinitiator) is required to have excited triplet energy higher than that of the barium salt.

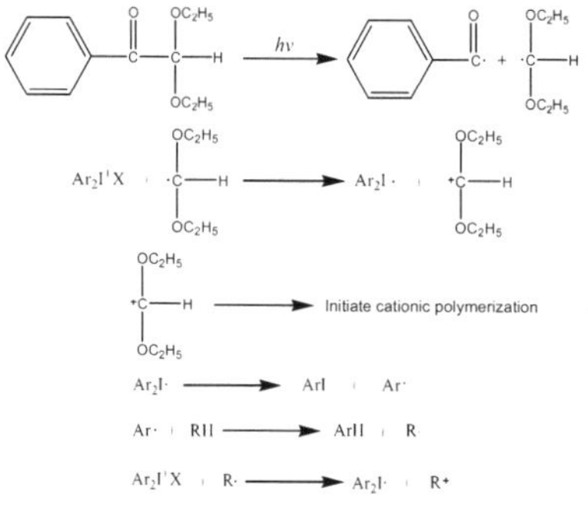

FIGURE 3.24 Equation of synergistic reaction initiated by DEAP and diphenyliodonium salt.

3.1.3 Characteristic parameters of photopolymer and ultraviolet light source

3.1.3.1 Characteristic parameters of photopolymer

Different photopolymer materials have different properties in SLA, that is, the photosensitivity of the resin is different. The photosensitivity of the photopolymer can be characterized by two characteristic parameters of the critical exposure amount E_c and the transmission depth D_p. The critical exposure amount of the photopolymer refers to the minimum energy that the resin liquid layer needs to obtain when the resin is photopolymerized under UV light to produce a gel. The photosensitivity of the photopolymer is also closely related to the transmission depth when the resin is cured. The transmission depth D_p of the photopolymer refers to the depth at which the UV light energy density in the resin is attenuated to $1/e$ of the incident energy density E_0, which is a performance indicator to measure the intensity of UV light energy absorbed by the photopolymer.

According to the polymerization kinetics, the mass fraction of initiator is an important parameter, and it is directly related to the characteristic parameters of the aforementioned photopolymer. In the photopolymer, the mass fraction of initiator is very low, generally less than 5% of the total mass of the system. Therefore, the photopolymer can be regarded as dilute solution of the photoinitiator. When the UV light is irradiated to the photopolymer, the absorption of the UV light by the photopolymer conforms to the Beer−Lambert rule, that is,

$$E(z) = E_0 \exp(-\varepsilon[I]z) \tag{3.1}$$

where $E(z)$ is the exposure energy at depth z, E_0 is the exposure energy of the incident UV light, ε is the molar extinction coefficient of the initiator, and $[I]$ is the mass fraction of initiator.

When the depth (z) is the maximum layer thickness (d) at the time of forming, then:

$$E(d) = E_0 \exp(-\varepsilon[I]d) = E_c \tag{3.2}$$

When the UV light reaches the resin depth (d), the energy of the UV light has been attenuated to be unable to continue to cure the resin at a depth greater than d, that is, it is attenuated to the minimum exposure energy when gel produces, so the exposure energy at the resin depth (d) is equal to the critical exposure.

The following equation can be obtained by mathematically transforming Eq. (3.2):

$$\ln\left(\frac{E_0}{E_c}\right) \varepsilon[I]d \tag{3.3}$$

$$d = \left(\frac{1}{\varepsilon[I]}\right) \ln\left(\frac{E_0}{E_c}\right) \qquad (3.4)$$

From the result of Eq. (3.4), it can be seen that the actual curing depth of the photopolymer is determined by the molar extinction coefficient, the mass fraction of initiator, the exposure amount of the incident UV light, and the critical exposure amount.

In addition, according to the definition of the transmission depth D_p of the photopolymer:

$$E(z) = E_0 \exp\left(\frac{-z}{D_p}\right) \qquad (3.5)$$

The following equation can be obtained by comparing Eq. (3.1) with Eq. (3.5):

$$D_p = \frac{1}{\varepsilon[I]} \qquad (3.6)$$

It can be seen from Eq. (3.6) that the transmission depth D_p of the photopolymer is inversely proportional to the mass fraction of initiator. From Eq. (3.4), it is known that E_c has a mathematical relationship with the mass fraction of initiator. Therefore, the resin characteristic parameter can be changed by changing the content of the initiator in the photopolymer.

For the photopolymer, the smaller the characteristic parameter (E_c) of resin is the higher photosensitivity the resin has; when the D_p is larger, the absorption of UV light by the resin is weaker. In the AM process, the smaller D_p is beneficial to improve the forming accuracy. The smaller D_p indicates that the photoinitiator in the resin absorbs more UV light, and the curing rate is faster, and the resulting laminate is thinner. However, this does not mean that it is the best if the D_p is smaller. If D_p is too small, it means that the mass fraction of the photoinitiator is large and the cost of resin is high. On the other hand, under the irradiation of same power of UV light, the energy of the incident UV light must be added to cure the resin thin layer of specified layer thickness. Therefore, the scanning rate of the UV lamp is lowered, which reduces the forming efficiency. Thus, when developing AM photopolymer, it is necessary to combine the characteristics of the AM rapid prototyping equipment and the forming process requirements to adjust the mass fraction of the relevant components in the photopolymer, and obtain reasonable resin characteristic parameters E_c and D_p.

3.1.3.2 Characteristic of ultraviolet light source

The most common source of radiation in SLA is the UV source. UV lamps are the core of UV light sources, so it is important to select the UV lamps rationally. There are many types of UV lamps to choose from. Technical requirements, as well as cost factors should be considered when selecting UV lamps. From a technical point of view, how to choose a UV lamp starts with two important factors that affect the curing speed and efficiency of the photopolymer material, namely the output spectrum and the light intensity. If the light intensity is unstable, it should be measured with the amount of exposure.

1. Output spectrum

    The output spectrum is the light intensity corresponding to each wavelength within the range of wavelength emitted by the UV lamp. UV curing is most efficient when the spectral output is consistent with the absorption spectrum required for the photoinitiator. Generally, the distribution law of high-intensity spectral output at different wavelengths is obtained by adjusting the materials (i.e., filler) and pressure in the UV lamp. Since different curing systems often require absorption spectra at specific wavelengths, it is important to consider the matching of UV lamps for a given photopolymer. The light source used for AM is metal halide lamp, in which a small amount of metal halide is added to the common mercury lamp tube to change the emission spectrum and improve the radiation efficiency of the ordinary high-pressure mercury lamp. The UV lamp of AM experimental prototype is a metal halide lamp with a main emission wavelength range of 360–390 nm and the strongest emission line of 365 nm, which accounts for about 40% of the total energy output.

2. Light intensity

    Light intensity is one of the most important parameters for characterizing light sources in photochemical research and SLA applications. The intensity of UV lamp is defined as the total output power of the UV lamp, also known as the power density spectrum, which is the sum of all output intensities of the UV lamp over the entire electromagnetic spectrum. Generally, the intensity of the UV lamp will affect the curing speed of the photopolymer because the intensity of the UV lamp will affect the actual light intensity of the UV light reaching the surface of the photopolymer, and the actual light intensity directly causes the photopolymer to produce chemical reaction and cure. Therefore, the intensity of the UV lamp and the actual light intensity received by the surface of the photopolymer are two separate concepts. The light intensity mentioned in the field of AM rapid prototyping refers to the sum of light intensities reaching the surface of the photopolymer in a specific UV wavelength range.

    The instrument for measuring light intensity is called a UV illuminance meter. After light energy is converted into electric energy by photoelectric tube, the relative value of light is obtained by measuring the photocurrent value, and the light intensity value is obtained by

calibration, which is generally expressed by W/cm^2. Due to the limitation of the spectrally sensitive region of the photovoltaic material (i.e., the probe) used in the phototube; the measured light intensity is only a value within the spectrally sensitive region.
3. Amount of exposure

   The amount of exposure refers to the integral of the light intensity over time. In the case where the light intensity does not change with time, the exposure amount is equal to the product of the light intensity and the exposure time. At this time, the exposure amount is easily obtained by only measuring the light intensity with photometer. However, in the case where the light intensity is unstable, it is necessary to use an exposure measuring instrument which measures the exposure amount of a specified spectral region. Generally, an exposure test meter can display both light intensity and exposure amount.
4. Light intensity loss

   In general, the output light intensity of electrode high-pressure mercury vapor lamp gradually decreases with increasing use time. For example, after being used for 1000 hours, a high-pressure mercury lamp will have its output light intensity dropped by 15%−25%. This loss occurs slowly and is not easily noticeable. Thus, regular monitoring of the actual received light intensity of the photopolymer surface is very important.

3.1.4 Characteristics of photopolymer materials and their stereolithography apparatus formability

3.1.4.1 Stereolithography apparatus properties of photopolymer

1. Cured shape of photopolymer

   The most common source of radiation for UV curing is UV lamp. The high precision requirements of the SLA technology determine that the most suitable source is the laser: the singularity of the laser allows it to concentrate the spot very small. In the forming process of SLA photopolymer, many factors affect the photopolymerization reaction, such as the composition of the material itself (especially the type and mass fraction of the initiating system), exposure energy, temperature, and the like. Among the many factors, the intensity of the laser light that is irradiated onto the material, that is, the exposure energy is the biggest factor affecting the curing behavior of the material. The intensity of the laser beam shows Gaussian distribution state along the radius of the spot (as shown in Fig. 3.25), where I represents the intensity of light per unit area, I_0 is the I value of the center portion of the beam. The Z-axis direction is the axial direction of the beam; it is a spatial distribution of light intensity. Taking the X and Y planes of the Cartesian coordinate system

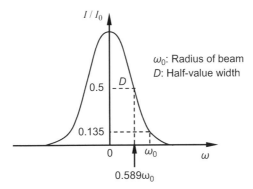

FIGURE 3.25 Light intensity distribution of a single mode laser.

perpendicular to the beam axis, the distribution of the light intensity in the X and Y planes can be expressed by Eq. (3.7).

$$I(x, y) = \left\{ \frac{2P_t}{\pi w_0^2} \right\} \exp\left(\frac{-2w^2}{-2w_0^2} \right) \quad (3.7)$$

where P_t is the full power of the laser, w is the distance from the origin of the optical axis origin (x_0, y_0), which can be expressed by Eq. (3.8).

$$w = \left\{ (x - x_0)^2 + (y - y_0)^2 \right\}^{1/2} \quad (3.8)$$

w_0 is the radius at light intensity value of $1/e^2$ (about 13.5%) of the laser beam.

Studies have shown that the transmission of light waves in SLA resin follows the Beer–Lambert theorem. Therefore, when the laser beam is vertically irradiated on the liquid surface of the resin, the liquid surface is located at the origin of the Z-axis, and the laser intensity $I(x,y,z)$ is distributed along the depth direction (Z) of the resin. The light intensity I decays along direction Z, that is,

$$I(x, y, z) = \left\{ \frac{2p_t}{\pi w_0^2} \right\} \exp\left(\frac{-2r^2}{r_0^2} \right) \exp\left(\frac{-z}{D_p} \right) \quad (3.9)$$

where D_p is the transmission depth of the laser in the resin.

When the laser beam irradiated on the resin is static, the exposure amount (E) of the resin at this point is a function of time (τ), which can be expressed as:

$$E(x, y, z) = I(x, y, z) \cdot \tau \quad (3.10)$$

At this time, the shape of the resin SLA is a rotating paraboloid (as shown in Fig. 3.26A).

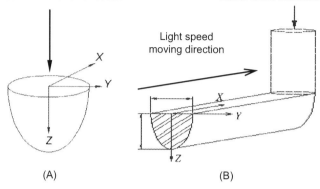

FIGURE 3.26 Cured shape of resin irradiated by laser beam. (A) Cured shape in static irradiation (B) Cured shape in moving irradiation.

The laser beam is scanned at a constant speed during SLA, and when it is scanned at a speed V in the X-axis direction, the light intensity at a certain point in the resin at a certain time (t) can be expressed as $I(x - Vt, y, z)$. When the scanning range is $-\infty < x < \infty$, the exposure amount of each part of the resin is:

$$E(x,y,z) = \int_{-\infty}^{\infty} I(x - Vt, y, z) \left\{ \frac{(\sqrt{2/\pi})pt}{(\omega_0 V)} \right\} \exp\left(\frac{-2Y^2}{\omega_0^2}\right) \exp\left(\frac{-z}{D_p}\right) \quad (3.11)$$

The resin starts to cure when $E = E_c$ (critical exposure amount), and when the resin cures in the space range of $E \geq E_c$ and $z \geq 0$, Eq. (3.11) is converted to Eq. (3.12):

$$2Y^2 \omega_0^2 + \frac{z}{D_p} = \ln\left\{ \frac{(\sqrt{2/\pi})P_t}{(\omega_0 V E_C)} \right\} \quad (3.12)$$

At this time, the cured shape is as shown in Fig. 3.26B, in which the (y,z) plane is a parabola about the Z-axis, and the cylinder is an equal section in the X direction.

When $Z = 0$, substitute it into Eq. (3.12), the y value is obtained, and the curing width L_w of a single scanning line is obtained, that is,

$$L_w = 2\omega_0 \left\{ \ln \frac{[\sqrt{2/\pi}]P_t}{(\omega_0 v E_c)} \right\}^{1/2} \quad (3.13)$$

As shown in Fig. 3.27, the shape and size of a single scan line section are determined by the total laser power P_t, the laser spot radius ω_0, scan speed v, and critical exposure amount E_c.

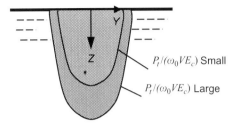

FIGURE 3.27 Curing factor and size.

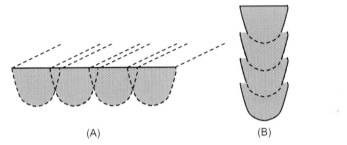

FIGURE 3.28 SLA-forming process. (A) Horizontal scanning forming (B) Vertical superposition forming.

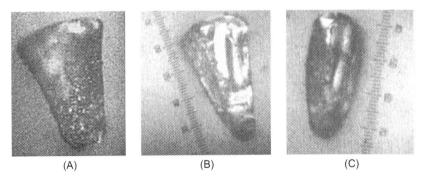

FIGURE 3.29 Micrographs of scanning line curing cross sections at different scanning speeds. (A) P: 160 mw, V: 100 mm/s (B) P: 160 mw, V: 200 mm/s (C) P: 160 mw, V: 400 mm/s.

Fig. 3.28 shows that the single resin cured scan line is overlapped and scanned in the horizontal direction and the vertical direction, respectively, so that the plurality of cured scan units is bonded to each other to form an integral shape. The aforementioned analysis is confirmed by experiments. Fig. 3.29 is a cross-sectional micrograph of a single scan line obtained with the same resin at the same laser power and resin temperature and different scanning speeds. It can be seen from the figure that the cross-sectional shape of the single curving line is similar to the distribution of the laser light intensity, and the outline is approximately parabolic.

2. SLA curve of photopolymer

The two most important indicators of SLA photopolymer are its photosensitivity and the lowest energy E_c (i.e., critical exposure) at which it causes photopolymerization to produce a gel. The photosensitivity of the resin is closely related to the transmission depth D_p of the cured resin during forming. Thus, D_p is a very important parameter when forming the SLA material.

1. SLA curve of photopolymer

Although different photopolymers have different characteristic parameters, namely, critical exposure amount and transmission depth, there is a necessary relationship between the critical exposure amount and the transmission depth of the same photopolymer, which can be described by SLA equation.

When a uniform laser beam is vertically irradiated from the above liquid to the liquid surface, the exposure amount is E at a certain depth (z) under the resin liquid surface, and the following equation is obtained by the exposure time τ multiplier (3.9), that is $I\tau = E$, $I_0\tau = E_0$

$$E(z) = E_0 \exp\left(-\frac{z}{D_p}\right) \qquad (3.14)$$

where E_0 is the exposure amount of the liquid surface.

In the curing process of free-radical type SLA resin represented by acrylate resin, there is a polymerization inhibition effect of oxygen, that is, the free radical generated by decomposition of photoinitiator under irradiation of light is initially consumed by the oxygen dissolved in the resin, so that photopolymerization cannot occur. However, when E exceeds a certain value, the free radical generated by the decomposition of the initiator is sufficient to completely consume the oxygen in the resin, and at this time, a polymerization reaction starts to occur, the critical value is set to E_c, and curing occurs in a certain depth range. When $E(z) \geq E_0$, the curing depth is

$$Z \leq D_p \ln\left(\frac{E_0}{E_c}\right) \qquad (3.15)$$

The z on the left side of the equal sign is the curing depth, and the value is set to C_d, then

$$C_d = D_p \ln\left(\frac{E_0}{E_c}\right) \qquad (3.16)$$

Plotting $\ln E_0$ with C_d, a straight line with a slope of D_p is obtained. The straight line is SLA curve, and the slope of the curve varies depending on the kind of the resin, which reflects the laser SLA property of the resin to some extent.

2. Resin parameters of photopolymer

The slope D_p of the SLA resin curing curve and the intercept E_c obtained from the straight line, the so-called resin parameters of the photopolymer, are important for determining the process parameters of the SLA-forming equipment and ensuring the high forming efficiency and forming precision of the SLA technology.

If Eq. (3.16) is rewritten into Eq. (3.17)

$$C_d = D_p \ln E_0 - D_p \ln E_c \tag{3.17}$$

The second term in Eq. (3.17) reflects the inhibition factor during the curing of the resin, and when E_0 is increased to make C_d becomes a positive value, the curing begins. In the curing process of free-radical SLA materials, the main factor of inhibition is oxygen, so if oxygen condensation is avoided during the curing process, E_c will be small and C_d will soon reach a positive value, that is, curing will start quickly. In the photopolymerization process of the cationic SLA-forming material, there is no inhibition of oxygen, and the second term in Eq. (3.17) will be significantly smaller than the curing of the radical type material, which leads to an increase in the value of C_d.

Since the characteristic parameter D_p of the SLA material, that is, the photopolymer, is inversely proportional to the initiator mass fraction, it can be seen from Eq. (3.3) that there E_c is also proportionally related to the initiator mass fraction. Therefore, the resin parameters can be changed by changing the content of the initiator in the photopolymer when studying the SLA-forming material.

The D_p of SLA photopolymer is an indicator for measuring the energy intensity of the photopolymer to absorb UV light, and refers to the depth at which the UV laser energy density in the resin is attenuated to $1/e$ of the incident energy density E_0. For a confirmed photopolymer, the larger the resin parameter D_p, the weaker the absorption of UV light by this value. In the SLA-forming process, small D_p is advantageous in terms of the accuracy requirements of the parts. Because the smaller D_p means that the laminate of the resin obtained by curing under laser irradiation is thinner. However, this does not mean that it is the best if the D_p is smaller, if D_p is too small, then under the same power of laser irradiation, it is necessary to reduce the laser scanning rate when curing the sheet with the specified layer thickness, which will reduce the forming efficiency, and, in order to ensure the adhesion between the layers after the resin is cured during the forming process, the curing depth must be greater than the layer thickness (d), that is:

$$D_p = d + h_0 \tag{3.18}$$

where h_0 is called the over curing depth. The smaller the D_p, the smaller the d is required, and the viscosity of the resin determines the extent of the thickness reduction of the layer. When the layer thickness is reduced to a certain extent, it will not only reduce the forming efficiency, but also makes the precision of the resin liquid level during the forming process difficult to control, or causes large internal stress of the part or even peeling between layers, resulting in failure of the part. Therefore, when developing SLA photopolymer, it is necessary to accurately control the resin parameter D_p by adjusting the mass fraction of the relevant components in the resin formulation in combination with the characteristics of the SLA equipment and the forming process requirements.

3. Determination of photopolymer parameters

It can be seen from the aforementioned analysis that as long as a series of curing depths and exposure amount are measured, the values of D_p and C_d can be obtained. The specific method for determining the resin parameters is as follows: placing a certain amount of photopolymer in a culture dish having a diameter of 20 cm, accurately measuring the power of the laser spot irradiated on the surface of the liquid resin, and then scanning and curing on the free surface of the resin at a series of scanning rates to obtain a series of sheets having the same layer thickness. Carefully clipping the sheets out with tweezers and cleaning them with isopropyl alcohol, and measuring the thickness (d) accurately with a micrometer. In this experimental method, the scanning rate is selected in the method that the cured sheet has certain strength. In the case where the laser power is determined, the higher the scanning rate, the smaller the exposure amount of the resin, and the thinner the thickness of the cured layer, when the thickness is too thin, the sheet will have insufficient strength, thus affecting the accuracy of thickness measurement.

When scanning a sheet by laser, the calculation equation of the exposure amount, that is, the energy E_0 irradiated on the surface of the resin is:

$$E_0 = \frac{P}{(v \cdot S)} \qquad (3.19)$$

where P represents the laser power irradiated on the surface of the liquid resin, v represents the scanning speed, and S represents the scanning pitch.

According to the thickness (d) of a series of sheets measured and the corresponding E_0 values obtained by Eq. (3.19), $\ln E_0$ is plotted with d, and the slope of the fitted straight line and the intercept on the x-axis are obtained by least squares method, thus the curing depth D_p of the resin and the critical exposure amount E_c are obtained.

3.1.4.2 Characterization of curing shrinkage of photopolymer materials

1. Curing shrinkage of photopolymer

    The reasons for the curing shrinkage of the photopolymer material have been described earlier. This section mainly discusses the characterization of the curing shrinkage of the photopolymer, and the method mainly includes the linear shrinkage rate and the volume shrinkage rate.

2. Measurement of volume shrinkage rate

    Measure the density ρ_1 of the resin system before curing and the density ρ_2 after the complete curing by using the pycnometer method at 25°C and obtain the volume shrinkage rate of the system:

$$\text{Volume shrinkage rate} = \frac{\rho_2 - \rho_1}{\rho_2} \times 100\% \qquad (3.20)$$

3. Measurement of linear shrinkage rate

    Pour a certain amount of resin into a mold of $L_0 \times W_0 \times H_0$ (length × width × height) and place it in an exposure chamber. After it is completely cured, measure the actual dimension (L) in the longitudinal direction. Then, the linear shrinkage rate is:

$$\text{Linear shrinkage rate} = \frac{L_0 - L}{L_0} \times 100\% \qquad (3.21)$$

4. Formation mechanism of warping deformation of SLA parts
1. Formation mechanism of warping deformation

    It can be seen from the aforementioned discussion that the fundamental reason for the curing shrinkage of photopolymer is the polymerization reaction between molecules, which is unavoidable, and this phenomenon directly leads to the warping deformation. There are two directions of warping deformation, one is a horizontal direction and the other is a vertical direction. Fig. 3.30 shows the shrinkage model of single scan line.

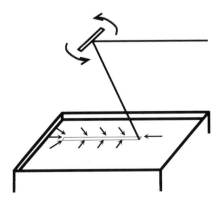

FIGURE 3.30 Shrinkage model of single scan line.

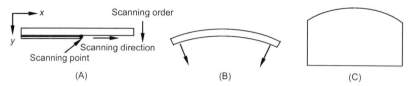

FIGURE 3.31 Plane scan deformation model. (A) Horizontal scanning. (B) Deformation when scanning. (C) Deformation after scanning.

When the UV laser is scanned to the liquid surface, the resin is excited by the laser to generate curing reaction, and the completion of the reaction takes a certain time. During this time, the intermolecular distance changes from the van der Waals distance in the liquid state to the covalent bond distance, and the shrinkage occurs. Since the scan line is filled with liquid resin, the portion where the scan line shrinks is sufficiently supplemented, and the size does not change in the direction of the scan line. In the plane of the scan line, since the shrinkage is symmetrically distributed in the scanning direction, the scan line is not deformed, and the deformation in the vertical direction can be ignored.

As shown in Fig. 3.31, it is a shrinkage model diagram of the UV laser when it is scanned in a certain order on the $X-Y$ plane. In the plane scanning process, the layer is cured by a plurality of curing scan lines, and two adjacent curing lines are embedded in one another. According to the shrinkage model of the above single scanning line, when the first curing line is scanned, although the second curing line shrinks, it is not restrained by the outside since the curing unit floats on the liquid surface, therefore, no deformation occurs. When the second curing line is scanned, the curing line shrinks, and because the curing line is partially embedded in the first curing line, the shrinkage is constrained by the first curing line, and deformation occurs, and the deformation direction is consistent with the scanning order. The subsequent third curing line is also deformed by the constraint from the front cured body. However, as the scanning progresses, when the strength of the cured body at the frontend is sufficient to resist the shrinkage deformation force of the single curing line, the curing line does not undergo significant deformation, as shown in Fig. 3.31C.

The deformation in the vertical direction is similar to the deformation in the horizontal plane, both due to the volume shrinkage of the resin, but the deformation unit in the vertical direction is each layer. The SLA-forming method requires that the layers of the parts must be cured and joined. The most typical deformation of the vertical direction is the warping deformation of cantilever beam, as shown in Fig. 3.32.

Fig. 3.32A shows that the first layer of the cantilever beam end of the part is originally produced on the liquid resin. Because there is no

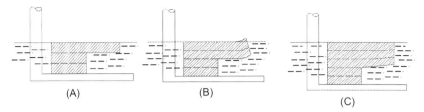

FIGURE 3.32 Warping deformation model of the cantilever beam in the vertical direction. (A) The first layer of the cantilever. (B) The second layer of the cantilever. (C) The multiple superpositions of the cured layers.

support at the bottom, it can shrink freely during the curing process without being restrained by force, and does not show warping deformation.

Fig. 3.32B shows that when the second layer is scanned and cured, the layer is accumulated on the first layer, and the two layers are partially embedded and cured into one body, so the second layer cannot freely shrink. This produces an upward tensile stress on the first layer, resulting in warping deformation.

Fig. 3.32C shows that with the multiple superposition of the cured layers, the effects of self-interlacing reactions occur, and the warping deformation gradually disappears due to the self-correcting effect of the curing formation.

Under normal circumstances, whether there are cantilever beam features or not in the unregulated parts, the shrinkage stress that causes warping deformation will exist, which is exhibited as warping deformation.

In extreme cases, warping deformation due to shrinkage stress can damage the connection between the parts and the lifting platform, causing cracking inside the parts or collision of the scraping plate with the parts to terminate the forming process.

2. Characterization of warping deformation of parts

The warping deformation of the SLA part during the forming process is closely related to the resin characteristics. The stress caused by the curing shrinkage of the resin is the most important factor for the warping deformation of the part during the forming process, but it is still impossible to establish the strict quantitative relationship between the curing volume shrinkage and the degree of warping deformation of the SLA parts for SLA photopolymer. Therefore, additional parameters must be used to evaluate the curing warping deformation of the photopolymer. Since the warping deformation of the cantilever beam is the most typical case in the SLA technique, people use the experimental method for characterizing the degree of warpage of the photopolymer in the SLA-forming process by determining the warpage factor (C'_f). In general, different photopolymers have different warpage factors, and the measurement of this factor can be determined by the following experiment:

As shown in Fig. 3.33, this experiment uses a part with a double-cantilever beam structure as warping deformation model. Although the

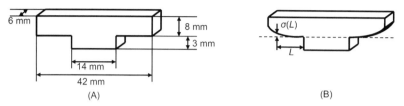

FIGURE 3.33 Double-cantilever test piece for testing warpage factor. (A) Size of the double-cantilever test piece. (B) Warpage of the double-cantilever test piece.

model is an experiment focusing on resin properties, it changes the scanning parameters of the laser, the way it is made, and the postcuring method. The relationship between warping deformation and the aforementioned factors can be obtained from the warping deformation rate of the test piece. In the experiment, the warping deformation rate (C_f) is defined as the ratio of the cantilever beam end displacement (σ) in the vertical cantilever direction to the length (L) of the cantilever beam end, and L is calculated from the root of the cantilever beam end. For convenience, this value is expressed as a percentage, that is:

$$C_f = \frac{\sigma(L)}{L} \times 100\% \qquad (3.22)$$

where the warpage factor is defined as the percentage of the amount of warpage at a distance of 6 mm from the root of the cantilever beam end divided by the distance, that is:

$$C_f' = \frac{\sigma(6)}{6} \times 100\% \qquad (3.23)$$

3.2 Research on stereolithography apparatus solid materials

SLA solid materials are mainly composed of monomers, oligomers, photoinitiators, and other additives. The study of solid materials involves the selection of raw material components, testing of resin properties, testing of jetting effects, and analysis of forming processes and performance analysis after curing. This section combines the AM experimental prototype requirements for solid resin to study the aforementioned contents.

The AM experimental prototype jointly developed by Huazhong University of Science and Technology and Shenzhen Yishan Digital Technology Co., Ltd. uses the Xaar XJ500 piezoelectric nozzle with a normal working temperature range of 20°C−60°C, and ink is supplied on demand. When performing AM and forming, the working temperature is generally set to 50°C−55°C to ensure long-term stable operation of the nozzle. The specific performance parameters of the XJ500 piezoelectric nozzle are shown in Table 3.5. The requirements for the ink are at a working

TABLE 3.5 Performance parameters of an XJ500 printhead.

Parameter	Value
Physical resolution of nozzle	180 dpi
Diameter of nozzle hole	50 μm
Droplet volume	80 pL
Droplet deviation accuracy	± 1 degree
Maximum ejection frequency	4 kHz
Droplet ejection speed	6 m/s
Maximum print line speed	5.1 m/s
Effective print nozzle hole number	500
Print bar width	70 mm
Optimal print distance	1 mm
Printhead net weight	290 g
Printhead size (length × width × height)	110 mm × 94 mm × 39.7 mm

FIGURE 3.34 AM experimental prototype.

temperature, the viscosity is 5−20 mPa s, and the surface tension is 25−40 N/m. The UV lamp of the AM experimental prototype is metal halide lamp. The main emission wavelength range is 360−390 nm, the strongest emission line is 365 nm, which accounts for about 40% of the total energy output, and the output power is 300 and 800 W (optional). The AM experimental prototype is shown in Fig. 3.34.

3.2.1 Study on benzyl alcohol accelerator in cationic stereolithography apparatus system

Firstly, let us discuss the study of cationic solid materials, which mainly involves the selection of benzyl alcohol promoters in cationic SLA systems, the study of the curing properties of trimethylene oxide monomers, the preparation of cationic solid materials, the ejection process and the properties after curing.

Compared with the free-radical curing system, the main disadvantage of the cationic SLA system is that the curing rate is relatively slow, and one way to increase the curing rate is to add a photosensitizer, which absorbs long-wave UV light and transfers energy to the cationic photoinitiator, so as to promote photolysis of photoinitiators and take advantage of UV light more efficiently. However, this method also has several major disadvantages: (1) the photosensitizer has low solubility in most monomers, sometimes it is necessary to add solvent; (2) the photosensitizer has high vapor pressure at room temperature, and it is easy to volatilize during SLA; (3) the photosensitizer is easy to oxidize, resulting in yellowing of the cured product; (4) the photosensitizer is generally toxic and easy to migrate out. In view of the shortcomings of photosensitizers, Crivello JV found that benzyl alcohol has the function of accelerating the polymerization of cationic system. It is a good SLA accelerator of cationic curing system, which has obvious acceleration effect and overcomes the aforementioned shortcomings of photosensitizer. However, Crivello JV did not mention which benzyl alcohol is better for AM by photopolymer and did not indicate detailed acceleration mechanism. This section discusses the acceleration of several common benzyl alcohols on cationic systems to guide the formulation and performance studies of subsequent cationic SLA resins.

Alicyclic epoxy and trimethylene oxide are monomers and oligomers commonly used in cationic SLA resins, of which UVR6105 and OXT221 are the most commonly used AM photopolymer monomers. The monomers have the characteristics of low viscosity, fast curing speed and excellent performance after curing, which are very suitably used as the cationic monomers for AM solid materials. This section mainly studies the SLA acceleration of benzyl alcohol on UVR6105 and OXT221.

3.2.1.1 Experimental part

1. Experimental materials

    Bis(3-ethyl-3-oxetanyl)methyl group ether OXT221 (DOX), industrial product, Japan TOAGOSEI company; UVR6105, industrial product, US DOW company; Polyol0301, industrial product, United States DOW company; UVI6976 (cationic photoinitiator), industrial products, US DOW company; 2-ethyl-9,10-dimethoxy anthracene, purity: 97%, industrial product, Shanghai Sigma-Aldrich company; 3,4-dimethoxybenzyl alcohol, chemically pure, Shanghai Sigma-Aldrich; 3,4-methylenedioxybenzyl alcohol, chemically pure, Shanghai Sigma-Aldrich; benzyl alcohol, chemically pure, Shanghai Sigma-Aldrich; 4-methoxybenzyl alcohol, chemically pure, Shanghai Sigma-Aldrich; 4-nitrobenzyl alcohol, chemically pure, Shanghai Sigma-Aldrich; 3,5-dimethoxybenzyl alcohol, chemically pure, Shanghai Sigma-Aldrich the company [3].

All raw materials are used directly without purification and the chemical structural formula of the main raw materials is shown in Fig. 3.35.

2. Experimental equipment

Real-time Fourier infrared spectrometer RT-FTIR (resolution: 4 cm^{-1}, data sampling interval: 1 time/s), Nicolet Magna-IR 850, Nicolet Company, United States; diode-emitting UV light source with a wavelength of 365 nm, Shenzhen Lamplic Technology Co., Ltd. Company; Con-Trol-Cure UV light meter, UV Process Supply, Inc., United States.

The diagram of experimental device is shown in Fig. 3.36.

3. Sample preparation and experimental methods

The alicyclic epoxy sample is prepared as follows: the proportion of basic raw materials is UVR6105: Polyol Tone 0301: UVI6976 = 100:15:4.6 (mass percent, the same below), adding benzyl alcohol equivalent to 10% of the total mass of the sample as needed, and stirring 30 minutes through the magnetic stirrer to prepare a uniform colorless liquid.

The trimethylene oxide sample is prepared as follows: the proportion of basic raw materials is OXT221: UVI6976 = 100:2, adding benzyl alcohol equivalent to 10% of the total mass of the sample as needed, and stirring 30 minutes through the magnetic stirrer to prepare a uniform colorless liquid.

Experimental method: Coat the prepared liquid on KBr wafer, and cover another KBr wafer on it, and place aluminum foil of different thickness between the two KBr wafers as a space to control the film thickness of the sample. Fix the sample on the special fixture, with a distance of 5 cm from the fiber to the sample and a light intensity of 6.0 mW/cm^2. Use the RT-FTIR to monitor the change of the absorption peak area at 789 or 981 cm^{-1} at the UV light with a wavelength of 365 nm, corresponding to the characteristic absorption peaks of the cycloalkoxy group and the trimethylene oxide group respectively, and calculate the conversion ratio (Y) of the corresponding groups by the following equation:

$$Y = \frac{(A_0 - A_t)}{A_0} \times 100\% \quad (3.24)$$

In the equation, A_0 and A_t are the areas corresponding to the absorption peaks when the light irradiation time is 0 and t seconds, respectively. The experimental condition is controlled at 25°C with 45% air relative humidity.

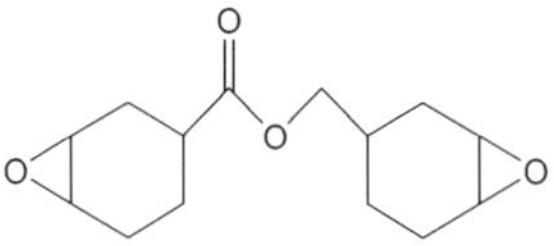

UVR-6105

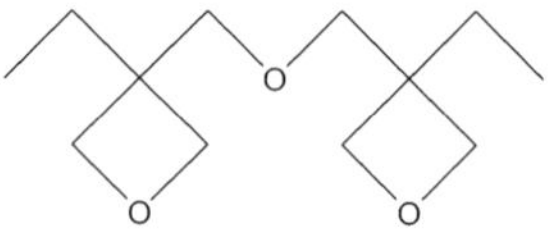

OXT-221

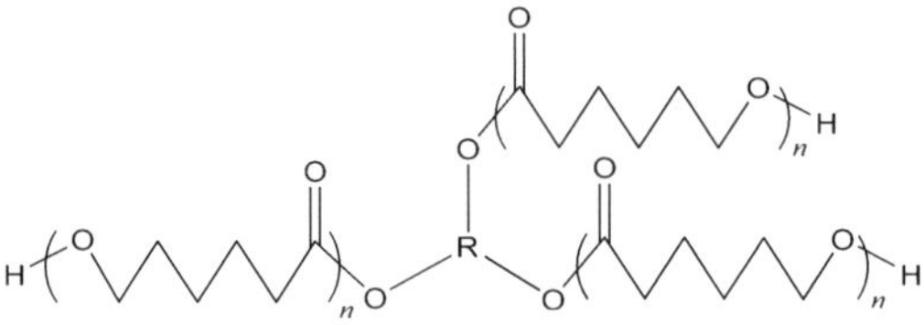

Polyol Tone 0301

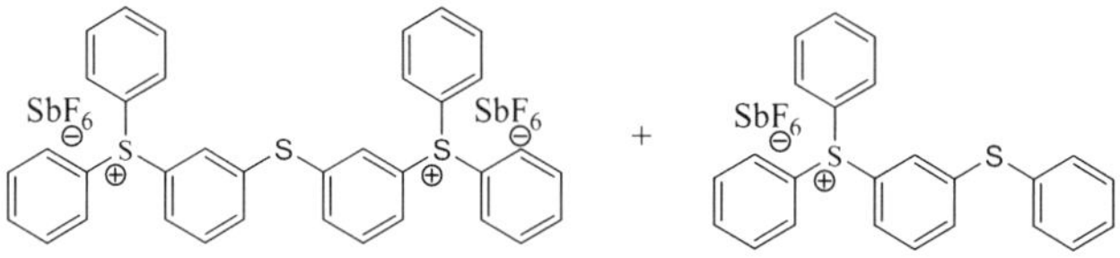

UVI-6976

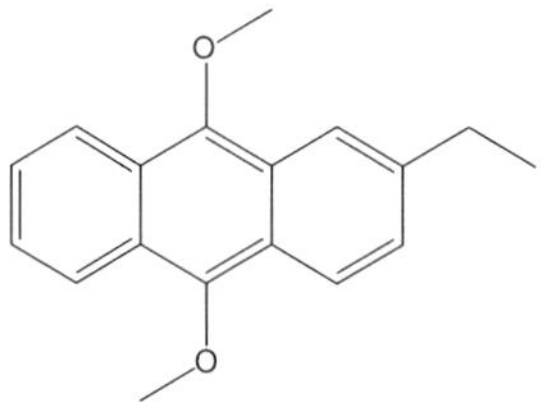

2-Ethyl-9,10-dimethoxy anthracene

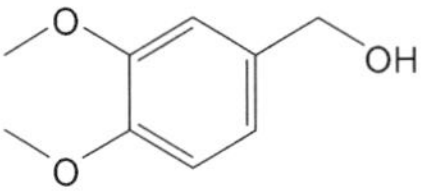

3,4-Dimethoxybenzyl alcohol

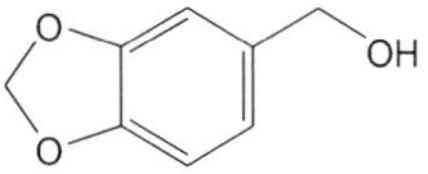

3,4-Methylenedioxybenzyl alcohol

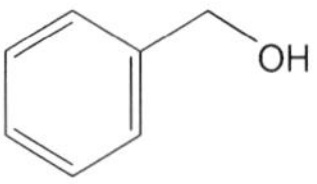

Benzyl alcohol

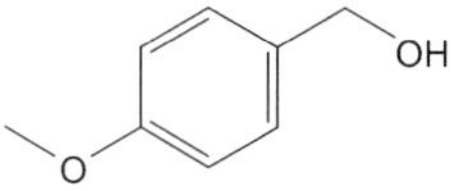

4-Methoxybenzyl alcohol

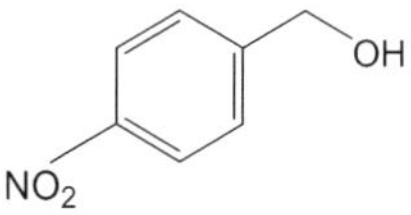

4-Nitrobenzyl alcohol

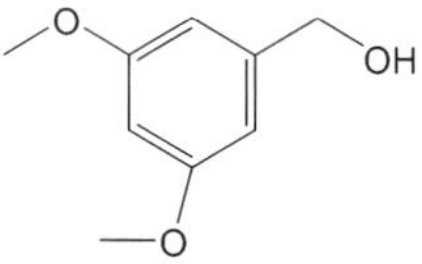

3,5-Dimethoxybenzyl alcohol

FIGURE 3.35 Chemical structural formula of the main raw materials.

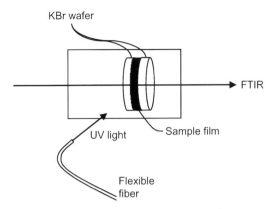

FIGURE 3.36 Schematic diagram of the real-time Fourier infrared spectrometer.

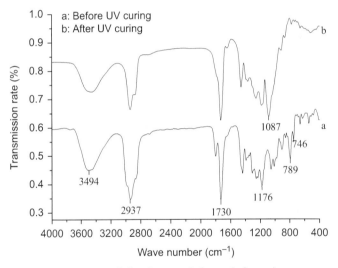

FIGURE 3.37 Infrared spectrum of alicyclic epoxy before and after curing.

3.2.1.2 Infrared spectroscopy analysis of UVR6105 and OXT221

Fig. 3.37 shows the infrared spectrum of the alicyclic epoxy UVR6105 alcohol before and after curing without adding benzyl. The 3494 and 2937 cm^{-1} are the stretching vibration peaks of —OH and —CH$_3$, respectively, and the 1730 cm^{-1} is the stretching vibration peak when C=O on the ester group. After photocuring of alicyclic epoxy UVR6105, the antisymmetric and symmetric deformation absorption peaks of 789 and 746 cm^{-1} of the original epoxy ring disappeared, forming a new peak, that is, the C—O—C stretching vibration peak after ring opening, the peak position is 1087 cm^{-1}. The 1087 cm^{-1} and C—O—C

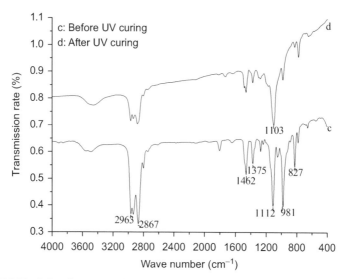

FIGURE 3.38 Infrared spectrum of trimethylene oxide before and after curing.

absorption peak of 1176 cm^{-1} on the ester group partially overlapped to form a broad peak. Both 789 and 746 cm^{-1} can be used as monitoring peaks for ring-opening reaction of epoxy groups, but the absorption peak of 789 cm^{-1} is stronger than that of 746 cm^{-1}, so the characteristic absorption peak of 789 cm^{-1} is selected as the monitoring peak of epoxy group [4].

Fig. 3.38 shows the infrared spectrum of trimethylene oxide OXT221 without adding benzyl alcohol before and after curing. 2963 and 2867 cm^{-1} are the antisymmetric stretching vibration peak and symmetric stretching vibration peak of C−H in CH$_3$, respectively; 1462 and 1375 cm^{-1} are the antisymmetric deformation vibration peak and symmetrical deformation vibration peak of C−H, respectively; 1112 cm^{-1} is the stretching vibration peak of C−O−C in the nonquaternary ring main chain, which combines the C−O−C stretching vibration peaks formed after ring-opening polymerization with trimethylene oxide to form a broad peak of 1103 cm^{-1}; 981 and 827 cm^{-1} are the antisymmetric deformation vibration peak and the symmetrical deformation vibration peak of the trimethylene oxide four-membered ring, respectively. After SLA, it can be seen that the peaks of 981 and 827 cm^{-1} drop sharply, indicating that the four-membered ring of trimethylene oxide undergoes ring-opening reaction, and both 981 and 827 cm^{-1} can be used as the monitoring peaks of the ring-opening reaction of trimethylene oxide group, but the 981 cm^{-1} absorption peak is stronger than the 827 cm^{-1} absorption peak, so 981 cm^{-1} is selected to detect the peak of the group reaction rate by real-time Fourier infrared spectrometer.

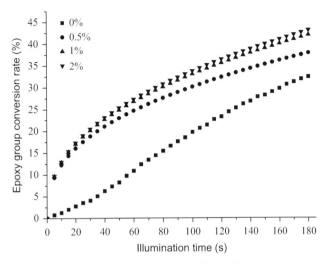

FIGURE 3.39 Effect of different proportions of photosensitizer on SLA rate of cycloaliphatic epoxy.

3.2.1.3 Acceleration of photosensitizer on UVR6105 SLA

2-Ethyl-9,10-dimethoxy anthracene is a relatively effective photosensitizer, so this experiment used it as a comparison with the accelerated effect of benzyl alcohol. Fig. 3.39 is a plot of epoxy functional group conversion over time during SLA of UVR6105 with the addition of 2-ethyl-9,10-dimethoxy anthracene in varying proportions. As can be seen from Fig. 3.39, as the mass fraction of the photosensitizer increases, the SLA curve becomes steeper and steeper, indicating that the SLA rate is faster (the slope of the curve is the SLA rate), when the photosensitizer is used in excess of the mass fraction. At 1%, the SLA rate is increased less. Of course, the mass fraction of the photosensitizer should not be too much, otherwise most of the UV light is absorbed by the surface of the resin, resulting in a small transmission depth of the resin.

3.2.1.4 Mechanism of benzyl alcohol to accelerate stereolithography apparatus rate of cationic system

Fig. 3.40 shows the effect of different kinds of benzyl alcohol on the SLA conversion of the alicyclic epoxy UVR6105. As shown in the figure, several types of benzyl alcohol have different acceleration effects. Benzyl alcohol and 4-methoxybenzyl alcohol increase the SLA rate of the epoxy group, while 4-nitrobenzyl alcohol reduces the SLA rate of the epoxy group, and benzyl alcohol and 4-methoxybenzyl alcohol The acceleration effect is better than 2-ethyl-9,10-dimethoxy anthracene.

In order to characterize the acceleration effect of various benzyl alcohols on the cation system, *IA* is defined as the acceleration index, which is the

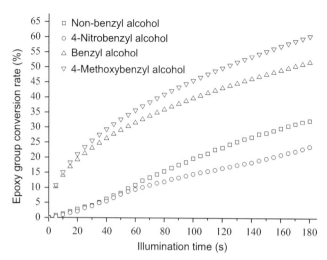

FIGURE 3.40 Effects of benzyl alcohol type on SLA rate of cycloaliphatic epoxy.

ratio of the SLA rate when benzyl alcohol is added to the SLA rate when benzyl alcohol is not added. The expression is as follows:

$$IA = \frac{(R_p/[M]0)_{alcohol}}{(R_p/[M]0)_{without}} \qquad (3.25)$$

where R_p is the SLA rate and $[M]_0$ is the starting monomer concentration. When various benzyl alcohols are added to the cationic system, the kinetic parameters of SLA are shown in Table 3.6. It can be seen that the *IA* values of benzyl alcohol and 4-methoxybenzyl alcohol are 1.7 and 2.3, respectively, and the *IA* value of 4-nitrobenzyl alcohol is 0.71.

There are two possible mechanisms by which benzyl alcohol accelerates the cationic system. The first type of acceleration is the activation monomer mechanism of the alcohol, as shown in Fig. 3.41. The cationic active center reacts with the alcohol to terminate the cation and release a proton to form a new active center, as shown in Fig. 3.41b and Fig. 3.41c. In a cationic SLA system having a large degree of crosslinking, this undoubtedly increases the mobility of the cationic active center, thereby increasing the SLA rate.

Another major acceleration principle is the photolysis mechanism of free radical−induced strontium salts, as shown in Fig. 3.42. After the benzyl alcohol participates in the cationic polymerization reaction, a benzyl ether group is formed, and the α-hydrogen on the benzyl ether group is easily taken up by the aryl radical generated by the photolysis of the sulfonium salt to form benzyl ether radical, as shown in Fig. 3.42a. The benzyl ether radical has low redox potential and is easily oxidized by the cerium salt to form benzyl ether cation, as shown in Fig. 3.42b; the benzyl group has a large conjugated system that can stabilize the benzyl ether free radical, and can also stabilize the benzyl ether cation, so that the

TABLE 3.6 Kinetic parameters of UVR6105 and OXT221 SLA.

Proportion	$Rp/[M]0$ $(10^{-2}\ s^{-1})$	IA	Group conversion rate (%)
UVR6105	1.15	–	32.4
UVR6105 + 4-nitrobenzyl alcohol	0.82	0.71	23.7
UVR6105 + benzyl alcohol	1.96	1.7	51.7
UVR6105 + 4-methoxybenzyl alcohol	2.65	2.3	60.5
UVR6105 + 35-dimethoxybenzyl alcohol	2.88	2.5	67.1
UVR6105 + 34-methylenedioxybenzyl alcohol	4.03	3.5	82.6
UVR6105 + 34-dimethoxybenzyl alcohol	4.26	3.7	87.0
OXT221	2.71	–	63.7
OXT221 + 34-methylenedioxybenzyl alcohol	7.32	2.7	91.6
OXT221 + 34- dimethoxybenzyl alcohol	8.40	3.1	96.5

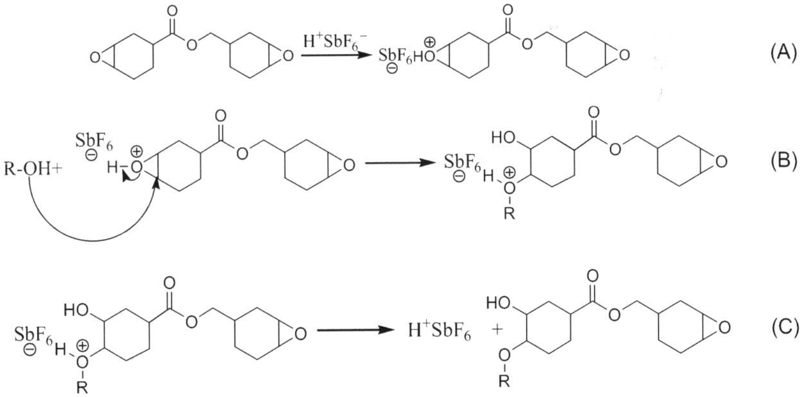

FIGURE 3.41 Activated monomer mechanism of alcohol.

reactions in aforementioned two steps are easy to carry out, and the electron-donating group on the benzene ring can enhance the stabilizing effect, and the electron-withdrawing group can reduce the stabilizing effect, so 4-nitrobenzyl alcohol has no acceleration effect; the triarylsulfonium radical formed in the

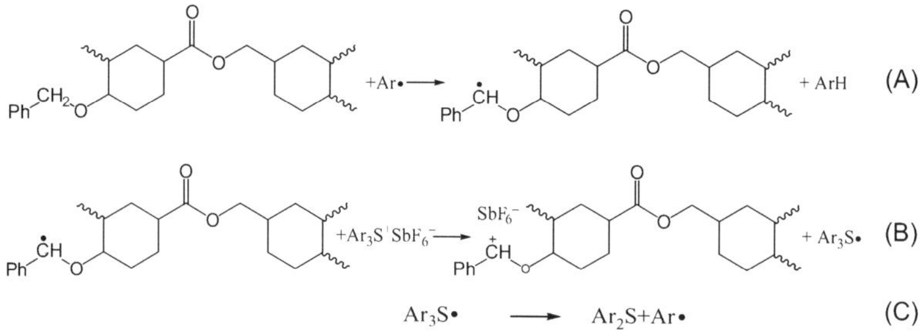

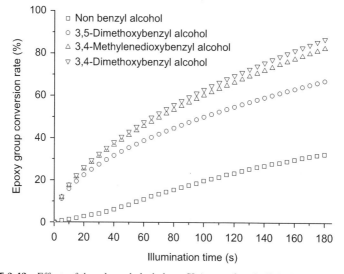

FIGURE 3.42 Photolysis mechanism of free radical–induced strontium salt.

FIGURE 3.43 Effects of three benzyl alcohols on SLA rate of cycloaliphatic epoxy.

second step is extremely unstable, and immediately decomposes into new aryl radical and diaryl sulfide, as shown in the Fig. 3.42c, this reaction is irreversible, forming a chain reaction that accelerating the photolysis process of the triarylsulfonium salt photoinitiator. Therefore, the SLA rate of the cationic system is increased.

3.2.1.5 Selection of benzyl alcohol

From the aforementioned analysis, it is known that the electron-donating group on the benzene ring of benzyl alcohol can enhance the stabilizing effect on the benzyl ether free radical and the benzyl ether cation, thereby accelerating the decomposition of the photoinitiator. Therefore, if the electron-donating property of the pendant group on the benzyl alcohol benzene ring is enhanced, the acceleration effect on the cationic system is more obvious, and the results are shown in Fig. 3.43. Among them, 3,4-dimethoxybenzyl alcohol and 3,4-

methylenedioxybenzyl alcohol have fast SLA rate, which is consistent with the mechanism analysis. The SLA kinetic parameters are listed in Table 3.6.

From Table 3.6, it is known that the acceleration index of 3,4-dimethoxybenzyl alcohol is 3.7, and the acceleration index of 3,5-dimethoxybenzyl alcohol is 2.5. This is because the methoxy coupling at the position 5 is less conjugated than the methoxy group at the 4-position, which weakens the electron-donating properties. The acceleration index of 3,4-methylenedioxybenzyl alcohol is 3.5, which is also relatively high because it contains an electron-rich methylenedioxy group.

3.2.1.6 Acceleration effect of benzyl alcohols on OXT221 stereolithography apparatus

Fig. 3.44 shows the acceleration effect of two benzyl alcohols on OXT221. It can be seen from the figure that the accelerated action of 3,4-dimethoxybenzyl alcohol and 3,4-methylenedioxybenzyl alcohol on the SLA rate of OXT221 is also more obvious. The kinetic parameters are listed in Table 3.6.

Benzyl alcohol accelerates the SLA rate of alicyclic epoxy and trimethylene oxide by activating monomer mechanism and free radical−induced photolysis mechanism of sulfonium salt, and the photolysis mechanism of free radical−induced strontium salt plays a major role. The electron-donating group on the benzyl alcohol more effectively stabilizes the benzyl ether radical and the benzyl ether cation, and accelerates the photolysis rate of the cationic photoinitiator, thereby accelerating the curing rate of the cationic system. 3,4-Dimethoxybenzyl alcohol and 3,4-methylenedioxybenzyl alcohol are two ideal cationic system curing accelerators.

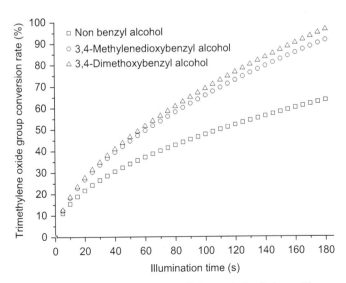

FIGURE 3.44 Effects of two benzyl alcohols on SLA rate of trimethylene oxide.

3.2.2 Study on iodonium salt and its photoinitiators

The initiator is the most important factor affecting the photopolymerization reaction. The effective initiator system must have strong absorption at the laser emission wavelength, and at the same time, it can produce active sites with high quantum yield after illumination. In addition, it should also be featured by low toxicity and stable storage. Most visible laser-initiated systems consist of initiator and sensitizer. The sensitizer, as an intermediate transmitter of energy, undergoes sensitization mainly through electron transfer after illumination. For some initiators that can absorb visible light, they can be used directly in the photopolymerization reaction.

Cationic photopolymerization can be initiated by various organic salts such as diaryliodonium salts, triarylsulfonium salts, aromatic diazonium salts, etc. When these organic salts are unable to absorb visible light, light sensitization can be obtained by adding appropriate dyes. Taking diphenyl iodonium salt as an example, it acts as an electron acceptor to carry out the following sensitization-initiating reaction with the dye in excited state:

$$Dye^* + Ar_2I^-X^- \rightarrow Dye^+ \cdot X^- + Ar_2I \cdot \quad (3.26)$$

$$Dye^+ \cdot X^- + RH \rightarrow H^+X^- + Dye + R \cdot \quad (3.27)$$

$$Ar_2I \rightarrow Ar_2I + Ar \quad (3.28)$$

The generated neutral iodonium salt radical is decomposed into alkyl group or aryl radical by dark reaction, and the generated dye cation radical can directly initiate cationic polymerization, and can further react with hydrogen-containing nucleophile to form Bronsted acid, so as to initiate cationic polymerization reaction. Both singlet and triplet dyes can react with sulfonium salt. Commonly used sensitizing dyes include methine compounds, P-aminobenzylidene derivatives, xanthene, and coumarin. The initiating system consisting of coumarin, diphenyl iodonium salt, and N-aminoglycine has the sensitivity up to 0.1 mJ/cm.

The photosensitizer can make the iodonium salt absorb UV light of 300–400 nm, even visible light. That is, a lower energy beam can be used to generate active cation. The mechanism of photosensitivity is mainly the electron transfer mechanism. The photoreaction is bimolecular reaction, that is, the reaction can proceed only when the photosensitizer molecule and the phosphonium salt collide with each other. When the viscosity of the SLA system increases with the progress of the reaction, the movement of the molecule becomes more and more difficult, and the chance of collision between the molecules becomes smaller and smaller, and the speed of the molecular expansion determines the speed of the sensitization reaction. Based on the aforementioned reasons, PapPas and Hong Xiaoyin designed some new bismuth salts with intramolecular photosensitization. Their sensitization occurs in the molecule, which has the characteristics of single molecule reaction and is not affected by the degree of diffusion, overcoming the aforementioned disadvantages. The structure of these new phosphonium salts can be expressed by the following formula:

$$\underset{R_1}{\overset{D}{\underset{R_2}{\overset{|}{S}}}}-R_3 \cdot X^- \qquad \underset{R}{\overset{I^+ X^-}{\diagdown}}\overset{D}{\diagup}_{R'}$$

In the sulfonium salt, R_1, R_2, and R_3 may be aryl group or alkyl group or benzoylmethyl group. In the iodonium salt, R and R' are aryl groups, and D is electron-donating group. When exposed to UV light, electron transfer can occur in the complex bismuth salt molecule, thereby greatly increasing the efficiency of photoinitiation.

3.2.2.1 Synthesis of iodonium salt and its photoinitiator
1. Synthetic route of iodonium salt

   Iodonium salt not containing substituent group or containing electronic substituent group •

$$X\text{-}\bigcirc + KIO_3 + 2H_2SO_4 + 2Ac_2O \longrightarrow$$

$$(X\text{-}\bigcirc)_2 I^+ HSO_4^- + KHSO_4 + 4AcOH + [O] \quad (3.29)$$
$$\text{I}$$

$$\begin{array}{l} NH_4Cl \\ I + NaBr \\ KI \end{array} \longrightarrow \begin{array}{l} (X\text{-}\bigcirc)_2 I^+ Cl^- \downarrow \quad NH_4HSO_4 \\ (X\text{-}\bigcirc)_2 I^+ Br^- \downarrow + NaHSO_4 \\ (X\text{-}\bigcirc)_2 I^+ I^- \downarrow \quad KHSO_4 \end{array} \quad (3.30)$$

$$Ph_2IHSO_4 + KPF_6Ph_2 \longrightarrow Ph_2IPF_6 \downarrow + KHSO_4 \quad (3.31)$$

$$\begin{array}{l}(X\text{-}\bigcirc)_2 I^+ Cl^- \\ (X\text{-}\bigcirc)_2 I^+ Br^-\end{array} + Ag_2O \xrightarrow{H_2O} (X\text{-}\bigcirc)_2 I^+ OH^-$$

$$\xrightarrow[-10^\circ C]{HBF_4} (X\text{-}\bigcirc)_2 I^+ BF_4^- \quad (3.32)$$

X = H, CH3, (CH3) 3C, etc.

When the iodonium salt not containing substituent group or containing electronic substituent group is synthesized on the benzene ring, since the system is sensitive to water, concentrated sulfuric acid is present as reactant and catalyst, and acetic anhydride is used as reactant and solvent, so that the impact of water on the reaction can be reduced. The $Ph_2IHSO_4(I)$ containing the electron-donating group substituent formed in the first step reaction is reacted with halogen ion to form precipitate, and precipitate the solution. If the anion portion of the final product is BF_4^-, the onium salt with the anion portion being the halogen reacts with silver oxide to form iodonium salt having an anion of OH$^-$, and then reacts with tetrafluoroboric acid to obtain iodonium salt having an anion of BF$_4^-$. If the desired anion of the onium salt is SbF$_6^-$ or PF$_6^-$ partially, the alkali metal salt of SbF$_6-$ or PF$_6-$ can be added in the solution of the product in the step directly to dissolve out the precipitate of onium salt from the solution.

Iodonium salt containing electron-withdrawing substituent group

$$2I_2 + 6KIO_3 + 11H_2SO_4 \longrightarrow 5(IO)_2SO_4 + 6KHSO_4 + 8H_2O \quad (3.33)$$

$$2X\text{-}\bigcirc + (IO)_2SO_4 + H_2SO_4 \xrightarrow{Fast} 2X\text{-}\bigcirc\text{-}I^+OH^- + 2HSO_4^- \quad (3.34)$$

$$X\text{-}\bigcirc + X\text{-}\bigcirc\text{-}I^+OH^- \xrightarrow{Slow} (X\text{-}\bigcirc\text{-})_2I^+ + H_2O \quad (3.35)$$

$$\begin{array}{c}(X\text{-}\bigcirc)_2I^+Cl^- \\ (X\text{-}\bigcirc)_2I^+Br^-\end{array} + Ag_2O \xrightarrow{H_2O} (X\text{-}\bigcirc)_2I^+OH^- \xrightarrow[-10°C]{HBF_4} (X\text{-}\bigcirc)_2I^+BF_4^- \quad (3.36)$$

X = Cl, Br, COOH, etc.

For the synthesis of iodonium salts containing electron-withdrawing substituent group, the intermediate product (IO) SO$_4$ is first synthesized. There are many ways to synthesize (IO) SO$_4$ intermediates, different oxidants are used in the process. Peroxyacetic acid can be selected as the oxidant and catalyst. Alternatively, peracetic acid and concentrated sulfuric acid can be used as catalyst to directly react with iodobenzene. We use a simpler and less stringent method. First, the concentrated sulfuric acid is used as the oxidant, the iodine and iodine in the potassium iodate are converted to +3 iodine, then benzene with electron-withdrawing substituent group (such as chlorobenzene or bromobenzene) is added dropwise, at which point there is a fast and a slow reaction in the solution. The iodonium salt cation containing electron-withdrawing substituent group is formed, and the subsequent reaction is the same as the synthesis of the iodonium salt containing the electron-donating group.

Synthesis of ethyleosin diphenyl iodonium salt

$$\text{(structure with Br, NaO, Br, Br substituents)} \xrightarrow[\text{r. t.}]{CH_2Cl_2} \text{(product structure with Br, O, Br, Br substituents)} \quad (3.37)$$

2. Synthesis of iodonium salt
   Preparation of Ph2I + Cl-(DPIOC)
   Add 10 g (0.118 mol) KIO_3, 10 m acetic anhydride and 22.5 mL benzene to 250 mL three-necked flask, cool the system to below 5°C (ice-sodium chloride bath), and then add a mixture of 10 mL acetic anhydride and 10 mL concentrated sulfuric acid to it dropwise with vigorous stirring. During the addition, the temperature does not exceed 5°C, and the system changes from colorless to yellow; allow the mixed system to return to room temperature, and bath in ice−NaCl−water and keep the water bath at room temperature for 12 hours; after standing for 24 hours, the system is cooled to 5°C, add 25 mL distilled water to it with stirring, the temperature does not exceed 10°C; the gas is released during the addition, and a white precipitate is produced; then add 10 mL anhydrous diethyl ether dropwise in the system at 10°C, remove the resulting $KHSO_4$ by filtration to obtain yellow solution. The solution is extracted twice with 40 mL diethyl ether, and then extracted once more with 20 mL petroleum ether. At this time, the water layer is pale yellow; add 7.2 g grinded NH_4Cl into it, the white precipitate is immediately produced; filter the precipitate and wash twice with diethyl ether and distilled water, and make it dry, the white solid is obtained; recrystallize it from methanol and dissolve in methanol, decolor with activated carbon, and carry out hot filtration and recrystallization once, wash three times with frozen methanol and drain it. The product obtained is colorless needle-like crystal $Ph_2I + Cl^-$, the infrared spectrum and UV absorption spectrum are respectively shown in Figs. 3.45 and 3.46.
   Preparation of Ph2I + BF4 − (DPIOF)
   Mix the 25 mL 2.0 M NaOH and 2 mL 2.0 M $AgNO_3$ uniformly to produce Ag_2O. Pour out the aqueous solution, and wash Ag_2O with distilled water for three times, make it suction filtered, and wash with distilled water for three times. Pour 7.9 g unrecrystallized $Ph_2I^+Cl^-$ and above 9.3 g Ag_2O and a small amount of water in the glass mortar for grinding until no pale yellow solid is observed. Filter the black slurry into glass sand funnel with 90 mL distilled water and make it suction filtered. Wash twice with distilled water, cool the filtrate to −5°C, and add 7.0 mL HBF_4 (> 40.0%) while stirring to produce a large amount of white precipitate. Let the system slowly return to room temperature and make it suction filtered. Dry at 60°C under vacuum (18 mm Hg). The product obtained is colorless needle crystal $Ph_2I^+BF_4^-$.

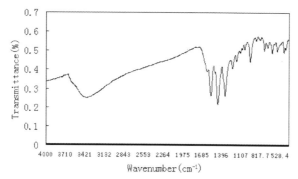

FIGURE 3.45 Infrared spectrum of diphenyliodonium salt $Ph_2I^+Cl^-$(DPIOC) (benzene ring skeleton: 1650, 1580, 1480, 1430 cm^{-1}; substituted by benzene ring: 740, 680 cm^{-1}; C–I: 480 cm^{-1}; C–H: 3050^{-1}).

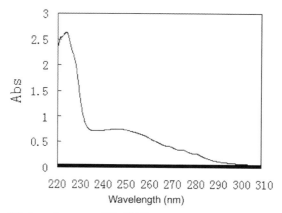

FIGURE 3.46 UV absorption curve of $Ph_2I^+Cl^-$(DPIOC) diphenyl iodonium salt.

Preparation of 34,4'-dimethyldiphenyliodonium salt(MePh)2I + Cl − + Cl-(DMPIOC)

Add 20.0 g KIO$_3$, 90.0 mL glacial acetic acid, 40.0 mL acetic anhydride, and 30.0 mL toluene in the 250 mL three-necked flask. Cool the system to 0°C (ice-sodium chloride-water bath), and add a mixture of 30.0 mL glacial acetic acid and 20.0 mL concentrated sulfuric acid in it dropwise with vigorous stirring. The temperature during the addition does not exceed 3°C. The system gradually changed from colorless to yellow. After the addition, continue stirring for 3 hours without withdrawing the ice-sodium chloride-water bath. Leave overnight and the solution is brown, in which there are white solids. Transfer the system to the funnel to filter with glacial acetic acid, making the total filtrate volume of 210 mL. Under ice cooling and stirring, add 210 mL distilled water to the filtrate, and as the distilled water is added, the solution becomes light and turbid. Finally, it becomes orange solution (lower layer, large amount) and brownish red solution (upper layer, small amount). The aforementioned solution is divided into two portions, each portion is extracted 4 times with 200 mL diethyl ether, and the two portions are combined to obtain a yellow solution. Add 11.0 g grinded NH_4Cl into it, after a while, colorless crystals appear, which are left overnight and filtered. Wash twice with ether and distilled water and drain. Dissolve the solid in 50 mL methanol, decolor with activated carbon and make it filtered. Evaporate the methanol in the filtrate to obtain pale yellow crystal. Redissolve the crystals with 20 mL methanol and then place them in the refrigerator, and after several hours of observation, the colorless precipitate appears. Filter by suction, wash the product twice with ice-cold methanol, drain and dry in vacuum. The product obtained is colorless granular crystal $(MePh)_2I^+Cl^-$. Figs. 3.47 and 3.48 show the infrared spectrum and UV absorption curve of DMPIOC, respectively.

Preparation of 44,4-dichlorodiphenyliodonium salt (ClPh)2I + Cl- (DCPIOC)

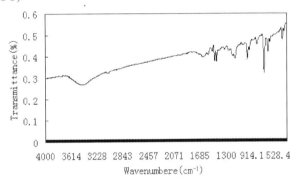

FIGURE 3.47 Preparation of 4,4'-dimethyldiphenyliodonium salt $(MePh)_2I + Cl^-$ (DMPIOC) (benzene ring skeleton: 1650, 1580, 1480, 1430 cm^{-1}; C-I: 480 cm^{-1}; substituted by benzene ring 1,4: 810 cm^{-1}; C—H: 3050 cm^{-1}).

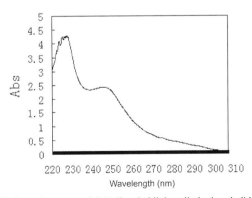

FIGURE 3.48 UV absorption curve of 4,4'-dimethyldiphenyliodonium iodide salt DMPIOC.

Add 12.0 g iodine, 31.2 g potassium iodate and 125.0 mL concentrated sulfuric acid to 500 mL three-necked flask in a cold water bath with vigorous stirring. After stirring for 3 hours, the solution turns from dark brown to yellow, stand it for one day and one night, and continue to stir for 3 hours, the brown color of the solution is lighter, the yellow solid increases; after standing for a while, cool the system to $-4°C$. Under vigorous stirring, add 50.8 mL chlorobenzene into it dropwise at a temperature not exceeding 7°C. During this process, the yellow solids become less. After the addition, the reaction system returns to room temperature, keep stirring for 12 hours. Stir the system at 45°C for 2 hours, the solution becomes dark brown at this time, then cool the system to 4°C, and add 200 mL distilled water dropwise and stir vigorously at a temperature not exceeding 7°C, then the gray-yellow solid is produced. At 4°C, add 20 mL diethyl ether to the system, followed by stirring for 10 minutes. Leave it overnight. After suction filtration, wash it with diethyl ether three times and dry to obtain 123.2 g pale solid, take 24.4 g of the solid, and add distill water to 150 mL. Heat to 72°C, the brown viscous material is produced, then it becomes solid. Carry out hot filtration to the system to obtain pale yellow filtrate, and add 4.5 g of NH_4Cl into it while stirring. The white precipitate is formed immediately. After suction filtration, wash the solid with distilled water for three times. Then, wash twice with diethyl ether and dry it, and apply vacuum drying, and the product obtained is colorless scaly crystal. The infrared spectrum and UV absorption spectrum are shown in Figs. 3.49 and 3.50 respectively.

Preparation of $(ClPh)_2I + Br - (DCPIOB)$

Heat the 48.8 g $(ClPh)_2I^+HSO_4$ crude product obtained above with water (total volume of about 300 mL) (temperature: 72°C) to produce brown viscous material, then it becomes solid and filter it while it is hot to obtain very pale yellow solution; added 13.0 g NaBr to it while stirring

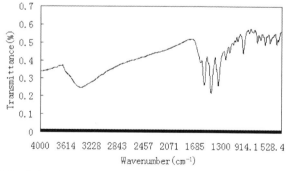

FIGURE 3.49 Infrared spectrum of 4,4-dichlorodiphenyl iodonium salt $(ClPh)_2I^+Cl^-(DCPIOC)$ (benzene ring skeleton: 1650, 1580, 1480, 1430 cm^{-1}; C—I: 480 cm^{-1}; substituted by benzene ring 1,4: 810 cm^{-1}; C—H: 3050 cm^{-1}).

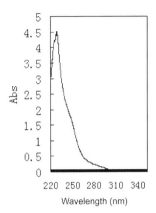

FIGURE 3.50 UV absorption curve of 4,4-dichlorodiphenyl iodonium iodide salt DCPIOC.

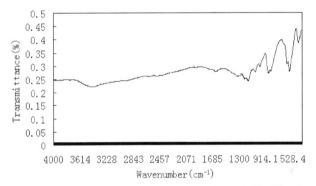

FIGURE 3.51 Infrared spectrum of eosin-bis(diphenyliodonium salt) [Eo(IPh$_2$)$_2$].

to immediately produce white precipitate. Carry out suction filtration, wash with water for three times, wash with diethyl ether for three times and dry it, recrystallize the product twice with methanol as solvent. The product obtained is white needle-like crystal.

Preparation of eosin-bis(DPIOC)[Eo(IPh2)2]

Add 1.70 g (2.45 mmol) eosin disodium salt and 1.61 g (5.09 mmol) diphenyl iodonium salt h$_2$I$^+$Cl$^-$ in 100 mL dichloromethane, and stir in dark environment under room temperature for 3 hours. After suction filtration, evaporate the solvent with filtrate to obtain the solid. Add 100 mL anhydrous diethyl ether in it, stirred and filter, and dry under vacuum at 45°C overnight to obtain eosin-bis(diphenyliodonium salt). The product is brownish red with lustrous crystal, the yield rate is 31%. Fig. 3.51 shows the infrared spectrum of eosin-bis(diphenyliodonium salt).

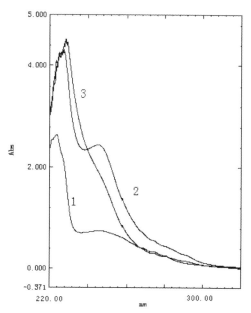

FIGURE 3.52 UV-visible spectra of iodonium salt with different substituents.

3.2.2.2 Iodonium salt-amine composite system

1. Spectroscopic properties of iodonium salt and ITS amine composite system

The absorption spectrum of the iodonium salt is generally below 300 nm. As shown in Fig. 3.52, the introduction of electron-donating or electron-withdrawing group in the para position of the phenyl group only causes limited red shift of the maximum absorption band. Curve 1 is the absorption curve of diphenyliodonium salt DPIOC; curve 2 is the absorption curve of 4,4'-dimethyldiphenyliodonium salt DMPIOC; curve 3 is the absorption curve of 4,4-dichlorodiphenyliodonium salt DCPIOC (Table 3.7).

However, we note that cerium salt is a good electron acceptor, and some suitable electron donors can be found. They form stable charge-transfer complexes (CTCs) in the ground state, which makes the absorption spectrum of the system have a large red shift. And a charge-transfer transition occurs in the excited state, thereby causing photolysis reaction to generate fragments having initiating activity. Amine is a good electron donor, which is susceptible to charge-transfer processes with some electron acceptors, especially tertiary amine which has high electron-donating ability. Therefore, we added some electron donor, trimethylamine, to the DPIOC ethanol solution. From the UV/visible light spectrum of the system (as shown in Fig. 3.53), it can be seen that a new wide absorption peak without vibration structure appears in the visible light region, the maximum absorption is located at 507 nm, and the triethylamine and iodonium salt can form CTC with a molar ratio of 1:1 by

TABLE 3.7 Maximum absorption peaks of iodonium salts with different substituents.

Structure	Anion	λ_{max} (nm)
⟨⟩-I⁺-⟨⟩	Cl⁻	225
H₃C-⟨⟩-I⁺-⟨⟩-CH₃	Cl⁻	230
Cl-⟨⟩-I⁺-⟨⟩-Cl	Cl⁻	238.2

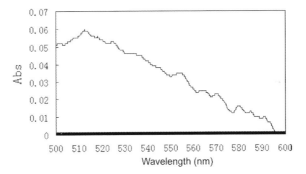

FIGURE 3.53 UV absorption curve of DPIOC/TEA complex system at 500–600 nm.

intermolecular interaction in the ground state. The complex visible light forms CTC in excited state, and then an electron transfer reaction occurs to form amine cation radical and diphenyl iodide radical, the latter one further decomposes to produce iodobenzene and benzene radicals, which can initiate polymerization. The rate at which the system initiates the polymerization of the ethylenic monomer is much higher than the rate of polymerization initiated by the iodonium salt alone.

From Fig. 3.54, we can see that the UV absorption of iodonium salt alone is below 300 nm, the absorption of triethylamine itself at 300–400 nm, and the UV absorption curve of DPIOC/TEA system relative to the UV absorption of iodonium salt and triethylamine has large red shift. Comparing Figs. 3.53 and 3.54, it can be seen that the DPIOC/TEA complex system forms a small absorption platform around 510 nm.

The interaction path between DPIOC and TEA is shown in Fig. 3.55.

2. Photochemical properties of DPIOC/TEA system

    The UV/visible light spectrum of the DPIOC/TEA complex system under illumination is shown in Fig. 3.56.

    From Fig. 3.56, we can see that the DPIOC/TEA complex system changes with time under illumination, and the absorption peak of the

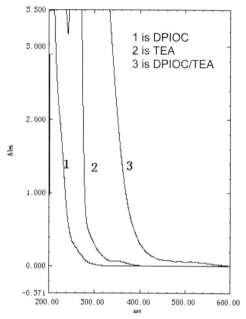

FIGURE 3.54 UV absorption spectra of DPIOC, TEA and DPIOC/TEA.

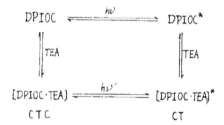

FIGURE 3.55 Interaction path between DPIOC and TEA.

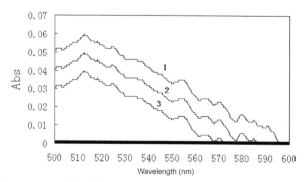

FIGURE 3.56 Curve of DPIOC/TEA complex system changing with time under illumination. Illumination time curve 1 (0 min); curve 2 (2 min); curve 3 (5 min).

CTC decreases continuously. It can be considered that the photochemical reaction of the sulfonium salt and the amine firstly generates CTC in excited state by photoexciting the CTC in ground state, and a single electron transfer occurs to form diphenyl iodine radical and ammonium cation radical. The diphenyl iodide radical is very unstable and is easily decomposed to produce iodobenzene and benzene radicals. The benzene radical and the hydrogen donor may undergo hydrogen transfer reaction or complex reaction, and the amine cation radical may easily react with the hydrogen donor to form proton and hydrogen donor radicals.

$$(Ph_2I \cdot NEt_3) + Cl^- \rightarrow Ph_2I + N + Et_3 + Cl^-$$

$$Ph_2I \rightarrow PhI + Ph$$

$$Ph + HR \rightarrow PhH + R$$

$$2Ph \rightarrow Ph - Ph$$

$$N + Et_3 + HR \rightarrow Et_3N + H + + R$$

3. Effects of free-radical polymerization inhibitor on polymerization

   Since DPIOC/TEA complex system has photodecomposition of iodobenzene and benzene radicals, and the proton and hydrogen donor radicals generated by the action of amine cation radicals and hydrogen donors, there is a significant induction period in the presence of oxygen, which is caused by the quenching of excited state by oxygen and the termination of free radicals. It can be seen from Table 3.8 that the free-radical polymerization inhibitor inhibits that the hydroquinone and DPIOC/TEA complex system photosensitivity induce MMA polymerization reaction. As the concentration of the inhibitor increases, the induction period extends. Therefore, the free-radical polymerization inhibitors such as oxygen and hydroquinone have great inhibition effect on the photoinitiated polymerization of the DPIOC/TEA complex system, and TEA itself is a cationic polymerization inhibitor. Therefore, most of the DPIOC/TEA complexation systems are carried out according to the free-radical mechanism.

TABLE 3.8 Inhibitory effect of free-radical polymerization inhibitor on polymerization of hydroquinone.

Inhibitor	Induction period (min)
Oxygen	5
Hydroquinone (9.78×10^{-6} M)	15
Hydroquinone (9.78×10^{-5} M)	30

3.2.2.3 Dye-onium salt photosensitive system

The dyes are featured by low cost, wide spectral absorption range, and coverage of the entire visible light region. In the molecule, especially in the ion pair, the light-induced electron transfer has the important characteristics of high quantum efficiency, fast speed, and small influence on the viscosity and rigidity of the medium, especially in the SLA speed of the polymer, the SLA speed will affect the nature of the entire system. Therefore, if a dye-onium salt compound is synthesized with these two characteristics combined, we can obtain a highly efficient photosensitive system that senses visible light. Therefore, this article discusses the synthesis of ionic compound by eosin dye and iodonium salt, compares the effects of their different bonding methods on the photoinduced electron transfer efficiency, and studies the photophysical and photochemical properties of the dye/onium salt ions on compounds and their influencing factors (such as the influence of the medium), photoinitiated allyl monomer polymerization reaction and its effects, compares the properties of ion pair compounds in solution and SLA systems, as well as the different efficiencies of ion pair electron transfer photosystems and intramolecular electron transfer photosystems.

1. Effects of dye/iodonium salt bonding method on electron transfer rate

    Eosin is a xanthene-based dye whose maximum absorption is in the visible region. Meanwhile, it is an anionic dye. Compared with the strong electron acceptor of iodonium salt, it can introduce onium salt through ion exchange reaction to form ion-pair photosensitive system. In this way, the spectral response range of the onium salt is effectively extended to the visible light region. Moreover, the disadvantage that the two-molecular electron transfer system is greatly affected by the viscosity and rigidity of the medium can be overcome. Table 3.9 compares the photophysical and photochemical properties of the dye/iodopidine salt system with different linkages, in which the eosin disodium salt and the diphenyliodonium salt undergo electron transfer reaction between the two molecules, and the photofading speed is slow. However, there are no inert sodium ions (Na^+)

TABLE 3.9 Photochemical properties of eosin and iodonium salts.

Compound	λ_{max} (nm)	Photobleaching speed
Eosin disodium salt + diphenyl iodonium salt (EoNa$_2$ + DPIOC)	483.6	Slow
Eosin-bis(disodium salt)(Eo(Ph$_2$I)$_2$)	519.5	Fast
Disodium salt (DPIOC)	225	–
Solvent: ethanol.		

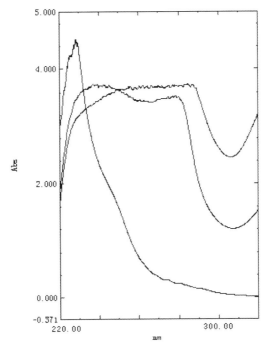

FIGURE 3.57 UV spectrum of eosin disodium salt, eosin-bis(diphenyliodonium salt), and diphenyliodonium salt.

$$A^+B^- \rightleftharpoons A^+//B^- \overset{K_d}{\rightleftharpoons} A^+ \; B^-$$

Contact ion pairs   Solvent-separated ion pairs   Free solvated ions

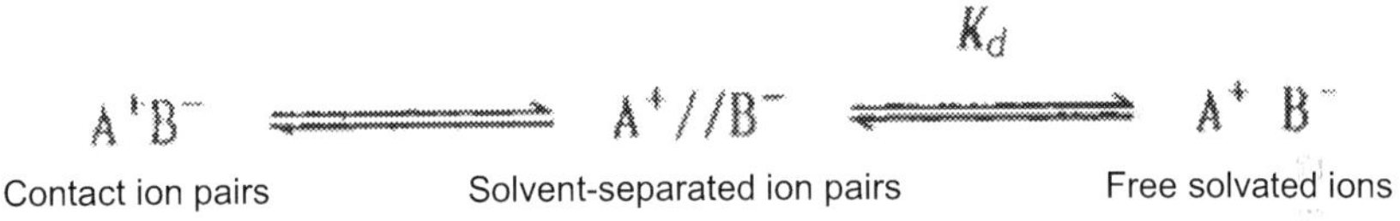

FIGURE 3.58 Binding form of ion pairs.

and chloride ions (Cl^-) in the eosin-bis(diphenyliodonium salt), and it is an intramolecular electron transfer reaction, and thus its photofading rate is much higher than that of the bimolecular system (Fig. 3.57).

2. Influence of medium polarity

Since the eosin diphenyl iodonium salt ($Eo(Ph_2I)_2$) is an ionic compound, its physicochemical properties in the medium are greatly affected by the polarity of the medium. The polarity of the solvent has a great influence on the tightness of the positive and negative ion pairs of the eosin diodide salt. As the polarity of the solvent increases, the ion-pair binding becomes loose and dissociation occurs. It directly affects the light-induced electron transfer efficiency and the subsequent free-radical generation reaction. Winstein has pointed out that many organic ionic compounds (A^+B^-) may have at least three forms of existence in a suitable solvent: contact (or tight) ion pairs, solvent-separated ion pairs, and free solvated ions, as shown in Fig. 3.58.

The value of K_d varies with the nature of A^+ and B^-, the solvation of the solvent used and the ability to dissociate ion pairs and temperature. Dennison and Ramsey quantitatively deal with the solvent effect and believe that the following relationship holds:

$$-\ln K_d = \frac{Z2}{aDKT} \qquad (3.38)$$

where Z is the charge of ion, A is the sum of ion radius of Van der Waals, D is the dielectric constant of solvent, K is Boltzmann constant, and T is temperature.

It can be seen that the smaller the polarity of the solvent, the lower the temperature, and the smaller the dissociation constant K_d of the A^+B^- ion pair, the system exists in the form of ion pairs to a larger extent. The main form of $Eo(Ph_2I)_2$ in solution is mainly the change of the absorption curve peak when reflected in the spectrum. When comparing dimethoxyethane (DME) of weak polar solvent with ethanol of strong polar solvent, DME has smaller polarity than that of ethanol, the tightness of positive and negative ion pairs increase, the maximum absorption wavelength is red-shifted, and the molar extinction coefficient decreases, as shown in Fig. 3.59.

When the solvent is strongly polar, the absorption spectrum of $Eo(Ph_2I)_2$ is (λ_{max} 537.5 nm and λ_{max} 500 nm), indicating that the bisiodonium salt of eosine is mainly present in the form of loose ion pair. As the

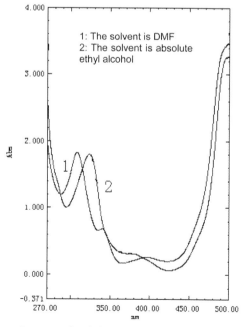

FIGURE 3.59 Absorption curve of eosin iodonium salt in DME and absolute ethyl alcohol.

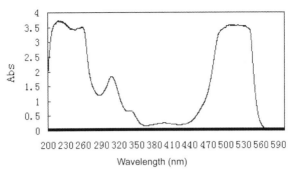

FIGURE 3.60 Ultraviolet absorption curve of diethyl eosin.

polarity of the solvent decreases, the loose ion pair decreases, and the close ion pair is generated at the same time, resulting in a decrease in absorption of absorption peak, while new absorption peaks (λ_{max} 475 nm and λ_{max} 380 nm) appear in the short-wave direction, which basically have the same absorption spectrum of ethyl eosin EoEt$_2$ connected to the covalent bond, indicating that Eo(Ph$_2$I)$_2$ exists mainly in the form of tight ion pairs (Fig. 3.60).

Of course, the concentration of the eosin iodonium salt itself also affects its presence form in solution. At low concentrations, Eo(Ph$_2$I)$_2$ is mainly present in the form of tight ion pairs, while at high concentrations, the contents of tight ion pairs and loose ions increase at the same time.

3. Photobleaching reaction of bisiodonium salt of eosine

The eosin-bisiodonium salt can undergo photobleaching reaction under the illumination of light, as shown in Fig. 3.61. In the polar solution, ethanol, the maximum absorption peak of Eo(Ph$_2$I)$_2$ decreases with the illumination time, and there is slight red shift phenomenon, as shown in Fig. 3.62. In the nonpolar solution, DME, the maximum absorption peak of Eo(Ph$_2$I)$_2$ decreases with the illumination time, there is almost no red shift, as shown in Fig. 3.63.

This is because the photobleaching reaction is caused by the charge-transfer between the eosin anion and the iodonium cation, and the nature of the solvent has great influence on the photofading reaction. The rate of photobleaching in nonpolar solvents is much higher than in polar solvents. This may be because in nonpolar solvents, the onium salt cations are closer to the eosin anion, that is, they are more likely to form tight pairs, so that even if the lifetime of the excited state is short, the collision and reaction between anion and cation can be guaranteed. Conversely, as the polarity of the solvent increases, the dissociation degree of ion pair also increases, and electron transfer is more difficult. This effect is called "inverted solvent polarity dependence." At the same time, the electron transfer of the ion pair itself can neutralize the positive and negative charges. The total charge

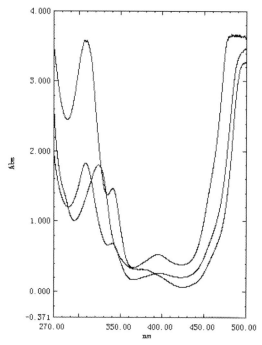

FIGURE 3.61 Absorption of eosin-bisiodonium salt in DMF and absolute ethyl alcohol and absorption spectrum of eosin sodium salt.

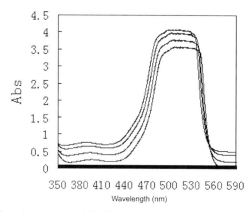

FIGURE 3.62 Time change curve of $Eo(Ph_2I)_2$ in ethyl alcohol. Illumination time from top to bottom: 0, 10, 20, 35 min.

changes easily in the nonpolar solvent during the electron transfer process, and there is no coulomb attractive force balanced with it in the residual repulsive effect after electron transfer, the rebound of free radicals can accelerate their separation. Moreover, the diphenyl iodide radicals generated by electron transfer are easily decomposed, thus preventing the electron reverse

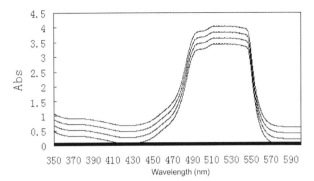

FIGURE 3.63 Time change curve of Eo(Ph$_2$I)$_2$ in DME. Illumination time from top to bottom: 0, 10, 20, 35 min.

transfer of wasted energy. In addition, there is also hydrogen bond in the solution. In the solvent containing groups such as $-$OH and $-$NCOH, the stable hydrogen bond and the solvent forming hydrogen bond in the eosin-bisiodonium salt dye precursor produce isolating effect on the ion pair, which will also slow down photobleaching.

4. Photoelectron transfer reaction of eosin diphenyl iodonium salt

The main volatile products of photobleaching in the photochemical reaction of eosin diphenyl iodonium salt are iodobenzene (PhI) (electron transfer reduction product) and benzene radical product, chlorobenzene (PhCl) and Bromobenzene (PhBr), as well as the photobleaching reaction product of eosin, which undergoes photoelectron transfer reaction according to the following step-by-step reaction mechanism:

$$\text{Eosin-Ph}_2\text{I} \xrightarrow{h\nu} S_1 \xrightarrow{isc} T_1 \quad (3.39)$$

$$S_1 \xrightarrow{et} \text{[radical intermediate]} + Ph^\cdot + PhI \quad (3.40)$$

$$Ph^\cdot + \text{(I)} \longrightarrow \text{(I)} \rightleftharpoons \text{(II)} \quad (3.41)$$

$$\text{I} \xrightarrow{h\nu} \text{[structure]} + \text{Ph}[3\text{-}42] \text{ PhI} \quad (3.42)$$

$$\text{II} \xrightarrow{h\nu} \text{[structure]} + \text{Ph}\cdot + \text{PhI}$$

$$\text{[structure]} \longrightarrow \text{[structure]} \quad (3.43)$$

$$\text{[structure]} \longrightarrow \text{[structure]} \quad (3.44)$$

5. Role of eosin iodonium salt in SLA system

As discussed earlier, the biggest disadvantage of the bimolecular photoinitiator system is that it is controlled by the diffusion rate. In the polymer SLA system, the movement of the molecule is greatly limited when the viscosity is large, which will directly affect the curing speed of the system, and is fatal to rapid prototyping. The ion pairing compound can undergo electron transfer reaction in the ion pair, which overcomes the shortcomings of bimolecular photosensitization, and its photoinitiation efficiency is greatly improved, especially in polymer SLA systems. Therefore, this section further discusses the role of eosin iodonium salts in SLA systems.

The prepolymer used herein is methyl group methacrylate.

The absorption spectrum of the eosin iodonium salt in the SLA system is basically similar to that in the solution, as shown in Fig. 3.64. Curve 1 is the absorption curve of the eosin iodonium salt in the SLA system, and curve 2 is the absorption curve of the eosin iodonium salt in the solution.

Add the fluorescent probe N,N-dimethylamino benzal malononitrile (structure is shown in Fig. 3.65) to the SLA system of the bisiodonium salt. When excited by light to form an excited state, the probe is easily

Polymer materials for additive manufacturing: liquid materials Chapter | 3 **259**

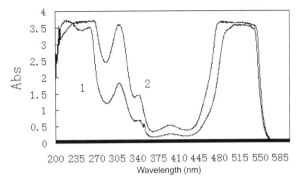

FIGURE 3.64 Absorption spectrum of eosin iodonium salt in SLA system and solution.

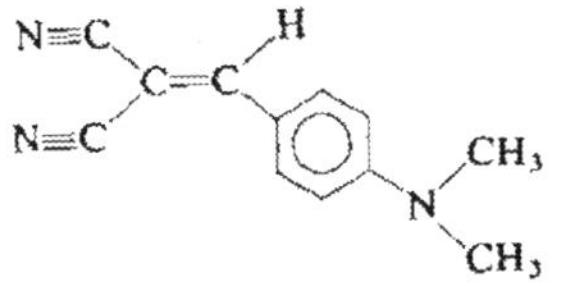

FIGURE 3.65 Fluorescent probe, N,N-dimethylamino benzal malononitrile.

inactivated by the rotation of the $(CH_3)_3N$ gene and the distortion between the donor-receptor. If environmental factors such as viscosity, rigidity, and temperature of the system change, such as increase in viscosity and rigidity, the rotation of the $(CH_3)_3N$ group and the distortion between the donor and the acceptor will be hindered, and the rate of nonradiative deactivation of the excited state will be reduced. In the SLA system, as the curing progresses, the viscosity and rigidity of the system increase, limiting the nonradiative deactivation of the fluorescent probe. Therefore, the fluorescence intensity will continue to increase. In this way, the change in fluorescence intensity can be utilized to track the SLA process. Experiments show that the photobleaching speed and SLA speed of bimolecular system of eosin iodonium salt are fast in the initial stage of the reaction, but with the change of time, the rigidity increases due to the increase of the viscosity of the SLA system, and the speed is slower and even to be zero; but there is almost no change in the photobleaching speed and SLA speed of the single molecule system of eosin diphenyl iodonium salt has in the same period of time.

6. Effects of oxygen polymerization inhibitor

In the photoinitiated MMA polymerization of eosin-bis-diphenyliodonium salt, the presence of oxygen also causes a significant induction period of the polymerization reaction. This is related to the mechanism of dye sensitized iodonium salt by Wang Wenzhi et al, in which the sensitized iodonium salt decomposes through the photoelectron transfer mechanism to produce cationic active center, and the sensitization of iodonium salts by photoelectron transfer to initiate cationic SLA is different.

3.2.3 Study on stereolithography apparatus kinetics of trimethylene oxide

Trimethylene oxide is a new monomer for cationic photopolymerization and is the main raw material for high-end UV cationic polymerization products. It can be applied to special occasions such as UV inks, UV coatings, UV adhesives and electronic encapsulants, SLA, and AM. It is featured by low viscosity, low curing shrinkage, high bond strength, fast polymerization speed, very low toxicity and volatility, and is not inhibited by oxygen. It also has good chemical resistance, thermal stability and excellent mechanical properties. The trimethylene oxide is similar to the alicyclic epoxy, the ring tension of the trimethylene oxide is 107 kJ/mol, and the ring tension of the epoxy group in the alicyclic epoxy is 114 kJ/mol; and their pK_a are 2.0 and 3.7, respectively. Therefore, trimethylene oxide has stronger alkalinity than alicyclic epoxy, which is more conducive to photoinitiated cationic ring-opening polymerization, and photopolymerization is faster.

In this section, bis(3-ethyl-3-oxetanyl)methyl group ether (OXT221) is used as a monomer, triarylsulfonium hexafluoroantimonate is used as an initiator, and the diode-emitting UV lamp with a wavelength of 365 nm is used as the UV light source to study the photopolymerization kinetics of trimethylene oxide monomers through UV-FTIR technique, so as to guide the preparation of cationic SLA resins in subsequent experiments.

3.2.3.1 Experimental part

The photoinitiator UVI6976 and polycaprolactone Polyol Tone 0301 of the monomer OXT221 and different amounts (mass percentage of monomers, the same below) are stirred by magnetic stirrer for 30 minutes to prepare uniform colorless liquid. The liquid is then coated on a KBr wafer, overlying another KBr wafer, and different thicknesses of aluminum foil are placed between the two KBr wafers as spacers to control the film thickness of sample.

The change of the absorption peak area at 981 and 827 cm^{-1} is monitored by RT-FTIR while irradiated with UV light. The conversion rate (Y) of trimethylene oxide group is calculated according to Eq. (3.27), where, A_0 and A_t are the sum of the areas of the absorption peaks at 981 and 827 cm^{-1} at 0 and t seconds of light irradiation, respectively. (Unlike the previous section, here we monitor the change in the absorption peak area at both locations simultaneously.)

3.2.3.2 OXT221 stereolithography apparatus reaction mechanism

The SLA reaction mechanism of trimethylene oxide OXT221 is shown in Fig. 3.66.

In reaction formula 1, UVI6976 is photolyzed under UV light to obtain protonic acid, $HSbF_6$. It has strong acidity, and then protonate the monomer to form the secondary oxonium cation (formula 2). The activation energy (Ea1) of this step is low and is an exothermic reaction. The secondary oxonium cation is extremely active, and it is ring opened after nucleophilic

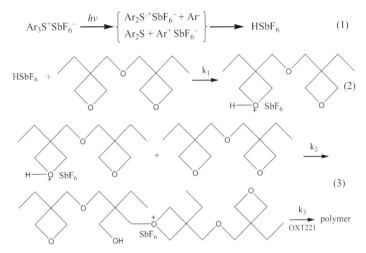

FIGURE 3.66 OXT221 SLA polymerization mechanism.

attack by another monomer, correspondingly generating a third-order oxonium cation (the first step in formula 3), and its activation energy (Ea2) is also low. The heat release is greater than formula 2. Finally, monomer OXT221 repeats the nucleophilic attack on the third-order oxonium cation to form polymer (the second step in formula 3). Since the third-order oxonium cation is relatively stable, the activation energy (Ea3) of the second step in the formula 3 is high, which is much larger than the first step in the formula 3, that is, Ea3 >> Ea2. A lot of heat will be released in the last step. Thus, once Ea3 is overcome, photopolymerization proceeds quickly.

3.2.3.3 Effects of various factors on the stereolithography apparatus conversion rate of OXT221

1. Effects of mass fraction of photoinitiator on curing conversion rate

    As can be seen from Fig. 3.67, there is a certain induction period during the SLA of OXT221. The induction period is defined as the time when the conversion rate is 5%. After the induction period, the photopolymerization reaction proceeds rapidly. The existence of the induction period is not due to the occurrence of oxygen inhibition, but the activation energy of k3 is high, and it is necessary to accumulate enough energy to overcome Ea3 in order to rapidly polymerize. As the mass fraction of initiator increases, the polymerization rate and conversion rate increase slowly. The increase of initiator mass fraction makes the photolysis rate of formula 1 faster while more monomers are protonated (formula 2). And the released photolytic heat and protonation heat provide activation energy for the polymerization of the monomer in the formula 3, and finally the total polymerization rate and the conversion rate are increased. The mass fraction of the photoinitiator should

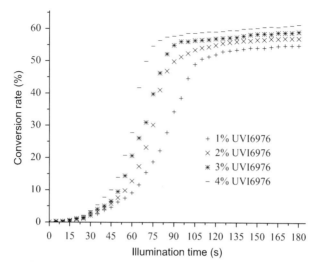

FIGURE 3.67 Monomer conversion rate curve at different initiator mass fraction (25°C, light intensity: 15 mW/cm^2).

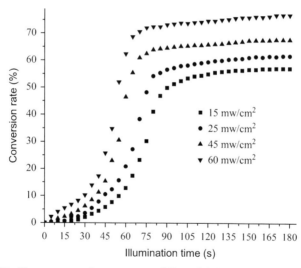

FIGURE 3.68 Monomer conversion rate curve at different light intensities (25°C, 2% UVI6976).

not be excessively high. First, the cationic photoinitiator is very expensive (2000−3000 RMB/kg), causing a significant increase in cost. Secondly, the surface of the monomer absorbs too much UV light, while the inner layer absorbs insufficient UV light, resulting in an insufficient curing depth.

2. Effects of light intensity on curing conversion rate

Fig. 3.68 shows the conversion curve of OXT221 at different light intensities. Obvious effects can be achieved by increasing the light intensity ratio and the mass fraction of initiator, and the induction period is shortened as the light intensity increases. This is mainly because as the light intensity is increased, more protonic acid is generated in the photolysis process of the formula 1, and more secondary oxonium protonated monomers are correspondingly obtained the formula 2, releasing a large amount of heat, so that the chain growth in the formula 3 advances rapidly, and the induction period is shortened. When the light intensity is increased to 60 mW/cm^2, the induction period is almost negligible. As the light intensity increases, the polymerization rate and conversion rate of the monomer continue to increase rapidly.

3. Effects of temperature on curing conversion rate

    Fig. 3.69 shows the conversion rate curve of monomer photopolymerization when the initial reaction temperature is controlled at 25°C, 35°C, 45°C, and 55°C, respectively. As can be seen from the figure, increasing the initial temperature can significantly shorten the induction period and increase the polymerization rate and conversion rate. Increasing the temperature can increase the kinetic energy of the starting reaction molecule rapidly, which is beneficial to overcome the activation energy of each step in the reaction mechanism formula, and also causes the formula 3 to proceed rapidly, causing a change in the polymerization rate. The initial reaction temperature should not be too high, depending on the application and the vapor pressure and boiling point of the monomer.

4. Effects of polyol content (mass fraction) on curing conversion rate

    Alcohol and moisture are terminators and chain transfer agents for cationic polymerization. According to Bednarek et al., polyols can promote

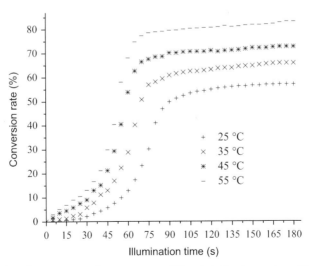

FIGURE 3.69 Monomer conversion rate curve at different starting temperatures (2% UVI6976, light intensity: 15 mW/cm^2).

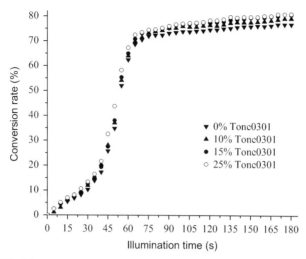

FIGURE 3.70 Monomer conversion curve at different polyol concentrations (25°C, 2% UVI6976, light intensity: 60 mW/cm^2).

cationic photopolymerization by a mechanism that activates monomers. This experiment analyzes the effect of polyol on the SLA of OXT221.

Fig. 3.70 shows the SLA conversion rate curve of OXT221 after Tone0301of a weight ratio of 10%, 15%, and 25% is added respectively. It can be seen that when the polyol is added, the conversion rate of trimethylene oxide is increased, but not significantly, and the content of the polyol does not have much influence on the conversion rate. The reason is that the polymerization process of OXT221 is a reaction in which the monomer conducts continuous nucleophilic attacks to the third-order oxonium cation. The viscosity of the monomer is small and its activity is strong, even if there is no chain transfer (activated monomer mechanism), it can also be polymerized to a system with a very high viscosity.

In summary, increasing the light irradiation intensity and the reaction temperature can effectively reduce the SLA induction period of the trimethylene oxide, thereby increasing the photopolymerization reaction rate and the conversion rate; increasing the amount of the initiator mass fraction can increase the polymerization rate and the conversion rate. However, polyols have little effect on the SLA conversion rate of trimethylene oxide.

3.2.4 Study on the preparation and properties of solid materials in cationic systems

Cationic SLA system is a commonly used resin type for SLA rapid prototyping technology. It has advantages of small volume shrinkage, good dimensional stability, high forming precision, easy recoating, good wear resistance, high modulus and high tensile strength. It is not easy to be deformed, and it is not affected by oxygen resistance when it is cured by light. Especially the last point is especially important. Since the size of the ink droplets is very small during the ink jet printing process, the surface area in contact with the air is relatively large, and a large amount of oxygen is inevitably dissolved. Therefore, the free-radical SLA system should be specially designed to prevent oxygen resistance, or increase the mass fraction of initiator, or increase the intensity of UV light irradiation. The pure cationic SLA system does not have this problem, which greatly simplifies the formulation and forming process.

3.2.4.1 Principles for selecting raw materials for cationic solid materials

1. Selection of monomers and oligomers

For solid materials, the choice of monomers and oligomers is of the utmost importance, and it is related to the ejection properties of the solid materials, the forming accuracy, and the mechanical properties of the cured molded parts. The alicyclic epoxy compound and trimethylene oxide are the preferred components of the solid material because the epoxy group of the alicyclic epoxy compound and the quaternary heterocyclic ring of the trimethylene oxide are electron-rich groups, which are easy to react with proton acid or Lewis acid generated by photolysis of cationic photoinitiator to initiate photopolymerization.

There are many types of alicyclic epoxy compounds, such as Sigma-Aldrich Reagent, Inc., which can provide bifunctional alicyclic epoxy compound as shown in Fig. 3.91.

In Figs. 3.71, (1) is vinylcyclohexene dioxide, (2) is dimethyl substituted ethylene cyclohexene, (3) is bicyclopentadiene dioxide, (4) is 3,4-epoxy caproyl methanoic acid-3',4'-epoxy caproyl methyl group ester, (5) is 3,4-epoxy-6-methyl caproyl methanoic acid-3',4'-epoxy-6'-methyl caproyl methyl ester, and (6) is adipate (3,4-epoxy caproyl methyl ester).

Among them, the viscosity of the variety (1) is the lowest, which can be as low as 1.385 mPa s at 25°C, but it has strong smell, and the SLA rate is not fast. The presence of methyl group in the varieties (2) and (4) forms steric hindrance to the ring-opening polymerization of the epoxy group. Industrial products of variety (3) are rare. The cured product of the variety (6) has good flexibility but high viscosity. The curing rate of the variety (5) is very fast, the cured product is wear-resistant, the toughness is good, and the viscosity is low, for example, the UVR6105 provided by DOW Chemical Co., Ltd. has a viscosity of 220–250 mPa s at 25°C, which is an ideal alicyclic epoxy compound prepolymer. Based on the aforementioned factors, 3,4-epoxy caproyl methanoic acid-3',4'-epoxy caproyl methyl group ester is mainly used as the alicyclic epoxy component in this article.

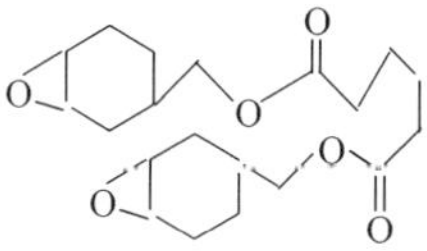

FIGURE 3.71 Main varieties of alicyclic epoxy compounds.

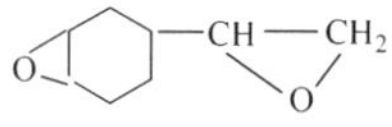

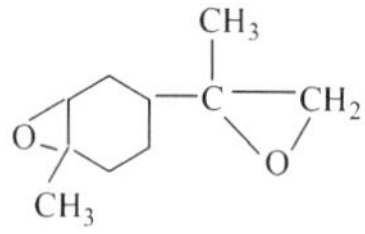

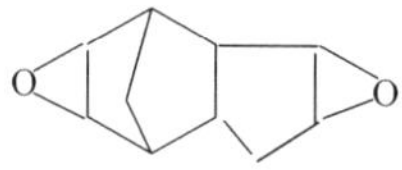

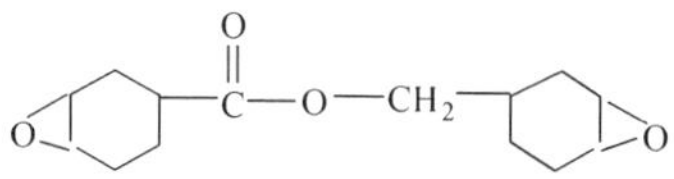

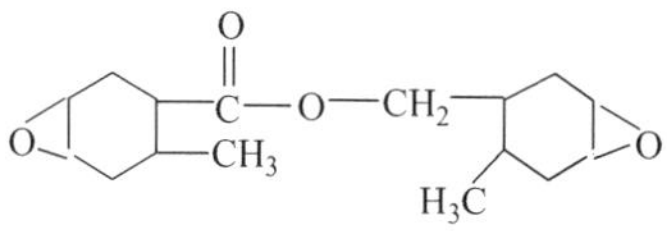

The main industrial product supplier of trimethylene oxide monomers and oligomers is TOAGOSEI CO. in Japan, which can provide the following trimethylene oxide monomers and oligomers, as shown in Fig. 3.72.

In Fig. 3.72, the viscosities of the six trimethylene oxides at 25°C are OXT101, 22.0 mPa s; OXT 212, 5.0 mPa s; OXT 211, 13.8 mPa s; OXT221, 12.8 mPa s; OXT121, 180 mPa s; and OXT 610, 8.0 mPa s, respectively.

It can be seen from the aforementioned data that trimethylene oxide is an ejection material which is very suitable for AM, and according to the study by Noutary et al., it is found that trimethylene oxide and alicyclic epoxy SLA have synergistic effects, which is an excellent inkjet printing material with corresponding patents. Trimethylene oxide not only has very low viscosity, but also has low curing shrinkage, high bonding strength, high polymerization rate, very low toxicity and volatility, and is not inhibited by oxygen. In addition, it has good chemical resistance and heat stability and excellent mechanical properties. Therefore, trimethylene oxide is the preferred raw material for the physical material of the cationic system. Among them, OXT211 (DOX) and OXT121 contain benzene ring, which not only has slightly higher viscosity, but also increases brittleness when the system crosslinks to a large extent; OXT610 is mainly used in light and moisture double curing systems;

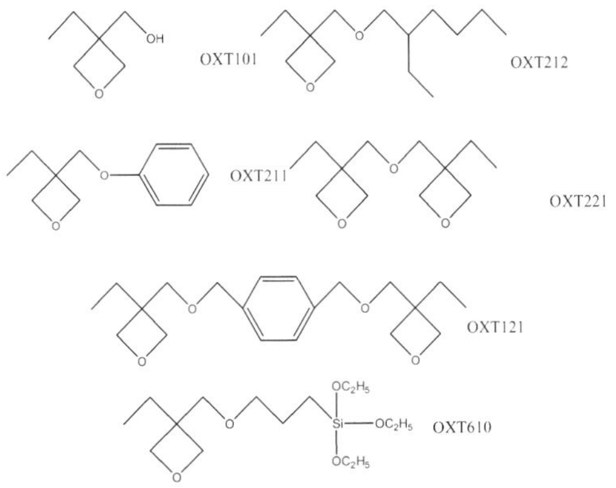

FIGURE 3.72 Main varieties of trimethylene oxide monomers and oligomers.

OXT212 and OXT101 are monofunctional monomers with slightly slow SLA reaction rate. Based on the aforementioned factors, OXT221 has a bifunctional group with fast SLA reaction rate, and contains ether bond with good flexibility, and its viscosity is only 12.8 mPa s. Therefore, this paper mainly uses it as the trimethylene oxide monomer of a cation system.

Vinyl ether is a monomer containing $CH_2 = CH-O-$ structural unit, and the lone pair electrons on the oxygen atom can be conjugated with carbon–carbon double bond to increase the electron cloud density of the double bond, and thus it is an electron-rich monomer that can be subjected to cationic polymerization and radical polymerization, which is particularly suitable for cationic polymerization and hybrid polymerization. Vinyl ether has excellent reactivity and dilution ability, and is low toxicity, low odor and low irritant monomer. It works well with alicyclic epoxy and trimethylene oxide. Therefore, this paper also discusses the use of vinyl ether as a monomer of the cationic system. Common vinyl ether monomers are shown in Fig. 3.73.

HBVE is a monofunctional vinyl ether with slightly slow curing rate; CHVE has great smell odor and unpleasant taste, with a viscosity of 1.2 mPa s at 25°C; DVE-3 has small smell with good flexibility, and the viscosity at 25°C is 2.67 mPa s, which is an ideal vinyl ether monomer, so it is selected as the third monomer of the cationic system.

2. Selection of photoinitiators

   The main considerations for the selection of UV photoinitiators are the main emission wavelength range of the UV light source, the photolysis efficiency and photoinitiation activity of the photoinitiator, the initiation efficiency of the photoinitiator, and the compounding of the photoinitiator. The main emission wavelength of the UV light source of the AM experimental prototype is 360–390 nm, and the strongest

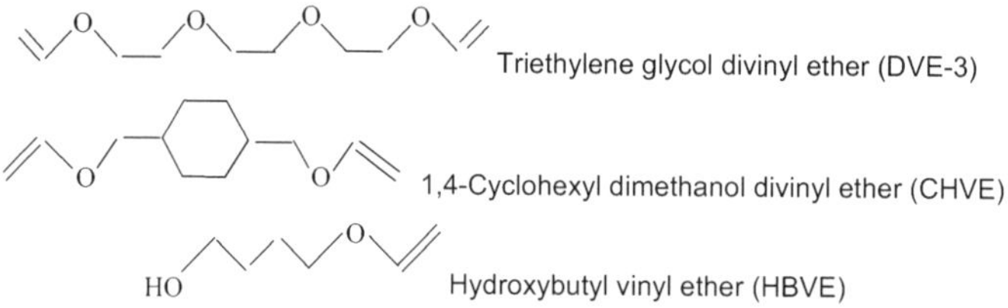

Triethylene glycol divinyl ether (DVE-3)

1,4-Cyclohexyl dimethanol divinyl ether (CHVE)

Hydroxybutyl vinyl ether (HBVE)

FIGURE 3.73 Main varieties of vinyl ether monomers.

emission line is 365 nm. Therefore, the absorption of the UV light by the cationic photoinitiator should be matched with the emission wavelength of the UV lamp as much as possible. Although the aromatic ferrocene salts cationic initiator has a large absorption at 390−400 nm, it also has a large absorption in the visible light region of 530−540 nm. If the solid photopolymer is prepared using aromatic ferrocene salts as an initiator, the photopolymer is easily cured and deteriorates after receiving visible light upon storage, which affects ejection stability, so that the aromatic ferrocene salts is not suitable for use as photoinitiator. The UV absorption wavelength of triarylsulfonium hexafluoroantimonate, triarylsulfonium hexafluoroarsenate salt, and triarylsulfonium hexafluorophosphate can be extended to 300−360 nm, and their absorption wavelength can be well matched with the UV light source, and the triarylsulfonium salt can be conjugated with three aromatic ring moieties because of the sulfur atom, the positive charge is dispersed, the molecular thermal stability is good, and it is not decomposed when heated to 300°C, so it is suitable for being used as ejection material under medium and high temperature. The initiating activity of triarylsulfonium hexafluoroantimonate is much greater than that of triarylsulfonium hexafluoroarsenate and triarylsulfonium hexafluorophosphate. Therefore, triarylsulfonium hexafluoroantimonate is used as the cationic initiator in the experiment. It is commercially available from DOW Chemical Company of the United States, with the trade name of UVI6976.

The absorption wavelength of UVI6976 is still too small, which cannot fully utilize the UV light emitted by the metal halide lamp. Therefore, it is compounded with Darocur ITX, a free-radical type photosensitizer that extends the absorption wavelength of UVI6976 through electron transfer activation and direct photosensitization to promote the photolysis process of UVI6976. Darocur ITX has a maximum absorption wavelength of 380−420 nm, and the corresponding molar absorption coefficient is also high, about 102 orders of magnitude, which can fully utilize the 365 nm emission line of the UV lamp.

3. Selection of alcohols

3,4-dimethoxybenzyl alcohol is cationic polymerization curing accelerator, and polycaprolactone triol (Polyol0301) mainly plays a role in increasing the flexibility of the cured product, reducing the crosslink density, and also accelerating the SLA speed of the alicyclic epoxy compound. The tripropylene glycol monomethyl ether (Dowanol TPM) is a relatively effective wetting agent that improves the wetting and fluidity of photopolymers and can also participate in cationic photopolymerization. The total mass fraction of the three alcohols generally cannot exceed 25%. Otherwise, the photopolymerization rate of the cationic system is lowered [5].

4. Selection of other additives

   Pigment: It is required to not contain the group with strong nucleophilicity as much as possible, such as amine group, otherwise, it is easy to stop the cationic active center, slow down the rate of cationic photopolymerization, and even prevent the SLA reaction. The particle size of the pigment is required to be 1.2 μm or less, otherwise, the nozzle is easily blocked. Therefore, the photopolymer to be prepared generally filters through the filter membrane of 0.8–1.2 μm, and even filters through three-stage nanofiltration sometimes. It is required to disperse the pigments by special wetting agent and ensure that the pigments are not easily agglomerated after dispersion.

   Surfactant: The surface tension of the photopolymer can be made in a reasonable range.

   Defoamer: It can reduce the bubbles generated during the jet printing process. Once the bubble problem occurs, the printer nozzle will be broken, that is, some nozzle holes do not discharge ink, which seriously affects the injection amount and printing accuracy during forming.

3.2.4.2 Preparation of solid material resin

Dissolve the photoinitiator Darocur ITX in monomer DOX, and add the solution to a three-necked bottle. Add appropriate amount of UVR6105, Polyol0301, benzyl alcohol, UVI6976, defoamer and leveling agent and stir it evenly; Then, mix the appropriate amount of tripropylene glycol monomethyl ether (TPM), pigment and pigment wetting and dispersing agent in a beaker to allow the pigment to be sufficiently wetted and evenly dispersed; pour the pigment component into a three-necked flask, light-shielded and heat it to 45°C, stir it for 4 hours, and then filter it for use.

3.2.4.3 Main injection parameters of cationic solid materials

1. Effects of viscosity on solid materials

   Viscosity is an important factor in determining whether ink can be ejected from a nozzle. When the viscosity is too high, the process of ink flowing and small droplets forming consume a large amount of kinetic energy, and the ink cannot be ejected from the print head; while when the viscosity is too low, the ink tailing occurs, it is easy to leak and

TABLE 3.10 Viscosity of main raw materials of solid materials.

Raw material	Viscosity (25°C, mPa s)
DOX	12.8
UVR6105	220–250
Polyol0301	2250
TPM	5.5
DVE-3	2.67
UVI6976	60–90

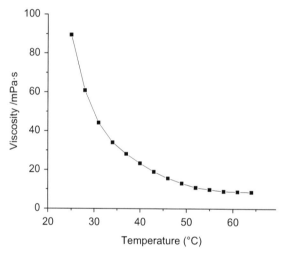

FIGURE 3.74 Viscosity of cationic solid materials at different temperatures.

splash ink. Only when the viscosity of the solid material resin is 5–20 mPa s, the resin can be continuously and stably ejected.

For SLA AM material, the ideal state is that the viscosity is high at room temperature, preferably above 150 mPa s, that is, when it is ejected from the nozzle and falls on the working surface, the temperature of droplet, subject to the working surface and the air heat transfer, has dropped to near room temperature, in order to prevent the droplets from flowing around the working surface, it is required to have high viscosity; at the working temperature (50°C–55°C), the viscosity is low. Below 20 mPa s, the material can be ejected smoothly from the nozzle. The viscosity of the main raw materials in the cationic solid material at 25°C is shown in Table 3.10.

As shown in Fig. 3.74, the cationic solid material has a viscosity of about 89.5 mPa s at 25°C. The effect of temperature on viscosity is obvious.

When the temperature rises, the viscosity gradually decreases, and finally reaches a critical value. Continued increase of temperature has little effect on reduction of viscosity. According to the operating temperature range of Xaar XJ500 piezoelectric nozzle, the operating temperature of the selected solid resin is 55°C, and the viscosity of the resin is about 10.0 mPa s.

2. Effects of surface tension on solid materials

    Surface tension is another important factor in determining whether ink can be ejected from the nozzle stably. When the surface tension is too high, the surface energy required to form the droplets is large, the ink is not easy to form small droplets, and it is difficult to eject; and when the surface tension is too low, the droplets are quickly ejected on the working surface. On the other hand, it is impossible to form an effective layer height, and on the other hand, it affects the dimensional accuracy of the part. Generally, it is suitable that the surface tension is 26–36 mN/m, and it is the best if it is 30 mN/m.

    The surface tension of the resin can be adjusted by surfactant, and a small amount of additives can be added to adjust the surface tension of the resin to an appropriate value. In addition, the temperature has slight influence on the surface tension, and as the temperature increases, the surface tension of the resin slightly decreases. The surface tension of the cationic solid material measured at 55°C is 29.2 mN/m.

3. Reynolds number and Weber number

    The Reynolds number (Re) represents the flow state (laminar flow or turbulent flow) of the sprayed liquid flowing in the nozzle. The flow state of the liquid in the piezoelectric print head is generally laminar flow, and the Weber number (We) represents the ratio of the inertial force of the sprayed liquid to the surface tension effect (the smaller the Weber number, the more important the surface tension is). The calculation equations of Reynolds number (Re) and Weber number (We) are:

    $$Re = \frac{\rho v d}{\eta} \tag{3.45}$$

    $$We = \frac{\rho v^2 d}{\sigma} \tag{3.46}$$

    where ρ is the density of sprayed liquid, η is the dynamic viscosity of the sprayed liquid, v is the tangential velocity of the jet droplet, σ is the surface tension of the sprayed liquid, and d is the diameter of nozzle hole. After calculation, the following results can be obtained:

    $$\frac{R}{W} = \frac{Re}{\sqrt{We}} \tag{3.47}$$

    The main conclusions about the influence of R/W value on the droplet ejection process and the shape of the droplet during ejection are as follows:
    1. Only when $1 < R/W < 10$, the liquid droplets can be normally and stably ejected from the piezoelectric nozzle.

2. If R/W is large, the viscosity at the operating temperature of the sprayed liquid is the main parameter affecting the injection process. At this time, the nozzle needs to generate great pressure to spray the liquid, the injection rate of the liquid will decrease, and the extension column before the liquid injection will become shorter.
3. If the R/W is small, the extension column before the liquid ejection becomes long, and the droplet is easily expanded to produce comets tail before the ejection.

According to the calculation equation of Reynolds number (3.49) and the calculation equation of Weber number (3.50), the Reynolds number and Weber number of cationic solid material at working temperature of 55°C are calculated. The solid material has a density of 1.08 g/cm^3, a viscosity of 10.0 mPa s, a surface tension of 29.2 mN/m, a nozzle jet velocity of 6.0 m/s, and a nozzle aperture of 50 μm, and the following results are obtained:

$$Re = \frac{1.08 \times 10^3 \times 6.0 \times 50 \times 10^{-6}}{10.0 \times 10^{-3}} = 32.4$$

$$We = \frac{1.08 \times 10^3 \times 6.0^2 \times 50 \times 10^{-6}}{29.2 \times 10^{-3}} = 66.6$$

From the aforementioned calculation results of Reynolds number and Weber number, according to Eq. (3.51), the value of stable injection parameter R/W of the solid material can be calculated as follows:

$$\frac{R}{W} \frac{Re}{\sqrt{We}} \frac{32.4}{\sqrt{66.6}} = 3.97$$

Therefore, $1 < R/W < 10$, which satisfies the requirement of stable injection index of solid materials.

3.2.4.4 Sputtering coefficient, contact angle, maximum spreading factor, and spreading time

1. Sputtering coefficient

   According to the calculation equation $K = We^{\frac{1}{2}}Re^{\frac{1}{4}}$ of the sputtering coefficient, the sputtering coefficient can be obtained as follows:

   $$K = We^{\frac{1}{2}}Re^{\frac{1}{4}} = 66.6^{\frac{1}{2}} \times 32.4^{\frac{1}{4}} = 19.5$$

   Therefore, the sputtering coefficient is smaller than the sputtering threshold K_C (57.7), and the sputtering phenomenon does not occur when the solid material is sprayed onto the working surface. The test straight line is printed and found to be clear, that is, no sputtering phenomenon exists.

2. Contact angle, maximum spreading factor

   The contact angle measurement photo of the cationic solid material on the cured surface of the cationic solid material is shown in Fig. 3.75.

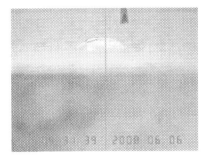

FIGURE 3.75 Contact angle measurement photo of solid material on its own cured surface.

The measured contact angle is 44 degrees and the contact angle with the cured surface of the support material is 52 degrees.

According to the aforementioned measurement results of the contact angle, the equation $\xi = \frac{d_{max}}{d_0} = \left[\frac{We+12}{3(1-\cos\theta)+4(We/\sqrt{Re})}\right]^{1/2}$ is used to calculate the spreading factor ξ_1 of the solid material on the cured surface of the solid material and the spreading coefficient ξ_2 on the cured surface of the support material, and the following results can be obtained:

$$\xi_1 = \left[\frac{We+12}{3(1-\cos\theta_1)+4(We/\sqrt{Re})}\right]^{1/2}$$

$$= \left[\frac{66.6+12}{3(1-\cos 44°)+4(66.6/\sqrt{32.4})}\right]^{1/2} = 1.28$$

$$\xi_2 = \left[\frac{We+12}{3(1-\cos\theta_2)+4(We/\sqrt{R_e})}\right]^{1/2}$$

$$= \left[\frac{66.6+12}{3(1-\cos 52°)+4(66.6/\sqrt{32.4})}\right]^{1/2} = 1.28$$

It can be seen from the aforementioned results that the cationic solid material has the same spreading diameter on the self-curing surface and the curing surface of the support material. The contact angle has a small effect on the spreading diameter.

3. Spreading time

According to the calculation equation $t_{spread} = \frac{\sqrt{\rho d_0^3}}{\sigma}$, the spreading time of the solid material droplets on the working surface can be obtained as follows:

$$t_{spread} = \frac{\sqrt{\rho d_0^3}}{\sigma} = \frac{\sqrt{1.08 \times 10^3 \times (53.5 \times 10^{-6})}}{29.2 \times 10^{-3}} = 7.5 \times 10^{-5} s$$

It can be seen here that the spreading speed of the droplets on the working surface is extremely fast and can be ignored.

3.2.4.5 Printing stability of cationic solid materials

In general, the fewer the number of nozzle plugging holes during printing, the more stable the ink, the more compatible the ink and the nozzle. When the print head is tested, the status of the print head is checked by printing a test strip of the specified shape. The shape of the printed test strip is shown in Fig. 3.76. Each nozzle hole corresponds to one print line, and one Xaar XJ500 nozzle has 500 nozzle holes, corresponding to 500 print lines. In Fig. 3.76, the mark 1 is the broken line, indicating that there is a nozzle hole where no ink is discharged; the mark 2 indicates that the nozzle hole is inked, but the spray path is inclined, that is, it is not vertically sprayed. Therefore, there are two broken lines and one oblique line in Fig. 3.76. (The printed test strips have blank space at each of the opposite corners. There are not a few broken lines here, but it is the design of the print strips of the print heads. The 500 print lines are not in completely rectangular arrangement.)

Table 3.11 shows the number of nozzle plugging holes (no ink discharge or oblique spray) when the nozzle is tested. It can be seen from Table 3.11

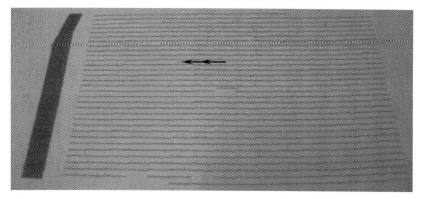

FIGURE 3.76 Print head test strip.

TABLE 3.11 Test results of printing stability of physical materials.

Printing duration printing temperature (°C)	0 h	6 h	12 h	18 h
45	21	76	No ejection	–
50	3	8	31	No ejection
55	0	1	3	3
60	0	1	15	–
65	0	6	–	–

that when the temperature is low, the resin cannot be continuously and stably ejected from the nozzle because the viscosity of the resin is too large to be sprayed; and the temperature is too high, the nozzle itself is unstable, so the ejection effect is not good. The results show that the resin can be continuously and stably ejected at a temperature of 55°C. From the case of the printed test strip, no sputtering occurs when the solid material is sprayed onto the work surface, verifying the theoretical calculation analysis of the sputtering phenomenon in the aforementioned section.

3.2.4.6 Stereolithography apparatus properties of cationic solid materials

1. Characteristic parameters

The SLA property parameters of the cationic solid material are measured. The data of the exposure amount E_0 and the curing depth C_d of the solid material under different exposure time are recorded, and C_d is used to plot $\ln E_0$, as shown in Fig. 3.77.

It can be seen from Fig. 3.77 that C_d has a linear relationship with $\ln E_0$, and the linear correlation coefficient is 0.97804, which is in accordance with the Beer–Lambert rule. The slope of the fitted straight line and the intercept of the abscissa axis are obtained by the least squares method, and thus, the E_C of the solid material is 22.4 mJ/cm^2 and D_p is 0.147 mm. In the AM process, the cured layer thickness of the solid material is set at 0.03–0.05 mm, so the transmission depth of the cationic solid material

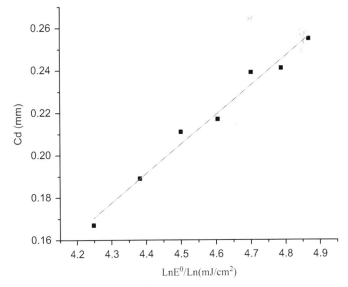

FIGURE 3.77 SLA curve of cationic solid materials.

satisfies the requirements of AM, and interlayer bonding can be achieved. The critical exposure gives the minimum curable absorbance of cationic solid material. This value is not very large. The 300 W metal halide lamp can fully meet the requirements of SLA of solid materials.

2. Study on SLA kinetics

Fig. 3.78 shows the curve of the conversion rate of three monomer functional groups changing with time. It can be seen from the figure that there is a certain induction period during the SLA of the resin at room temperature. The conversion rate of DOX monomer is the highest, followed by UVR6105, and the conversion rate of DVE-3 monomer is the lowest. As can be seen from this sequence, the relative activities of the three resins are DOX > UVR6105 > DVE-3. Both DOX and UVR6105 are highly reactive SLA monomers with fast polymerization rates, and they can be used together to promote mutual conversion. UVR6105 has relatively large molecular volume and high viscosity, and molecular motion is difficult. Thus, the conversion rate is lower than DOX.

As can be seen from Fig. 3.79, when the resin is SLA at 55°C, the induction period disappears and the conversion rate is remarkably improved. The DOX monomer conversion rate reaches 97.4% at 60 seconds of light irradiation; at this time, the UVR6105 monomer conversion rate reaches 77.8%, and the DVE-3 conversion rate reaches 70.8%. As the initial reaction temperature of photopolymerization increases, the kinetic energy of the monomer molecule increases, and the activation energy of

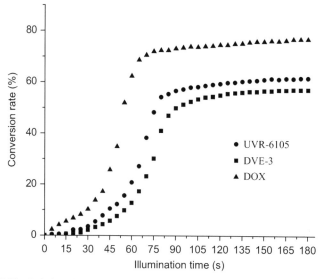

FIGURE 3.78 Relationship of conversion rate of three monomer functional groups varying with illumination time (25°C, 6.0 mW/cm^2).

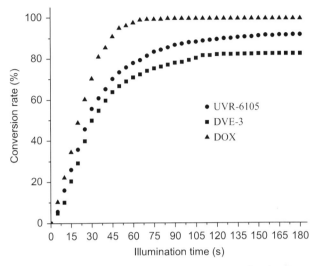

FIGURE 3.79 Relationship of conversion rate of three monomer functional groups changing with illumination time (55°C, 6.0 mW/cm²).

the polymerization reaction is easily overcome, so that the induction period disappears. In addition, the viscosity of the resin decreases, the thermal motion of the monomer molecule is free, and the conversion rate increases. The actual machine test of 3D printer shows that the solid resin sprays when heated to 55°C. After UV illumination, the curing effect is better, and the instantaneous surface drying can be achieved (lighting speed: 0.8 m/s, lamp power: 300 W).

3.2.4.7 Stereolithography apparatus properties of cationic solid materials after stereolithography apparatus

1. Shrinkage coefficient and warpage factor

    According to the test method in Section 3.3.2, the volume shrinkage rate and linear shrinkage rate of the solid material are respectively tested. The density of the liquid solid material is 1.08 g/cm³, and the density after complete curing is 1.12 g/cm³, according to the Eq. (3.20), the volume shrinkage rate is:

$$\text{shrinkage} = \frac{\rho_2 - \rho_1}{\rho_2} \times 100\% = \frac{1.12 - 1.08}{1.12} \times 100\% = 3.57\%$$

In the linear shrinkage test of the solid material, the actual length of the resin after curing is 120.01 mm. According to Eq. (3.21), the linear shrinkage rate is:

$$\text{Linear shrinkage rate \%} = \frac{120.06 - L}{120.06} \times 100\% = \frac{120.06 - 120.01}{120.06} \times 100\% = 0.042\%$$

It can be seen from the aforementioned calculation results that the curing shrinkage rate of the cationic solid material is extremely low, which is smaller than the shrinkage rate of the photopolymer of the hybrid system reported in some literatures, and can fully meet the requirements of AM dimensional accuracy. Through the actual machine test of 3D printer, the warpage factor of the cationic solid material is 1.26%, and the deformation amount during forming is small.

2. Mechanical properties after curing

    The mechanical properties of the solid materials after forming using 3D printer are listed in Table 3.12. It can be seen from the test results in the table that the photopolymer has large tensile strength, good toughness and high hardness, and satisfies the mechanical properties of AM parts. Fig. 3.80 shows the 3D print forming test piece for performing hardness test.

3. Glass transition temperature after curing

    The glass transition temperature of the cationic solid material after curing is 54.5°C, as shown in Fig. 3.81. For the use of general AM parts, it can meet the thermal performance requirements.

4. Picture of cationic solid material forming parts

    As shown in Fig. 3.82, it shows a blue cationic solid material molded part.

TABLE 3.12 Mechanical properties of cationic solid materials after curing.

Tensile strength (MPa)	Tensile modulus (MPa)	Elongation at break (%)	Bending strength (MPa)	Flexural modulus (MPa)	Impact strength (J/m^2)	Hardness (HRM)
59.4	1074	11.5	70.2	1451	50.3	81

FIGURE 3.80 Hardness test sample for AM forming.

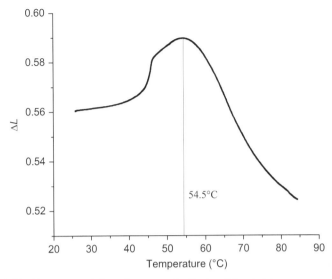

FIGURE 3.81 Temperature-deformation curve of cationic solid materials.

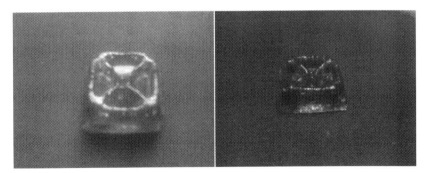

FIGURE 3.82 Dimensional test piece.

3.2.5 Study on the preparation and properties of hybrid system solid materials

The hybrid SLA system has the advantages of free-radical curing system featured by fast SLA rate, insensitivity to moisture and adjustable performance, as well as cationic curing system featured by low volume shrinkage rate, good performance after curing, high forming precision, easy recoating, no oxygen resistance and dark reaction. Therefore, hybrid SLA system is a major class of products for rapid prototyping of SLA materials. This section focuses on the hybrid solid material for SLA AM.

3.2.5.1 Preparation of hybrid solid materials

Dissolve the photoinitiator Irgacure 184 in tripropylene glycol diacrylate, and add surfactant and defoamer into the three-necked flask with the cationic system, shield and heat the solution to 45°C. Then, stir it for 4 hours, and then filter it for future use.

3.2.5.2 Properties of hybrid solid materials

1. Main physical properties and injection parameters of hybrid solid materials

   Table 3.13 lists the measured or theoretically calculated value of viscosity, surface tension, density before and after curing, Reynolds number and Weber number of the hybrid solid materials.

   From the data in Table 3.13, it can be calculated that $R/W = 2.83$, and $1 < R/W < 10$, which satisfies the requirement of stable injection index of solid materials, and the working temperature of the nozzle is set to 55°C.

2. Sputtering coefficient, contact angle, spreading factor and spreading time of hybrid solid materials

   Table 3.14 lists the measured or theoretically calculated value of the sputtering coefficient, contact angle, spreading factor, and spreading time of the hybrid solid materials.

TABLE 3.13 Main physical properties and injection parameters of hybrid solid materials.

Viscosity (mPa s)		Density (g/cm³)		Surface tension (55°C, mN/m)	Reynolds number (Re)	Weber number (We)
25°C	55°C	Before curing	After curing	27.4	24.3	73.6
115	13.8	1.12	1.17			

TABLE 3.14 Sputtering coefficient, contact angle, spreading factor and spreading time of hybrid solid materials.

Contact angle		Spreading factor		Spreading time (s)	Sputtering coefficient (K)
With itself	With support material	With itself	With support material		
32 degrees	46 degrees	1.19	1.19	7.9×10^{-5}	19.0

From the data in Table 3.14, it is known that the sputtering coefficient is smaller than the sputtering threshold K_C (57.7), and the spattering phenomenon does not occur when the hybrid solid material is sprayed onto the working surface. In fact, no sputtering phenomenon is found in the actual machine test. The contact angle of the hybrid solid material is small, which may be related to its small surface tension. The spreading coefficient is smaller than that of the cationic solid material, and seems to contradict the contact angle data. Actually, the contact angle is the static contact angle, and the spreading factor is the dynamic spreading size change obtained by the ejection force during the spraying process. The density and viscosity of the hybrid solid material are slightly larger, and the surface tension is slightly smaller, which makes the difference between the Reynolds number and the Weber number of the hybrid solid material larger, which is the main factor in the spreading coefficient. Therefore, the spreading factor is smaller in spraying.

3. Printing stability experiment

   Table 3.15 shows the number of nozzle plugging holes in the printing test of hybrid solid materials.

   As can be seen from the data in Table 3.15, the jet stability of the hybrid solid material is not as good as that of the cationic solid material. At 50°C, the injection is stopped after 12 hours of spraying. The main reason is that the viscosity is high at 50°C. The spray force of the Xaar XJ500 nozzle is not as large as that of the SPECTRA nozzle, so the adaptability of the nozzle to the ink is poor. At 55°C, a total of 22 nozzle holes cannot normally discharge ink in 18 hours of spraying, and the bubbles are found in the experiment, which is related to the use of the defoamer. The operating temperature of 60°C is slightly high for the Xaar XJ500 nozzle, and the oblique spray phenomenon is found after 6 hours of spraying.

4. Characteristic parameters and SLA rate

   The SLA curve of the hybrid solid material is shown in Fig. 3.83. It can be seen that C_d has a linear relationship with $\ln E_0$, and the linear correlation coefficient is 0.98729, which is in accordance with the Beer–Lambert rule. The slope of the fitted straight line and the intercept of the abscissa axis are obtained by the least squares method, so that the

TABLE 3.15 The number of nozzle plugging holes in the printing test of hybrid solid materials.

Printing temperature	0 h	6 h	12 h	18 h
50°C	5	47	No spray	—
55°C	0	4	7	22
60°C	0	3	19	—

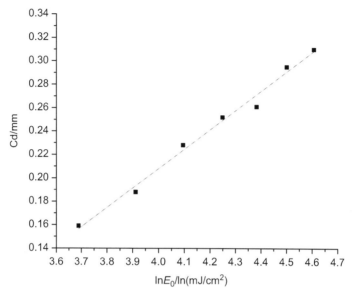

FIGURE 3.83 SLA curve of hybrid solid materials.

E_C of the hybrid solid material is 15.6 mJ/cm^2 and D_p is 0.165 mm. The critical exposure is slightly lower than that of the cationic solid material, indicating that the hybrid solid material is more photosensitive than the cationic solid material. There are two main reasons. First, the SLA rate of polypropylene glycol diacrylate is higher. Second, there is no pigment added in the hybrid solid material. Thus, the critical exposure is small, and the transmission depth is slightly deeper.

Fig. 3.84 is the curve of the gel content of cationic solid materials, hybrid solid materials, and free-radical solid materials changing with the exposure time. It can be seen that the hybrid solid material has the highest SLA rate, followed by the cationic solid material, and the free-radical solid material has the slowest SLA rate. When the hybrid solid material is SLA, the radical and cationic polymerization reactions proceed simultaneously, and form IPN structure. The SLA gel content is high, and the corresponding overall reaction conversion rate is high. The free-radical solid material uses urethane acrylate as the main oligomer, and its SLA rate is slower than that of epoxy acrylate, and free-radical photoinitiated polymerization has oxygen inhibition, the inhibition is more obvious for inkjet printing. Thus, the overall polymerization rate is slow.

5. Shrinkage rate and warpage factor

   The shrinkage rate and warpage factors of hybrid solid materials are listed in Table 3.16. Compared with cationic solid materials, the shrinkage rate of hybrid solid materials is large. The main reason is that two types of

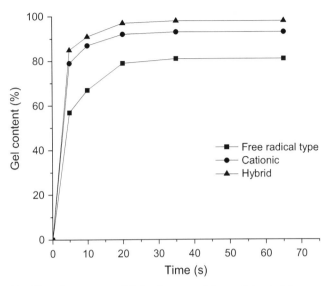

FIGURE 3.84 Time change curve of SLA gel content of three solid materials.

TABLE 3.16 Shrinkage rate, warpage factor, and glass transition temperature of hybrid solid materials.

Volume shrinkage rate	Linear shrinkage rate	Warpage factor	Glass transition temperature
4.27%	0.092%	1.43%	58°C

monomers or oligomers react with each other in the hybrid solid materials and form IPN structure, the system is denser after curing. Thus, the amount of shrinkage is slightly larger.

6. Mechanical properties and glass transition temperature

Table 3.17 shows the mechanical properties of the hybrid solid material after curing, and Table 3.16 lists the glass transition temperature of the hybrid solid material after curing. It can be seen from Table 3.17 that the hybrid solid material has higher tensile strength after curing, lower elongation at break and impact strength, and higher hardness than those of the cationic solid material. It shows that the mechanical strength of the hybrid solid material is improved, and a little toughness is sacrificed. Although polypropylene glycol diacrylate has good flexibility, its viscosity is large, the small amount added cannot remarkably improve the toughness. Hybrid solid materials also have a slightly higher glass transition temperature than cationic solid materials.

TABLE 3.17 Mechanical properties of hybrid solid materials after curing.

Tensile strength (MPa)	Tensile modulus (MPa)	Elongation at break (%)	Bending strength (MPa)	Flexural modulus (MPa)	Impact strength (J/m^2)	Hardness (HRM)
64.5	1121	9.7	67.8	1487	46.7	85

3.2.6 Study on preparation and properties of solid materials in free-radical system

Free-radical systems are still relatively rare in the field of rapid prototyping. The main disadvantages are that the curing shrinkage rate is large and oxygen resistance is likely to occur. However, the performance of the free-radical system is adjustable, and it is widely used in UV inkjet printing inks. The urethane acrylate has the advantages of low temperature resistance, good flexibility, and high bonding strength. In this section, the urethane acrylate is used as the main oligomer to prepare the AM solid material of the free-radical system, and the acrylate system where an oxygen resistance occurs easily is improved by designing the initiation system.

3.2.6.1 Principles for selecting raw materials

1. Selection of monomers

    There are many types of acrylate monomers, which are used for AM. Monomers with lower viscosity and faster curing rate should be used. The viscosity of 2-(2-ethoxyethoxy) ethyl acrylate (SR256) is very low, which is 6 mPa s at 25°C; the curing rate of 1,6-hexanediol diacrylate (SR238) is fast but viscosity is low, which is 9 mPa s at 25°C; tripropylene glycol diacrylate (SR306) and triethyl oxide trimethylolpropane triacrylate (SR454) have viscosities of 15 and 60 mPa s, respectively, and their curing rates are fast. Generally, monofunctional, difunctional, and trifunctional monomers are used in combination, which has a lower viscosity, a higher curing rate, and a lower curing shrinkage.

2. Selection of the initiation system

    The free-radical polymerization has the biggest disadvantage of easy oxygen resistance. Therefore, the initiation system should be specially designed, and the cracking type and hydrogen-extraction type initiators should be used together, which has good oxidation resistance effect. For example, when Irgacure 184 and benzophenone series are used in combination, the excited triplet of benzophenone series can effectively promote the decomposition of hydroperoxide, while the free-radical generated by photolysis of Irgacure 184 reacts with oxygen to consume oxygen, making the quenching effect of oxygen to the excited triplet state of benzophenone series inhibited, and the two components have synergistic effects. Reactive amine coinitiators can also inhibit oxygen resistance and can participate in double bond reactions without surface migration. Acrylates are used at slightly elevated temperatures and are prone to thermal polymerization. In order to ensure the stability of the resin, it is generally necessary to add some polymerization inhibitor.

3.2.6.2 Preparation of physical materials

Dissolve the photoinitiator Irgacure 184 in acrylate monomer, and pour the surfactant and defoamer into the three-necked flask with the aforementioned components, shield and heat the solution to 45°C. Then, stir it for 4 hours, and then filter it for future use.

3.2.6.3 Properties of free-radical solid materials

1. Main physical properties and injection parameters of free-radical solid materials

    Table 3.18 lists the measured value or theoretical calculation value of the viscosity, surface tension, density before and after curing, Reynolds number and Weber number of the radical type solid materials.

    Based on the data in Table 3.18, it can be calculated that $R/W = 4.74$, and $1 < R/W < 10$, which meets the requirement of stable injection index of solid materials. The working temperature of the nozzle is set to 50°C, and the XJ500 nozzle always works at high temperature. Its stability is very poor when working, and it often causes oblique spray and printing failure. Therefore, the section discusses the use of solid material prepared with low-viscosity urethane acrylate at a slightly lower temperature to reduce the instability of the working nozzle.

2. Sputtering coefficient, contact angle, spreading factor and spreading time of free-radical solid materials

    Table 3.19 lists the measured value or theoretical calculation value of the sputtering coefficient, contact angle, spreading factor, and spreading time of the radical type solid materials.

    From the data in Table 3.19, it is known that the sputtering coefficient is smaller than the sputtering threshold K_C (57.7), and the sputtering

TABLE 3.18 Main physical properties and injection parameters of free-radical solid materials.

Viscosity (mPa s)		Density (g/cm³)		Surface tension (50°C mN/m)	Reynolds number (Re)	Weber number (We)
25°C	50°C	Before curing	After curing			
51.2	8.5	1.04	1.11	31.2	36.7	60.0

TABLE 3.19 Sputtering coefficient, contact angle, spreading factor and spreading time of free-radical solid materials.

Contact angle		Spreading factor		Spreading time (s)	Sputtering coefficient (K)
With itself	With support material	With itself	With support material		
41 degrees	49 degrees	1.34	1.33	7.1×10^{-5}	19.1

TABLE 3.20 Printing stability test results of free-radical solid materials.

Printing duration printing temperature (°C)	0 h	6 h	12 h	18 h
50	0	1	2	2
55	0	1	2	2

phenomenon does not occur when the radical type solid material is sprayed onto the working surface. In fact, no sputtering phenomenon is found in the actual machine test. The free-radical solid material has large spreading diameter during the spraying process, and the viscosity at room temperature is too low. Therefore, in the printing process, the vertical surface of the printing formed part becomes slope surface, and it is considered to add thixotropic agents to solve this problem, but thixotropic agents can easily cause plugging.

3. Printing stability experiment

    Table 3.20 shows the number of nozzle plugging holes when printing free-radical solid materials.

    It can be seen from the data in Table 3.20 that the free-radical solid material has good jet stability and few plugging happens.

4. Characteristic parameters

    According to the test method of the characteristic parameters in Section 3.1.4, the E_C of the free-radical solid material is $E_C = 31.5$ mJ/cm², and $D_P = 0.102$ mm. The critical exposure amount is relatively high, mainly because the free-radical solid material is prone to oxygen inhibition. It has been discussed earlier that the curing rate of free-radical solid materials is relatively slow, and oxygen inhibition is one of the reasons. In addition to the special design of the free-radical initiating system, the 800 W metal halogen lamp instead of the original 300 W metal halogen lamp is used in the experiment, and the problem is completely solved by improving the light intensity. The experiment proves that the instantaneous dryness can be achieved when the lamp illumination speed is 0.8 m/s.

TABLE 3.22 Mechanical properties of free-radical solid materials after curing.

Tensile strength (MPa)	Tensile modulus (MPa)	Elongation at break (%)	Bending strength (MPa)	Flexural modulus (MPa)	Impact strength (J/m^2)	Hardness (HRM)
60.5	1091	10.8	74.9	1391	50.6	78

TABLE 3.21 Shrinkage rate, warpage factor and glass transition temperature of free-radical solid materials.

Volume shrinkage rate	Linear shrinkage rate	Warpage factor	Glass transition temperature
6.3%	0.175%	1.70%	55°C

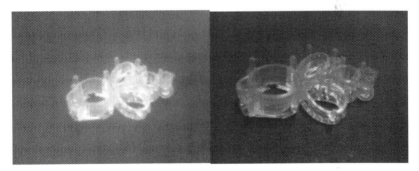

FIGURE 3.85 Jewelry inserts.

5. Shrinkage rate and warpage factor

    The shrinkage rate and warpage factors of free-radical solid materials are listed in Table 3.21. Compared with the other two solid materials, the shrinkage rate of free-radical solid materials is larger, which is also one of the main disadvantages of free-radical solid materials.

6. Mechanical properties and glass transition temperature

    Table 3.22 shows the mechanical properties of the free-radical solid material after curing, and Table 3.21 lists the glass transition temperature of the free-radical solid material after curing. It can be seen from Tables 3.21 and 3.22 that the mechanical properties and thermal properties of the free-radical solid materials satisfy the general mechanical and thermal properties of the AM parts.

7. Pictures of formed parts of free-radical solid materials

    Fig. 3.85 shows the pictures of the formed parts of free-radical solid materials.

3.3 Oligomers in stereolithography apparatus solid materials

The photopolymer of AM requires a low viscosity, a low volatility, good rheological property and photoreaction characteristics, and requires a high ejection performance, no sedimentation, flocculation, and plugging. The photopolymer after curing is required to have good accuracy and mechanical properties. It is therefore particularly important to synthesize solid material oligomers with specified performance requirements. This section mainly discusses the methods and conditions of oligomer synthesis, and aligns the partial performance characterization of the polymer as a reference for the subsequent study of photopolymer of solid materials.

In SLA, oligomers of free-radical systems or hybrid system photopolymers often use bisphenol A epoxy acrylate, phenolic epoxy acrylate or toughened modified epoxy acrylate, which have fast UV curing speed, and the cured molded part has high mechanical strength and high temperature resistance. If used in 3D printers, the high viscosity and high surface tension of these two types of epoxy acrylates are obvious disadvantages. When formulated, a large amount of reactive diluent is required, which affects the SLA speed and mechanical properties and accuracy after curing. Therefore, the synthesis of low-viscosity and high-performance oligomers is an idea choice for the preparation of AM solid materials [6].

3.3.1 Polypropylene glycol diglycidylether diacrylate

The oligomer of polypropylene glycol diglycidylether diacrylate contains ether bond with excellent flexibility, which has a low viscosity, a high SLA speed and excellent mechanical properties, and can be combined with UV cationic raw material (such as alicyclic epoxy, trimethylene oxide) to prepare hybrid solid materials, and can obtain 3D printed solid materials with excellent flexibility and mechanical properties. In this section, polypropylene glycol diglycidylether is used as raw material to react with acrylic acid to prepare polypropylene glycol diglycidylether diacrylate, and the synthesis conditions and product properties are discussed.

3.3.1.1 Synthesis of oligomers

In a four-necked flask equipped with stirrer, dropping funnel, reflux condensing tube and thermometer, add 0.5 mol of polypropylene glycol diglycidylether, hydroquinone or p-methoxy with a total mass fraction of 0.2%−0.4%, and N,N-dimethylbenzylamine with a total mass fraction of 0.40%−1.00%. While stirring, bath the mixture in the oil and heat it. Then, add the acrylic acid dropwise at 80°C−120°C, after that and keep it reacting at the reaction temperature for a while until the acid value is less than 6 mgKOH/g.

3.3.1.2 Reaction mechanism

First, the catalyst N,N-dimethylbenzylamine attacks the epoxy group of the polypropylene glycol diglycidylether to positively charge the epoxy group,

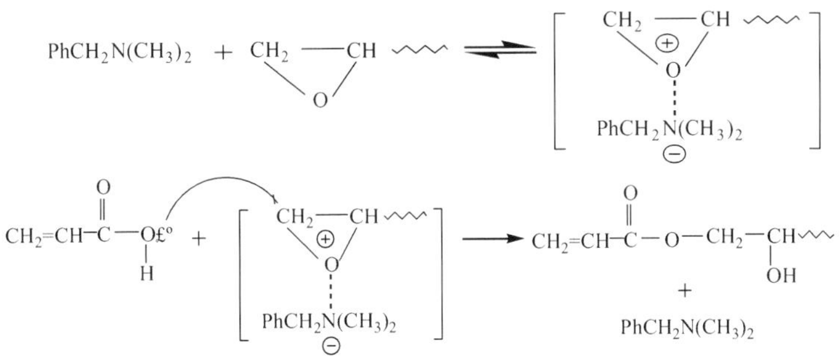

FIGURE 3.86 Synthesis reaction mechanism of polypropylene glycol diglycidyl ether diacrylate.

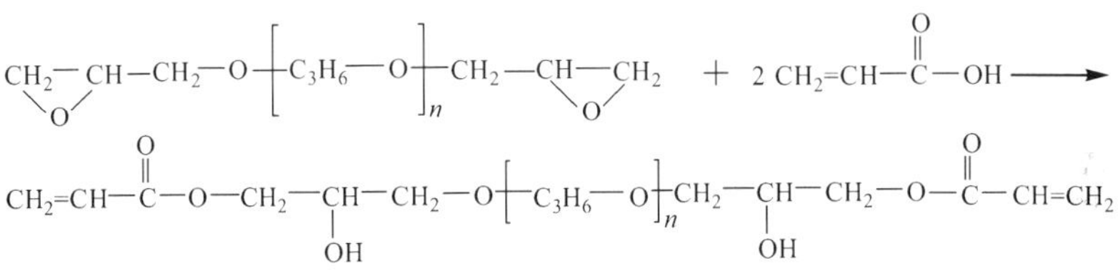

FIGURE 3.87 Chemical reaction formula of polypropylene glycol diglycidylether diacrylate.

so that the nucleophilic acrylic acid easily reacts with it, resulting in ring opening of the epoxy group and generating polypropylene glycol diglycidylether diacrylate. The reaction mechanism and reaction formula are shown in Figs. 3.86 and 3.87.

3.3.1.3 Influencing factors

1. Effects of catalyst mass fraction

While feeding in a molar ratio of polypropylene glycol diglycidylether: acrylic acid = 1:2, add p-methoxyphenol with a mass fraction of 0.30% as polymerization inhibitor, and keep the system at the reaction temperature of 110°C. Under the reaction conditions, consider the effects of adding N,N-dimethylbenzylamine with different mass fractions on the conversion rate of the product, polypropylene glycol diglycidylether diacrylate during the reaction, as shown in Fig. 3.88.

It can be seen from Fig. 3.88 that when the mass fraction of the catalyst N,N-dimethylbenzylamine is 0.4%–0.8%, the conversion rate increases as the amount of the catalyst mass fraction increases, especially in the initial stage; When the mass fraction of N,N-dimethylbenzylamine is 0.4%, and the reaction time is 6 hours, the conversion rate of polypropylene glycol diglycidylether diacrylate is only 68%; when the mass fraction of N,N-dimethylbenzylamine is 0.6%, and the reaction time is 6 hours, the conversion rate is increased to 85%; when the mass fraction of N,N-dimethylbenzylamine is

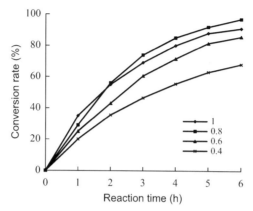

FIGURE 3.88 Effects of catalyst mass fraction on reaction conversion rate.

0.8%, and the reaction time is 6 hours, the conversion rate is 97%; when the mass fraction of N,N-dimethylbenzylamine is 1.0%, and the reaction time is 6 hours, the conversion rate is reduced to 91%, indicating that the mass fraction of good catalyst is 0.8%.

2. Selection of inhibitor and mass fraction

   When no polymerization inhibitor is added to the reaction system, the viscosity of the system increases rapidly with the extension of the reaction time, and the temperature inside the reaction bottle rises sharply, which causes gelation. When hydroquinone or p-methoxyphenol is added, it can function to prevent the double bond of the acrylate, and gelation is less likely to occur. In the experiment, the polymerization inhibition effect of hydroquinone and p-methoxyphenol is very good. When the mass fraction of hydroquinone is 0.20%, the inhibition effect is good, but the product has very deep color, because quinoid structure is generated after oxidation of hydroquinone. When the mass fraction of p-methoxyphenol is 0.20%–0.40%, it has a better inhibition effect and the product has a lighter color.

3. Effects of reaction temperature

   Fig. 3.89 shows the conversion rate of polypropylene glycol diglycidylether diacrylate at different reaction temperatures when the mass fraction of the catalyst N,N-dimethylbenzylamine is 0.8% and the mass fraction of polymerization inhibitor p-methoxyphenol content is 0.30%. As can be seen from Fig. 3.37, the temperature has great influence on the reaction rate. When the reaction temperature is 90°C, the reaction time is 6 hours, and the conversion rate is only 81%; when the reaction temperature is 100°C, the reaction time is 6 hours, and the conversion rate is increased to 89%; when the reaction temperature is 110°C, the reaction time is 6 hours, and the conversion rate is increased to 97%, indicating that increasing the temperature can speed up the reaction speed. When the reaction temperature is 120°C, the gelation phenomenon occurs,

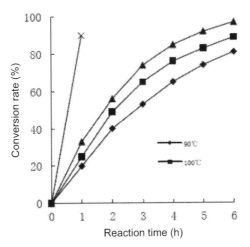

FIGURE 3.89 Effects of reaction temperature on reaction conversion rate.

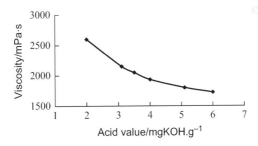

FIGURE 3.90 Relationship between acid value and viscosity of the reaction product.

which is mainly caused by self-polymerization reaction between $C=C$ double bonds.

4. Effects of residual acid value on viscosity

The polypropylene glycol diglycidylether diacrylate is synthesized when the reaction temperature is 110°C, the mass fraction of N,N-dimethylbenzylamine is 0.8%, and the mass fraction of p-methoxyphenol is 0.30%. At 25°C, the relationship between viscosity and acid value is shown in Fig. 3.90.

As can be seen from Fig. 3.90, when the acid value is 5.0 mgKOH/g at 25°C, the product viscosity is 1810 mPa s, it indicates that the synthesized polypropylene glycol diglycidylether diacrylate has a moderate viscosity as a raw material for formulating solid material.

3.3.1.4 Characterization of the infrared spectrum of the product

In the oligomer of polypropylene glycol diglycidylether diacrylate with a residual acid value of 5.0 mg KOH/g, the photoinitiator 1-hydroxycyclohexyl

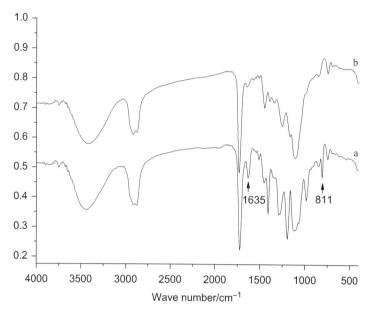

FIGURE 3.91 Infrared spectrum of polypropylene glycol diglycidylether diacrylate before and after SLA.

benzophenone with a mass fraction of 2% is added to test the infrared spectrum before and after curing, as shown in Fig. 3.91, where a is the infrared spectrum before curing and b is the infrared spectrum after curing. In Fig. 3.91a, 811 cm^{-1} is the bending vibration absorption peak of =C–H, 1635 cm^{-1} is the stretching vibration absorption peak of C=C, and 1725 cm^{-1} is the characteristic absorption peak of acrylate –C=O. 1105 cm^{-1} is the stretching vibration absorption peak of C–O, and 3469 cm^{-1} is the characteristic absorption peak of hydroxyl group. In the infrared spectrum of the oligomer of polypropylene glycol diglycidylether diacrylate after curing in Fig. 3.91b, the three distinct characteristic absorption peaks of 1725, 1105, and 3469 cm^{-1} are unchanged, while the two characteristic absorption peaks of the carbon–carbon double bond are substantially disappeared at 1635 and 811 cm^{-1}. The aforementioned infrared spectrum analysis showed that the polymerization reaction is complete under the initiation of 1-hydroxycyclohexyl benzophenone of polypropylene glycol diglycidylether diacrylate, and the curing effect is good.

3.3.1.5 Other properties of the product

1. Surface tension

For inkjet printing, the surface tension of the photopolymer oligomer is an important parameter. The oligomer of polypropylene glycol diglycidylether diacrylate weighing at 25 g and having acid value of 5.0 mgKOH/g is placed in the glass, and its surface tension change curve in the range of 20°C–60°C is measured, as shown in Fig. 3.92.

As can be seen from Fig. 3.92, the surface tension of polypropylene glycol diglycidylether diacrylate is 33.7 mN/m at 25°C, and it is 31.6–31.1 mN/m when the working temperature of the nozzle is 50°C–55°C. As an oligomer, its surface tension range is relatively close to that of the piezoelectric nozzle

2. Density

The density of the oligomer at 25°C is measured by the pycnometer method, and the measured value is 1.16 g/cm^3.

3. Glass transition temperature

The glass transition temperature (T_g) is the transformation temperature of the polymer segment from freezing to movement. Its height is related to the flexibility of the molecular chain. The better the molecular chain flexibility, the lower the glass transition temperature, the greater the rigidity of the molecular chain, and the higher the glass transition temperature. The cured product of 10–15 mg is taken to measure its glass transition temperature by differential scanning calorimeter at a linear heating rate of 5°C/min and a nitrogen flow rate of 100 mL/min. The measurement results are shown in Fig. 3.93. As can be seen from the figure, the glass transition temperature of polypropylene glycol diglycidylether diacrylate is 86°C, indicating that the polypropylene glycol diglycidylether diacrylate has a high flexibility.

3.3.2 Low-viscosity urethane acrylate

Urethane acrylate is a kind of important SLA oligomer. The synthesis process of urethane acrylate is simple and flexible. It can be structurally designed and synthesized to obtain oligomers of various structures, and its molecular structure is very

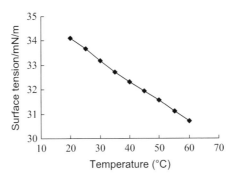

FIGURE 3.92 Relationship between surface tension and temperature of polypropylene glycol diglycidylether diacrylate.

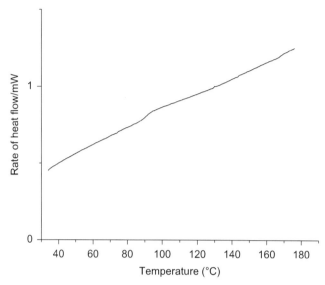

FIGURE 3.93 DSC curve of cured product of polypropylene glycol diglycidylether diacrylate.

adjustable to a large extent from the aspects of relative molecular mass, reaction functionality, flexibility and modulus. Therefore, urethane acrylate is the oligomer with the most product marks currently.

The urethane acrylate molecular chain contains functional groups such as ether bond (−O−), ester bond (−COO−), and urea bond (−NH−) in addition to the carbamate (−NH−COO−) segment. Wherein, the long-chain diol unit provides softness to form a soft segment; and the aryl group and the urethane to form a hard segment. The thermodynamic incompatibility of the two segments produces microphase separation, the soft segment region plays the role of toughening, and the hard segment region plays the role of cross-linking and reinforcing to make the urethane acrylate have high modulus and elasticity, which is an important reason which explains why urethane acrylate has excellent mechanical properties.

The viscosity of urethane acrylic resin is generally large, which is mainly due to the formation of by-products in the reaction process that results in a substantial increase in viscosity, while urethane acrylic resin of lower viscosity are often required in AM applications. Therefore, it is very important to strictly control the reaction conditions, carry out structural design to the urethane acrylate, and synthesize low-viscosity urethane acrylate.

Generally, the hybrid SLA system cannot use urethane acrylate because the curing rate is very slow when it forms hybrid SLA system with cationic monomer. The main reason is that the singly paired electrons on the N in the urethane acrylate structure have neutralizing effects on the Bronsted acid or Lewis acid produced by the photolysis of the cation, consuming the active species which initiate the photopolymerization of the cation, thereby affecting the cationic photopolymerization and reducing the SLA rate of the hybrid SLA system. This section synthesizes low-viscosity urethane acrylate and uses it as an oligomer to prepare free-radical solid material to meet the low viscosity and high performance required for AM materials.

3.3.2.1 Synthesis experiment

1. Purification and dewatering of raw materials

   A small amount of water can react with isocyanate to form amine, and then continue to react with isocyanate to form urerea structure. Excessive urea group will cause excessive hydrogen bonding of the cured resin, thus increasing the brittleness of the cured product, and N—H group on the urea group can undergo branching reaction with the isocyanate, resulting in an increase in the viscosity of the polyurethane oligomer, and even the scrapping of gel. Polyethylene glycol (PEG) is particularly easy to absorb moisture and contains more water. Therefore, it is necessary to purify PEG and other raw materials to remove moisture. The chemical formula for the reaction of water with isocyanate is shown in Fig. 3.94.

   Purify and dewater the PEG, toluene diisocyanate (TDI) and hydroxyethyl acrylate (HEA) by using reduced pressure distillation at a suitable temperature and vacuum; dewater the dibutyltin dilaurate and p-methoxyphenol by using the molecular sieve; and reflux the di-n-butylamine with flaky potassium hydroxide and then resteamed it.

2. Synthesis experiment

   In a four-necked flask equipped with heating device, stirring device, condenser tube (with drying tube), and thermometer, first add 1 mol TDI, and an appropriate amount of catalyst, polymerization inhibitor, and protect with nitrogen gas, and add 1 mol of HEA dropwise while stirring, control the reaction temperature at 40°C, and the reaction is about 2—3 hours. At the same time, detect the content of

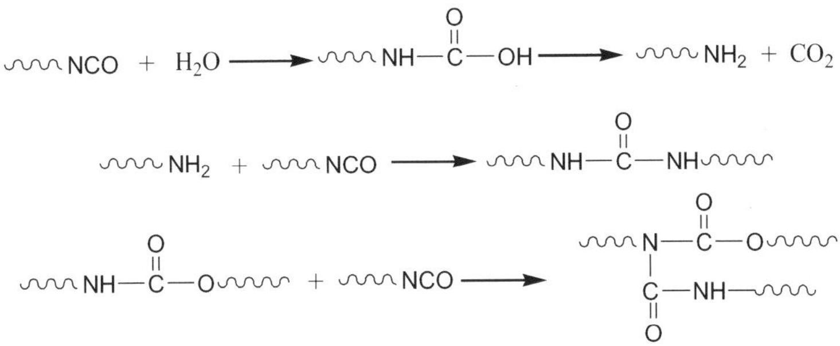

FIGURE 3.94 Chemical formula for the reaction of water with isocyanate.

the isocyanate in the reaction flask (mass fraction). When the isocyanate is reacted for 50%, it is considered that the reaction in the first step is complete. The temperature is raised to 55–75°C, while stirring, gradually add 0.5 mol PEG200, and fully react until the isocyanate is not detected (when the residual content of isocyanate is too high, an appropriate amount of ethanol can be added to remove it). Cool the mixture to room temperature and discharge and finally purify the synthesized oligomer.

3. Determination of free-NCO content (mass fraction)

Determine the content of isocyanate in the system by the di-n-butylamine method. The principle of di-n-butylamine countertitration is as follows: the isocyanate in the oligomer or intermediate product reacts with excessive amount of di-n-butylamine in solvent. After the reaction is completed, the excessive amount of di-n-butylamine is titrated with hydrochloric acid standard solution, and then the amount of isocyanate is determined. The methods are as mentioned subsequently.

Accurately weigh a small amount of the sample to be tested in a triangular flask, add 15 mL N,N-dimethylacetamide, absorb 10 mL di-n-butylamine anhydrous toluene solution with pipette into a triangular flask, and shake to make the sample completely dissolved. The operation speed should be as fast as possible and is required to be completed within 5 minutes. After shaking for 30 minutes, add 30 mL isopropanol, drip three drops of the indicator bromophenol blue, and titrate the solution with hydrochloric acid until the solution turns from blue to yellow, and a blank experiment is simultaneously carried out according to the method described earlier. The calculation equation is:

$$X_1 = \frac{(V_1 - V_2) \times C_1 \times 4.2}{M_1} \qquad (3.48)$$

where X_1, the mass percentage of residual isocyanate in the reaction flask, %; V_1, the volume of hydrochloric acid standard titration solution consumed by the blank experiment, mL; V_2, the volume of hydrochloric acid standard titration solution consumed by the sample, mL; C_1, the actual concentration of hydrochloric acid standard titration solution, mol/L; and M_1, the mass of the sample to be measured, g.

4. Determination of iodine value

Morpholine addition method is used to determine the content of double bonds in the product.

Accurately weigh a small amount of the sample to be placed in a triangular flask, add 10 mL morpholine reagent and 7 mL of 50% acetic acid aqueous solution, shake it, and then place it at room temperature for 30 minutes, add 50 mL ethylene glycol monomethyl ether and 20 mL acetic anhydride, respectively; keep stirring constantly, after cooling to room temperature, add indicator methyl group orange. The mixture is titrated to green color with 0.5 mol/L HCl methanol standard solution, and a blank experiment is simultaneously performed as described earlier. The calculation equation is:

$$X_2 = \frac{(V_3 - V_4) \times C_2 \times 1.1269 \times 200}{M_2} \quad (3.49)$$

where X_2, iodine value; V_3, the milliliters of 0.5 mol/L HCl methanol standard solution consumed by the sample; V_4, the number of milliliters of 0.5 mol/L HCl methanol standard solution consumed by the blank test; C_2, molar concentration of HCl methanol standard solution, mol/L; and M_2, the mass of the sample to be measured, g.

5. Determination of viscosity

    Take a certain amount of the final reaction product and measure it at 25°C by using an SNB-2 rotary viscometer.

6. Determination of SLA speed

    The gel content of the oligomer at different SLA times is determined to characterize the SLA rate of the oligomer.

    Gel content determination method: first weigh the mass (m_1) of the glass slide, apply the oligomer resin on the glass slide, weigh the total mass (m_2) of the resin and the glass slide, cure the resin with UV light, and place the cured product and glass slide in the Soxhlet extractor to extract with isopropyl alcohol for 4 hours, then place it in an oven to bake at 50°C for 10 hours, and finally weigh the total mass (m_3) of it. Then, the gel rate is:

$$\text{Gel rate \%} = \frac{m_3 - m_1}{m_2 - m_1} \times 100\% \quad (3.50)$$

3.3.2.2 Principle of preparation of urethane acrylate

The low-viscosity urethane acrylate synthesized in this section is a bifunctional acrylate synthesized by a two-step process. The first step in the synthesis reaction is that TDI is first reacted with HEA and reacted at the position 4 of 2,4-TDI. Generally, the position 2 does not participate in the reaction because the reaction activity of two isocyanates of 2,4-TDI is very different, when reacting with alcoholic hydroxyl groups, the relative rate constant of the reaction at the position 4 is 400, and the relative rate constant of the reaction at the position 2 is 33, an order of magnitude difference. In the second step of the synthesis reaction, the isocyanate at the position 2 of 2,4-TDI starts to react with PEG under the action of the catalyst dibutyltin dilaurate and the temperature is appropriately raised. The reaction formula is shown in Fig. 3.95.

3.3.2.3 Synthesis conditions of urethane acrylate

1. Reaction time of the second step at different temperatures

    In the preparation process of the low-viscosity urethane acrylate oligomer, the position 4 on 2,4-TDI has high reaction activity in the first

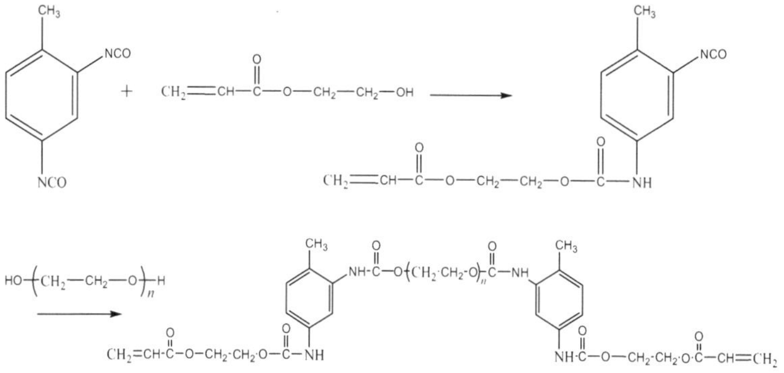

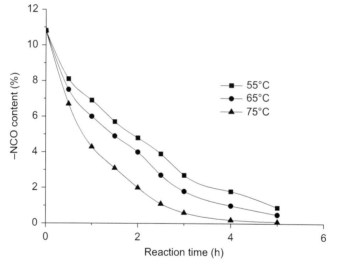

FIGURE 3.95 Reaction formula of synthesis of low-viscosity urethane acrylate with two-step method.

FIGURE 3.96 Relationship between reaction time and $-NCO\%$ at different temperatures.

step, and the molecular weight of HEA is small, the movement rate of HEA is fast, so the first step reaction is generally completed within 2–3 hours. In the second step, the reaction with PEG200 has relatively slow reaction rate, so the reaction temperature is raised. Fig. 3.96 shows the curve of the relationship between reaction time and $-NCO\%$ at different temperatures.

It can be seen from Fig. 3.96 that the higher the reaction temperature, the faster the isocyanate concentration decreases, indicating that as the reaction temperature increases, the reaction rate increases and the corresponding reaction time decreases. In general, when the residual content of the isocyanate is less than 1%, the reaction is considered to be nearly complete. As can be seen from Fig. 3.96, it takes about 5 hours to reduce to 1% under the reaction temperature of 55°C, about 5 hours at 65°C, and about 3 hours at 75°C.

2. Effects of reaction temperature on the degree of double bond destruction in the second step

The acrylate group on the oligomer is a functional group participating in photoinitiated radical polymerization. In the synthesis process of the low-viscosity urethane acrylate oligomer, the high temperature reaction easily causes the double bond thermal polymerization of the acrylate. On one hand, the viscosity of the oligomer is increased, and on the other hand, the photoreactivity of the oligomer is lowered. Therefore, the loss of the acryloyloxy double bond should be minimized in the synthesis of the low-viscosity urethane acrylate oligomer, thereby improving the SLA activity of the oligomer and lowering the viscosity of the oligomer. When the isocyanate reaction is complete, the degree of decline in the iodine value of the oligomer, that is, the degree of destruction of the acryloyloxy double bond at different reaction temperatures is shown in Fig. 3.97.

It can be seen from Fig. 3.97 that the higher the reaction temperature of the second step, the more serious the iodine value of the oligomer decreases, indicating that the acryloyloxy group of the urethane acrylate is destroyed to large extent, which can also be seen from the viscosity of oligomer product. One of the main reasons for the increase in viscosity of

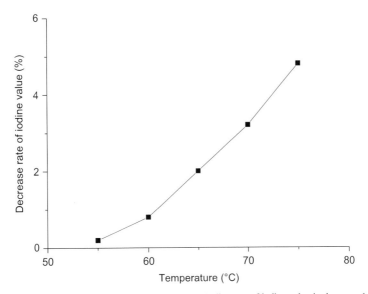

FIGURE 3.97 Effects of reaction temperature on the decline rate of iodine value in the second step.

the product is that the double bond is destroyed and the side reaction is increased. In the experiment, the second step reaction temperature is controlled at 55°C–60°C, and the iodine value decrease rate is less than 1%.
3. Mass fraction of the catalyst

In the first and second steps of the low-viscosity urethane acrylate synthesis process, in order to accelerate the reaction rate of HEA with isocyanate and PEG 200 with isocyanate to a reasonable level, and to ensure double bond is not thermally polymerized as much as possible, and a more selective catalyst is required. The addition of the catalyst dibutyltin dilaurate facilitates the selective reaction, such that the hydroxyl group of HEA in the first step is added at the position 4 of the TDI, and the hydroxyl group of PEG in the second step is added at the position 2 of the TDI. At the same time, the catalyst can effectively prevent the crosslinking, so that the whole reaction process proceeds according to the set route, and the specified structure product is obtained.

In the synthesis, when the first step of the reaction is carried out for 2 hours, the effect of the mass fraction of the catalyst on the isocyanate concentration in the first step is shown in Fig. 3.98. The theoretical isocyanate concentration after the first step of the reaction is 14.5%. As can be seen from the figure, when the mass fraction of catalyst is increased, it is getting closer to the theoretical isocyanate concentration.

When molar number TDI: PEG is 2:1, and the reaction proceeds for 5 hours when the second step reaction temperature is fixed at 60°C, the effects of the mass fraction of the catalyst on the isocyanate concentration in the second step is shown in Fig. 3.99. As shown, as the amount of

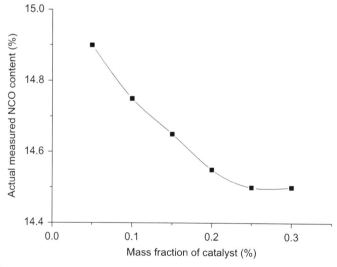

FIGURE 3.98 Effects of the catalyst on the first step reaction.

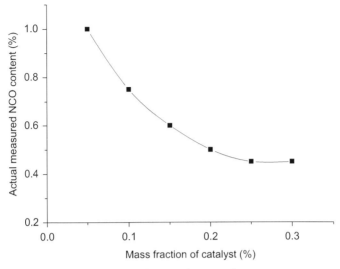

FIGURE 3.99 Effects of the catalyst on the second step reaction.

mass fraction of the catalyst increases, the isocyanate concentration becomes lower and lower. Figs. 3.98 and 3.99 also show that the catalyst mass fraction has an optimum value. When the mass fraction exceeds 0.25%, the final conversion of isocyanate tends to be stable after the specified reaction time.

3.3.2.4 Analysis of the product of carbamate acrylate

1. Characterization of infrared spectrum

    It can be seen from the infrared spectrum of the urethane acrylate in Fig. 3.100 that the absorption peak at $2250\ cm^{-1}$ disappears, indicating that the isocyanate reaction in the system is complete. The hydroxyl stretching vibration peak at $3571\ cm^{-1}$ disappears, indicating that the hydroxyl groups on HEA and PEG have reacted. The stretching vibration peak of saturated $-C-H$ at $2944\ cm^{-1}$ and the $-C-O-C-$ stretching vibration peak at $1197\ cm^{-1}$ indicate that the PEG segment is added to the product. The N−H stretching vibration absorption peak is at $3341\ cm^{-1}$, and the C-N characteristic absorption peak is at $1535\ cm^{-1}$, indicating that a carbamate group is formed during the reaction. The stretching vibration peak of $C=O$ is at $1730\ cm^{-1}$, the stretching vibration peak of $C=C$ is at $1601\ cm^{-1}$, and the out-of-plane vibration peak of $=C-H$ is at 811 and $768\ cm^{-1}$, indicating the double bonds of HEA have been inserted into the oligomer.

2. Viscosity

    The 3D printer uses a piezoelectric nozzle to spray solid material. The viscosity of the solid material at the working temperature is required

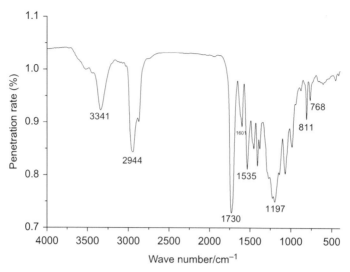

FIGURE 3.100 Infrared spectrum of carbamate acrylate.

to be between 8 and 15 mPa s. The control of the viscosity of the solid material requires dilution of the monomer, but the low viscosity of the oligomer itself is more important to obtain solid materials of a low viscosity and high performance.

In this experiment, the effects of the mass fraction of the catalyst, the reaction temperature and the ratio of the reactants on the viscosity of the product are investigated. Fig. 3.49 shows the effect of the mass fraction of catalyst on the viscosity of the product (the reaction temperature in the first step is 40°C, and the reaction temperature in the second step is 60°C). It can be seen that the viscosity of the product increases as the mass fraction of catalyst increases. Because when the amount of catalyst mass fraction is increased, the reaction rate of isocyanate and hydroxyl group is increased, and the rate of heat release is faster. When the heat is accumulated to a certain extent, various side reactions are correspondingly increased, which inevitably increases the viscosity of the product. However, the mass fraction of the catalyst should not be too low, otherwise the conversion of isocyanate will be low, and the residual isocyanate will affect the stability of the oligomer. Therefore, the optimum catalyst mass fraction in this experiment is 0.20%.

In the first step, the reaction temperature is low, and HEA reacts easily with isocyanate, which has little effect on the viscosity of the product. Fig. 3.101 shows the effects of the reaction temperature on the viscosity of the product in the second step when the catalyst mass fraction is 0.20% of the total mass fraction of the reactants, When the reaction temperature is 55°C, the viscosity is 150 mPa s; when the reaction temperature is 60°C,

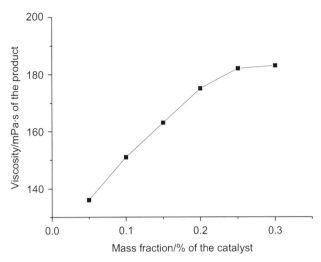

FIGURE 3.101 Effects of catalyst mass fraction on product viscosity.

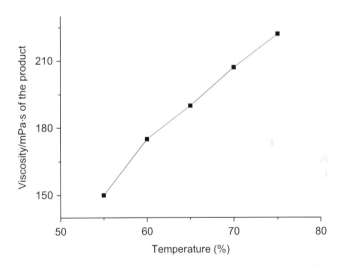

FIGURE 3.102 Effects of reaction temperature on product viscosity in the second step.

the viscosity is 175 mPa s; the higher the temperature, the larger the viscosity. Here is the same reason. When the temperature rises, the isocyanate reacts with the hydroxyl group faster, and the reaction heat release rate is faster, when the heat accumulates to a certain extent, various side reactions increase correspondingly, which inevitably increases the viscosity of the product.

Fig. 3.102 shows that when the mass fraction of the catalyst is 0.20% of the total mass fraction of the reactants, and the reaction temperature in the second step is 60°C, the mass fraction of TDI is constant. When the sum of the mole number of hydroxyl groups of PEG and HEA is equal to the mole number of isocyanate, and the viscosity of the product varies with the molar ratio of PEG to HEA.

It can be seen from Fig. 3.103 that when the proportion of PEG is small, the viscosity of the product is large, because the molecular segment of PEG is flexible, and the larger the concentration of PEG, the more easily the oligomer molecules are spheroidal, so the viscosity is lower. When PEG:HEA > 0.5, due to the slow addition reaction rate of PEG, it takes a long time for PEG to complete the reaction, resulting in excessive cross-linking reaction of acryloyloxy double bonds, and the viscosity of the product increases again. When PEG:HEA = 0.5, the viscosity is the lowest.

3. Research on SLA speed

The synthesis condition of the urethane acrylate is that the reactant n (TDI): n (HEA): n (PEG) is 2: 2: 1, the reaction temperature in the first step is 40°C, the reaction temperature in the second step is 60°C, and the mass fraction of catalyst is 0.2%, so that the product has a viscosity of 175 mPa s. A certain amount of the reaction product is taken, and a different ratio of the photoinitiator 1-hydroxycyclohexyl benzophenone series is added and stirred uniformly to obtain a SLA material. The SLA material is uniformly coated on a glass slide, and the gel content at a temperature of 25°C irradiated with 1000 W high-pressure mercury lamp (vertical distance of the glass slide and the high-pressure mercury lamp is 15 cm) is measured at different times, and the results are shown in Fig. 3.52.

It can be seen from Fig. 3.104 that as the mass fraction of 1-hydroxycyclohexyl benzophenone series increases, the SLA speed of the

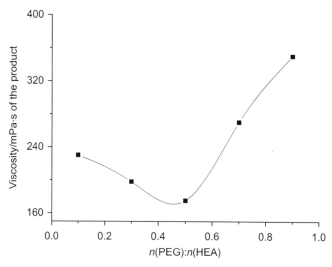

FIGURE 3.103 Product viscosity at different PEG to HEA molar ratios.

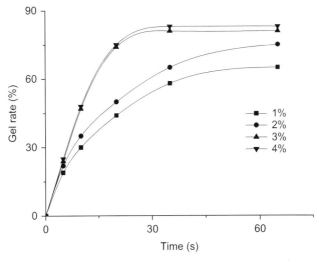

FIGURE 3.104 Effects of the mass fraction of photoinitiator Irgacure 184 on the SLA speed of the product.

oligomer increases. When the mass fraction of the initiator is more than 3%, increasing the mass fraction of the photoinitiator has little effect on the SLA speed of the aligned polymer. In the later stage of the reaction of the oligomer, as the SLA crosslinking reaction proceeds, the viscosity of the oligomer is already large, and the reaction progress of the oligomer is hindered. A gel rate close to 90% indicates that the oligomer is cured firmly.

4. Density and surface tension

    The synthesis condition of the urethane acrylate is that the reactant n (TDI):n(HEA):n (PEG) is 2:2:1, the reaction temperature in the first step is 40°C, the reaction temperature in the second step is 60°C, and the mass fraction of catalyst is 0.2%, so that the product has a viscosity of 175 mPa s. The density at 25°C is measured by the pycnometer method, and the measured value is 1.09 g/cm^3. The curve of the surface tension of the oligomer changing with temperature is shown in Fig. 3.105. It can be seen that the surface tension is 40.95 mN/m at 40°C, and is 39.4−39.1 mN/m at 50°C−55°C.

3.3.3 Oligomer of stereolithography apparatus support material

The support material for SLA AM can be rapidly cured under the irradiation of UV light, and the SLA support material is required to be partially dissolved in water or easily peeled off. The strength of the support material is not required to be high, but it should not be too low, otherwise the solid material cannot be supported and the solid material will sink. There are two main types of water-soluble SLA oligomers, one is polyethylene glycol acrylate, and the

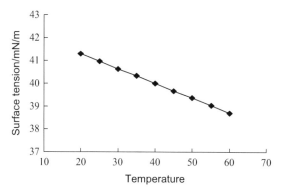

FIGURE 3.105 Surface tension of low-viscosity urethane acrylate.

other is water-based urethane acrylate. Both of the two raw materials have good water solubility, the former one has no high strength, and is a nonionic water-soluble oligomer, and the latter one has adjustable curing strength, and is an anionic water-emulsion oligomer. This chapter first discusses the synthesis of two oligomers in the support material, and then uses these two oligomers as the main raw materials to prepare the SLA support material.

3.3.4 Synthesis and properties of waterborne urethane acrylate

The aqueous waterborne SLA system is divided into two types: water dispersion and water solubility. The water dispersion system is an unstable emulsion which is dispersed and stabilized by emulsifier and is likely to be deposited in AM head, and block the head. The water-soluble resin is stable in storage and uniform in curing under UV light reaction. The water-soluble resin support material prepared by the water-soluble oligomer cannot form hydrogel due to the water has no sufficient time to evaporate when curing, so that the support material is easily peeled off and removed. The waterborne urethane acrylate oligomer prepared in this section uses polyethoxylated chain and carboxylic acid anion as hydrophilic group and belongs to water-soluble oligomer. It has small average molecular weight and high crosslink density, and is intended to be used as an oligomer in water-soluble support material, mainly giving mechanical strength to the support material.

Waterborne urethane acrylate oligomers require sufficient hydrophilic groups if they are to be water soluble. In order to obtain oligomers with good hydrophilic properties, this experiment synthesizes the high carboxyl content urethane acrylate containing hydrophilic polyethoxy soft segment. By changing the amount of the carboxyl group-containing monomer, the effects of the carboxyl group content on the water solubility of the alignment polymer are compared to obtain several urethane acrylates of different carboxyl group.

3.3.4.1 Synthesis experiment

1. Principle of synthesis design

    The waterborne urethane acrylate oligomer is prepared using TDI, PEG, dimethylolpropionic acid (DMPA) and HEA as raw materials. DMPA can introduce carboxyl group on the side chain of the oligomer, and become hydrophilic anion after being neutralized to form salt; PEG introduces hydrophilic nonionic soft segment on the main chain of oligomer; HEA is end capped and introduces SLA acryloyl group.

2. Purification and dewatering of raw materials

    Conduct decompression and distillation for PEG400, TDI, HEA, N, and N-dimethylacetamide, and place them in a desiccator for use; dewater the dibutyltin dilaurate and hydroquinone by using a molecular sieve; dry the DMPA at 80°C for 2 hours; and reflux the butylamine with flaky potassium hydroxide, and then resteam it.

3. Synthesis of waterborne urethane acrylate

    1. Add a TDI and N,N-dimethylacetamide solvent to a four-necked flask equipped with stirrer, condenser (with a drying tube) and thermometer with a heating device, and protect it with nitrogen. at 30°C. DMPA is added dropwise at a stirring speed for 1.5 hours. After the detection of the -NCO content in the reaction system reaches the theoretical calculation value, the reaction in the first step is thought to have been completed.
    2. At a suitable stirring speed, gradually add the PEG400 containing the catalyst dibutyltin dilaurate (0.2% of the total amount) dropwise, and control the temperature at 45°C–50°C for 3.5 hours. After the detection of the -NCO content in the reaction system reaches the theoretical calculation value, the reaction in the second step is thought to have been completed.
    3. Raise the temperature to 50°C–55°C. Under stirring, add the HEA containing appropriate amount of polymerization inhibitor dropwise until no − NCO is detected (when the residual content of − NCO is too high, an appropriate amount of ethanol can be added to remove it). Cool the mixture to room temperature to discharge, and conduct extraction and purification for the synthesized oligomer.
    4. Neutralization reaction: Dilute with water at room temperature to obtain aqueous solution of the oligomer. When stirring, add the neutralizer triethylamine (or sodium hydroxide) to the aqueous solution of the oligomer to form salt reaction. After the reaction is completed, detect PH value.

    The synthetic reaction formula of waterborne urethane acrylate is shown in Fig. 3.106.

3.3.4.2 Analysis of the test method

1. Determination of free-NCO value: di-n-butylamine countertitration method, same as the method in Section 3.2.2.

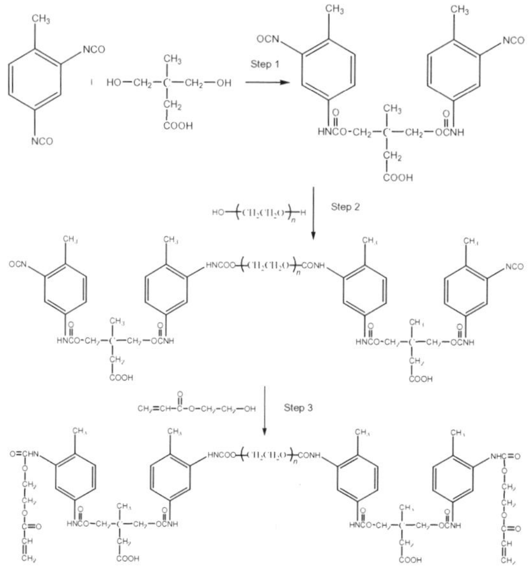

FIGURE 3.106 Synthetic reaction formula of waterborne urethane acrylate.

2. Purification of the oligomer: Weigh a certain amount of the sample, dissolve it in the appropriate amount of N,N-dimethylacetamide, and then quickly pour into the appropriate amount of petroleum ether to precipitate and layer, repeat aforementioned steps for several times, then the obtained precipitate is removed by vacuum drying to remove the solvent.
3. Analysis of infrared spectroscopy: After purification, the sample is measured by a Bruker FTIR EQUINOX 55 infrared spectrometer.
4. Product acid value: KOH titration method.
5. Determination of iodine value: morpholine addition method, same as the method in Section 3.2.2.
6. Measurement of viscosity: The viscosity of the aqueous solution is measured using SNB-2 type rotational viscometer.
7. Measurement of average particle diameter: it is measured by ZETAIZZER particle size tester.
8. Determination of gel content: first weigh the mass (m_1) of the slide glass, take a certain amount of oligomer and dilute with 30% deionized water, add alkali to neutralize, then add photoinitiator and apply it on the slide glass, make it dried and SLA, weigh the mass of cured product and the slide glass (m_2). After extraction with isopropyl-ketone for 4 hours, place it in an oven and bake at 50°C for 10 hours. Finally, take out the remaining cured product and slide glass to weight he mass (m_3), and the gel fraction is calculated by the Eq. (3.26).

3.3.4.3 Synthesis process and product analysis of waterborne polyurethane acrylate

1. Reaction process of aqueous urethane acrylate

    The synthetic route in Fig. 3.107 is a route in ideal conditions, and each step in the actual synthesis may generate a series of homologues of different molecular weights. As shown in Fig. 3.107, after diisocyanate oligomer is generated through the reaction between TDI and DMPA, it can also react with DMPA to form the oligomer of longer molecular chain, making the final product a mixture of a series of homologues. The drip addition method is used to control the drip speed of DMPA, which can make the polymerization process closer to the ideal condition and the molecular weight distribution more concentrated. In the second step, the diisocyanate oligomer reacts with PEG, which also produces a mixture of different molecular chain lengths.

    The first step in the synthesis reaction is to react at the position 4 of 2,4-TDI. Generally, the position 2 does not participate in the reaction. In the second step of the synthesis reaction, when the catalyst dibutyltin dilaurate is added, and the temperature is raised appropriately, the isocyanate at the position 2 in 2,4-TDI begins to react with PEG. In the third step of the synthesis reaction, since HEA is thermally polymerized under heating, hydrothermal polymerization agent, hydroquinone, is added to prevent HEA from self-polymerization.

2. Infrared spectrum of waterborne urethane acrylate product

    It can be seen from Fig. 3.108 that the characteristic absorption peak at 2273 cm^{-1} disappears, indicating that the isocyanate has completely reacted in the system. At 2872 cm^{-1}, it is a stretching vibration peak of saturated $-C-H$ bond. At 3380 cm^{-1}, it is a combined absorption peak

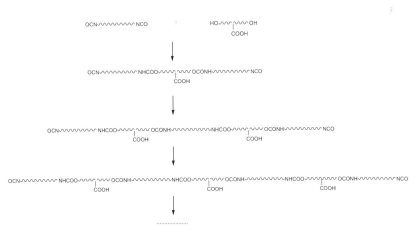

FIGURE 3.107 Formation of homologues in the first step of waterborne urethane acrylate synthesis.

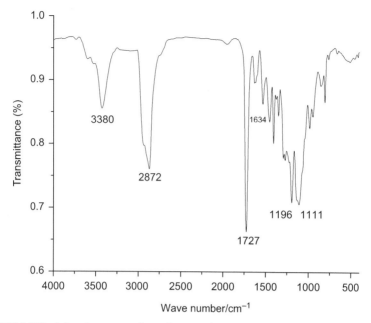

FIGURE 3.108 Infrared spectrum of waterborne urethane acrylate.

of N−H and carboxyl−COOH, and at 1727 cm^{-1}, it is an absorption peak of C=O, indicating that carbamate is formed and carboxyl group is attached to the product. A stretching vibration peak of C=C at 1634 cm^{-1} indicates that the acrylate structure has been attached to the product. The stretching vibration peak of C−O−C at 1196 and 1111 cm^{-1} indicates that there is an ethoxylated chain on the product. The aforementioned analysis proves that the reaction product is consistent with the theoretical structure.

3. Changes in isocyanate during the synthesis of waterborne urethane acrylate

When the feeding molar ratio of TDI: DMPA: PEG is 4:2:1, the content of isocyanate during and after each reaction in the synthesis process of waterborne urethane acrylate is detected by di-*n*-butylamine reverse titration. The rate of change of isocyanate content during the reaction is shown in Fig. 3.109.

It can be seen from Fig. 3.109 that the reaction in the first step (the reaction time of the first 1.5 hours) is close to a straight line, which indicates that the dripping rate is less than the polymerization rate. Since the reactivity at the position 4 is much higher than the position 2, the dripping rate at this time is advantageous for preferentially consuming the isocyanate at the position 4 of the TDI, so that the reaction continues as far as possible according to the set molecular structure.

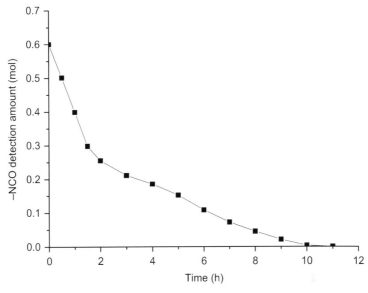

FIGURE 3.109 Change rate of isocyanate content during the reaction.

TABLE 3.23 Detected value of isocyanate after each step.

Reaction step	Reaction time (h)	Detection amount of isocyanate (mol)
Initial state	0	0.600
First step	1.5	0.298
Second step	3.5	0.152
Third step	5	0

The values of isocyanate content after the end of each reaction are shown in Table 3.23. It can be seen from the results that the concentration of isocyanate is high at the beginning of the reaction, and the reactivity of the isocyanate at the position 4 of TDI is relatively high, and the reaction rate in the first step is relatively fast, and the theoretical value isocyanate content is reached within 1.5 hours (about half of the initial value). The detected value of isocyanate is slightly lower than 0.300 mol, and it is possible that a very small amount of isocyanate at the position 2 participates in the reaction. The reaction in the second step lasts 3.5 hours. After the isocyanate content reaches the theoretical value (0.150 mol) roughly, the reaction is stopped, and then the temperature is raised. After adding HEA and an appropriate amount of polymerization inhibitor to dilute benzenediol, −NCO cannot be detected after about five hours of reaction. Due to the high reactivity of TDI, the concentration of isocyanate at the end of the reaction is still gradually reduced. As the reaction time is prolonged, the isocyanate gradually reacts with the hydroxyl group.

4. Acid value of waterborne urethane acrylate

Generally, hydrophilic group of high concentration can make the polymer water-soluble, and the waterborne urethane acrylate has hydrophilic PEG segment and two carboxyl groups, thus making it water-soluble. The content of the carboxyl group can be measured by the acid value, and the theoretical acid value of the carboxyl group in the waterborne urethane acrylate is represented by the Eq. (3.51):

$$y = \frac{2 \times 56.1 \times 1000}{M} \quad (3.51)$$

where y, acid value, mgKOH/g; M, theoretical molecular weight of waterborne urethane acrylate (calculated by theoretical structural formula).

When the feeding molar ratio of TDI:DMPA:PEG is 4:2:1, the theoretical molecular weight of the waterborne urethane acrylate is about 1656.0 g/mol, and the theoretical acid value calculated is 67.75 mgKOH/g, and the acid value measured by KOH titration is 66.5 mgKOH/g, which is very slightly different from the theoretical calculation value. This indicates that the DMPA monomer has been polymerized into the oligomer, and the carboxyl group does not participate in the reaction in the polymerization.

5. Iodine value of waterborne urethane acrylate

The iodine value of the waterborne oligomer indicates the number of reactive double bond groups which are SLA, and the high double bond content provides fast curing rate upon UV curing. In the synthesis of waterborne urethane acrylate, the loss of acryloyloxy double bond should be minimized to improve the SLA reactivity of the oligomer. When the basic reaction of isocyanate is complete, the effects on the decline rate of the product iodine value at different reaction temperatures in the third step, that is, the degree of destruction of the acryloyloxy double bond, are shown in Fig. 3.110.

It can be seen from Fig. 3.90 that the higher the reaction temperature of the HEA added in the third step, the iodine value of the final oligomer decreases to a larger extent, indicating that the lower the acryloyloxy group content in the oligomer, therefore, the reaction temperature should be controlled at 50°C–55°C.

3.3.4.4 Study on the hydrophilic properties of waterborne urethane acrylate

1. Effects of neutralization degree on water solubility of waterborne urethane acrylate

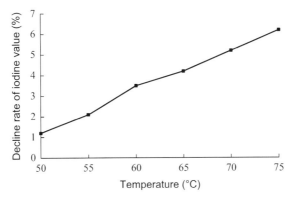

FIGURE 3.110 Effects of reaction temperature on the decline rate of iodine value in the third step.

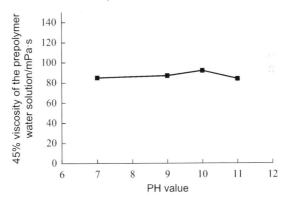

FIGURE 3.111 Curve of the relationship between viscosity of oligomer aqueous solution and neutralization degree when the solid content is 45%.

Before the waterborne urethane acrylate is neutralized, the hydrogen bond interaction between the carboxyl groups is strong, the molecular chain is not easy to be hydrated with water, and the water dispersibility is poor. After the carboxyl group is neutralized, the carboxylic acid group is ionized, the degree of ionization is increased, the hydrogen bond interaction is weakened, and the hydrophilicity of molecular chain is improved. Therefore, the degree of neutralization has great influence on the water solubility, and the water solubility of the waterborne urethane acrylate is not good when there is no 100% neutralization. Fig. 3.111 shows the effect of pH increase on viscosity after 100% neutralization at 25°C. The viscosity change is not significant after 100% neutralization, indicating that the oligomer already has good hydrophilicity. Thus, 100% neutralization is sufficient.

2. Effects of carboxyl content (mass fraction) on the hydrophilicity of the product

    The introduction of carboxyl group imparts hydrophilicity to the oligomer, and the hydrophilic group and the hydrophobic group is simultaneously

TABLE 3.24 Appearance and stability of oligomer aqueous solution of 70% solid contents at different acid values.

Acid value of the oligomer mgKOH (g)	23.9	49.6	66.5	72.8
Appearance	Milky white	Milky white	Transparent	Transparent
Stability	Have precipitation	Stable	Stable	Stable

TABLE 3.25 Appearance of aqueous solutions of oligomers with different solid contents.

Solid content%	70	50	30	15
Appearance	Transparent	Transparent	Milky white	Milky white

present, so that the oligomer has the function of polymer emulsifier. When the content of carboxyl group is low, the hydrophilicity is poor, and the system is white emulsion; when the content of carboxyl group is high, the compatibility with water is improved, and the transparency and water absorption rate of the system are increased. In the experiment, different ratios of DMPA are added to obtain oligomers having different carboxyl groups, which have different acid values, and the results are shown in Table 3.24.

As can be seen from Table 3.24, when the content of carboxyl group is low, the water-soluble emulsion formed is milky white and unstable. When the content of carboxyl group is 49.6−72.8 mg KOH/g, the water dispersibility of the oligomer gradually becomes better, and its appearance changes from milky white to transparent.

3. State of the aqueous solution of the oligomer at different solid contents

Table 3.25 shows the appearance of the oligomer of aqueous solution dispersion system at different solid contents when the content of carboxyl group is 66.5 mg KOH/g. In the experiment, when the solid content is high, the system is transparent; when the solid content is low, the system is milky white. When the solid content is high, the oligomer is a continuous phase. Due to the good hydrophilicity of the oligomer molecules, water is dispersed in the oligomer by hydrogen bonding or by solubilization, and the appearance is nearly transparent. As the amount of water increases, the phase transition of the system occurs, that is, the aqueous phase gradually transforms into a continuous phase, and the oligomer is dispersed in water in the form of emulsoid particles. At this time, aqueous emulsion is formed, and the appearance of the aqueous solution is

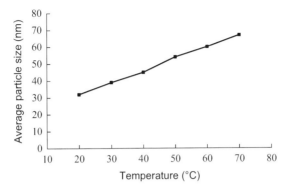

FIGURE 3.112 Relationship between average particle size and temperature of emulsoid particles in aqueous solution of oligomer.

milky white. This reflects from one side that the oligomer of high carboxyl content has good hydrophilicity.

4. Effects of temperature on aqueous solution of oligomer

The temperature affects the viscosity, particle size and appearance of the aqueous solution of the oligomer. When the temperature rises, the molecular chain of the oligomer is more easily stretched, which is manifested as the improved hydrophilicity, and the appearance of the aqueous solution tends to be transparent, and the oligomer is uniformly dispersed. Fig. 3.112 shows the average particle size distribution of oligomer emulsoid particle when the aqueous solution of oligomer changes with the temperature when the hydrophilic group has a carboxyl group content of 66.5 mg KOH/g and solid content of 30%. When the temperature in the figure is higher, the average particle size of the oligomer latex will be larger, indicating that the faster the molecules of the oligomer move, the easier the molecular chain stretches.

Fig. 3.113 shows the change of the viscosity of oligomer aqueous solution with the temperature when the solid content is 30%. As the temperature increases, the kinetic energy of the thermal movement of the oligomer molecules increases, the flow resistance decreases, and the viscosity of the aqueous solution of the oligomer decreases.

In addition, the relationship between viscosity and temperature conforms to the Arrhenius equation:

$$\eta = A e^{\Delta E_\eta / RT}$$
$$\ln \eta = \ln A + \frac{\Delta E_\eta}{RT} \tag{3.52}$$

From this equation, it can also be seen that as the temperature increases, the viscosity of the aqueous solution of the oligomer decreases. Therefore, the viscosity of aqueous support material will be lower when working above room temperature, which can facilitate inkjet printing.

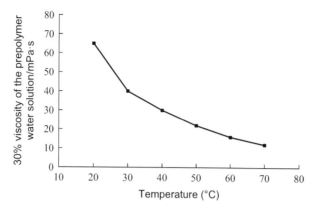

FIGURE 3.113 Relationship between the viscosity of oligomer aqueous solution and the temperature when the solid content is 30%.

3.3.4.5 Stereolithography apparatus speed of waterborne urethane acrylate

In this subsection, the factors affecting the SLA speed of the oligomer are discussed from the aspects of the type of photoinitiator and its mass fraction, neutralizer, light intensity and temperature. Two high-pressure mercury lamps with 500 W power and one high-pressure mercury lamp with 10,000 W power are used as the curing conditions in the experiment. Unless otherwise specified, the distance between the high-pressure mercury lamp and the oligomer is 10 cm, and the coating thickness of the oligomer is about 50 μm. The SLA speed of the oligomer is measured according to the gel content.

1. Types of photoinitiator

    Photoinitiators are important components in waterborne SLA support materials. When the type of photoinitiator is different, the curing rate of waterborne urethane acrylate is also different. In waterborne SLA systems, it mainly considers the UV absorption wavelength of the photoinitiator, the solubility of the photoinitiator in the oligomer, the water solubility of the photoinitiator, the photoinitiator activity and the initiation efficiency, and the loss (volatility) of the photoinitiator in the process that the droplet ejects from the 3D printer nozzle to the working surface in choosing the photoinitiator.

    Fig. 3.114 shows the curve of gel content of oligomer changing with the irradiation time when using different photoinitiators at 25°C under the conditions that the total pressure of the high-pressure mercury lamp is 2000 W, the mass fraction of photoinitiator is 2%, the neutralizer is triethylamine, and the degree of neutralization is 100%. As can be seen from the figure, Irgacure 2959 has good compatibility with oligomers due to good water solubility, and has high initiation efficiency, and the oligomer has the highest curing rate.

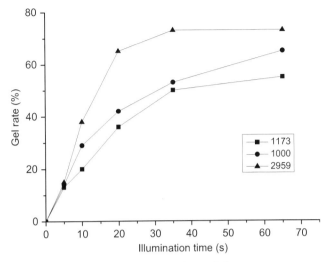

FIGURE 3.114 Effects of different initiators on gel rate of oligomer.

2. Mass fraction of photoinitiator

   Generally, when the mass fraction of photoinitiator increases, the curing rate of the SLA system will also increase. However, it is not always the case that the higher the mass fraction of the photoinitiator, the better. First, when the mass fraction of the photoinitiator is too high, the surface of the aqueous support absorbs too much UV light energy, so that the UV light entering the bottom layer is reduced, which is disadvantageous to the curing in the deep layer; secondly, when the mass fraction of the photoinitiator is large, more active radicals will be generated, the generated free radicals are easily terminated by mutual coupling, resulting in decline in photoinitiation efficiency; finally, when the mass fraction of the photoinitiator is increased, the cost of waterborne support materials will be inevitably increased. Fig. 3.115 shows the curve of gel content of oligomer changing with the irradiation time when using Irgacure 2959 photoinitiators of different percentages at 25°C under the conditions that the total pressure of the high-pressure mercury lamp is 2000 W, the neutralizer is triethylamine, and the degree of neutralization is 100%.

   It can be seen from Fig. 3.115 that Irgacure 2959 with 2% mass fraction has a faster curing rate of the oligomer, and when the mass fraction of the photoinitiator is further increased, the curing rate does not change much.

3. Types of neutralizer

   The neutralizer mainly affects the water solubility of the oligomer, which in turn affects the SLA rate of the oligomer. Two neutralizers, triethylamine and sodium hydroxide, are mainly tested here. The effects of the neutralizer

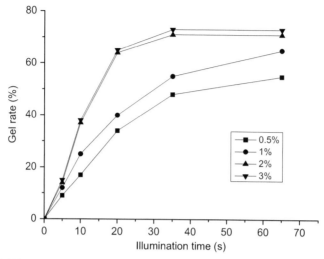

FIGURE 3.115 Effects of initiator mass fraction on the gel rate of oligomer.

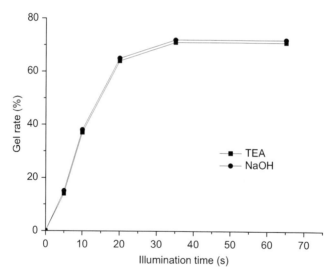

FIGURE 3.116 Effects of the neutralizer type on the gel rate of the polymer.

type on the SLA rate of oligomer is shown in Fig. 3.116, in which the basicity of sodium hydroxide is stronger, and the neutralizer has little effect on the SLA rate of the oligomer as can be seen from the gel content in the SLA process. The main reason is that when two different neutralizers are used, the water solubility of the oligomer is similar, except that the pH of the aqueous solution after 100% neutralization is different. The curing is carried out at 25°C under the conditions that the total pressure of high-pressure pump lamp is 2000 W, the neutralization degree is 100%, and the mass fraction of photoinitiator, Irgacure 2959, is 2%.

4. Light intensity

Fig. 3.117 shows the effects of different light intensities on the SLA rate of the oligomer. The light intensity is changed by adjusting the number of high-pressure mercury lamps. The high-pressure mercury lamps used are arranged from small power to large power. They are one 500 W, one 1000 W, one 500 W and one 1000 W, two 500 W, and one 1000 W. The curing is carried out at 25°C under the conditions that the total pressure of high-pressure pump lamp is 2000 W, the neutralizer is trimethylamine, the neutralization degree is 100%, and the mass fraction of photoinitiator, Irgacure 2959, is 2%. It can be seen from the figure that as the light intensity increases, the gel content of the oligomer increases rapidly. It is proved that the SLA rate is significantly affected by the light intensity, but the light intensity cannot be too high, otherwise the yellowing phenomenon of the oligomer will be serious.

5. SLA temperature

Fig. 3.118 shows the curve of the gel content of oligomers changing over time at different temperatures. The curing is carried out at under the conditions that the total pressure of high-pressure pump lamp is 2000 W, the neutralizer is trimethylamine, the neutralization degree is 100%, and the mass fraction of photoinitiator, Irgacure 2959, is 2%. It can be seen from the figure that as the temperature rises, the molecular segment of the oligomer moves faster, and the molecular segments of the oligomer are more easily rearranged, making the double bonds on the oligomer easier to move closer to each other, and the contact of double bond to increase. Thus, the rate of SLA increases.

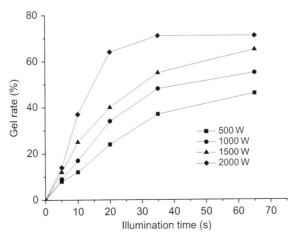

FIGURE 3.117 Effects of light intensity on gel rate of oligomer.

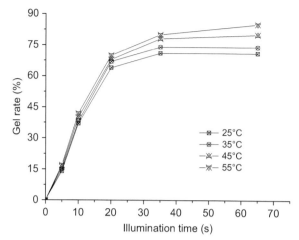

FIGURE 3.118 Effects of temperature on gel rate of oligomer.

3.3.5 Synthesis and properties of polyethylene glycol diacrylate

Polyethylene glycol diacrylate is a nonionic aqueous SLA oligomer. After curing, the mechanical strength is not high due to weak interaction force between the ether bonds, and the cohesive force is poor. It is this oligomer that is easy to peel off, has low mechanical properties and poor water resistance, and is very suitable for use as an SLA support material for AM. The ether chain network structure formed by SLA of polyethylene glycol diacrylate is also an effective water retention system. Therefore, many people have studied to prepare SLA hydrogel using aqueous solution of polyethylene glycol diacrylate. In fact, the SLA AM support material can also be regarded as an SLA hydrogel after curing, but the support material also requires a certain strength, which can play a role of supporting the solid material. Therefore, this section synthesizes the polyethylene glycol diacrylates of three different molecular weights, and selects the oligomer of the optimal molecular weight as another important component in the support material.

3.3.5.1 *Synthesis experiment*

In a 500 mL three-necked flask equipped with stirrer, dropping funnel and thermometer, add 0.25 mol PEG, 0.5 mol sodium and 200 mL tetrahydrofuran solvent, and make it react for 2 hours under the protection of nitrogen, add 0.6 mol acryloyl chloride slowly dropwise, and then react for 2 hours continuously. After filtration of the product, the solvent of the filtrate is removed by rotary evaporation, and PH value is adjusted to 7 with potassium carbonate solution. Wash with water repeatedly, and dry with anhydrous magnesium sulfate, filter it, and the filtrate is dried in vacuum to give a product after removing the solvent.

3.3.5.2 Synthesis reaction formula of polyethylene glycol diacrylate

In the synthesis of polyethylene glycol diacrylate by sodium alkoxide-acid chloride method, sodium first reacts with PEG to form sodium polyethylene glycol, and then reacts with acryloyl chloride to obtain polyethylene glycol diacrylate. The reaction rate between sodium and PEG at both terminal hydroxyl groups is fast, and the reactivity between sodium glycolate and acryloyl chloride is higher than that of PEG and acryloyl chloride. The reactivity of acryloyl chloride is significantly higher than that of acrylic acid, so the total reaction rate is fast, and the reaction can be carried out smoothly at room temperature, and a higher conversion rate can be achieved in a certain period of time. It only takes four hours for reaction, and the purity of the product is high. The main chemical reaction formula is:

$$HO(CH_2CH_2O)_nH + Na \longrightarrow NaO(CH_2CH_2O)_nNa$$

$$NaO(CH_2CH_2O)_nNa + CH_2=CH-\overset{O}{\underset{\|}{C}}Cl \longrightarrow CH_2=CH-\overset{O}{\underset{\|}{C}}O(CH_2CH_2O)_n\overset{O}{\underset{\|}{C}}-CH=CH_2$$

3.3.5.3 Analysis of the product of polyethylene glycol diacrylate

1. Infrared spectroscopy analysis of polyethylene glycol diacrylate

    Fig. 3.119 shows the infrared spectrum of polyethylene glycol 400 diacrylate. The infrared spectrum of the synthetic product, polyethylene glycol 400 diacrylate, shows that the characteristic absorption peak of hydroxyl group is 3500 cm^{-1}, which is still obvious, indicating that the product also

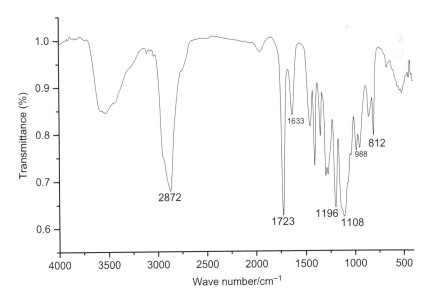

FIGURE 3.119 Infrared spectrum of the product of polyethylene glycol 400 diacrylate.

contains a small amount of water. 2872 cm^{-1} is the stretching vibration peak of −C−H, the characteristic absorption peak of 1723 cm^{-1} is the stretching vibration absorption peak of acrylate carbonyl −C = O, and the characteristic absorption peak of 1633 cm^{-1} is the stretching vibration of acrylate C = C. 988 cm^{-1} is the δ bond out-of-plane bending vibration peak of acrylate C = C, and 812 cm^{-1} is the denatured vibration absorption peak of acrylate olefin hydrogen bond = C−H. The above absorption peak indicates that the acrylate structure has been connected to the product. There is no change in the −C−O−C-antisymmetric and symmetric stretching vibration peaks at 1196 and 1108 cm^{-1}, indicating that the internal ether bond structure of PEG is not destroyed. The results show that PEG forms polyethylene glycol diacrylate after reaction in two steps.

2. Viscosity and surface tension of polyethylene glycol diacrylates with different molecular weights

    In this section, polyethylene glycol 400 diacrylate, polyethylene glycol 600 diacrylate and polyethylene glycol 1000 diacrylate are synthesized respectively, and the viscosity and surface tension of three polyethylene glycol diacrylates are compared and analyzed. Polyethylene glycol diacrylate increases in viscosity and surface tension at 25°C as the molecular weight of the polyethoxylated chain increases. The results are shown in Table 3.26.

    When the molecular segment of the PEG is longer, the content of the ethoxylated chain in the molecular chain of the product increases, causing the increase of the surface tension. The greater the molecular weight of the polyethylene glycol diacrylate is, the greater the corresponding viscosity is.

3. SLA speed of polyethylene glycol diacrylate with different molecular weights

    Take a certain amount of polyethylene glycol diacrylate with different molecular weights, and add 2% mass fraction of photoinitiator, Irgacure 2959, in it. The light intensity is obtained with 2000 W high-pressure mercury lamp, and the vertical irradiation distance is 15 cm. The SLA gel-time curve of polyethylene glycol diacrylate with different molecular weights is shown in Fig. 3.120. It can be seen from Fig. 3.120 that the initial SLA speed of polyethylene glycol 400 diacrylate is relatively fast.

TABLE 3.26 Viscosity and surface tension of polyethylene glycol diacrylates with different molecular weights.

Product (25°C)	PEG400 diacrylate	PEG600 diacrylate	PEG1000 diacrylate
Viscosity (mPa s)	61	102	198
Surface tension (mN/m)	42.6	43.7	44.5

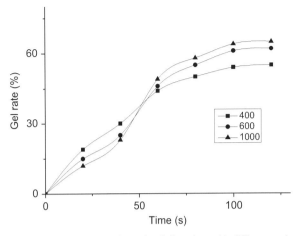

FIGURE 3.120 SLA speed of polyethylene glycol diacrylate with different molecular weights.

With the extension of reaction time, the functional density of polyethylene glycol 400 diacrylate is relatively large, and the crosslinking density is high, so the SLA speed in the later stage is lowered. The initial SLA speed of polyethylene glycol 1000 diacrylate is slower, and the conversion rate in the late SLA is higher than that of polyethylene glycol 400 diacrylate. This is because the flexible molecular chain of polyethylene glycol 1000 diacrylate is longer, at higher reaction levels, the acrylate functional groups can still rearrange together, and continue to participate in the addition reaction, and the conversion rate increases. Since the oligomer does not contain reactive diluent and water, the viscosity of the system is too large when the oligomer is crosslinked to a certain extent, and the molecular motion of the oligomer is hindered. Thus, the final conversion rate of the double bond of the oligomer is not high.

3.3.6 Study on the preparation of support materials by oligomers

In this section, waterborne urethane acrylate and polyethylene glycol 1000 diacrylate are used as the main oligomers to prepare support material for SLA hydrogel. After the support material is formed, it can be easily removed, and the compressive strength is appropriate, which is not easy to produce compression deformation, ensuring the forming accuracy of the solid material.

3.3.6.1 Selection of raw materials for stereolithography apparatus support materials

1. Selection of monomers and oligomers

    On the one hand, the SLA support material is required to be SLA and crosslinked, and has certain mechanical properties; on the other hand, it is required to have low curing strength, so that it can be relatively easily removed, and is preferably capable of being removed by water dissolution. To achieve the two points at the same time, it must start with the selection of oligomers and monomers. As mentioned earlier, waterborne urethane acrylate can impart mechanical strength to the support material, while polyethylene glycol 1000 diacrylate has low cure strength and is easy to remove. In order to allow the cured support material to be easily removed in alkaline water, β-carboxyethyl acrylate (β-CEA) is also added. β-CEA is a monofunctional acrylate which does not increase the crosslink density, and at the same time, carboxyl group is added to the polymer after the support material is cured, and when the content is relatively large, it is easily dissolved in the alkali water.

2. Selection of temperature-sensitive materials

The so-called temperature-sensitive material refers to the materials whose properties vary with the temperature. The nonionic material is a temperature-sensitive material with a cloud point (lower critical solution temperature, LCST). When the temperature is above the cloud point, nonionic material is insoluble in water and sediments; when the temperature is below the cloud point, it is water soluble. The principle is that at low temperatures, the hydrogen bonding between the hydrophilic segment and the water molecule in the polymer takes the predominating position, the polymer molecular chain is in an extended state, macroscopically appearing as an aqueous solution; when the temperature rises, the hydrophobic interaction between the hydrophobic segments of the polymer chain is enhanced, and the hydrogen bonding is weakened, the polymer segments are associated by hydrophobic action to form a gel. By using this property, the temperature-sensitive material is added to the support material. When the support is formed, the temperature is relatively high, and the temperature-sensitive material is solid, which can serve as supporting the solid material; when the forming is completed, the formed part is cooled. The temperature drops below the cloud point of the temperature-sensitive material in the support material, making it water soluble and can remove the support material in the water. PEO-PPO-PEO triblock oligomer is a nonionic temperature-sensitive material and its commercial products are mainly BASF's Pluronic series products, among which F127 has suitable cloud point and is most suitable for being used as a support material.

3. Humectants and wetting agents

The humectant generally uses water-soluble organic solvent, such as alcohol or polyhydric alcohol, and its function is to improve the stability of the supporting resin, so that the viscosity and the surface tension are not easily changed with temperature, and it is difficult to block the nozzle. The wetting agent delays the drying of the aqueous supporting resin and allows the resin to easily wet the nozzle holes, so that the nozzle can be quickly restarted after the ejection is stopped. Commonly used wetting agents are propylene glycol, ethylene glycol and the like, and their mass fraction is 1%–10% of the aqueous support material. In this experiment, Dowanol TPM is used as a humectant, and Rcy101 is used as a wetting agent.

4. Photoinitiator

The photoinitiator is selected for the water-based supporting resin mainly because it has thermal stability, water solubility, photoinitiating activity, and maximum UV absorption peak of the photoinitiator. Irgacure 2959 is a white crystal with very good thermal stability, so there is no problem with the spray temperature of 50°C–55°C. The solubility in water is much higher than the aqueous photoinitiator 1173. At 276 and 331 nm, the absorption wavelength of λ_{max} can be extended to 380 nm. Both the hydroxyethylether benzoyl radical and the α-hydroxyisopropyl radical generated by photolysis are highly reactive free radicals. In combination with the aforementioned characteristics, Irgacure 2959 is used as the photoinitiator for the supporting resin.

5. Other additives

The surface tension of water is very strong, so the proper surfactant should be used to adjust the surface tension. In order to prevent the acrylate from being thermally polymerized at a high temperature to block the nozzle, the polymerization inhibitor is generally added. Keeping the aqueous supporting resin alkaline (PH = 8–9) can reduce the corrosion to the metal nozzle, and when the PH value of the resin is low, it is likely to cause nozzle clogging. The commonly used pH adjusting agent is triethanolamine, etc., and the mass fraction thereof is generally 1%–5% of the aqueous supporting resin, and at the same time, it can prevent oxygen inhibition. In order to prevent the aqueous supporting resin from generating bubbles during the spraying process, a certain amount of defoamer may be added.

3.3.6.2 Support material resin

Dissolve F127 in deionized water, add monomers and oligomers into it, and finally add photoinitiator and other additives. Shade the mixture and heat it to 45°C. Stir the mixture for 4 hours, and then filter it for use. The density of the liquid support material is measured by the pycnometer method to be 1.03 g/cm^3, and the density of the support gel after curing is 0.99–1.01 g/cm^3. The solid material liquid has a contact angle of 44 degrees on the cured surface of the solid material and a contact angle of 52 degrees on the cured surface of the support material.

3.3.6.3 Main injection parameters of aqueous support materials

1. Viscosity and surface tension

Table 3.27 shows the viscosity and surface tension of the aqueous support material at different temperatures. As can be seen from the table, the viscosity of the support resin at room temperature is low, which is

TABLE 3.27 Viscosity and surface tension of supporting resin at different temperatures.

Temperature (°C)	25	30	35	40	45	50	55
Viscosity (mPa s)	31.7	26.1	21.6	17.4	13.5	10.8	9.4
Surface tension (mN/m)	32.3	31.8	31.2	30.7	30.2	29.5	29.0

TABLE 3.28 Reynolds number, Weber number, R/W value, sputtering coefficient of supporting resin.

	Reynolds number (Re)	Weber number (We)	R/W value	Sputtering coefficient (K)
50°C	28.6	62.8	3.6	18.3
55°C	32.9	63.9	4.1	19.1

only 31.7 mPa s. The working temperature of the support material can be set between 50°C and 55°C, which is generally set at 50°C or 55°C.

2. Reynolds number, Weber number, R/W value, sputtering coefficient

According to the calculation formula of Reynolds number and the calculation formula of Weber number, the Reynolds number and Weber number of the support material at working temperature 50°C or 55°C are calculated. The density of the support material is 1.03 g/cm³, the jet velocity of the nozzle is 6.0 m/s, and the nozzle aperture is 50 μm. The results of Table 3.28 are obtained:

It can be seen from Table 3.28 that the support material can be sprayed normally at 50°C and 55°C without sputtering on the working surface. In fact, during the printing process with an actual machine, no sputtering phenomenon is detected.

3.3.6.4 Contact angle, maximum spreading factor, and spreading time of support material

The contact angle, maximum spreading factor and spreading time of the support material on its cured surface and the solid surface of the cationic solid material are shown in Table 3.29 (50°C); the contact angle, spreading factor and spreading time of the support material with the hybrid solid material and the free-radical solid material are shown in Tables 3.30 and 3.31 (50°C), respectively. It can be seen from the table that the spreading of the support material on the cured surface of various solid materials is generally the same, there is no big difference, and the spreading factor is not very large compared with the spreading factor of the solid material.

TABLE 3.29 Contact angle, spreading factor and spreading time of support materials on cationic solid materials.

Contact angle		Spreading factor		Spreading time (s)
With itself	With solid material	With itself	With solid material	
45 degrees	33 degrees	1.25	1.26	7.3×10^{-5}

TABLE 3.30 Contact angle, spreading factor and spreading time of support materials on hybrid solid materials.

Contact angle		Spreading factor		Spreading time (s)
With itself	With solid material	With itself	With solid material	
45 degrees	38 degrees	1.25	1.25	7.3×10^{-5}

TABLE 3.31 Contact angle, spreading factor and spreading time of support materials on free-radical solid materials.

Contact angle		Spreading factor		Spreading time (s)
With itself	With solid material	With itself	With solid material	
45 degrees	41 degrees	1.25	1.25	7.3×10^{-5}

3.3.6.5 Printing stability of the support material

Table 3.32 shows the number of nozzle plugging holes during the printing test of the aqueous support material. It can be seen from the data in Table 3.32 that the support material has good ejection stability at operating temperatures of 50°C and 55°C, and plugged holes are seldom found.

3.3.6.6 Stereolithography apparatus and removal of support materials

1. Characteristic parameters

    According to the test method of the characteristic parameters, the support material is obtained with $E_C = 36.5$ mJ/cm^2 and $D_P = 0.222$ mm.

TABLE 3.32 Printing stability test results of the support material.

Printing duration	0 h	3 h	6 h	9 h	12 h
Printing temperature (°C)					
50	0	1	1	2	3
55	0	1	1	3	4

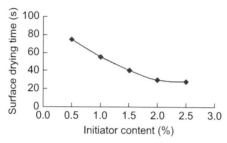

FIGURE 3.121 Effect of photoinitiator on the surface drying time of support materials.

The critical exposure is slightly higher than that of the solid material, mainly because the aqueous system cures slightly slower and requires sufficient energy to initiate polymerization; since the photoinitiator 2959 has a low mass fraction, its transmission depth is slightly higher than that of the solid material.

2. Volume shrinkage rate

    The density of the support gel after curing is 0.99–1.01 g/cm^3, which is less than 1.03 g/cm^3 before curing. Therefore, after curing, the support actually expands slightly. In the support forming process, on the one hand, the volatilization of water causes loss of the forming material; on the other hand, part of the water which is not volatilized is swollen in the cured supporting gel to lower the density of the supporting gel. The combined effect lowers the volume shrinkage rate of the support [7].

3. SLA speed

    Fig. 3.121 shows the effects of photoinitiator Irgacure 2959 on the surface drying time of the SLA gel of support material. It can be seen from the figure that as the mass fraction of photoinitiator increases, the surface drying time becomes shorter, indicating that the curing rate is increased; when the mass fraction of photoinitiator is greater than 2%, the effect on the SLA surface drying time is very small. Moreover, when the photoinitiator content is too high, excessive radicals are generated, and biradical termination is liable to occur, thus the SLA rate is lowered. When tested with AM test machine, the power of the UV metal halide lamp is 800 W and the movement rate of the lamp is 0.8 m/s, the support material can achieve instant surface drying.

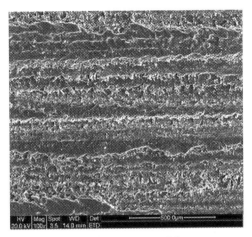

FIGURE 3.122 SEM of the surface of the support material.

4. Performance after removal of support

   After the solid material and the support material are formed together, cool them and then place them in water, and the support will be able to be easily removed by peeling. Place the support in the water alone, and soak it for 1−2 hours, and you will find that the support becomes soft obviously and can be destroyed with a slight force.

   Fig. 3.122 shows the SEM of the support material after the AM is formed by separating the cationic solid material from the support material and scanning the vertical joint surface of the support material and the solid material by SEM. It can be seen from the SEM that many gullies are formed on the surface of the support material, which is extremely advantageous for the support material to be separated from the solid material without damaging the surface quality of the solid material. The main reason for the formation of surface gullies is that the cationic solid material is completely oil-soluble, and the support material is water-soluble, so the two materials are incompatible. Therefore, before SLA, the two materials are repelled, and after SLA, due to various factors such as shrinkage stress, polymer molecular repulsion, layer pressure, etc., the joint of the two materials is easy to isolate, and many gullies are formed after stacking multiple layers.

   Based on the aforementioned studies, the support has higher removal performance.

3.4 Modified stereolithography apparatus forming materials

The SLA-forming material has the advantages of fast curing rate, little environmental pollution, high utilization rate, short forming period, etc., but there are still many shortcomings that need further research, such as poor transparency, poor surface quality, low hardness, poor toughness, and poor heat resistance of formed parts. These defects make the formed parts unable to be used directly as functional parts, thus greatly reducing its application range. Researchers try to improve its performance by changing the formulation of the material, but this often increases the performance at the expense of another performance. Therefore, traditional methods are difficult to improve strength, toughness, and heat resistance at the same time, and it is of great significance to develop new materials and processes.

3.4.1 Nano-SiO$_2$ modified stereolithography apparatus forming material

3.4.1.1 SiO$_2$ surface treatment

The SiO$_2$ is surface treated with different silane coupling agents and titanate coupling agents, and the treatment effects are evaluated by sedimentation experiments. If the particles have good dispersibility in the liquid, the particles are less likely to aggregate and bond, the particle size is small and the settling time is long. On the contrary, if the particle dispersibility is not good, the particles tend to aggregate, the particle size is large, and the sedimentation speed is fast. Since the viscosity of HDDA is the smallest among the components used for the UV curable resin, the sedimentation time is measured using HDDA as a medium. The surface-treated SiO$_2$ powder is incorporated into HDDA at a mass fraction of 3%, shaken in an ultrasonic wave to make it uniformly mixed, and then allowed to stand for a while to observe the sedimentation time and dispersion degree, as shown in Table 3.33 [8].

In general, silane coupling agents work better than titanate coupling agents. The transparency of modified SiO$_2$ of silane coupling agent in HDDA is significantly better than the modified SiO$_2$ of titanate. This is because the silane coupling agent has strong interaction with the acidic substance, and the titanate coupling agent has strong interaction with the alkaline substance, while the SiO$_2$ is acidic, the silane coupling agent should be selected as the modifier. KH-560, abbreviated from γ-glycidyloxypropyl trimethoxysilane, has the best modification effect in the silane coupling agent. Its molecular formula is shown in Fig. 3.123.

TABLE 3.33 Settlement experiment results of surface-treated samples.

Coupling agent type	Settling time and simple description
KH-570	Small amount of sedimentation at 80 min
KH-560	Small amount of sedimentation at 140 min
NDZ-101	Precipitation at 60 min
NDZ-102	Precipitation at 60 min
NDZ-201	Small amount of sedimentation at 30 min

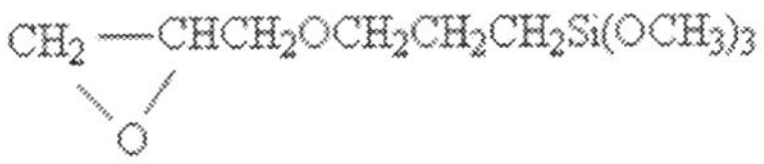

$$CH_2\!\!-\!\!\underset{\underset{O}{\diagdown\;\diagup}}{CH}CH_2OCH_2CH_2CH_2Si(OCH_3)_3$$

FIGURE 3.123 KH-560 molecular formula.

TABLE 3.34 Viscosity of photopolymers containing different mass fractions of SiO_2.

SiO_2 content (%)	Viscosity(mPa s) (28°C)	
	Mixing under heating temperature	Mixing under normal temperature
0	400	400
1	415	490
2	431	525
3	488	560
4	512	665
5	663	754

3.4.1.2 Nano-SiO_2 modified stereolithography apparatus forming material

1. Preparation of nano-SiO_2-photopolymer composite material

    Dry the KH-560-modified nano-SiO_2 in a vacuum oven at 100°C for 2 hours, and store it for standby use.

    Proportionally disperse the free radical and cationic initiator in the reactive diluent, stir and dissolve completely, add modified nano-SiO_2 of different masses, stir with the high-speed dispersing mixer for 15 minutes, ultrasonically disperse for 1 hour, and then add the bis(3,4-epoxycyclohexylmethyl) adipate, toughen the modified bisphenol an epoxy resin, polypropylene glycol diglycidyl ether diacrylate and additives respectively according to the ratio. The mixture is stirred at a high-speed dispersing mixer at 60°C for two hours until the resin system is uniformly translucent, and then vacuumed in a vacuum drying oven to remove bubbles generated during the stirring process to prepare UV curing nanocomposite.
2. Properties of nano-SiO_2 modified SLA-forming materials
1. Effects of nano-SiO_2 on the viscosity of SLA-forming materials

    Measure the viscosity of the photopolymer containing 0%, 1% (mass fraction, the same below), 2%, 3%, 4%, and 5% of the modified SiO_2 by using a rotary viscometer. The viscosity is largely affected by temperature. The test temperature is 28°C and the results are shown in Table 3.34.

It can be seen from Table 3.34 that the effect of heating and mixing of the filler and the resin is obviously better than that of the nonheating mixing, because under the heating state, various molecules move faster, the contact is more sufficient, the mixing is more uniform, and the viscosity of the resin is less affected.

In the heating and mixing, as the mass fraction of the filler increases, the viscosity of the resin increases. The reason is that the surface of the nano-SiO_2 treated with the coupling agent is attached with organic group, and interacts with the organic matrix to form a network structure, resulting in a significant increase in the viscosity of the system. With the increase of the content of nano-SiO_2, the interaction between the nano-SiO_2 particles and the organic matrix per unit volume is further enhanced, and the viscosity of the photopolymer is correspondingly increased, which further promotes the stability of the system, and finally achieves a balanced state of stable dispersion and uniformity. However, the content of nano-SiO_2 should not be too large. On the one hand, it will exceed the maximum viscosity that the curing equipment can withstand. On the other hand, the uneven dispersion of the filler will cause a large amount of agglomeration, thereby losing the nanoeffect that it should have.

It can be seen from Fig. 3.124 that when the amount of filler added is less than 2%, the viscosity of the resin increases slightly, and the viscosity increases when the amount of addition is between 2% and 4%, and the viscosity increases sharply when the amount is more than 4%. When the amount of addition is less than 2%, the nano-SiO_2 is uniformly dispersed in the resin, and has little effect on the viscosity of the resin. The larger the amount of addition, the more difficult it is to disperse uniformly, and the viscosity change of the resin is more obvious. When it is more than 4%, SiO_2 has a significant thickening effect on the resin.

2. Dispersion of nano-SiO_2

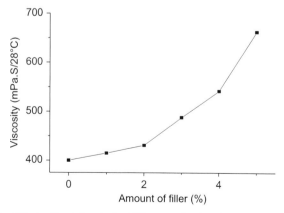

FIGURE 3.124 Effects of heated and mixed filler on resin viscosity.

Add the nano-SiO$_2$ of 3% mass fraction before modification to the photopolymer, disperse and cure it, and conduct field emission scanning for the cross section of the cured part. The result is shown in Fig. 3.125. It can be clearly seen that the nano-SiO$_2$ agglomerates very seriously, with the largest particle size reaching 10 μm, and the nano-SiO$_2$ also disperses very unevenly.

Fill the modified nano-SiO$_2$ into the photopolymer, and cure it by using an SLA device, and then conduct a field emission scanning for the cross section of the formed part Fig. 3.126 shows the SEM photo of a cured part with 3% filling amount after 16,000-fold magnification. It can be seen from the photo that the modified nano-SiO$_2$ is uniformly dispersed in the resin matrix, which has substantially no agglomeration, and the particle size remains in the nanometer range, which is about 80 nm, and the compatibility between the dispersed phase and the continuous phase interface is good. It indicates that the surface-treated nano-SiO$_2$ can be uniformly dispersed in the resin by the dispersion method employed herein. However, it can be seen from Fig. 3.126 that the filling amount of 3% nano-SiO$_2$ is too large.

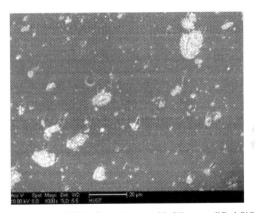

FIGURE 3.125 Cross-sectional SEM of cured parts with 3% unmodified SiO$_2$.

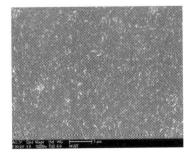

Element	Mass Fraction %	Atomic Fraction %
CK	58.29	70.00
OK	22.11	19.94
SiK	19.60	10.07
Matrix	Correction	ZAF

FIGURE 3.126 Cross-sectional SEM of the formed part with 3% SiO$_2$.

In order to better prove that the white part in Fig. 3.126 is nano-SiO_2, an energy spectrum analysis is performed for one of them. As can be seen in Fig. 3.127, the main elements of the white point are C, O, and Si, among which the ratio of the amount of the substance of Si and O is 1:2, indicating that SiO_2 is mainly contained therein, and the reason why the element C is contained is because the resin coats the nanoparticles, and the main element in the resin is C. This further demonstrates the uniform dispersion of nano-SiO_2 in the resin.

3. Effects of nano-SiO_2 on curing rate of photopolymer

Fig. 3.128 shows the relationship between the nano-SiO_2 and the gelation rate of the photopolymer. In the figure, C, C1, C2, C3, and C4

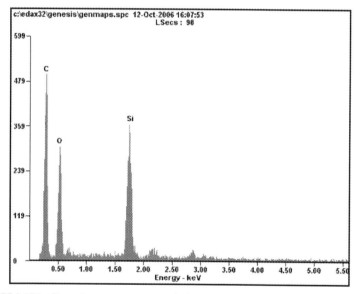

FIGURE 3.127 Chart of energy spectrum analysis.

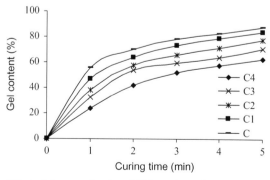

FIGURE 3.128 Effects of the mass fraction of SiO_2 on the curing rate of the resin.

indicate that the content of SiO_2 added to the photopolymer is 0%, 1%, 2%, 3%, and 4%, respectively. The experimental results show that the addition of SiO_2 reduces the curing rate of the resin, and as the amount of SiO_2 increases, the curing rate decreases.

4. Effects of nano-SiO_2 on curing shrinkage rate of photopolymer

As can be seen from Figs. 3.129 and 3.130, the addition of nano-SiO_2 reduces the linear shrinkage rate and volume shrinkage rate of the resin system, and the more the added amount, the more the shrinkage decreases. SiO_2 is a nonshrinking component which is added to fill voids caused by shrinkage of organic components, thereby reducing the shrinkage rate of the material.

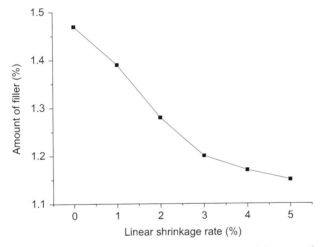

FIGURE 3.129 Effects of the mass fraction of SiO_2 on the linear shrinkage rate of resin.

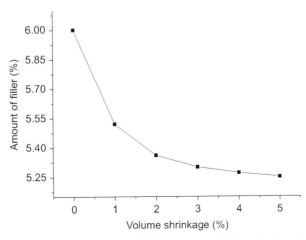

FIGURE 3.130 Effects of mass fraction of SiO_2 on volume shrinkage of photopolymer body.

5. Thermal properties

The thermal stability of the resin to which nano-SiO_2 is added is measured by a Shimadzu DTA-50 thermal analyzer. The N_2 atmosphere has a flow rate of 20.00 mL/min, a heating rate of 20.0°C/min, and a temperature range of (25°C–500°C), as shown in Fig. 3.131.

It can be seen from Fig. 3.112 that the thermogravimetric curve of the composite is obviously shifted to the high temperature direction compared with the thermogravimetric curve of the SLA material without filler, and the curve with the 1% filler has the largest offset, but the curves with 3%, 5%, and 1% filler are less obvious, indicating that the thermal performance improvement of the cured material has been optimized with the addition of 1% filler. The temperature at which the weight loss rate of the four samples is 5% is 120°C, 165°C, 149°C, and 144°C, and the temperature at the weight loss rate of 15% is 282°C, 299°C, 295°C, and 295°C, respectively. The addition of the nanofiller increases the heat resistance of the SLA material, and the heat resistance is optimal when 1% is added. The main reason is that the added rigid nano-SiO_2 forms chemical and physical crosslinking point with the polymer chain in the matrix, which limits the free movement of the polymer chain and inhibits the diffusion of oxygen into the interior when the composite is thermally decomposed, thereby delaying the decomposition process and increasing the thermal decomposition temperature of the material. When the addition ratio is further increased, it is difficult to disperse uniformly, and the nanofiller becomes agglomerated and micron-sized, so that its performance is affected.

6. Mechanical properties

Test the cured parts added with different amounts of filler in terms of tensile strength, impact strength, bending, and hardness. The test results are shown in Table 3.35.

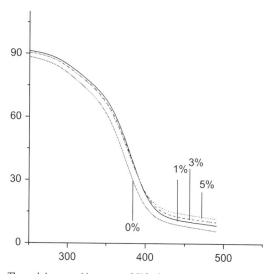

FIGURE 3.131 Thermal decomposition curve of SiO_2 forming materials with different mass fractions.

TABLE 3.35 Mechanical properties of formed parts containing different SiO_2.

SiO_2 (%)	Hardness (HRM)	Elongation (%)	Tensile strength (MPa)	Flexural modulus (MPa)	Impact strength (J/m^2)
0	88	5.50	89.98	2700	3.56
1	98	7.09	110.00	7813	4.02
2	95	6.84	129.10	5938	4.00
3	94	6.60	100.30	5313	3.42
4	90	4.83	85.20	4200	3.30
5	86	3.66	78.50	2300	3.01

It can be seen from the experimental results in Table 3.35 that the addition of nano-SiO_2 has a certain improvement in the properties of the material, and the addition ratio is preferably 1%–2%. When the filling amount is 1%, the hardness is increased by 16% and the flexural modulus is increased by more than 200%; when the filling amount is 2%, the tensile strength is increased by 60% and the elongation and impact strength are also improved. However, the filling amount of 4%–5% cannot obviously improve the system performance.

The following reasons are analyzed:

1. There are a large number of highly reactive hydroxyl groups on the surface of nano-SiO_2, and the role of the hydroxyl functional group in the photopolymer is to increase the strength of the extended chain, prevent the epoxy chain from an excessive crosslinking reaction, and increase the toughness of the part;
2. The silane coupling agent attached to the surface of the nano-SiO_2 contains epoxy group, which can participate in the polymerization reaction in the curing process and functions as a crosslinking point;
3. It can be seen from the SEM photo that the filler is dispersed in the matrix in nanostate, so that the active surface of the nanofiller can be combined with several polymer chains through physical and chemical reactions to form a crosslinked network structure and strengthen the interaction between the chains. When one of the molecular chains is stressed, the stress dispersion can be transmitted to other molecular chains through the crosslinking point. If one of the chains breaks, the other chains can act as they are, without damaging the whole part, so the filler pair has a good reinforcement effect on the matrix and the rigidity is also improved.
4. When the material is impacted, the cross section of the filler and the matrix can absorb a part of the impact energy and play a damping role, so that the toughness of the composite material can be improved.

However, the rigid nano-SiO$_2$ particles do not break or deform in themselves, and the introduction of rigid particles reduces the volume fraction of the matrix, thereby reducing the contribution of the matrix to toughness. At the same time, too many crosslinking points have great limits on the matrix molecular chain, reducing the flexibility of the molecular chain. The combination of the two factors makes the toughening effect of the nano-SiO$_2$ particles not obvious. If the filling ratio is too large, it cannot toughen, but increase the brittleness of the material.

It can be seen from Table 3.35 that when filling 1%–2% of nano-SiO$_2$, the performance indexes reach the maximum, and when it is more than 3%, it starts to decrease. It can be seen from the SEM that although the mass fraction of the filler is only 3%, the volume ratio of the filler in the matrix is large. When the filler content is increased, the nanoparticles will agglomerate several times, and it is difficult to disperse uniformly, and the microstructure of the material nonuniformity occurs, stress concentration occurs, and the material loses the nanoeffect and cannot achieve the modification effect.

1. Effects of nano-SiO$_2$ on critical exposure (E_c) and transmission depth (D_p) of photopolymer

    Prepare the photopolymer containing no SiO$_2$ and containing 1% SiO$_2$. Determine the resin parameters by using the HRPL-I type equipment developed by the Rapid Prototyping Center of Huazhong University of Science and Technology.

    The resin parameters are determined as follows: place a certain amount of photopolymer, a petri dish having a diameter of 20 cm, and accurately measure the power of the laser spot irradiated on the surface of the liquid resin, and then scan and cure the free surface of the resin with the scanning rate v of 800–1600 mm/s to obtain a series of sheets having a uniform layer thickness of 30.0 mm × 60.0 mm. Carefully clip the sheets with tweezers and clean with isopropyl alcohol. Each small piece of the test piece is divided into two pieces along the center line, and the thickness C_d is accurately measured with a micrometer. In this experimental method, the scanning rate is selected such that the cured sheet has certain strength. In the case where the laser power is determined, the higher the scanning rate, the smaller the resin exposure amount, and the thinner the cured layer thickness. When the layer thickness is too thin, the sheet will be too soft and thus affect the accuracy of thickness measurement.

    When conducting laser scanning for a sheet, the calculation equation of the energy E_0 which is irradiated on the surface of the resin, that is, the exposure amount, is:

$$E_0 = \frac{P_L}{v_2 S} \quad (3.53)$$

where P_L represents the laser power irradiated on the surface of the liquid resin, v_s represents the scanning speed, and S represents the scanning pitch.

The SLA equation is:

$$C_d = D_p(\ln E_0 - \ln E_c) \tag{3.54}$$

According to the thickness C_d of a series of sheets measured and the corresponding E_0 value obtained by Eq. (3.53), $\ln E_0$ is plotted against C_d, and the slope of the fitted straight line and the intercept on the x-axis are obtained by least squares method. The curing depth D_P of the resin and the critical exposure amount E_c are obtained. Taking $\ln E_0$ as the abscissa and C_d as the ordinate, the SLA equation is a straight line, the slope of the line is D_p, and the intersection of the line and the $\ln E_0$ axis is $\ln E_c$. Figs. 3.132 and 3.133 show the SLA curves of the photopolymer without filler and with 1% SiO_2, respectively.

It can be seen from Figs. 3.132 and 3.133 that C_d is linear with $\ln E_0$, and the linear correlation coefficients are 0.99697 and 0.99789, respectively. It can be seen that it conforms to the Beer−Lambert rule within the scanning speed range of the experiment. For filler-free photopolymer, $E_c = 11.446$ mJ/cm^2, and $D_p = 0.106$ mm; for photopolymer containing 1% SiO_2, $E_c = 12.941$ mJ/cm$_2$, and $D_p = 0.103$ mm, indicating that the addition of nano-SiO_2 reduces the

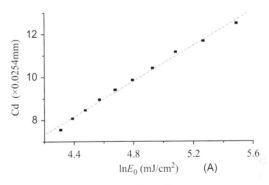

FIGURE 3.132 SLA curve of photopolymer without filler.

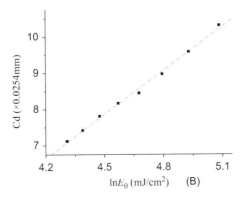

FIGURE 3.133 SLA curve of photopolymer containing 1% SiO_2.

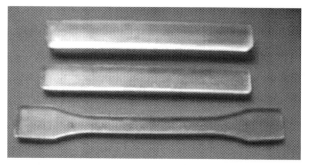

FIGURE 3.134 Sample of cured parts of resin containing 1% SiO_2.

transmission depth of the resin, and increases the critical exposure amount increase, this is because the nano-SiO_2 has a certain absorption effect on UV light, but it is weak and does not affect the curing and forming of the resin.

The sample of photopolymer cured part containing 1% SiO_2 is shown in Fig. 3.134.

3.4.2 Toughened photopolymer material of epoxy acrylate

The main component of SLA photopolymer is epoxy resin and its acrylate, and its cured product is a 3D network structure with high crosslink density. The movement of the main chain is very difficult, so the SLA formed parts has low impact strength and poor flexibility. The cured products will crack due to deformation and stress concentration when the temperature changes rapidly. As the SLA formed parts are widely used, the current trend is increasing the use of functional parts, meaning that the formed parts are directly used as parts, which requires SLA photopolymer to have good toughness and impact resistance.

3.4.2.1 Overview of toughening methods for photopolymers

Improvement of the toughness of the SLA photopolymer can be achieved by adding filler and adjusting the component having toughening effect in the photopolymer.

1. Adding filler

    When using the filler, it must be properly selected according to the requirements. From a chemical point of view, the filler must be free of bound water, inert to the SLA photopolymer system, nonadsorptive or less adsorptive to liquids and gases. From the operation point of view, the particles of the filler are 0.1 μm or more, have good affinity with the resin, and have small settleability in the resin. It is desirable that the mass fraction of the filler has little effect on the viscosity of the resin.
2. Inactive filler

The amount of inactive filler to be added is determined by three factors: (1) Control the resin to a certain viscosity. Too much amount will increase the viscosity of the resin, which is not conducive to the process; (2) ensure that each particle of the filler can be wetted by the resin, so the amount of filler should not be added excessively; (3) ensure that the part can meet various performance requirements.

Selectable fillers include SiO_2 crystallites, silicates, $Al(OH)_3$, mica or feldspar, and kaolinite.

Generally speaking, lightweight fillers such as asbestos powder are generally used in an amount of 25% or less. As the specific gravity of the filler increases, the amount can be increased accordingly. For example, the amount of mica powder and aluminum powder can reach more than 200%, and the amount of iron powder can exceed 300%. It can be seen that the amount of filler used varies very differently, which should be determined specifically.

It is not suggested to apply excessive filler to the photopolymer, because it is formed after accumulation by layer-by-layer scanning and curing, and the added inactive filler itself does not participate in the chemical reaction, but only acts as physical filling. Excess filler may cause bonding between layers or in the layers lose, thus affecting the mechanical properties of the parts.

3. Active filler

The active filler is a substance that can participate in the reaction of the system. If an active toughening agent is used, it can react with the epoxy resin of the system to form a part of the compound of the network structure of the curing system.

The active filler has active groups and directly participates in the curing reaction, which can greatly reduce the brittleness and easy-cracking of the cured parts of the SLA resin, and improve the impact strength and elongation of the resin. Generally, the active toughening agent commonly used in epoxy resin is mainly the monofunctional epoxidized vegetable oil and polyfunctional thermoplastic polyamide resin, polysulfide rubber, nitrile rubber, and polyester resin.

The experiment uses the acrylic modified solid epoxy resin (E-06) as active filler, and the modified solid bisphenol A epoxy acrylate has a higher molecular weight than the original resin, it is ground into fine powder and added in photopolymer. Since the added solid powder has high molecular weight and long molecular chain, the formed material has a shrinkage ratio lower than that of the raw material.

4. Method of adjusting some components in the photopolymer having toughening effect

1. Toughened epoxy resin

Such epoxy resin has a long fatty chain in the molecular structure and thus impart toughness to the cured product. For example, a side-chain epoxy resin (ADK-EP 4000) is shown in Fig. 3.135.

It is a resin with good toughness cured under room temperature and generally has good compatibility with liquid bisphenol A type epoxy

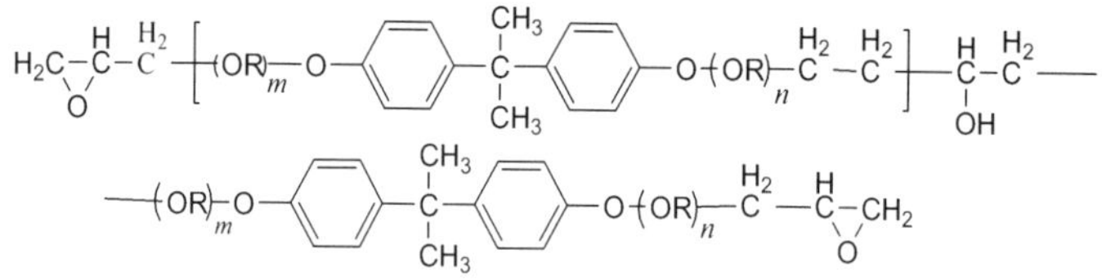

FIGURE 3.135 Side-chain epoxy resin.

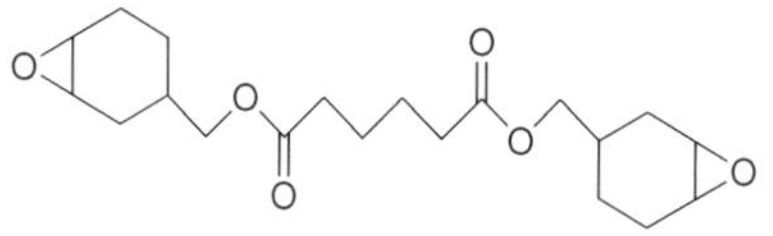

FIGURE 3.136 Bis-(3,4-epoxycyclohexyl) adipate.

resin. The longer the side chain on the bisphenol is, the more advantageous it is to increase the toughness of the cured resin.

There is also an alicyclic epoxy resin which can be cationically cured: bis-(3,4-epoxycyclohexyl) adipate, since it has a long molecular chain and can improve the toughness of the cured product during polymerization, so it is often used to adjust the performance of SLA photopolymer systems. Its molecular formula is shown in Fig. 3.136.

2. Toughened polyol compound

   The polyol compound can also exert certain toughening effects on the SLA photopolymer. In this study, the bisphenol A epoxy acrylate resin solid is synthesized from a long-chain bisphenol A epoxy resin, which is ground and added to the earlier-mentioned hybrid system SL-01 to discuss the impact of its mass fraction on all aspects of the resin. It is used as the toughening material SL-02 to produce the formed part.

3.4.2.2 Synthesis of epoxy acrylate toughening filler

1. Synthesis method

   There are two kinds of acrylate methods for epoxy resin: one is esterification of epoxy resin directly with acrylic acid under the action of catalyst; the other is the reaction between hydroxyalkyl acrylate, maleate or other anhydride intermediate and epoxy resin. In this experiment, epoxy resin and acrylic acid are reacted to synthesize epoxy resin acrylate. The reaction formula is as follows:

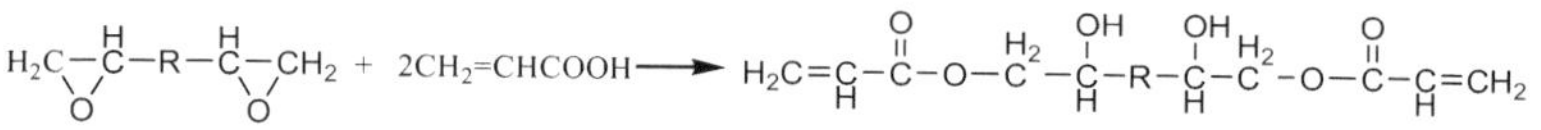

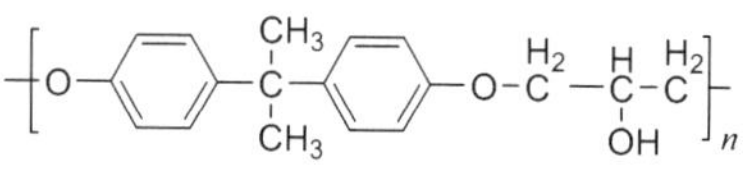

For bisphenol A epoxy resin, the smaller the epoxy value (the amount of the epoxy group-containing substance per 100 g of the resin), the longer the chain length, the better the flexibility, and the better flexibility of the product after curing. Since the purpose of this experiment is to improve the toughness of the SLA photopolymer, E-06 is thus selected as the experimental raw material to synthesize epoxy acrylate.

Add a certain amount of epoxy resin E-06 to a ground four-necked flask equipped with stirrer, dropping funnel, condenser and thermometer, and then add an appropriate amount of xylene (based on dissolved E-06) in it, add a certain amount of polymerization inhibitor, heat and stir to a certain temperature (90°C), then slowly add a mixture of acrylic acid and catalyst dropwise, maintain the reaction temperature of 100°C–110°C for 2–3 hours, raise the temperature to 110°C–115°C to react one hour further until the acid value is less than 8. The xylene is distilled under reduced pressure, and is subjected to purification, that is, to add an appropriate amount of hot water of 60°C–70°C and stir 10 minutes, and remove the upper aqueous solution by tilting the flask after it is layered, so as to remove unreacted acrylic acid, polymerization inhibitor, and catalyst. After repeating three times, the water remaining in the system is removed to obtain the pale yellow liquid. Pour out the hot yellow liquid to get solid substance.

2. Influencing factors

1. Polymerization inhibitor

The temperature of the synthesis reaction of bisphenol A epoxy acrylate resin (EA) is high, which is easy to trigger self-polymerization reaction of acrylic acid, thereby affecting the SLA property and even leading to the explosion. Therefore, the polymerization inhibitor is indispensable in the reaction. By comparing hydroquinone and *p*-hydroxyanisole, it is found that the color of EA resin synthesized by using hydroquinone as polymerization inhibitor is reddish, while the color of EA resin synthesized by using *p*-hydroxyanisole as polymerization inhibitor is yellowish. Therefore, *p*-hydroxyanisole is used as polymerization inhibitor. When the mass fraction is more than 0.3%, the color of the product becomes darker and even brownish red; when the mass fraction is less than 0.2%, the reaction time is prolonged and the reaction degree is small, so a mass fraction of 0.25% is selected.

2. Catalyst

The ring-opening reaction of acrylic acid and epoxy resin must adopt a catalyst with high selectivity for the reaction of hydroxyl groups with epoxy groups. Generally, tertiary amines and quaternary amines are used. *N,N*-dimethylbenzylamine is selected as synthetic catalyst in the experiment, the degree of reaction is determined by observing the change in the residual acid value, P value (P is defined as the ratio of the acid value measured at a certain time to the initial acid value) after changing its mass fraction. Fig. 3.137 shows the relationship between P value and reaction time for different catalyst contents.

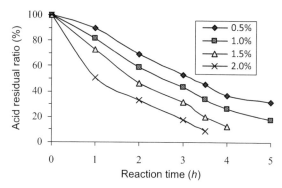

FIGURE 3.137 Effects of the mass fraction of the catalyst on the reaction.

Fig. 3.137 shows that as the mass fraction of catalyst increases, the reaction rate increases significantly. When the mass fraction of catalyst is increased to 2%, the synthesis reaction time is only 3.5 hours. This is mainly because the catalyst is bound to the epoxy group in the esterification reaction, and the epoxy group is positively charged, which is advantageous for the attack of acrylic acid as nucleophile. Therefore, the more the catalyst content, the more positively charged epoxy groups formed in the reaction, the faster the reaction rate, and the lower the P value; but when the reaction proceeds to a certain extent, the positively charged epoxy group gradually decreases, and the excess catalyst does not have much influence on the reaction, and the P value decreases relatively slowly. However, since the catalyst is not eliminated in the synthesized product, the excessive mass fraction affects certain properties of the synthesized product, such as product color and storage time. Taking into account the aforementioned factors, the mass fraction of catalyst should be 1.5%.

3. Reaction temperature

In the esterification reaction of epoxy resin with acrylic acid, the reaction temperature is one of the most important factors. The effects of different temperatures on the esterification of epoxy resin with acrylate are shown in Fig. 3.138.

The figure shows that the higher the reaction temperature is, the faster the reaction proceeds. When the reaction temperature exceeds 120°C, the reaction can be completed in a shorter period of time, and the final reaction degree is also higher. However, too high reaction temperature easily causes thermal polymerization or other side reactions of the double bond group contained in the material, resulting in an increase in viscosity of the product and affecting the color and purity of the product. In addition, due to the existence of exothermic heat peak in the synthesis reaction, if the reaction temperature is too high, it is easy to cause the temperature to

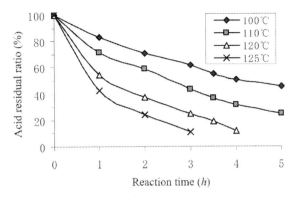

FIGURE 3.138 Effects of temperature on the reaction.

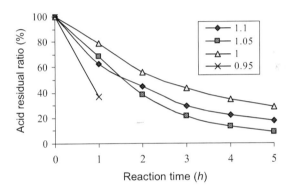

FIGURE 3.139 Effects of the ratio of different acrylic groups to epoxy groups on the reaction.

be uncontrollable to produce gel. Therefore, the temperature of the synthesis reaction is suitably in the range of 110°C–115°C.

4. Ratio of acrylic group to the epoxy group

Since some side reactions may occur during the acylation process, and the reaction may be incomplete, the ratio of the amount of the acrylic group to the epoxy group substance is controlled during the synthesis, and only under appropriate conditions, the needs of synthesis can be met. The effects of their difference ratio on the degree of reaction are shown in Fig. 3.139.

It can be seen from the figure that when the ratio is 0.95, the gel is generated during the synthesis of the system, causing the reaction to fail; and when the ratio is increased, the greater the ratio in the initial stage, the greater the degree of reaction, but after two hours, the reaction rate with a ratio of 1.10 is not as fast as a ratio of 1.05, and excessive acrylic acid undergoes self-polymerization reaction during the reaction, so that impurities in the reaction product are difficult to remove. To sum up, the ratio of the acrylic group to the epoxy group is selected to be 1.05.

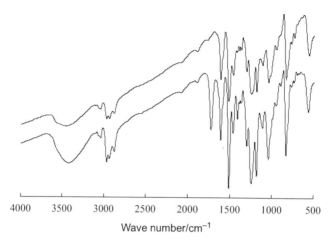

FIGURE 3.140 FTIR spectrum before and after modification.

Fig. 3.140 shows the infrared spectrum before and after modification of bisphenol A epoxy resin (E-06).

In the spectrum before resin modification, 1292.17, 1172.60, and 819.66 cm^{-1} are the characteristic absorption peaks of stretching vibration of $-CH$, and 1226.60 cm^{-1} is the characteristic absorption peak of stretching vibration of C-O, 1602.68 and 1502.39 cm^{-1} are the characteristic absorption peaks of the stretching vibration of the benzene ring, and 2858.20 and 2293.21 cm^{-1} are the characteristic absorption peaks of the symmetric and asymmetric stretching vibration of CH_3, respectively.

Compared with the infrared spectrum of the bisphenol A epoxy acrylate before and after modification, a distinct absorption peak appears at 1718.39 cm^{-1}, which is caused by the stretching vibration of $C=O$. It indicates that carbonyl group is introduced into the product to carry out an esterification reaction.

3.4.2.3 Effects of epoxy acrylate toughening filler on photopolymer

Grind the aforementioned synthesized EA solid into powders and add the powders as filler to the SLA photopolymer system. Investigate the influence on the curing system of the photopolymer to find a suitable ratio to toughen the SLA photopolymer.

1. Preparation and curing of resin

    In a 5000 mL special three-necked flask equipped with stirrer, reflux condenser and thermometer, add a certain amount of SLA photopolymer SL-01 sequentially, and the modified toughened bisphenol A epoxy resin of different mass fractions, heat and stir until it becomes pale yellow transparent liquid, then pour the photopolymer (SL-02) into the opaque container for storage for experimental use. (Note: During the preparation process, the three-necked flask should be protected from light as much as possible.)

Coat the prepared resin SL-02 evenly on a clean slide glass and cure it in an UV exposure chamber.

2. Effects of the mass fraction of EA on the curing rate of SLA photopolymer

Fig. 3.141 shows the relationship between the mass fraction of EA and the gel rate of SLA photopolymer. In the figure, C, C1, C2, C3, and C4 indicate that the content of EA added in the SLA photopolymer is 0%, 2%, 4%, 6%, and 8%, respectively. The experimental results show that the addition of EA reduces the curing rate of the resin, and the rate decreases as the amount of EA increases. The reason may be that EA has large molecular weight, small functional group density, and low reactivity, so that its addition causes a decrease in the curing rate of the SLA photopolymer.

3. Effects of the content of EA (mass fraction) on the shrinkage rate of SLA photopolymer

Figs. 3.142 and 3.143 show the effects of different contents of EA on the linear shrinkage rate and on volume shrinkage rate of SLA photopolymer, respectively.

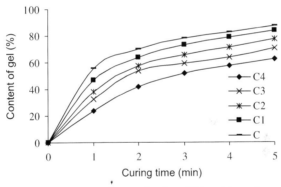

FIGURE 3.141 Relationship between the mass fraction of EA and the curing rate of the resin.

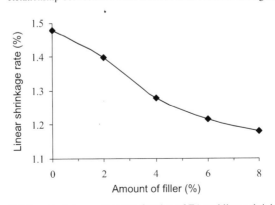

FIGURE 3.142 Relationship between the mass fraction of EA and linear shrinkage rate.

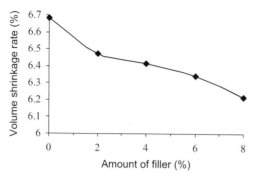

FIGURE 3.143 Relationship between mass fraction of EA and volume shrinkage rate of EA.

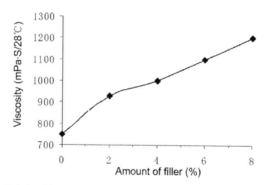

FIGURE 3.144 Relationship between the mass fraction of EA and the viscosity of the resin.

It can be seen from the figures that the larger the content of EA is, the lower the linear shrinkage rate and volume shrinkage rate of the material will be, but it does not mean that it would be better if the EA content is larger, and it should also be considered that the high EA content will affect other aspects of the curing system, for example, the viscosity.

4. Effects of the mass fraction of EA on the viscosity of SLA photopolymer

Fig. 3.144 shows the effects of the content of EA on the viscosity of SLA photopolymer system. The results show that the viscosity of the system increases with the increase of EA, so although the shrinkage rate of the curing system can be improved with the addition of EA, the curing rate and the viscosity of the system become poor. Therefore, a variety of factors should be considered when selecting the mass fraction.

5. Effects of the mass fraction of EA on the flexibility of SLA photopolymer cured film

The flexibility of the cured film is tested according to the GB/T1731-93 standard. The resin is coated on a special tinplate and placed in an UV curing box. After it is fully cured, the iron piece is bent to a certain radius, and

TABLE 3.36 Relationship between the mass fraction of EA and the flexibility of cured film.

EA (%)	0	2	4	6	8
Flexibility (mm)	<5	<4	<3	<2	<1

the flexibility is beyond a certain radius when there is no break after coating (usually expressed in less than a certain radius, mm).

Table 3.36 shows the effects of the mass fraction of epoxy acrylate on the flexibility of the cured film. The flexibility of the cured film increases with the increase of EA. The reason may be that the long molecular chain of EA provides flexibility to the cured film, so as the EA increases, the flexibility of the cured film becomes higher.

Considering all the aforementioned factors, the EA to be added in the SLA photopolymer should be 4%.

3.4.2.4 Preparation and properties of photopolymer toughening system

1. Preparation of photopolymer toughening system

    Take an appropriate amount of SL-01 photopolymer in a 5000 mL special three-necked flask equipped with stirrer and thermometer, and weigh the synthetic resin EA powder of 4% SL-01 mass fraction into it, stir it at high speed under the action of ultrasonic wave, and stir it evenly. After that, take it out and store in the dark, then the resin prepared is SL-02.

2. Performance index of photopolymer toughening system

    Table 3.37 lists the main performance indicators of SL-02 resin for parts made on the rapid prototyping equipment HRPL-I.

    It can be seen from the table that SL-02 resin is greatly improved in elongation at break and impact strength than SL-01, although the tensile strength is decreased, it can meet the requirements for SLA forming.

3.4.3 Synthesis and application of a novel alicyclic epoxy acrylate

In order to obtain good performance of SLA photopolymer, a radical−cation hybrid polymerization system can be used, which simultaneously generates two kinds of polymerization reactions: free radicals and cations, exhibiting good synergistic effect. It can complement each other, and expand the applications of UV curing system.

In this study, the alicyclic epoxy resin is used as the raw material to synthesize into acrylate resin. A simple SLA hybrid system is formed after adding the cationic photoinitiator and free-radical initiator in the resin.

TABLE 3.37 Main properties of SL-02 material.

Resin	Density (g/cm³)	Viscosity MPs 30°C	Curing depth (mm)	Critical exposure amount (mJ/cm²)	Tensile strength (MPa)	Elastic modulus (MPa)	Elongation at break (%)	Impact strength (J/m²)	Hardness	Glass transition temperature (°C)
SL-02	1.19	510	0.145	6.14	51	1580	8.3	30	D87	62
SL-01	1.15	420	0.125	5.62	53	1740	5.1	24	D86	58

3.4.3.1 Synthesis of alicyclic epoxy acrylate

1. Synthesis method

   EA is synthesized by direct reaction of alicyclic epoxy resin with acrylic acid. The reaction diagram is shown in Fig. 3.145.

   Since the main purpose of the experiment is to synthesize an EA containing both epoxy group and C=C double bond, and it can be used for UV curing, so the bis-(3,4-epoxycyclohexyl) which has good flexibility and low-viscosity adipate is selected as the raw material for synthesis [9].

   Add the alicyclic epoxy resin to a ground four-necked flask equipped with a stirrer, dropping funnel, condenser and thermometer, add a certain amount of polymerization inhibitor, stir evenly, and then warm up to a certain temperature (80°C), then add the mixture of acrylic acid and catalyst dropwise slowly, maintain the reaction temperature at about 80°C–90°C for 2.5–3.5 hours, and react for another 1.5 hours when the temperature is raised to about 90°C–95°C until the acid value is less than 8. Purify the product by adding an appropriate amount of water of 60°C–70°C and stir 10 minutes, and remove the upper aqueous solution by tilting the flask after it is layered, so as to remove unreacted acrylic acid, polymerization inhibitor and catalyst. After repeating this for three times, the water remaining in the system is removed to obtain the pale yellow liquid. Pour out the hot yellow liquid and store it in a brown bottle.

2. Influencing factors

1. Polymerization inhibitor

   The temperature of the synthesis of the alicyclic epoxy resin acrylate is low, but the self-polymerization reaction of acrylic acid is also prone to occur, which also affects the SLA property, so the polymerization inhibitor is also added in the reaction. p-Hydroxyanisole is used as polymerization inhibitor, and the mass fraction is 0.1%.

2. Catalyst

   N,N-dimethylbenzylamine is selected as a catalyst. Fig. 3.127 shows the relationship between P value and the reaction time for different catalyst contents.

   Fig. 3.146 shows that as the mass fraction of catalyst increases, the reaction rate increases significantly. If the catalyst mass fraction is

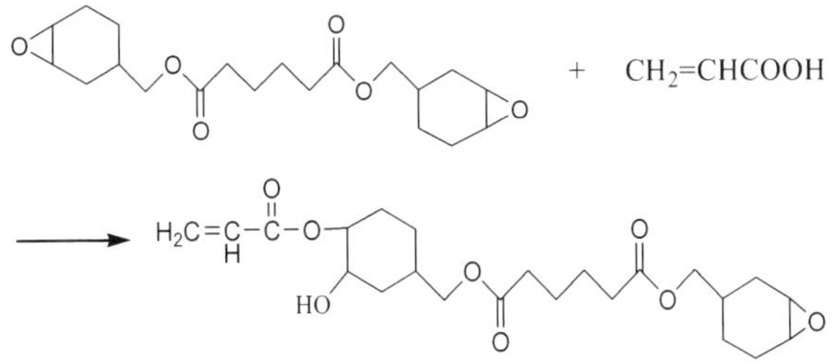

FIGURE 3.145 Synthetic route of alicyclic EA.

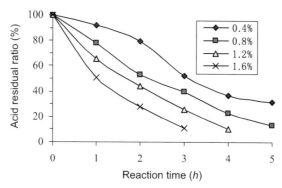

FIGURE 3.146 Effects of the mass fraction of the catalyst on the reaction.

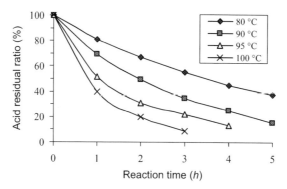

FIGURE 3.147 Effects of temperature on reaction.

increased to 1.6%, the synthesis reaction time is only 3 hours. However, while pursuing the decrease of reaction time, it should also consider that excessive catalyst is not easily eliminated in the synthesized product, and too large mass fraction will affect certain properties of the synthesized product, such as product color and storage time. For comprehensive consideration, the catalyst mass fraction is chosen to be 1.2%.

3. Reaction temperature

In the esterification reaction of epoxy resin with acrylic acid, the reaction temperature is one of the most important factors. The epoxy resin and acrylate esterification reactions are compared at several different temperatures. The results are shown in Fig. 3.147. The figure shows that the higher the reaction temperature, the faster the reaction rate. When the reaction temperature exceeds 100°C, the reaction can be completed in a short period of time, and the final reaction degree is also high. However, too high reaction temperature easily causes thermal polymerization or other side reactions of the double bond group contained in the material, resulting in an increase in viscosity of the product and affecting the color and purity of the product. In addition, due to the existence of an exothermic peak in the synthesis reaction, if the reaction temperature is too high, it is easy to cause the temperature to be uncontrollable to produce a gel. Therefore, the temperature of the synthesis reaction is suitably in the range of 90°C–95°C.

4. Ratio of acrylic group to epoxy group

In the acrylation reaction, only 1 mol of the epoxy group per 2 mol of the epoxy group is esterified with acrylic acid, so that the ratio of the amount of the acrylic group to the epoxy group should be controlled well to achieve successful synthesis. The effects of the difference between the two compounds on the degree of reaction are shown in Fig. 3.148.

It can be seen from the figure that under the aforementioned experimental conditions, the epoxy group appears to be absolutely excessive relative to the acrylic group, so the acid value conversion rate at the end of the reaction is very low, but when the ratio is 1:1.2, the reaction rate reaches the end point in the fastest speed. Since only one of the two epoxy groups is expected to participate in the reaction, the ratio of 1.1:2 is selected considering all factors, which satisfies the requirements.

The infrared spectrum of the epoxy resin before and after the reaction is shown in Fig. 3.149.

The three absorption peaks of the alicyclic epoxy resin before the reaction at 1722.25, 1145.60, 896.81, 1220.81, and 1054.95 cm^{-1} are characteristic absorption peaks of C=O stretching vibration, C-O stretching vibration, =C–H deformation vibration, C–O–C asymmetric stretching vibration, and C–O–C symmetric stretching vibration respectively in the alicyclic epoxy resin. The absorption peaks appearing at 2916.06, 1427.17, and 744.44 cm^{-1} are the characteristic absorption peaks of the stretching vibration and bending vibration of the CH bond in the $(CH_2)_6$ group, respectively.

In the alicyclic epoxy acrylate spectrum after the reaction, it can be seen that the stretching vibration absorption peak of C=C chain appears

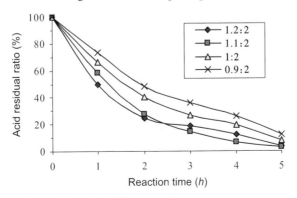

FIGURE 3.148 Effects of the ratio of different acrylic groups to epoxy groups on the reaction.

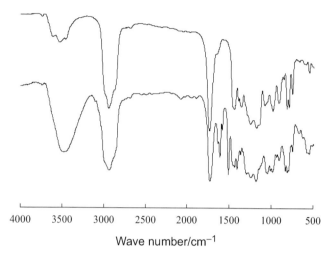

FIGURE 3.149 FTIR spectrum of epoxy resin before and after reaction.

TABLE 3.38 Different formulations of pure cationic systems.

Series	C1	C2	C3	C4
Resin A	100	100	100	100
Initiator a	5	3	2	1

at 1604.61 cm^{-1}, and the deformation vibration of —OH appears at 1404.03 cm^{-1}. This demonstrates the acrylic acid esterification of alicyclic epoxy resin is successful.

3.4.3.2 Stereolithography apparatus properties of resin

Accurately weigh the various resins and initiators in a 100 mL conical flask with cover according to the designed formula, magnetically stir for a certain period of time, and mix them evenly. Spread the prepared resin evenly on a clean slide glass and cure it in an UV exposure chamber.

1. Effects of the mass fraction of initiator on the initiation rate of pure cationic system

    According to Table 3.38, four kinds of SLA systems ranging from C1 to C4 are prepared, and the gel rates are measured. The results are shown in Fig. 3.130.

    Theoretically, due to the absence of oxygen inhibition in cationic polymerization, as the mass fraction of cationic photoinitiator increases, the more active species are produced by photolysis, the initiation rate of the photopolymer system will be faster, so that the degree of reaction will be greater.

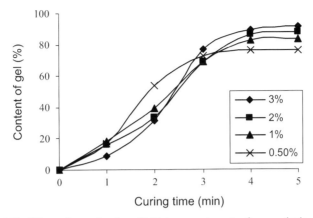

FIGURE 3.150 Effects of mass fraction of initiator on curing rate of pure cationic system.

TABLE 3.39 Initiator contents of different systems.

Series	C5	C6	C7	C8
Resin A	100	0	0	0
Resin B	0	100	100	100
Initiator a	3	3	3	0
Initiator b	0	3	0	3

However, it can be seen from Fig. 3.150 that the reaction rate of the photopolymer system decreases as the mass fraction of photoinitiator increases in the initial stage of reaction. It may be because the mass fraction of the initiator in the system is so large that the surface layer of the resin is rapidly cured, thereby affecting the penetration of the light wave into the resin, which is not conducive to the deep polymerization of the resin, and exhibits low gel rate. As the reaction continues, that is, prolonging the illumination time, the deep photopolymer is less affected by the surface gel, which is reflected as faster reaction rate and higher conversion rate with the increase of the mass fraction of initiator in the middle and later stage of the reaction,

Since the cationic initiator is expensive, increasing its mass fraction will undoubtedly increase the cost of the material. Therefore, in combination with various factors, the mass fraction of the initiator is selected to be 3%.

2. Effects of different initiation systems on the reaction rate of alicyclic epoxy acrylate

According to Table 3.39, four kinds of hybrid systems ranging from C5 to C8 are prepared, and the gel rates are measured. As shown in Fig. 3.151, the

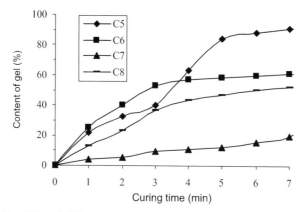

FIGURE 3.151 Effects of different initiation systems on the curing rate of alicyclic epoxy acrylate.

reaction rate of pure cationic system C5 is higher than that of the other three systems with resin B added. It may be because the density of functional groups in pure cationic system C5 is relatively large, the number of groups in which reaction occurs is relatively large, and the gel rate is high.

The reaction rate of the system C7 with the addition of the cationic initiator only is the lowest, followed by the system C8 with the addition of the radical initiator, while the reaction rate of the system C6 with the addition of the radical and the cationic initiator at the same time is the fastest. The reason is that the free-radical initiator first photolyzes to obtain a radical, as shown formula (3.55); then a charge-transfer reaction occurs between the radical and the sulfonium salt, eventually forming strong acid, as shown in the formula (3.56).

$$\text{Ph-C(=O)-CH(OH)-Cy} \xrightarrow{hv} \text{Ph-C(=O)} \cdot + \cdot \text{CH(OH)-Cy} \qquad (3.55)$$

$$\text{Cy-CH(OH)} \cdot + \text{Ar}_3\text{S}^+\text{SbF}_6^- \longrightarrow \text{Ar}_2\text{S} + \text{Ar} \cdot + \text{Cy-CH(OH)}^+ \text{SbF}_6^- \qquad (3.56)$$

Since the rate of polymerization increases as the strength of the co-initiator acid during cationic polymerization, the formation of a strong acid is advantageous for cationic polymerization, which greatly improves the initiation efficiency of the radical–cation hybrid SLA system, thereby initiating polymerization faster and more thoroughly, and exhibiting higher gel rate.

3. Volume shrinkage rate and linear shrinkage rate of different curing systems

For SLA formed parts, when the linear shrinkage rate is reduced to a certain extent, the shrinkage stress is much smaller than the adhesion between and in the layers, which no longer causes the warping deformation of the parts. The volume shrinkage rate and linear shrinkage rate of the C6–C8 systems after complete curing are shown in Table 3.40.

As can be seen from Table 3.40, the curing shrinkage rate of the system C6 in which both the radical and the cationic initiator are simultaneously added is smaller than the systems C7 and C8, and it is significantly lower than the radical system. It indicates that the resin has a low curing shrinkage rate when both the free-radical and cationic initiator are added at the same time, which can better meet the needs of the SLA process.

4. Infrared spectrums of cured films and resins of different systems

Fig. 3.152 shows the infrared spectrums of the cured films of different reaction systems. It can be seen from the figure that the spectrum of resin B shows three absorption peaks at 1610.12, 1247.02 and 830.90 cm^{-1}, which are the characteristic absorption peaks of $C=C$ stretching vibration, C–O–C symmetric stretching vibration, and C–O–C asymmetric stretching vibration conjugated with the carboxyl group, respectively. In

TABLE 3.40 Linear shrinkage rate and volume shrinkage rate of different resin systems.

Series	C6	C7	C8
Volume shrinkage rate (%)	5.73	5.97	7.08
Linear shrinkage rate (%)	0.092	0.059	0.473

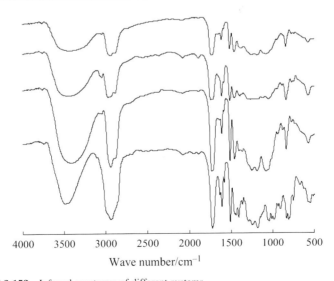

FIGURE 3.152 Infrared spectrums of different systems.

the C8 spectrum, the characteristic absorption peak of the C=C double bond is slightly weakened, indicating that the degree of reaction of the system is small when only the free-radical initiator is added. In the C7 spectrum, the characteristic absorption peak of the epoxy group is greatly weakened, and at the same time, the characteristic peak of the C=C double bond also appears to be largely weakened. This indicates that only the addition of cationic initiator allows the system to undergo both radical polymerization and cationic polymerization though the degree of reaction is not large. This is because the cationic initiator can generate both free radicals and cations upon photolysis, so that the system can be promiscuously polymerized to a lesser extent. In the C6 spectrum, the aforementioned three absorption peaks almost completely disappear, which fully demonstrates that the addition of the radical initiator can effectively sensitize the cationic initiator, so that the system can sufficiently carry out hybrid polymerization.

From the aforementioned analysis, it is found that the synthesized resin B contains epoxy group and acrylate group, and is able to cause hybrid polymerization under a suitable initiator.

5. Thermal stability analysis of cured films of different systems

In the SLA rapid prototyping material, the glass transition temperature (T_g) is an important index, when the T_g gets higher, the high temperature resistance will be strengthened, and the performance of the part will be more stable.

From the DSC analysis curves in Fig. 3.153, it can be seen that the material obtained after the curing of the system C6 with two initiators added has higher glass transition temperature than the system C7 in which only the radical initiator is added and the system C8 in which only the cationic initiator is added. This is because both the epoxy group and the acrylate group in the system participate in the reaction, so the density of the

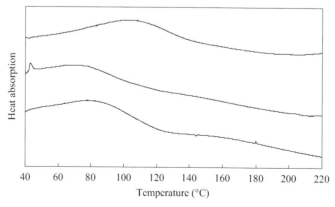

FIGURE 3.153 DSC curves of cured films of different systems.

crosslinking point in the system increases, the free volume of the polymer decreases, and the degree of molecular chain activity is also restricted. The average chain length between the crosslinking points becomes small, so that the glass transition temperature of the product is high.

References

[1] D. Dimitrov, W. van Wijck, N. de Beer, et al., Development, evaluation, and selection of rapid tooling process chains for sand casting of functional prototypes, Proc. Inst. Mech. Eng. Part B-J. Eng. Manuf. 221 (9) (2007) 1441–1450.

[2] L. Haitao, Research and application of SLA AM forming materials, Huazhong Univ. Sci. Technol. (2009).

[3] L. Haitao, M. Jianhua, L. Houcai, Study on SLA kinetics of trimethylene oxide, J. Huazhong Univ. Sci. Technol. 36 (11) (2008) 129–132.

[4] L. Haitao, M. Jianhua, H. Xiaomao, Research on the mechanism of accelerator for SLA resin in AM, J. Donghua Univ. 26 (1) (2009) 16–20.

[5] B.Y. Tay, S.X. Zhang, M.H. Myint, et al., Processing of polycaprolactone porous structure for scaffold development, J. Mater. Process. Technol. 182 (1–3) (2006) 1.

[6] L. Haitao, H. Shuhuai, M. Jianhua, T. Fulan, H. Biwu, Effects of photopolymer on surface quality of rapid prototyping parts, Polym. Mater. Sci. Eng. 23 (5) (2007) 170–173.

[7] D. Yugang, W. Suqin, L. Bingheng, Study on shrinkage of SLA resin for stereoscopic modeling, J. Xi'an Jiaotong Univ. 34 (3) (2000) 45–48.

[8] L. Haitao, M. Jianhua, Study on the nanosilica reinforcing stereolithography resin, J. Reinf. Plast. Comp. (2009).

[9] H. Zhang, J.L. Massingill, J.T.K. Woo, V.O.C. Low, Low viscosity UV cationic radiation-cured ink-jet ink system, J. Coat. Technol. 72 (905) (2000) 45–52.

Chapter 4

Polymer material for additive manufacturing—filament materials

4.1 Fused deposition modeling principle and process of polymer filament materials

4.1.1 Fused deposition modeling principle

Fused deposition modeling (FDM) is a typical forming method in three-dimensional (3D) printing technology, which involves mechanical, numerical control, materials, and computer technologies. The forming principle is shown in Fig. 4.1 [1].

During the FDM process, the filamentous (typically 1.75 or 3 mm in diameter) thermoplastic material is melted by heating the extrusion head, and the bottom of the extrusion head is provided with a fine nozzle (typically 0.2–0.6 mm in diameter) to allow the material to be extruded at a certain pressure. At the same time, the extrusion head moves two-dimensionally along the horizontal plane, and the workbench moves in the vertical direction. The material thus extruded is welded to the previous layer. After deposition of one layer is completed, the workbench is lowered in thickness by one layer according to the predetermined increment, and then further fused deposition is continued until the entire solid model is completed.

4.1.2 Analysis of material modeling process

At present, the materials used in FDM process are basically thermoplastics. Granular thermoplastics need to be extruded and drawn into filaments by an extruder before they can be used in FDM. The extrusion process and FDM process are specifically analyzed below.

4.1.2.1 Screw extrusion process

The filament used in FDM has a diameter of 1.75 mm, and is required to have uniform diameter (accuracy is ±0.05 mm), smooth surface, compact interior, good flexibility, and has no defects such as hollowness and surface

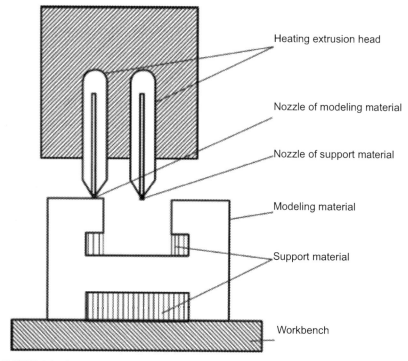

FIGURE 4.1 Schematic diagram of FDM technology.

knot. The forming method of FDM filament can be called precision extrusion.

Precision extrusion is a molding method that significantly improves the precision of extruded products by controlling the product process, optimizing the design of the extruder, applying new auxiliary molding machines, and controlling the mechanical and electrical precision. Product precision mainly refers to geometric precision, repetition precision, and functional precision. Geometric precision includes dimensional precision, shape, and position precision, which is the main problem to be solved in precision molding. Repetitive precision mainly reflects the dimensional stability of extruded products in axial direction. Functional precision refers to the mechanical properties, optical properties, thermal properties, and surface quality of the molded products. Different products have different requirements for the functional precision.

When the plastic is continuously extruded by a single screw extruder, the discharging is sometimes fast or slow, causing yield fluctuation. Fluctuation in yield is one of the main problems in extrusion production. It will cause unstable cross-sectional size of the product, resulting in uneven diameter of the filament. Assuming that the plastic is homogeneous Newtonian fluid under isothermal condition in the spiral groove, the theoretical calculation equation of the volume flow rate of the extruder is

$$Q = \frac{\pi^2 D^2 nh\sin\theta\cos\theta}{2} - \frac{\pi Dh^3\sin^2\theta \times P}{12\eta_1 L} - \frac{\pi^2 D^2 \delta^3 \varepsilon\tan\theta \times P}{12\eta_2 EL} \quad (4.1)$$

where Q is the volume flow rate (cm^3/s); D is the outer diameter of the screw (cm); n is the rotating speed of screw (r/s); h is the depth of the spiral groove in the metering section (cm); θ is the helix angle (°); E is the width of screw edge (cm); L is the length of screw metering section (cm); δ is the clearance between screw and cylinder (cm); ε is the check factor of screw eccentricity, usually take 1.2; P is the material pressure at the end of the screw metering section (Pa); η_1 is the viscosity of the molten material in the spiral groove (Pa s); and η_2 is the viscosity of the molten material in the gap (Pa s).

From this, it can be seen that the volume flow increases with the increase of rotational speed and decreases with the increase of pressure. The viscosity is affected by temperature, therefore, under the condition of fixed equipment, temperature, pressure, and rotational speed are the main factors affecting the flow rate.

1. Influence of temperature

    The quality of extruded products depends to a large extent on the temperature fluctuation ΔT_m of the melt at the head of the extruder screw. ΔT_m depends on the material's own performance and product requirements. It is found that some defects can still be detected in the final product when ΔT_m reaches $\pm 1°C$. To make the plastic extrusion process smoothly and improve the efficiency, one of the key factors is to control the temperature of each section of the cylinder. Higher cylinder temperature reduces the viscosity of the material. The relationship between the viscosity of polymer and temperature can be expressed by Arrhenius equation:

$$\eta = Ae^{-E_a/RT} \quad (4.2)$$

    where E_a is viscous flow activation energy (kJ/mol); R is gas constant (8.314 J/(mol K)); A is constant; and T is the absolute temperature.

2. Influence of screw rotational speed

    Screw rotational speed is one of the most important process parameters in extrusion molding. As the rotating speed increases, the shear rate increases, which is beneficial to material homogenization. However, the load capacity of the main machine and the melt pressure range should be considered when increasing the rotating speed. Otherwise, the material will be delivered to the machine head without being plasticized, which will lead to quality degradation. Therefore, when the machine is just started, the screw rotational speed is usually adjusted to be smaller and the charging amount is also smaller. After the material is extruded from the machine head, the rotating speed is slowly increased and the charging amount is increased. At the same time, the current and melt pressure of the main machine should be closely observed until the set rotating speed is reached. Increasing the rotating speed can greatly improve the productivity of the extruder.

3. Influence of pressure

    During the extrusion process, a certain pressure will be produced inside the material due to the resistance of the material flow and the change of the screw groove depth, the resistance of the filter screen, filter plate, and die. Pressure is one of the important conditions for plastic to turn into a molten state to obtain a uniform melt and finally to extrude the compact product. When the melt pressure is increased, the volume compresses and the molecular chain accumulates closely, so that the viscosity increases and fluidity decreases, resulting in decreased extrusion but dense product, which is conducive to improving product quality.

4. Interaction among temperature, pressure and flow fluctuation

    For precision extrusion process, the system is required to have stable flow rate, constant pressure, and constant temperature, and the extrudate has uniform physical and chemical properties. Thus, the quality can be assessed by any parameter change therein. However, these variables are not independent. For example, pressure fluctuation will cause flow fluctuation, temperature fluctuation will cause viscosity fluctuation, and viscosity fluctuation will cause pressure fluctuation and flow fluctuation. To ensure the quality of extrudate, the fluctuation of the above parameters must be reduced as much as possible.

    a. Interaction between pressure fluctuation and flow fluctuation

       The flow rate through the die depends on the pressure established in front of the die. As shown in Table 4.1, small changes in pressure can cause great changes in flow rate.

    b. Interaction between temperature fluctuation and pressure fluctuation

       Consider the assumption that the flow rate remains constant and the temperature fluctuates. For Newtonian fluid, temperature change will cause viscosity change, while for power-law fluid, it will cause the change of non-Newtonian index n. In order to maintain a constant flow rate, the pressure must change. If there is pressure fluctuation but no temperature fluctuation, it is almost certain that there is flow fluctuation at the same time.

5. Viscoelastic effect in extrusion process

    During extrusion, the viscoelasticity of polymer will lead to entrance effect and die swelling.

    a. Entrance effect

       At the entrance of the die, the melt enters the smaller diameter path from the larger diameter path, and the streamline shrinks. This

TABLE 4.1 Flow fluctuation of plastics when the pressure fluctuation is 1%.

Plastic name	Temperature (°C)	Shear rate range (γ)	Non-Newtonian index (n)	Flow fluctuation (%)
High-impact polystyrene	170	100–7000	0.21	5.00
	190	100–7000	0.20	4.76
	210	100–7000	0.19	5.26
ABS resin	170	100–5500	0.25	4.00
	190	100–6000	0.25	4.00
	210	100–7000	0.25	4.00

process has the characteristic of transition state. Its disturbance will not disappear until it flows through the length equivalent to several times the diameter of the pipe. At this time, the flow of polymer melt can be regarded as steady shear flow.

b. Die swelling

During extrusion, the phenomenon that the cross-sectional size of the extrudate after leaving the die is larger than that of the die due to elastic recovery is called die swelling, which is the expansion immediately after contraction. The die swelling will affect the dimensional accuracy and shape of the product. This kind of influence must be solved in the die process parameter design and process condition control. Properly increasing the melt temperature is beneficial to reduce the degree of die swelling. In actual production, as long as production can proceed smoothly and qualified products can be produced, the temperature should be kept as low as possible. Appropriate traction speed can offset the influence of die swelling. However, the traction speed should not be too high, otherwise, the product will have anisotropy.

4.1.2.2 Fuse deposition modeling process

The FDM feeding system uses a pair of motor-driven feed rollers to advance the filament with a diameter of 1.75 mm into the inlet of the heating chamber. Before the temperature reaches the softening point of the filament, there is an area with constant gap between the filament and the heating chamber, which is called the feeding section. As the surface temperature of the filament rises, the material begins to fuse, forming an area where the diameter of the filament gradually decreases until it is completely melted, called the

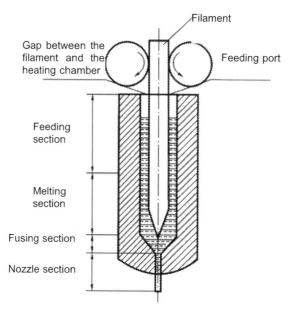

FIGURE 4.2 Structural diagram of FDM feeding system.

melting section. Before the material is extruded out of the nozzle, there is an area where the cylinder is completely filled with molten material, called the fusing section. In this process, the filament itself is not only the raw material, but also acts as a piston to extrude the molten material from the nozzle. The structure of FDM feeding system is shown in Fig. 4.2.

4.1.3 Thermodynamic transformation of polymer processing

4.1.3.1 Mechanical state and thermal transformation of polymers

Polymers have undergone morphological changes as shown in Fig. 4.3 throughout the processing.

The physical state of a polymer is reflected by its mechanical properties, which is usually expressed by a temperature-deformation (or modulus) curve (also known as a thermo-mechanical characteristic curve). This curve reflects the relationship between deformation characteristics and the physical state of high molecular materials, as shown in Fig. 4.4.

First of all, for polymers, temperature is the main factor that causes the transition of aggregation state. The development of deformation is continuous, which indicates that the transition of three aggregation states of amorphous polymers is not phase transition. When the temperature is lower than the glass transition temperature T_g, the polymer is in a glassy state and appears as a rigid solid. In this case, the cohesive force formed by the main

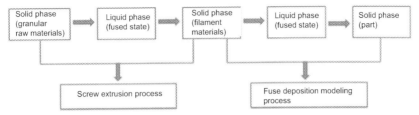

FIGURE 4.3 Morphological changes of polymers during processing.

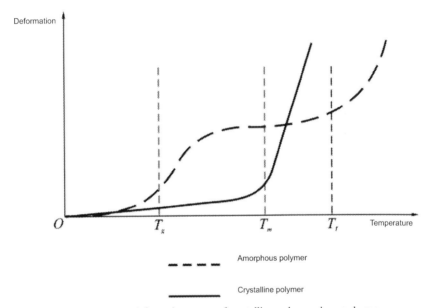

FIGURE 4.4 Temperature-deformation curves of crystalline and amorphous polymers.

valence bond and the secondary valence bond of the polymer enables the material to have considerable mechanical strength, small thermal motion energy, large intermolecular force, frozen single bond internal rotation of macromolecules, only thermal vibration of atoms or groups, and external force action is not enough to make macromolecules or chain segments do orientation displacement motion. Therefore, deformation is mainly caused by bond angle deformation, with small deformation value, reversibility of deformation within ultimate stress, large internal stress and elastic modulus. The deformation and deformation recovery are time-independent (instantaneous), and change little with temperature. Therefore, the deformation of glassy solid belongs to ordinary elastic deformation, which is called instantaneous elastic deformation. However, if the temperature is low to a certain extent, a small external force can break the macromolecular chain, and the corresponding temperature is brittle temperature, which will make the material lose its use value [2].

When the temperature is between the glass transition temperature T_g and the viscous flow temperature T_f, the polymer is in a highly elastic state and exhibits rubber-like properties. At this time, the temperature is higher, the segment movement has been activated (i.e., thawed), but the relative sliding movement between the molecular chain is still blocked, and the external force can only make the segment do orientation displacement movement. Therefore, deformation is caused by conformational stretching of macromolecules, with large deformation value and small internal stress and elastic modulus. After the external force is removed, the coiled conformation of the macromolecule is restored due to the irregular thermal movement of the segment, that is, the maximum conformation entropy state is restored, and the deformation is still reversible. Moreover, polymer is very viscous between the glass transition temperature T_g and the viscous flow temperature T_f (or melting temperature T_m), close to the viscous flow temperature T_f (or melting temperature T_m)-side [3].

When the temperature reaches or exceeds the viscous flow temperature T_f (amorphous polymer) or melting temperature T_m (crystalline polymer) of the polymer, the polymer is in a viscous flow state and presents a high-viscosity melt (liquid phase). In this state, the intermolecular force energy is of the same order of magnitude as the thermal motion energy, and the thermal energy further intensifies the relative sliding motion between the chain-like molecules. At this moment, the two motion units of the polymer appear simultaneously, making the properties of the aggregated state (liquid state) and the phase state (liquid-phase structure) consistent. External force not only makes macromolecular chains do orientation displacement movement, but also makes chains slide relatively. Therefore, high-viscosity melt shows continuous irreversible deformation under the action of force, which is called viscous flow, also known as plastic deformation. At this time, the deformation can be permanently maintained by cooling the polymer.

When the temperature rises to the vicinity of the decomposition temperature T_d of the polymer, the polymer will be decomposed to reduce the physical and mechanical properties of the product or cause poor appearance, etc.

The three aggregation states of amorphous polymers are only differences in dynamic properties (due to different forms of molecular thermal motion), rather than differences in physical phase states or thermodynamic properties, so they are often called mechanical three states. All dynamic factors, such as temperature, force magnitude and acting time, will lead to corresponding changes in their properties.

In FDM process, the temperature in the nozzle section is always controlled above the viscous flow temperature T_f (or melting temperature T_m) and below the decomposition temperature T_d.

4.1.3.2 Viscoelasticity of polymers

In FDM process, polymers undergo multiple changes from solid phase to liquid phase (melting and flowing), and then from liquid phase to solid phase (cooling and curing). Polymers exhibit different viscoelastic behaviors in different mechanical states.

According to the classical viscoelasticity theory, the total deformation $\varepsilon(t)$ of the linear polymer during processing can be regarded as a combination of the instantaneous elastic deformation ε_1, the delayed high elastic deformation ε_2 and the viscous deformation ε_3, which can be expressed by the following equation:

$$\varepsilon(t) = \varepsilon_1 + \varepsilon_2 + \varepsilon_3 = \frac{\sigma}{E_1} + \frac{\sigma}{E_2}\left(1 - e^{-(E_2/\eta_2)t}\right) + \frac{\sigma}{\eta_3}t \quad (4.3)$$

where σ is the external force; t is the acting time of external force; E_1 and E_2 are the instantaneous elastic deformation modulus and delayed high elastic deformation modulus of the polymer respectively. η_2 and η_3 represent the viscosity of the polymer at delayed high elastic deformation and viscous deformation, respectively. Fig. 4.5 shows the development process of deformation of the polymer with time under external force.

At the moment t_1, the instantaneous elastic deformation of the polymer caused by external force is shown in ab section in the figure, ε_1 is very small; when the external force is released at the time t_2, the instantaneous elastic deformation immediately recovers (cd section in the figure). Within the external force acting time t ($t = t_2 - t_1$), the delayed high elastic deformation and viscous deformation are shown in bc section in the figure. The external force is released at time t_2, after a certain period of time, the delayed high elastic deformation ε_2 is completely recovered, as shown in

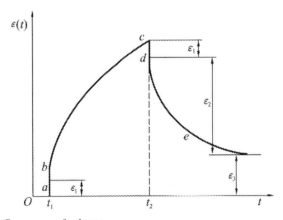

FIGURE 4.5 Creep curve of polymers.

the de section in the figure, while the viscous deformation ε_3 remains as permanent deformation in the polymer.

Under normal processing conditions, the deformation of polymers mainly consists of delayed high elastic deformation and viscous deformation (or plastic deformation), which includes reversible deformation and irreversible deformation in terms of deformation properties.

The relationship between deformation characteristics of polymers and processing temperature and time is investigated below.

It is easy to see from Eq. (4.3) that with the increase of temperature, the viscosity of the polymer decreases, that is, η_2 and η_3 both decrease, and then the deformation values of ε_2 and ε_3 both increase. When the processing temperature is higher than T_f (or T_m), the polymer is in a viscous flow state, it mainly exhibits viscous deformation. At this time, on the one hand, the polymer has low viscosity, good fluidity and easy molding; on the other hand, due to the irreversibility of viscous deformation, the elastic shrinkage of the product is reduced, and the stability of the shape and geometric dimensions of the product is improved.

It is worth noting that in Eq. (4.1), as the external force action time (t) is prolonged, the delayed high elastic deformation ε_2 and the viscous deformation ε_3 are increased, but ε_3 increases proportionally with time, while the increasing trend of ε_2 is gradually weakened. It can be seen that the reversible deformation energy is partially converted into irreversible deformation by prolonging the external force action time, thereby reducing the shrinkage deformation of the product.

Furthermore, the deformation of polymers during processing is the result of macromolecular deformation and rearrangement under the combined action of external force and temperature. Due to the long-chain structure of polymer macromolecules and the gradual nature of macromolecules' motion, any deformation of polymer molecules in response to stress under external force cannot be completed in an instant. Generally, at a certain temperature, the transition of macromolecular deformation from a series of intermediate states to an equilibrium state suitable for external forces is regarded as a relaxation process, and the corresponding relaxation time is denoted as τ.

Then, Eq. (4.3) can be written as:

$$\varepsilon(t) = \frac{\sigma}{E_1} + \frac{\sigma}{E_2}\left(1 - e^{-t/\tau}\right) + \frac{\sigma}{\eta_3}t \tag{4.4}$$

where $\tau = \eta_2/E_2$, the value is equal to the time required for stress relaxation to the initial stress value of $1/e$ (i.e., 36.79%).

Due to the relaxation process, the deformation of the material will inevitably lag behind the change of stress. This hysteresis phenomenon of polymers' response to external forces is called "hysteresis effect" or "elastic hysteresis."

During extrusion, the hysteresis effect will cause the die swell.

During fuse deposition modeling, the hysteresis effect will cause shrinkage and deformation of the part. Because the filaments are quenched after coming out of the nozzle, the macromolecules are piled up loosely, and then the rearrangement movement of the macromolecules chain is further carried out, so that the piling up is gradually tight, resulting in increased density and volume shrinkage. The degree of volume shrinkage of the part becomes serious with the increase of cooling speed, so quenching in the processing is generally unfavorable to the quality of the part. Volume shrinkage not only causes instability of the shape and geometric dimensions of the part, but also causes excessive internal stress in the part due to serious uneven shrinkage, resulting in warping deformation and even cracking of the part. For crystalline polymers (e.g., nylon and polypropylene), the part will have a larger volume shrinkage due to the gradual formation of crystalline structure of polymers. Therefore, the amorphous polymer is preferred in the selecting material of FDM, thus eliminating the influence of crystallization shrinkage.

In the process of fuse deposition modeling, increasing the forming temperature and prolonging the residence time of the modeling material in a molten state in the compression extrusion process are beneficial to reducing the reversible deformation of the material and increasing the irreversible deformation component, thereby reducing the shrinkage of the material in the molding process and improving the stability of the shape and the geometric dimension of the part. At the same time, slowing down the cooling speed after the melt is deposited on the forming surface is beneficial to reducing the volume shrinkage of the part at room temperature.

4.1.4 Performance requirements for fuse deposition modeling polymer materials

No matter the modeling material or the support material, before FDM process, the filament with a diameter of 1.75 mm must be extruded by an extruder, so the extrusion molding requirements must be met. In addition, according to the process characteristics of FDM, polymers should also meet the following requirements.

1. Melt viscosity

    The low melt viscosity and good fluidity of the material are helpful for the smooth extrusion of the material, but the material with too much fluidity will cause the occurrence of salivation. However, the material with poor fluidity requires a large feeding pressure to extrude, which will increase the start-stop response time of the nozzle, thus affecting the molding accuracy.

2. Melting temperature

    Low melting temperature allows the material to be extruded at a lower temperature, which is beneficial to prolonging the service life of the nozzle and the whole mechanical system, reducing the temperature difference between the material before and after extrusion, reducing the thermal stress, and thus improving the precision of the part. However, polymers with low melting temperature usually have poor heat resistance. If there are high requirements on the heat resistance of the part, polymers with good heat resistance and thermal stability should be selected. If the melting temperature is too close to the decomposition temperature, it will become extremely difficult to control the molding temperature.

3. Molding shrinkage

A large molding shrinkage of modeling materials will cause internal stress in the part during FDM process, which will lead to warpage and even interlayer stripping. The shrinkage of the support material will cause the support structure to deform without support effect. Therefore, when the molding shrinkage of the material is small, the dimensional accuracy of the model or product can be increased, and the linear molding shrinkage is generally required to be less than 1%.

4. Mechanical properties

From the perspective of FDM process, it requires the filament to have good flexural strength, compressive strength, tensile strength, and good flexibility by using the filament feeding method, so that filament breakage and bending will not occur under the traction and driving force of the driving friction wheel. The mechanical properties of the modeling materials are the main factors that affect the mechanical properties of FDM part, while it only needs to ensure that the support materials are not easily broken during the FDM process.

5. Bond properties

The FDM process is a process based on layered manufacturing. The weakest site of the part is between the layers, and the adhesion between the layers determines the strength of parts. If the bond strength is too low, sometimes, the thermal stress will lead to layer-to-layer cracking in the FDM process. For the break-away support materials, a weak bond should be formed with the modeling materials.

6. Hygroscopicity

A high hygroscopicity of the material will affect the part quality due to water volatilization when it is molten at a high temperature. Therefore, the filaments should be stored dry.

Judging from the influence of the above material characteristics on FDM process, the key requirements of FDM for modeling materials are appropriate melt viscosity, low melting temperature, good mechanical properties, good adhesion and low shrinkage. For water-soluble support materials, good water solubility shall be guaranteed, and it should be able to dissolve in water or alkaline aqueous solution within a certain period of time.

4.2 Modeling materials in fuse deposition modeling

At present, the modeling materials used for FDM mainly include ABS resin and polylactic acid, and nylon, polycarbonate, thermoplastic polyurethane, polyphenyl sulfone, polyether ether ketone, etc. Different materials are applied in different fields due to their unique properties.

4.2.1 ABS filament

ABS refers to the copolymer of butadiene, acrylonitrile and styrene, which has the advantages of high strength, high toughness, impact resistance, easy processing, good electrical insulation performance, corrosion resistance, low-temperature resistance and coloring performance, etc. It is widely used in household appliances, automobile industry, toy industry, and other fields. ABS resin is the earliest polymer consumable used in FDM printing due to its good hot melt and extrudability. At present, ABS resin is still one of the most commonly used modeling materials for FDM, with the advantages of stable printing process, high strength and high flexibility. Stratasys Company in the United States has developed a variety of ABS modified materials for FDM. ABS-plus is a special 3D printing material, which has the advantages of good environmental stability, low shrinkage, and water absorption. The strength of ABS-plus FDM parts is 20%−40% of that of standard ABS resin−molded parts. The strength of ABS-M30 FDM parts is 25%−70% of that of standard ABS resin−molded parts. It has strong interlayer adhesion ability and can print functional parts. ABS-M30i is a biocompatible 3D printing material widely used in medical, pharmaceutical, and food packaging industries. Its parts can be sterilized by γ rays or ethylene oxide. ABSi is a semitransparent material, which can be used as modeling material for automobile taillight protective cover and other products.

The application of pure ABS filament is limited due to its disadvantages such as large shrinkage rate, easy warpage and deformation of the parts and easy interlayer stripping. At the same time, as the printing temperature of ABS filament is as high as 230°C, it can cause partial decomposition of the material and generate odor, and the printing process consumes energy. In order to obtain ABS filament with excellent printing performance, it is very important to choose the brand of ABS resin. Blending modification, filling modification, and other means can be used to further improve various performance indexes of the filament and obtain functional ABS filament.

4.2.1.1 Selection of the ABS resin brand

ABS resin is generally composed of more than 50% (mass fraction, the same below) styrene, 25%−35% acrylonitrile and an appropriate amount of butadiene. Each of the three components has its own performance, making it have excellent comprehensive properties. As shown in Fig. 4.6, acrylonitrile

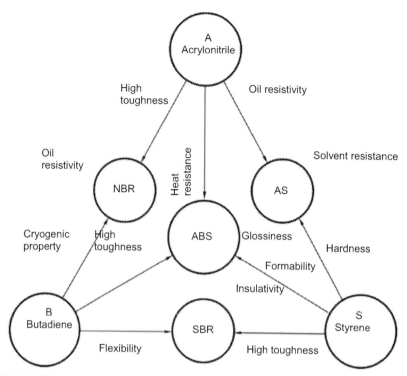

FIGURE 4.6 Component characteristics of ABS terpolymer.

gives ABS resin good chemical corrosion resistance, oil resistance, certain rigidity and surface hardness, butadiene improves toughness, impact resistance and cold resistance of ABS resin, while styrene gives ABS resin good dielectric property, processing fluidity, high smooth finish, and high strength.

Polymerization of the three monomers produces a ternary copolymer with a complex two-phase structure, in which styrene-acrylonitrile copolymer forms a continuous phase and polybutadiene is a dispersed phase. The properties of ABS are mainly determined by the ratio of the three monomers and the phase structure. This allows great flexibility in product design, and thus produces hundreds of ABS materials with different qualities. These different quality materials provide different characteristics, such as medium to high-impact resistance, low to high surface roughness and fluidity.

The rubber content in ABS resin, that is, the mass fraction of rubber particles in ABS resin, is an important factor affecting the performance of ABS resin. Increasing the mass fraction of rubber will reduce the tensile modulus, yield strength and hardness. Fig. 4.7 shows the relation curves of tensile strength, tensile modulus and rubber mass fraction. The mass fraction of rubber

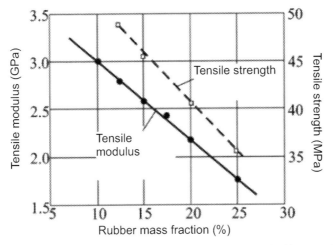

FIGURE 4.7 Relationship between tensile strength, tensile modulus and rubber mass fraction.

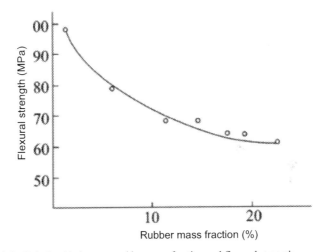

FIGURE 4.8 Relationship between rubber mass fraction and flexural strength.

in ABS resin is generally 10%–30%. Fig. 4.8 shows the relationship between rubber mass fraction and flexural strength, and Fig. 4.9 shows the relationship between rubber mass fraction and melt flow rate. As can be seen from Figs. 4.7–4.9, the tensile strength, tensile modulus, flexural strength and melt flow rate of ABS resin all decrease with the increase of rubber mass fraction. ABS resin with high rubber mass fraction has good flexibility and easy diffusion of molecular chains, which is conducive to improving interlayer bonding strength in FDM parts. Therefore, ABS resin with high rubber mass fraction is usually selected to prepare filament for FDM.

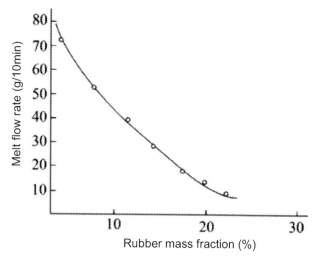

FIGURE 4.9 Relationship between rubber mass fraction and melt flow rate.

ABS resin has different brands suitable for various processing methods, of which injection molding is the most commonly used process, followed by extrusion. The melt flow rate of resin suitable for extrusion is lower than that of injection molding.

4.2.1.2 Preparation of ABS filament

ABS resin is an amorphous polymer with no obvious melting point and no crystallization after molding. ABS has a wide melting range, and its molding temperature is generally controlled at 180°C–230°C. If the temperature exceeds 250°C, it will degrade and even produce toxic volatile substances. The melt viscosity of ABS resin is moderate, and its rheological property in molten state is non-Newtonian. The fluidity of ABS resin is not sensitive to temperature during molding, so the molding temperature is easier to control. According to different types of resin, the linear expansion coefficient of ABS resin is $(6.2-9.5) \times 10^{-5}$ °C^{-1} and the molding shrinkage is 0.3%–0.8%.

TABLE 4.2 Setting of extrusion temperature for ABS756.

Experiment no.	Feeding section (°C)	Compression section (°C)	Metering section (°C)	Die section (°C)
1	185	190	200	215
2	190	195	200	215
3	195	200	215	220

TABLE 4.3 Performance indexes of ABS filament for FDM.

Serial number	Test item	Unit	Test standard	Test conditions	Indicators
1	Density	g/cm³	–	25°C ± 2°C	1.02–1.08
2	Water content	%	GB/T 6284—2006	–	≤ 0.2
3	Glass transition temperature	°C	GB/T 19466.2—2004	–	≥ 90
4	Tensile strength (XY direction printing)	MPa	GB/T 1040.2—2006	50 mm/min	≥ 25
5	Tensile strength (XZ direction printing)	MPa	GB/T 1040.2—2006	50 mm/min	≥ 22
6	Flexural strength (XY direction printing)	MPa	GB/T 9341—2008	2 mm/min	≥ 50
7	Flexural strength (XZ direction printing)	MPa	GB/T 9341—2008	2 mm/min	≥ 40
8	Notch impact strength (XY direction printing)	kJ/m²	GB/T 1843—2008	–	≥ 30
9	Notch impact strength (XZ direction printing)	kJ/m²	GB/T 1843—2008	–	≥ 10
10	Thermal deformation temperature (XY direction printing)	°C	GB/T 1634.2—2004	1.8 MPa	≥ 75
11	Thermal deformation temperature (XZ direction printing)	°C	GB/T 1634.2—2004	1.8 MPa	≥ 70

FDM filament was prepared from ABS756 and ABS757 produced by Taiwan Chimei Company of China. The filament made from ABS757 by extrusion has good filament diameter uniformity, surface quality, and strength. However, when the filament is applied to an open FDM system, the filament is quickly solidified and cooled after extrusion by FDM nozzle, delamination occurs during stacking, and the printing process performance is poor.

When ABS756 performs extrusion, the extrusion temperature has a certain influence on the surface quality and precision of the filament. Table 4.2 shows the temperatures of various sections during extrusion of ABS756.

The experiment found that the third condition has the best extrusion effect for ABS756 and the printing performance of filament was good. Typical performance indexes of ABS filament for FDM are shown in Table 4.3.

The sample printing direction is shown in Fig. 4.10.

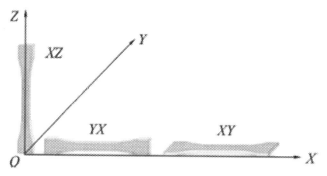

FIGURE 4.10 Schematic diagram of sample printing direction.

4.2.2 Polylactic acid filament

4.2.2.1 Characteristics of polylactic acid

Polylactic acid is a new type of biodegradable thermoplastic resin. Its raw material lactic acid has a wide range of sources and can be obtained by fermentation of agricultural products such as corn and starch. Polylactic acid has the advantages of high strength, good biocompatibility, environmentally friendly, and low odor, and is suitable for office environment. At the same time, its low shrinkage makes it not warp when printing large-sized parts. However, polylactic acid also has the disadvantages of poor toughness, low melt strength, poor thermal stability, etc. Unmodified polylactic acid is easy to degrade in the melting process, resulting in an increase in melt flow rate. During the printing process, the melting material will leak out from the nozzle due to a decrease in melt strength, and the leaked material will stick to the part to form burrs, affecting the surface quality of the part.

4.2.2.2 Preparation of polylactic acid filament

ADR 4370S, a compound with epoxy active groups, is used as a chain extender to undergo chemical reactions such as crosslinking, branching, chain extension and the like with active groups (such as carboxyl groups) generated in the fusing process of polylactic acid, so that the melt strength of polylactic acid can be remarkably improved, and the leakage phenomenon of materials at nozzles in the printing process can be improved.

The polylactic acid was hot-air dried at 80°C for 4 h, then the polylactic acid and chain extender ADR 4370S were mixed and melt blended by a twin-screw extruder, extruded and granulated, and the temperatures from the feeding area to the die were 160°C, 190°C, 200°C, 200°C, and 200°C, respectively. Then the pure polylactic acid granules and the chain-extended modified polylactic acid granules were subjected to a 3D consumables extruder respectively to obtain polylactic acid filaments with a diameter of 1.75 ± 0.05 mm, and the extrusion temperature was 175°C−190°C.

The influence of chain extender ADR 4370S on polylactic acid melt flow rate is shown in Tables 4.4 and 4.5.

Table 4.4 shows the melt flow rates of pure polylactic acid and modified polylactic acid granules. From the table, it can be seen that the melt flow rate of polylactic acid decreases with the increase of the mass fraction of chain extender ADR 4370S, and the melt flow rate changes little after the mass fraction of chain extender increases to 0.6%. Table 4.5 shows the melt flow rates of pure polylactic acid filament and modified polylactic acid

TABLE 4.4 Melt flow rates of pure polylactic acid and modified polylactic acid granules.

Granules	Pure polylactic acid	Polylactic acid/0.2% (mass fraction) ADR 4370S	Polylactic acid/0.4% (mass fraction) ADR 4370S	Polylactic acid/0.6% (mass fraction) ADR 4370S	Polylactic acid/0.8% (mass fraction) ADR 4370S
Melt flow rate (g/10 min)	7.24	5.28	2.88	1.44	1.2

TABLE 4.5 Melt flow rates of pure polylactic acid filament and modified polylactic acid filament.

Filament	Pure polylactic acid	Polylactic acid/0.2% (mass fraction) ADR 4370S	Polylactic acid/0.4% (mass fraction) ADR 4370S	Polylactic acid/0.6% (mass fraction) ADR 4370S	Polylactic acid/0.8% (mass fraction) ADR 4370S
Melt flow rate (g/10 min)	12.06	6.96	3.42	1.50	1.44

filament. Compared with Table 4.4, the melt flow rate of polylactic acid after extrusion increases, in which the melt flow rate of pure polylactic acid increases from 7.24 g/10 min before extrusion to 12.06 g/10 min after extrusion, with an increase of 66%. This is caused by the degradation of polylactic acid molecular chain during high temperature extrusion. The melt flow rate of polylactic acid filament modified by chain extender increases slightly, which indicates that the addition of chain extender reduces the degradation degree of polylactic acid to a certain extent during processing.

Table 4.6 shows the influence of printing temperature on the printing effect of different polylactic acid filaments. For pure polylactic acid filaments, when the printing temperature exceeds 200°C, material will leak from the nozzle, forming burrs on the finished part. Increasing the mass fraction of chain extender can increase the leak temperature, effectively improve the leakage phenomenon and expand the printing temperature range. However, when the mass fraction of chain extender is greater than 0.6%, the material will not flow smoothly during printing, which may be caused by the chain extension branching reaction that causes a too high melt viscosity and poor melt fluidity to the material.

TABLE 4.6 Influence of printing temperature on printing effect of different polylactic acid filaments.

Filament	Printing temperature (°C)	Printing situation
Pure polylactic acid	195, 200, 210, 220, 230	Able to print, above 200°C materials are easy to leak from the nozzle, and are easy to stick on the part to form burrs
Polylactic acid/0.2% (mass fraction) ADR 4370S	195, 200, 210, 220, 230	Able to print, above 210°C materials are easy to leak from the nozzle, and are easy to stick on the part to form burrs
Polylactic acid/0.4% (mass fraction) ADR 4370S	195, 200, 210, 220, 230	Able to print, above 230°C materials are easy to leak from the nozzle, and are easy to stick on the part to form burrs
Polylactic acid/0.6% (mass fraction) ADR 4370S	195, 200, 210, 220, 230	Poor melt fluidity below 210°C, and printing is available when above 210°C, little material leakage
Polylactic acid/0.8% (mass fraction) ADR 4370S	195, 200, 210, 220, 230	Poor melt fluidity below 230°C, and though the temperature is raised, and the nozzle basically does not leak material

When FDM process is used to mold complex parts, if the parts are directly molded on the base plate, the parts will be easily damaged when the parts are removed from the base plate while the parts completed. In order to avoid this situation, the base is usually designed. Table 4.7 shows the bonding between the base and the base plate during the FDM process of different polylactic acid filaments.

TABLE 4.7 Bonding of base and base plate in FDM process of different polylactic acid filaments.

Filament	Printing temperature (°C)	Bonding between base and base plate
Pure polylactic acid	195–230	Bonding is firm and difficult to remove
Polylactic acid/0.2% (mass fraction) ADR 4370S°	195–230	Bonding is firm and relatively difficult to remove
Polylactic acid/0.4% (mass fraction) ADR 4370S°	195–230	Bonding is firm and can be removed
Polylactic acid/0.6% (mass fraction) ADR 4370S°	210–230	Bonding is relatively firm and easy to remove
Polylactic acid/0.8% (mass fraction) ADR 4370S°	≥230	Bonding is not firm enough and the extruded materials are not easy to adhere to the base plate and are easy to be scraped off by the nozzle

Table 4.7 shows that when pure polylactic acid filament is used as modeling material, the base and the base plate are firmly bonded and it is difficult to separate the base from the base plate. The polylactic acid filaments with the mass fraction of chain extender of 0.4% and 0.6% have good adhesion between the base and the base plate during FDM process, and can better fix the parts. After parts completed, the base and the base plate are easy to separate, and the comprehensive molding effect is the best.

Adding toughening agent to polylactic acid can improve its impact strength. The addition of inorganic filler can reduce the molding shrinkage and improve the dimensional accuracy of the parts. Adding reinforcing agents such as carbon fiber powder and glass fiber powder can improve their mechanical properties and heat resistance.

The addition of nanomaterials such as carbon nanotubes and graphene can produce functional filaments with conductive properties. Table 4.8 shows the performance indexes of the three modified polylactic acid filaments [4].

TABLE 4.8 Performance indicators of polylactic acid filament for FDM.

Serial number	Test item	High-impact type	Filling modified type	Enhanced modified type
1	Density (g/cm^3)	1.15–1.35	1.25–1.55	1.30–1.60
2	Water content (%)	≤0.05	≤0.08	≤0.08
3	Melting point (°C)	≥140		
4	Glass transition temperature (°C)	≥90		
5	Tensile strength (X-Y direction printing) (MPa)	≥40	≥25	≥60
6	Tensile strength (X-Z direction printing) (MPa)	≥30	≥18	≥40
7	Flexural strength (X-Y direction printing) (MPa)	≥55	≥40	≥70
8	Flexural strength (X-Z direction printing) (MPa)	≥40	≥35	≥50
9	Notch impact strength (X-Y direction printing) (kJ/m^2)	≥3	≥2	≥3
10	Notch impact strength (X-Z direction printing) (kJ/m^2)	≥1	≥1	≥1
11	Thermal deformation temperature (X-Y direction printing) (°C)	≥45	≥55	≥90
12	Thermal deformation temperature (X-Z direction printing) (°C)	≥35	≥40	≥70

4.2.3 Polycarbonate and its composites filaments

Polycarbonate is a kind of thermoplastic engineering plastic with excellent performance, and is also one of the engineering plastics with the largest consumption at present. Polycarbonate has almost all the excellent characteristics of engineering plastics, tasteless and nontoxic, high strength, good impact resistance and low shrinkage, in addition, it has good flame retardant properties and pollution resistance. Polycarbonate is made into 3D printing filament, its strength is about 60% higher than ABS resin, and it has super engineering material properties. However, polycarbonate also have some disadvantages, such as high notch sensitivity, single color, high price, etc. The composite material made of polycarbonate and ABS resin can combine the strength of polycarbonate with the toughness of ABS to obtain a material with better mechanical properties [5].

Due to the poor compatibility between polycarbonate and ABS resin, it is necessary to add compatilizer when preparing polymer composites. Using ABS-g-MAH as compatilizer, polycarbonate, ABS resin, and ABS-g-MAH were dried at 80°C for 8 h, and then mixed in a high-speed mixer for 10 min. Adding the mixed materials into a twin-screw extruder for melt blending and extrusion, granulation, wherein the blending temperature is 210°C−240°C. The blended granules were dried in an oven at 80°C for 8 h, and then extruded filament by a 3D consumables extruder, with the extrusion temperature of 220°C−240°C, and the cooling water temperature of 55°C, and the filament with a diameter of 1.75 mm was obtained.

Table 4.9 lists the comprehensive properties of polycarbonate, ABS resin, and polycarbonate/ABS composites. The mechanical property indicators and Vicat softening temperature data of each material in the table are all obtained by testing injection molding samples. As shown in Table 4.9, the tensile strength and flexural strength of polycarbonate/ABS composites are equivalent to those of polycarbonate. As ABS resin effectively improves the notch impact sensitivity of polycarbonate,

TABLE 4.9 Properties of polycarbonate, ABS resin and polycarbonate/ABS composites.

Performance indicators	Pure polycarbonate	Pure ABS resin	Polycarbonate/ABS composites
Tensile strength (MPa)	51.82	47.71	54.84
Flexural strength (MPa)	89.97	58.49	82.56
Notch impact strength (kJ/m^2)	7.77	20.30	33.92
Melt flow rate (g/10 min)	32.76	16.08	26.00
Vicat softening temperature (°C)	114.2	94.4	107.97

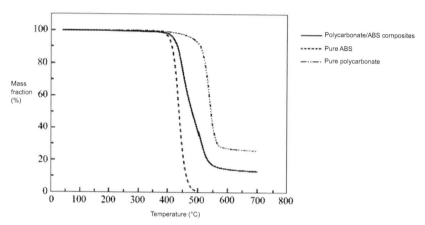

FIGURE 4.11 Thermal weight loss curves of polycarbonate, ABS resin, and polycarbonate/ABS composites.

the notch impact strength of polycarbonate/ABS composites is greatly increased, which is much larger than the notch impact strength of polycarbonate. Therefore, polycarbonate/ABS composites are superior to polycarbonate in terms of mechanical properties. The melt flow rate in Table 4.9 is measured at a load of 1.2 kg and a temperature of 300°C. As the melt flow rate of ABS resin used is small, the melt flow rate of polycarbonate/ABS composites is lower than that of polycarbonate, but the decrease degree is not large. Similarly, the Vicat softening temperature of polycarbonate/ABS composites is slightly lower than that of polycarbonate because the heat resistance of ABS resin is worse than that of polycarbonate.

Fig. 4.11 shows the thermal weight loss curves of polycarbonate, ABS resin, and polycarbonate/ABS composites [6].

As can be seen from Fig. 4.11, the thermal weight loss temperature of polycarbonate is 470°C−500°C, the starting temperature of thermal weight loss of ABS resin is lower than 400°C, and the thermal stability of ABS resin and polycarbonate are quite different. The thermal weight loss starting temperature of polycarbonate/ABS composites is about 420°C, and its thermal stability is between polycarbonate and ABS resin. When the temperature rises to about 500°C, the residual mass fraction of ABS resin is close to 0, while the residual mass fraction of polycarbonate/ABS composites is still about 50%, indicating that the composites have good thermal stability.

The dimensional stability of the filament diameter of polycarbonate during extrusion and drawing is not as good as that of ABS resin, and it is difficult to control the filament diameter within the standard range of 1.75 ± 0.25 mm. However, ABS resin has excellent dimensional stability, and the filament diameter can be accurately controlled within 1.75 ± 0.05 mm during extrusion. The diameter of polycarbonate/ABS composites filament can be controlled within the range of 1.75 ± 0.20 mm, which indicates that the ABS component in the composites effectively improves the dimensional stability of polycarbonate filament.

TABLE 4.10 FDM printing conditions of polycarbonate filament and polycarbonate/ABS composites filament.

Printing conditions	Polycarbonate filament	Polycarbonate/ABS composites filament
Printing speed (mm/s)	30	30
Nozzle temperature (°C)	235–240	220–230
Base plate temperature (°C)	110	90

TABLE 4.11 Comparison of mechanical properties of printed parts between polycarbonate filament and polycarbonate/ABS composites filament.

Mechanical properties	Polycarbonate filament printed parts	Polycarbonate/ABS composite filament printed parts
Tensile strength (MPa)	86.3	76.0
Flexural strength (MPa)	122.2	87.5
Notch impact strength (kJ/m^2)	6.2	21.5

During FDM process, polycarbonate extruded from the nozzle is easy to adhere to the nozzle, but difficult to adhere to the base plate, which easily leads to building failure. Polycarbonate/ABS composites filament is smoother than polycarbonate filament, has better adhesion with the base plate and better process performance. FDM printing conditions of polycarbonate filament and polycarbonate/ABS composites filament are shown in Table 4.10.

As can be seen from Table 4.10, during FDM process, the nozzle temperature and the base plate temperature for polycarbonate/ABS composites filament are lower than those for polycarbonate filament, and the printing process is easy to control.

Table 4.11 shows the mechanical properties of FDM parts made of polycarbonate filament and polycarbonate/ABS composites filament, and the printing direction is X-Y direction.

As can be seen from Table 4.11, although the tensile strength and flexural strength of polycarbonate/ABS composites filament printed parts are lower than those of polycarbonate filament printed parts, the notch impact strength is greatly increased, which indicates that ABS resin greatly improves the notch impact sensitivity of polycarbonate. Compared with the mechanical properties of the injection-molded parts in Table 4.9, the tensile strength and flexural strength of the two kinds of filaments printed parts are higher than those of the injection-molded parts, which may be related to the orientation alignment of macromolecular chains during printing.

4.2.4 Nylon filaments

Nylon molecules have a large number of hydrogen bonds with extremely strong acting force, which make it have the advantages of good mechanical properties, good wear resistance and strong corrosion resistance. Nylon is widely used in the fields of automobile industry, electronics, medical treatment, machinery, military, and aerospace. Nylon is a crystalline polymer with large internal molecular stress and large molding shrinkage. The depositing layer of pure nylon filament is prone to warp deformation in FDM process, so it is the focus of research to improve its warp deformation [7].

There are many varieties of nylon, including nylon 6, nylon 66, nylon 11, nylon 12, nylon 1010, and various nylon copolymer. Table 4.12 lists the

TABLE 4.12 Filament drawing properties and printing properties of nylon of different grades.

Serial number	Nylon grade	Melt flow rate (g/10 min)	Filament drawing and printing
1	PA6 1013B	60.72	Too good fluidity and difficult filament drawing
2	PA6 1022B	6.00	Smooth filament drawing, too high printing temperature of 260°C, large shrinkage, high requirements for equipment
3	PA6 1030B	2.64	Melt is difficult to extrude and cannot be drawn
4	PA12 L1670	48.32	Too good fluidity and filament drawing fails
5	PA12 3020U	15.35	Filament drawing and printing are very successful, but the bonding between layers of formed parts is not good
6	COPA6/66 5033B	2.76	Smooth filament drawing and warpage of formed parts

filament drawing properties and printing properties of several different grades of nylon.

As can be seen from Table 4.12, three grades of nylons, PA6 1022B, PA12 3020U, and COPA6/66 5033B can be extruded to filaments, while the other three grades of nylons are not suitable for FDM because filament cannot be drawn or drawing is difficult. Considering the cost and FDM equipment, the forming temperature of PA6 1022B filament can reach 260°C or higher, and the common FDM equipment on the market cannot meet the requirements. Moreover, the molding shrinkage is large and the process performance is poor. For PA12 3020U, filament drawing and printing are relatively smooth, but the cost of raw materials is high, the bonding between layers of formed parts is poor, and delamination is easy to occur. COPA6/66 5033B has a printing temperature of about 235°C, and good interlayer adhesion, and good toughness and strength of the parts, but there is warping phenomenon when forming large-sized parts. Taking COPA6/66 as the base material, the molding shrinkage can be reduced by filling modification and the warping phenomenon can be improved in the FDM process. Experiments show that COPA6/66 is melt blended with glass microbeads, glass fiber powder, carbon fiber powder, aluminum powder, and other fillers through a twin-screw extruder, extruded and drawn to prepare nylon composite filaments which can be used in FDM process.

4.3 Support materials in fuse deposition modeling

4.3.1 Overview of support materials

According to the process characteristics of FDM, the system must support the 3D CAD model of the product. Otherwise, in the layered manufacturing process, when the upper section is larger than the lower section, the excess part of the upper section will be suspended, causing the section to collapse or deform, affecting the molding accuracy of the formed parts, and even the parts cannot be formed. Another important role of support is to establish a base layer and a buffer layer between the working platform and the bottom layer of the prototype so that the prototype can be easily detached from the working platform after the prototype is completed. In addition, the support can also provide a base plane for the manufacturing process.

The support can be made of the same material, only one extrusion head is needed, and the material density is controlled by controlling the filling degree of the material at the support site and the modeling site to distinguish the molded part from the support structure. Nowadays, many FDM equipment use double nozzles for independent heating, and one for extruding modeling materials to make parts; the other is used to extrude the support material to make the support structure. The characteristics of the two materials are different, so it is easy to remove the support after the part fabricated.

At present, the commonly used support materials for FDM process include break-away support materials and water-soluble support materials.

When forming parts with simple structure and few hollow structures, it is convenient to use break-away supports. Water-soluble support provides a better solution for integrally formed assemblies. Because the water-soluble support can be dissolved in water and automatically removed, the assembly can be constructed integrally in one step. If the gear set is manufactured by FDM technology, it can be completed without manual peeling of the support structure, thus the production efficiency is high and the appearance quality of the manufactured parts is good. In addition, in the process of product development, it is very important to evaluate the design and functionality of the assembly. With water-soluble support, CAD data of the entire assembly can be treated as data of a workpiece, and the assembly of the workpiece is not required.

FDM also has certain requirements for support materials based on the role of support materials in the process and the removal steps after forming. The break-away support material and the modeling material have the same requirements in terms of shrinkage and hygroscopicity, and other special aspects are specifically described as follows.

1. Withstand a certain high temperature

    Since the support material is in contact with the modeling material on the support surface, the support material must be able to withstand the high temperature of the modeling material, at which no decomposition and melting occur. Because the materials extruded by FDM process are relatively thin and can be cooled relatively quickly in air, the support material is required to withstand a temperature of about $100°C$ under normal circumstances.

2. Mechanical properties

    FDM does not require high mechanical properties of support materials, but it must have certain strength to facilitate the transmission of filament. The break-away support material needs certain brittleness to be easy to break during stripping, and at the same time, it is necessary to ensure that the filament does not easily bend or break under the action of traction and driving force when driving the friction wheel.

3. Fluidity

    Because the molding precision of the support material is not high, in order to improve the printing speed, the support material is required to have good fluidity.

4. Bonding

    The support structure is an auxiliary means adopted in the processing, and must be removed after the processing is finished, so the bonding of the break-away support material may be inferior to that of the modeling material.

5. Filament drawing requirements

    Filaments used in FDM are usually 1.75 mm in diameter, which requires smooth surface, uniform diameter, internal compactness, no surface pimples, and other defects. In addition, certain toughness is required in performance to ensure smooth filament feeding during 3D printing. Therefore, for break-away support materials, the raw materials that are brittle at room temperature should be modified to improve their toughness appropriately.

6. Strippability

    The most critical performance requirement for break-away support materials is to ensure that the materials can be easily peeled off under certain force, and the support materials can be conveniently removed from the molded parts without damaging the surface accuracy of the molded parts. In this way, it is beneficial to process complicated parts with cavity or cantilever structure.

    For water-soluble support materials, besides the general properties of modeling materials, they are also required to dissolve rapidly in water or alkaline aqueous solution. This kind of support material is especially suitable for manufacturing hollow and microfeature parts, solving the problem that the support is not easy to dismantle manually or is torn down due to too fragile structure, and reducing the surface roughness of the support contact surface.

4.3.2 Break-away support materials

4.3.2.1 Bonding model

In FDM, the bonding between layers is accomplished by bonding after melting of the material. The interlayer bond is similar to the bond between the two polymers, and the spun filamentous material is bonded to the former layer, except that the bond contact area is very small and the bond strength is limited. Fig. 4.12 shows an FDM interlayer bonding model [8].

4.3.2.2 Bonding mechanism and material factors affecting bond strength

There are various physical models about the mechanism of bonding between two objects, including mechanical bonding theory, adsorption theory, diffusion theory and electrostatic theory. The bonding process between FDM layers is realized by molecular diffusion process driven by temperature potential after the melting of the material itself.

1. Diffusion theory

    Diffusion theory is also called molecular penetration theory, which holds that the bonding between polymers is caused by molecular diffusion.

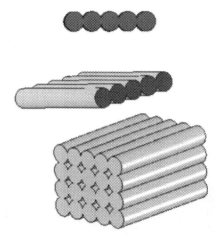

FIGURE 4.12 FDM interlayer bonding model.

The molecular chains of polymers are flexible, and the molecules between the layers are in constant thermal motion. Due to the mutual diffusion of long-chain segments, the interface between the layers disappears, forming a firm combination of mutual "interweaving." The strength of the bond increases to the maximum value with the increase of time.

Diffusion theory is based on the basic characteristics of polymers (chain structure and flexibility of molecules, micro-Brownian motion) and the existence of polar groups in polymer molecules. The bonding strength of polymers in glassy and crystalline state is low due to the influence of mutual diffusion.

2. Material factors affecting bond strength

The bond strength increases with the increase of contact time, bonding temperature, bonding pressure and flexibility of molecular chain, and decrease of relative molecular weight. In addition, the disappearance of bulky short side groups will also increase the bond strength. These phenomena can be used as examples of diffusion playing an important role in the bonding process.

Influence of relative molecular weight: The relationship between relative molecular weight and bonding strength is generally discussed, and it is limited to linear chain polymers. Increasing the relative molecular weight will increase the self-adhesive strength of the material, thus being conducive to obtaining high adhesive strength. However, an increase in viscosity will hinder diffusion. In this way, the influence of relative molecular weight on bond strength is uncertain, which depends on the nature of the material.

Influence of molecular structure: Increasing the rigidity of the chain tends to reduce the bond strength, which is not always the case. The stiffness of the chain will increase the self-adhesive strength of the material, but it will also increase the viscosity and retard the diffusion. Therefore, the influence of chain stiffness on bonding is uncertain. However, under normal circumstances, the bond strength tends to decrease with the increase of chain stiffness. The polymer molecular-containing benzene ring is not easy to diffuse due to the influence of chain flexibility, so the bond strength is usually low.

4.3.2.3 Bonding process

Fig. 4.13 shows an enlarged view of the bonding process between the two layers during FDM process. It can be seen that with the increase of time, molecules between layers are continuously diffused. From Fig. 4.13AD, macromolecular chains are gradually bonded into a whole and finally become parts.

Fig. 4.14 shows the schematic diagrams of the bond between the support material and the modeling material. Interface a is an interface layer formed by interdiffusion of the two materials.

When a certain external force is applied, the fracture always occurs at a portion where the mechanical strength is weak. If the fracture occurs at b (modeling material), the surface of the part is liable to appear small pits. If the fracture occurs at c (support material), some burrs are easily left on the surface of the part. In both cases, the surface of the part must be smooth to obtain the desired surface roughness. When removing the support, it is desirable to break at the interface a. In order to facilitate the removal of the support material, relatively weak bonding force should be formed between the support material and the modeling material relative to each layer of the modeling material, and certain bonding strength should also be ensured

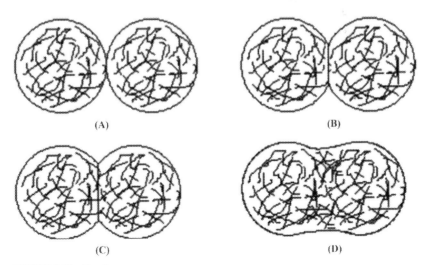

FIGURE 4.13 Bonding process.

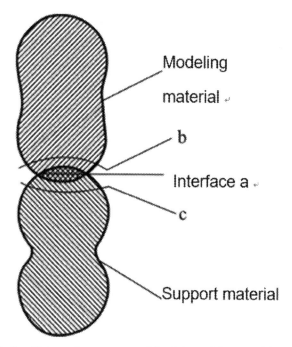

FIGURE 4.14 Bond between the support material and the modeling material.

between each layer of the support to avoid delamination. Therefore, bonding is the focus of development of break-away support materials.

4.3.2.4 Preparation of break-away support filament

1. Selection of base materials

   In the FDM process, the support filament is used together with the modeling filament. As ABS resin is the most commonly used modeling filament, the support material matching with ABS resin is mainly discussed. According to the performance requirements of the break-away support filament, polystyrene is selected as the base material.

   a. General purpose polystyrene

      General purpose polystyrene (GPPS) is a linear polymers with high relative molecular weight. Commercially available polystyrene is usually polymerized from more than 1000 styrene monomers. Its structural formula is:

$$\text{---}(CH_2\text{---}CH)_n\text{---}$$
$$|$$
$$C_6H_5$$

The existence of a large number of side-group benzene rings in the GPPS molecular chain makes its molecular structure irregular, and the steric hindrance of benzene rings is large. The increase of the internal rotation barrier of the chain segments leads to the rigidity and poor flexibility of the molecular chain, making it an amorphous brittle material with a glass transition temperature as high as 100°C.

b. High-impact polystyrene

High-impact polystyrene (HIPS) is a graft copolymer prepared by copolymerization of styrene monomer and polybutadiene or butadiene styrene rubber. Its structural formula is:

$$\mathrm{\{CH_2-CH=CH-CH_2\}_m\{CH-CH_2\}_n}$$
$$\qquad\qquad\qquad\qquad\qquad |$$
$$\qquad\qquad\qquad\qquad\bigcirc$$

The blending rubber into polystyrene is mainly to improve its impact strength. The mass fraction of rubber has a significant impact on the properties of HIPS products. The elongation at break and impact strength of the products have been improved to varying degrees with the increase of rubber content, but the tensile strength and melt flow rate have decreased. However, when the rubber content increases to a certain value, the elongation at break of the product decreases instead, because with the increase of the rubber content, the grafting site of the product increases, forming an approximate network structure, while the impact resistance of the product is greatly improved.

Considering that the GPPS molecular chain contains a large number of benzene rings with poor flexibility and large brittleness, and the molecules are not easy to diffuse and bond when the temperature is slightly lower in the fuse deposition modeling process, it is not adopted. The impact strength of HIPS can be up to seven times higher than that of GPPS. Its processing properties and chemical properties are similar to those of GPPS, and the tensile strength, hardness and thermal stability are lower than those of GPPS. HIPS is used as the base material of the support material because it contains rubber components, has a long molecular chain, is easy to diffuse, and is easy to bond when used as the support material.

Table 4.13 shows the technical parameters of HIPS PH-88 produced by Taiwan Chimei, China. It can be seen that this material has relatively low water absorption and molding shrinkage, slightly poor mechanical properties and relatively high softening point (the softening point is close to the glass transition temperature) and can withstand the high temperature of the modeling material; and with good fluidity, it can basically meet the requirements of the substrate of support material.

TABLE 4.13 Technical parameters of high-impact polystyrene PH-88.

Characteristics	Test method	Unit	Data
Tensile strength	D-638	MPa	25
Elongation at break	D-638	%	40
Flexural modulus	D-790	10^4 MPa	2.0
Flexural strength	D-790	MPa	38
Rockwell hardness	D-785	–	L-75
Softening point	D-1525	°C	99
Thermal deformation temperature	D-648	°C	82
Relative density	D-792	(23/23°C)	1.05
Melt flow rate	D-1238	g/10 min (Cond. G)	5.0
		g/10 min (Cond. I)	15.0
Water absorption	–	%	0.10–0.14
Molding shrinkage	–	%	0.02–0.006

Note: The processing temperature range is 180°C–210°C.

2. Blending modification

During fuse deposition modeling, HIPS contains long-chain molecules, with deep interlayer diffusion, and it is easy to bond but difficult to strip, so it needs to be modified.

To reduce the bonding of HIPS, the filling modification method is adopted. After the filler is added, it can bring about two effects. On the good side, the melt viscosity of the material is increased, the extent of diffusion of the long-chain molecules is reduced, and the bonding is deteriorated; moreover, the addition of the filler brings certain brittleness to the material and is convenient for stripping; at the same time, it increases the hardness and provides the driving force for the transmission of filament. On the bad side, when the melt viscosity of the material increases, the fluidity becomes poor. Therefore, in the process of blending modification, the key is to reasonably adjust the ratio of HIPS and filler to meet the requirements of bonding and stripping balance.

Considering that the support material does not require high rigidity and also needs to reduce toughness, ordinary calcium carbonate is selected as the filler.

The break-away support material needs certain brittleness to be easily broken when stripping, and at the same time, it also needs to ensure that

TABLE 4.14 Effect of calcium carbonate addition on performance of high-impact polystyrene.

Calcium carbonate mass fraction (%)	Tensile strength (MPa)	Elongation at break (%)	Flexural strength (MPa)	Glass transition temperature (°C)
0	24.5	40	37.2	95.86
20	29.5	20.4	39.7	95.94
25	31.3	12.8	43.8	98.2
30	34.1	7.5	47.2	99.43

the extruded filament cannot be easily bent or broken under the traction and driving force of the driving friction wheel. Therefore, it is necessary to inspect the brittleness of materials.

The HIPS and calcium carbonate are melt blended in proportion and extruded and granulated. The experiments show that when the mass fraction of calcium carbonate is below 20%, the blended material still has high toughness. When the mass fraction of calcium carbonate exceeds 30%, the blended material is brittle and can be easily broken, so the mass fraction of calcium carbonate should be 20%–30%.

The effect of calcium carbonate addition on the properties of HIPS is shown in Table 4.14.

As can be seen from Table 4.14, with the increase of calcium carbonate mass fraction, the tensile strength, and flexural strength of the blended material increase, the elongation at break decreases, and the material changes from ductile fracture to brittle fracture. The glass transition temperature of the material is increased, but the range is small, which has little influence on the processing temperature of the material.

4.3.3 Water-soluble support materials

For FDM technology, it is a great challenge to remove the support without damaging the feature parts. The water-soluble polymer as the support material provides a good solution. Since no mechanical removal method is required, the water-soluble support material can be located in any area of the model or contact with the fine feature parts, thus protecting the fine feature parts from damage during removal of the support. At present, the research on water-soluble support materials at home and abroad is still in the primary stage. In foreign countries, Stratasys Company has developed acrylate copolymer, which can be partially dissolved in alkaline water by ultrasonic cleaner. The support material is particularly suitable for manufacturing hollow and fine characteristic parts, which is also very advantageous for forming an assembly composed of a plurality of elements. In China, Yu Meng of Huazhong University of Science and Technology and Chen Wei of Wuhan Institute of Technology have respectively studied acrylates copolymer, polyvinyl alcohol (PVA) and other water-soluble support materials, and have made certain progress. Due to its outstanding advantages, it is inevitable that water-soluble support materials will replace break-away support materials to become mainstream support materials in the future. At present, there are two main types of water-soluble support materials that can be used in FDM process: one is polyvinyl alcohol water-soluble support materials, and the other is acrylic copolymer water-soluble support materials [9].

4.3.3.1 Polyvinyl alcohol water-soluble support material

Polyvinyl alcohol is a widely used synthetic water-soluble polymer. Due to its molecular chain contains a large number of hydroxyl groups, polyvinyl alcohol has good water solubility, coupled with its good mechanical properties and bonding properties, so from the performance requirements, polyvinyl alcohol can be selected as FDM support material. However, since the melting temperature of polyvinyl alcohol is higher than its decomposition temperature, it cannot be melted processed, so it needs to be modified to improve the melt processing performance of polyvinyl alcohol. In addition, to improve the stripping efficiency of water-soluble supporting materials and reduce the loss of energy sources, polyvinyl alcohol is required to dissolve rapidly in low-temperature solvents, so water-soluble modification of polyvinyl alcohol is also required [10].

1. Modification principle and method of polyvinyl alcohol melt processing performance

    Polyvinyl alcohol has a high degree of polymerization and alcoholysis. It contains a large number of hydroxyl groups in the flexible main chain, and there are a large number of intramolecular and intermolecular hydrogen bonds. It has many physical crosslinking points and high density, which leads to difficulties in melt processing of polyvinyl alcohol. Therefore, lowering the melting temperature and improving the thermal stability are the necessary conditions to realize the melt processing of polyvinyl alcohol. To improve the melt processability of polyvinyl alcohol, the intermolecular force must be weakened. Generally, the melt processability of polyvinyl alcohol can be improved by adding plasticizer to destroy its intermolecular force or reduce the hydroxyl content and increase the hydroxyl spacing. Common modification methods are as follows.

    a. Blending modification. Adding plasticizer can reduce intermolecular force, thus reducing melting temperature and improving melt processability.

b. Copolymerization modification. By introducing copolymerization components with other monomers, the chemical structure and regularity of the polyvinyl alcohol molecular chain are changed to reduce the intramolecular and intermolecular hydroxyl group actions to improve the melt processability.
c. Postreaction modification. The melt processability of polyvinyl alcohol can be improved by chemical reaction of partial hydroxyl groups of polyvinyl alcohol, introduction of other groups, and reduction of intramolecular and intermolecular hydrogen bonds.
d. Controlling the polymerization degree and alcoholysis degree. The lower the degree of polymerization of polyvinyl alcohol, the lower the degree of alcoholysis is, and the lower melting temperature it has.

Among the above modification methods, blending modification is widely used because of its simple operation and low cost, which is suitable for industrial production. Considering the water solubility requirement of the modified polyvinyl alcohol, the plasticizer used should preferably have greater solubility in water or be mutually soluble with water in any proportion, and should not be volatile in the processing temperature range. Using glycerol and polyethylene glycol as compound plasticizer can effectively reduce the melting temperature of polyvinyl alcohol and improve the processing fluidity of polyvinyl alcohol.

2. Principle and method of water-soluble modification of polyvinyl alcohol

Polyvinyl alcohol has intramolecular and intermolecular hydrogen bonds. Therefore, although polyvinyl alcohol has water solubility, its solubility is not very good. Some polyvinyl alcohols need to be heated and stirred in water for several hours before dissolving. To make polyvinyl alcohol have the effect of instant dissolution at low temperature, two methods can be adopted, that is, appropriately reducing hydroxyl groups or increasing molecular spacing, to introduce anionic groups with water solubility into polyvinyl alcohol molecular chains to enhance solubility. The modification method is as follows.

a. Copolymerization modification. The water solubility of polyvinyl alcohol can be greatly improved by introducing copolymerization components, changing the chemical structure and regularity of polyvinyl alcohol molecular chains, and reducing intermolecular and intramolecular hydrogen bonds.
b. Postreaction modification. Polyvinyl alcohol with a certain degree of polymerization and alcoholysis and methacrylic acid compounds are used to carry out a Michael addition reaction under alkaline conditions, after full reaction, partial or complete hydrolysis is carried out under alkaline conditions to obtain carboxylic acid modified polyvinyl alcohol.

The water solubility of polyvinyl alcohol is related to the degree of alcoholysis. The polyvinyl alcohol with an alcoholysis degree of 87%–89% has the best water solubility, while the polyvinyl alcohol with a low average polymerization degree has good water solubility and processing fluidity. Therefore, polyvinyl alcohol with an alcoholysis degree of 87%–89% and a low average polymerization degree should be selected for making the water-soluble support filament.

Since polyvinyl alcohol can be precipitated from aqueous solution when the low concentration solutions such as sodium hydroxide, calcium sulfate, sodium sulfate, or potassium sulfate are used as the precipitants, and thus can be recycled, the polyvinyl alcohol water-soluble support material is a green and environment-friendly material.

3. Preparation technology of polyvinyl alcohol water-soluble support filament

Cooling is an important link in extruded filament. The cooling effect directly determines the filament diameter stability and apparent quality of the filament. Since polyvinyl alcohol filament is soluble in water, conventional water cooling method cannot be adopted. However, the traditional air cooling equipment has slow cooling speed, long process response time, and poor cooling effect, which easily leads to the problems of low roundness precision of filament and poor stability of filament diameter. Therefore, researchers have developed a multistage variable pitch double helix air cooling device for preparing water-soluble filament. The device is shown in Fig. 4.15 and comprises a variable pitch double helix spoiler 1, annular air shield 2 (consisting of an annular inner shell, an annular outer shell and an insulating layer), hollow annular air duct 3, air inlet 4 (air inlet a and air inlet b), filament inlet 5, filament outlet 6, and guide roller 7. On the one hand, the double spiral spoiler enables gas at constant temperature to form spiral airflow in the hood, thus increasing the heat exchange time between the cooling gas and the filament (the direct cooling mode compared with the traditional cooling fan), and improving the cooling efficiency; on the other hand, the spiral

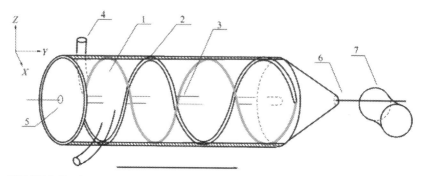

FIGURE 4.15 Structural diagram of double helix air cooling device.

airflow ensures the uniform heat exchange between the gas and the round filament, so that the filament solidifies and shrinks more uniformly along the radial direction, and the roundness precision and the filament diameter stability of the filament are improved.

4.3.3.2 Acrylic copolymer water-soluble support material

Polyacrylic acid, polymethacrylic acid and their copolymers are important water-soluble high molecular materials. Acrylic copolymers have many excellent properties. Copolymers with different relative molecular weights have very different water solubility, strength, hardness, adhesion and other properties. Acrylic acid and methacrylic acid are easy to copolymerize with other monomers and can be designed to meet the requirements of users.

There are a variety of monomers for synthesizing water-soluble acrylic copolymers, and their effects on the properties of copolymers are also very different. Specifically, monomers that affect the hardness of the copolymer include methyl methacrylate, styrene, vinyl toluene, acrylonitrile, etc. The monomers that affect the water-soluble acid value of the copolymer include acrylic acid, maleic anhydride, methylene succinic acid, etc. The monomers that affect the flexibility of the copolymer include ethyl acrylate, butyl acrylate, 2-ethylhexyl acrylate, ethylene (or butyl)methacrylate, 2-ethylhexyl methacrylate, etc. The monomers that affect the crosslinking of the copolymer include β-hydroxyethyl acrylate, β-hydroxypropyl methacrylate, glycidyl acrylate, acrylamide, N-butoxymethyl acrylamide, etc. The influence of various monomers on the copolymer is further introduced as follows.

(1) Methyl methacrylate is a hard monomer and butyl acrylate is a soft monomer. By adjusting the ratio of the two, the glass transition temperature of the copolymer can be changed in a wider range, thus greatly affecting the hardness, flexibility and impact resistance of the material.

(2) If the proportion of acrylic acid and methacrylic acid in the copolymer is too small, the water solubility of the copolymer is poor; with the increase of the dosage, the water solubility of the copolymer becomes better and the adhesive force increases, but it has a bad effect on the comprehensive properties of the material. Therefore, on the premise of full water solubility of the polymer, the amount of $-COOH$ introduced into the polymer needs to be controlled.

(3) The introduction of hydroxyl-containing alkyl (meth) acrylate monomer can increase the water solubility of the copolymer, and can crosslink the polymer to improve the strength of the material. It is generally believed that the mass fraction of 8%–12% is the best, and β-hydroxyethyl acrylate is better than β-hydroxypropyl acrylate.

(4) To obtain acrylic polymers with higher hardness, styrene and other monomers can also be added for modification, and the cost can be reduced.

1. Polymerization method of acrylic copolymers

   Two methods, solution polymerization and emulsion polymerization, are often used to prepare water-soluble acrylic copolymers. Acrylic copolymers prepared by emulsion polymerization have relatively high molecular weight, and the copolymers can be made to have certain water solubility through salinization treatment. However, the water solubility of copolymers synthesized by emulsion polymerization is not as good as that synthesized by solution polymerization, so water-soluble acrylic copolymers are generally prepared by solution polymerization.

2. Hydration method of acrylic copolymers

   In order to improve the water solubility of the copolymer, the prepared copolymer needs to be further hydrated. There are two hydration methods for water-soluble acrylic copolymers: (1) alcoholysis, that is, acrylic esters are copolymerized into viscous polyacrylate in solution, followed by partial alcoholysis; (2) salification method: acrylic esters and carboxylic acid monomers containing unsaturated double bonds such as acrylic acid, methacrylic acid, maleic anhydride, etc. are copolymerized in solution, and then amine is added to neutralize and form salt. Among them, salification method is the most commonly used.

   As there are many kinds of monomers to choose from for acrylic copolymer water-soluble support materials, and the performance of different monomers varies greatly, water-soluble support materials suitable for different kinds of modeling materials can be developed, but the disadvantages lie in that it is difficult to balance the mechanical properties, melt processing properties and water solubility, and it takes a long period to develop water-soluble support materials with stable performance.

References

[1] W. Guangchun, Z. Guoqun, Rapid Prototyping and Rapid Mold Manufacturing Technology and Its Application, China Machine Press, Beijing, 2004.
[2] X. Xianghong, Analysis of FDM 3D printer and its applicable materials, Guangdong Print. (1) (2015) 46−48.
[3] Y. Xue, P. Gu, A review of rapid prototyping technologies and systems, Comput.-Aided Des. 28 (4) (1996) 307−318.
[4] C. Wei, W. Yan, F. Yi, Preparation and research of modified polylactic acid filament for 3D printing, Eng. Plast. Appl. 43 (8) (2015) 21−24.
[5] K. Tiantian, K. Xinyu, X. Ping, et al., Research progress in preparation of polymer composites by FDM technology, China Plast. Indus. 45 (03) (2017) 45−49. 69.
[6] Y. Yuan, W. Yan, Performance of PC/ABS alloy for 3D printing, Eng. Plast. Appl. 46 (8) (2018) 34−39.
[7] C. Shengxing, Research on Nylon and Its Composite Filament for 3D Printing, Wuhan Institute of Engineering, Wuhan, 2018.

[8] C. Bellehumeur, L. Longmei, S. Qian, G. Peihua, Modeling of bond formation between polymer filaments in the fused deposition modeling process, J. Manuf. Process. 6 (2) (2004) 170–178.
[9] C. Wei, Preparation and Industrialization of Polyvinyl Alcohol Support Filament for Fuse Deposition Molding, Wuhan Institute of Technology, Wuhan, 2015.
[10] Y. Meng, Research on forming materials and supporting materials of fused deposition modeling, Huazhong University of science and technology, Wuhan, 2007.

Chapter 5

Metal materials for additive manufacturing

5.1 Additive manufacturing technologies for metal materials and the principles

The direct manufacturing of metal parts is one of the important signs that additive manufacturing (AM) technology has changed from "rapid prototyping" to "rapid manufacturing." In 2002, Germany successfully developed selective laser melting (SLM) equipment, which can form nearly fully dense fine metal parts and molds, and its performance can reach the level of homogenous forgings. At the same time, direct metal manufacturing technologies and equipment such as electron beam manufacturing and laser engineering net shaping have emerged. These technologies are aimed at superior manufacturing fields such as aerospace, weaponry, automotive molds and biomedical, directly forming complex and high-performance metal parts, and solving manufacturing problems such as difficult processing and even inability to process in traditional manufacturing processes. This section mainly introduces the AM technologies and forming principles of metal materials.

5.1.1 Selective laser melting technology

The SLM technology is developed based on selective laser sintering (SLS) technology. In the early days, due to the lack of support of powerful computer systems and high-powered lasers, metal parts were indirectly formed by coating metal powder in form of indirect SLS technology. With the development of computers and the gradual maturity of laser manufacturing technology, the Fraunhofer Institute for Laser Technology in Germany first explored the SLM technology by complete laser melting metal powder.

Based on the principle of discrete-layer-superposition and with the help of computer-aided design (CAD) and manufacturing, SLM technology uses high-energy laser beams to directly form metal powder materials into dense three-dimensional (3D) solid parts without any tooling during the forming process and not subject to the complexity of the shape of the part. It is one of the most advanced and fastest growing metal AM technologies in the world. SLM technology is actually the opposite of the traditional processing

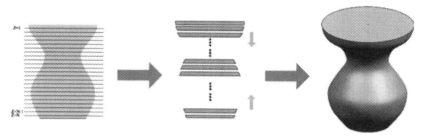

FIGURE 5.1 Schematic diagram of AM.

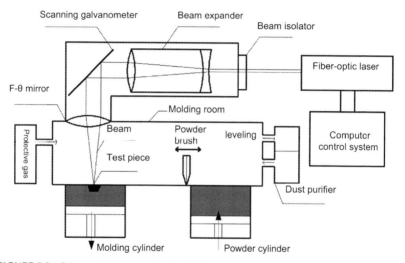

FIGURE 5.2 Schematic diagram of typical SLM process.

of removing materials for metal parts. Based on the concept of AM, it starts from the 3D part model with CAD software to make the model layered and discrete, and use the CNC forming system to transform the complex 3D manufacturing into the superstition of a series of planar two-dimensional (2D) manufacturing. The high-energy beam can be used to melt and accumulate forming materials (such as powder, strips, plates, etc.) without the fixture or tools, and quickly produce 3D metal parts with arbitrary complex shapes and functions. The schematic diagram is shown in Fig. 5.1.

SLM technology utilizes high-energy laser beam to selectively melt metal powder line by line, layer by layer, and ultimately achieve the purpose of manufacturing metal parts. The typical forming process is shown in Fig. 5.2.

1. Before the laser beam starts scanning, the substrate required for metal part stacking is installed on the working plane, and after the substrate is adjusted to the horizontal position with the work surface, the powder

cylinder is first lifted to a certain height above the powder leveling roller. The powder is brought onto the substrate of the working plane by rolling the powder leveling roller to form a uniform and even powder layer.
2. Under the control of the computer, the laser beam selectively melts the powder in a certain area of the powder layer according to the first layer data information of the CAD model of the part to form a 2D cross-section of the formed part in a horizontal direction.
3. After the scanning of the forming area of the layer is completed, the forming cylinder is lowered by a slice layer thickness, and the powder cylinder is further raised to a certain height. The powder leveling roller rolls to feed the powder to the upper part of the molten metal layer to form a uniform powder layer of one layer of thickness, and the computer imports the 2D shape information of the next layer to further processing.
4. The processing is performed layer by layer until the entire 3D part body is manufactured.

The direct formation of metal parts with practical industrial applications by using the SLM technology is one of the ultimate goals. Compared with traditional processing technology, the SLM technology has the following advantages:

1. A wide range of forming materials.
   In theory, any metal powder can be melted by high-energy laser beam; therefore, metal parts having certain functions can be directly formed by SLM technology through preparing the metal material into metal powder.
2. A simple manufacturing process and a short cycle for complex parts.
   The manufacture of traditional complex metal parts requires cooperation of a variety of processes. For example, the manufacture of artificial joints requires parallel manufacturing of various processes such as molding, precision casting, cutting, punching, etc., and the parts manufacturing requires a variety of professional technicians, in which not only the process is cumbersome, but also the cycle of the parts manufacturing is long. The SLM technology is a direct process of forming the final part from the metal powder material, which is independent of the complexity of the part, thus simplifying the manufacturing process of complex metal parts, shortening the manufacturing time, and improving the manufacturing efficiency.
3. A high utilization rate for parts materials, which saves materials.
   Traditionally machined metal parts are manufactured by removing the excess material from the blank to obtain the desired metal parts. The material used for manufacturing parts by SLM technology is basically equal to the actual size of the parts. The unused powder materials can be reused during the processing, and the material utilization rate is generally over 90%. Especially for some precious metal materials (such as gold), the cost of materials accounts for the majority of the entire processing cost. Wasting a large amount of materials leads to the increase of several

times of costs in the processing and manufacturing, so that the advantages of SLM technology to save materials are often more prominent.

4. Excellent comprehensive mechanical properties of the parts.

    The mechanical properties of metal parts are determined by their internal structure. When the grains are finer, the overall mechanical properties will be better. Compared with casting and forging, SLM parts use high-energy laser beam to selectively melt metal powder, which has small laser spot, high energy, and few internal defects. The internal structure of the part is formed under the condition of rapid melting/solidification. The microstructure often has the characteristics of small grain size, fine structure, and dispersed distribution of reinforced phase, so that the part exhibits excellent comprehensive mechanical properties. Usually most of its mechanical properties are superior to those of the forgings of the same material.

5. Suitable for the manufacturing of lightweight porous parts.

    For some porous parts with complex microstructures, the traditional method cannot process the complex porous structure inside the part. The SLM process can be used to achieve the above purpose by adjusting process parameters or data models, so as to achieve the requirements for lightweight and porous parts, as shown in Fig. 5.3. For example, artificial joints often require pores of a certain size internally to meet the requirements of biomechanics and cell growth, but the traditional manufacturing method cannot produce porous artificial joints that meet the design requirements. For SLM technology, the porous structure of any shape and complexity can be formed only by modifying the data model and the process parameters, so that it can better meet the actual needs.

6. Meeting the needs of individualized metal parts manufacturing.

    The SLM technology can easily meet the production of some personalized metal parts, and get rid of the dependence of traditional metal parts manufacturing on the mold. For example, for some personalized artificial metal prosthesis, as depicted in Fig. 5.4, designers only need to design their own products, and then use SLM technology to directly form their own products, without the need for professional manufacturing technology, so as to meet the individual needs of modern people.

FIGURE 5.3 Lightweight porous parts manufactured by SLM.

FIGURE 5.4 Personalized artificial prosthesis manufactured by SLM.

5.1.2 Wire and arc additive manufacture

With the electric arc as energy-carrying beam, the wire arc additive manufacture (WAAM) is an advanced digital manufacturing technology using layer-by-layer cladding principle to gradually form metal solid components from the line-surface-volume through the addition of the wire material according to the 3D digital model under the control of the program, which adopts electric arc produced from welding machines such as metal inert-gas welding (MIG) and tungsten inert-gas welding (TIG) as the heat source. The technology is mainly developed based on welding technologies such as TIG, MIG, and SAW, and the formed parts are composed of all welded joints with uniform chemical composition and high density.

The energy-carrying beam of WAAM has the characteristics of low heat flux density, large heating radius and high heat source intensity, coupled with the factors that the instantaneous point heat source reciprocating during the forming process strongly interacts with the forming environment, and the thermal boundary conditions have nonlinear time-varying characteristics, the stability control in the forming process is a difficulty in obtaining a consistent forming shape. The more stable the arc are, the more favorable it is to the forming process control, that is, the more favorable it is to the dimensional accuracy control of the formed shape. Therefore, arc-stabilized and spatter-free nonmolten gas shielded welding and cold metal transfer (CMT) technology developed based on melt inert/active gas shielded welding have become the main heat source supply methods currently used.

The WAAM technology, with the same principle of other AM technologies, first performs slicing along a certain coordinate direction through the STL point cloud data model to generate a discrete virtual layer; and then accumulates through the drops molten by the metal wire material wire from point to line and from line to surface, so as to print out the solid layers and form the final parts with pieces stacked, as shown in Fig. 5.5.

Compared with several other AM technologies (such as SLS, FDM, LOM, etc.), the WAAM technology uses a method of metal wire melting to

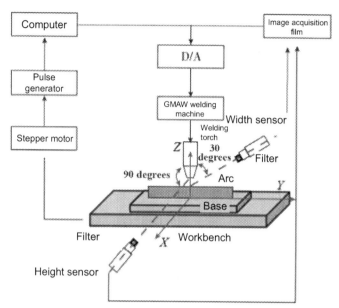

FIGURE 5.5 Wire and arc additive manufacture.

stack materials based on the metal deposition method in welding metallurgy, so the microstructure of the formed parts is compact and has good mechanical properties, printing and manufacturing costs are much lower than the same type of AM technologies, so this technology is expected to become a new manufacturing technology for mass production in the civilian market.

In addition to having the advantages common to AM technology such as forming without the need for traditional tools, reducing the number of processes and shortening the product cycle, WAAM technology also has the following four advantages:

1. The manufacturing cost is low. More than 90% of the welding materials can be utilized, and the material utilization rate is high. There is a large number of general-purpose equipment on the market for WAAM, and the investment cost is low. While the powder-based metal raw materials used in metal powder-based laser and electron beam AM technology have high preparation cost, vulnerability to contamination, low utilization rate, etc., which increase the cost of raw materials.
2. The stacking speed is high. The arc printing has a fast wire feeding speed and high stacking efficiency, which has obvious advantages in forming large-sized parts. The forming rate can reach several kilograms per hour. However, the metal AM with the laser as heat source has low forming rate, besides, the aluminum alloy has low absorption rate to the laser, and the laser is sensitive to some metal materials.

3. Manufacturing size and shape are not limited. The open forming environment has no restrictions on the size of the formed part. In the field of AM, the formed parts of WAAM are not limited by the mold, and the manufacturing size and shape are flexible. However, the metal AM with the electron beam as heat source has a certain limit on the component volume due to the size of its vacuum furnace.
4. It is not sensitive to metal materials. It is suitable for any metal material. Some materials have high reflectivity to the laser and are not suitable for AM using laser as a heat source.

In addition to the low surface quality of the solid components, the WAAM technology has a lower ability to shape the geometry than the laser. This is because the arc forming position is determined by the positions of the welding torch, the welding wire and the robot. However, the forming position of laser is controlled by the galvanometer, the accessibility and accuracy of the arc are lower than that of the laser.

In order to meet the needs of industrial production and batch repair of metal materials, arc AM technology has broad prospects and applications, and has become a hot spot of AM. Many scholars at home and abroad are dedicated to it, which has given this technology a broad development.

5.2 Forming mechanisms of metal materials

5.2.1 Laser energy transfer

5.2.1.1 Changes in physical state caused by laser and metal action

SLM formation mainly utilizes the photothermal effect. When the laser irradiates the metal surface, the metal surface area will undergo various changes at different power densities, including surface temperature rise, melting and vaporization. Moreover, changes in the physical state of the metal surface affect the absorption of the laser by the metal. As the power density and the action time increase, the following changes in the physical state of the metal will occur:

1. When the laser power density is low ($<10^4$ W/cm^2) and the irradiation time is short, the laser energy absorbed by the metal can only cause the temperature of the metal to rise from the surface to the inside, but the solid phase is kept unchanged. It is mainly used for part annealing and phase transformation hardening.
2. With the increase of laser power density (10^4-10^6 W/cm^2) and the irradiation time, the surface of the metal gradually melts. As the input energy increases, the liquid−solid interface gradually moves to the depth. This physical process is mainly used for metal surface remelting, alloying, cladding, and thermal conductivity welding.

3. When the power density ($> 10^6$ W/cm^2) is increased and the action time is lengthened further, the surface of the material not only melts but also vaporizes, and the vapor gather near the metal surface and weakly ionizes to form a plasma. This thin plasma contributes to the absorption of laser by metal. Under the vaporization expansion pressure, the liquid surface is deformed to form pits. This stage can be used for laser welding.
4. When the laser power density ($> 10^7$ W/cm^2) is increased and the irradiation time is lengthened further, the surface of the material is strongly vaporized to form a plasma with high ionization degree, which has a shielding effect on the laser and greatly reduces the energy density of laser incident into the inside of the metal. This stage can be used for laser drilling, cutting, and surface strengthening.

Vaporization is an important boundary when metal absorbs the laser. When the metal does not vaporize, whether it is solid phase or liquid phase, its absorption of laser only changes slowly with increasing surface temperature; and once the metal is vaporized and forms plasma, the absorption of metal to the laser will suddenly change. Therefore, the energy density of laser is required to exceed 10^6 W/cm^2 in the SLM forming process to ensure the remelting of the former forming layer, which facilitates metallurgical bonding.

5.2.1.2 Energy balance between laser and metal action

The interaction between laser and metal includes complex microscopic quantum processes, as well as macroscopic phenomena such as laser reflection, absorption, refraction, diffraction, polarization, photoelectric effects, and gas breakdown.

When the laser interacts with the metal, the energy conversion follows the law of conservation of energy. Assuming E_0 denotes the laser energy incident on the metal surface, $E_{\text{reflection}}$ denotes the laser energy reflected by the metal surface, $E_{\text{absorption}}$ denotes the laser energy absorbed by the metal surface, and $E_{\text{transmission}}$ denotes the laser energy transmitted through the metal, then

$$E_0 = E_{\text{reflection}} + E_{\text{absorption}} + E_{\text{transmission}} \tag{5.1}$$

$$1 = \frac{E_{\text{Reflection}}}{E_0} + \frac{E_{\text{absorption}}}{E_0} + \frac{E_{\text{transmission}}}{E_0} = \rho_R + \alpha_A + \tau_T \tag{5.2}$$

ρ_R indicates the reflectance, α_A indicates the absorbance and τ_T indicates the transmittance.

For opaque metals, the transmitted light is also absorbed, that is, $E_{\text{transmission}} = 0$, then

$$1 = \rho_R + \alpha_A \tag{5.3}$$

When the laser illuminates the metal surface, part of it is reflected by the metal, and part of it enters the inside of the metal, and the part entering into the inside is completely absorbed. During the propagation of the absorbed laser in the inside metal, according to Lambert's law, the laser intensity is exponentially attenuated, and the intensity I of light incident on the surface x from the laser is:

$$I = I_0 e^{-Ax} \tag{5.4}$$

where I_0 indicates the intensity of the laser incident on the metal surface ($x = 0$), and the unit is W/cm^2; A indicates the metal absorption coefficient, and the unit is cm^{-1}. If the penetration depth of the laser in the material is defined as the depth at which the light intensity drops to I_0/e, the penetration depth is $1/A$. This indicates that the laser intensity is reduced to $1/e$ of the original intensity after the laser passes through the metal with the thickness of $1/A$, indicating that the absorption capacity of the metal by the laser should be attributed to the absorption coefficient.

The absorption coefficient A of the metal to the laser depends on the type of metal material and the wavelength of the laser. The characteristic value of the metal material which the absorption coefficient A corresponds to is the absorption index K, and there is a relationship between them: $A = 4\pi K/\lambda$, therefore (Eq. 5.4) can be expressed as

$$I = I_0 e^{-4\pi Kx/\lambda} \tag{5.5}$$

The values of ρ_R, α_A, and A can be calculated from the optical constant of the metal or the measured value of the complex refractive index. The absorption index K is the imaginary part of the complex refractive index n of the metal material. The complex refractive index of the metal is

$$n = n_1 + iK \tag{5.6}$$

When the laser is incident perpendicularly to the metal, the reflectance (ρ_R) of the laser is:

$$\rho_R = \left|\frac{n-1}{n+1}\right|^2 = \frac{(n_1-1)^2 + K^2}{(n_1+1)^2 + K^2} \tag{5.7}$$

If it is opaque metal material, then $\alpha_A = 1 - \rho_R$

$$\alpha_A = \frac{4n_1}{(n_1+1)^2 + K^2} \tag{5.8}$$

Therefore, the absorption coefficient A can be obtained as:

$$A = \frac{4\pi K}{\lambda} \tag{5.9}$$

Therefore, the absorption of laser by metal depends mainly on the type of metal and the wavelength of the laser.

5.2.2 Absorption of laser energy by metal

The absorption of laser by metals is related to many factors such as metal properties, laser wavelength, metal temperature, metal surface condition, and laser polarization characteristics.

5.2.2.1 Effect of metal properties

The absorption coefficient of metal to particular laser wavelength can also be calculated from the electric resistivity of the metal (Eq. 5.10).

$$A = 0.365\sqrt{\frac{\rho_0}{\lambda}} \qquad (5.10)$$

A indicates the absorption coefficient of the metal surface, and the unit is cm^{-1}; ρ_0 indicates the DC resistivity of the metal, and the unit is $\Omega\, cm^{-1}$; λ indicates the laser wavelength, and the unit is cm.

Table 5.1 lists the electric resistivity of several metals. From this, it can be seen that the electric resistivity of silver, copper, and aluminum is small. Therefore, they have small absorption coefficient for laser of specific wavelength.

5.2.2.2 Effects of laser wavelength

Fig. 5.6 shows the relationship between reflectance and wavelength for a typical metal at room temperature. In the infrared region, as the wavelength increases, the reflectance increases. Most metals have strong reflections on infrared at a wavelength of 10.64 μm (10640 nm), while weak reflections on infrared at a wavelength of 1.064 μm (1064 nm).

5.2.2.3 Effects of metal temperature

The relationship between absorption rate and temperature is different at different bands. When $\lambda < 1$ μm (1000 nm), the relationship is complicated, but the overall change is relatively small. The intrinsic absorption rate of metal generally increases with increasing temperature. A laser beam having a wavelength of 10.64 μm (10640 nm) acts on the polished aluminum, and its reflectance to the laser is 98.6% at room temperature, and the reflectance of the liquid aluminum to the laser is 91%–96%. When $\lambda = 1$ μm (1000 nm),

TABLE 5.1 Electrical resistivity of several metals.

Metal	Ag	Cu	Al	Fe	Pt	Pb
$\rho_0/(\Omega/m)$	1.65×10^{-8}	1.75×10^{-8}	2.83×10^{-8}	9.78×10^{-8}	2.22×10^{-7}	2.08×10^{-7}

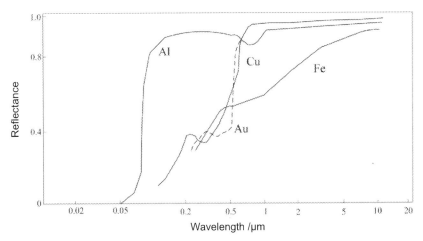

FIGURE 5.6 Relationship between reflectance and wavelength for a typical metal at room temperature.

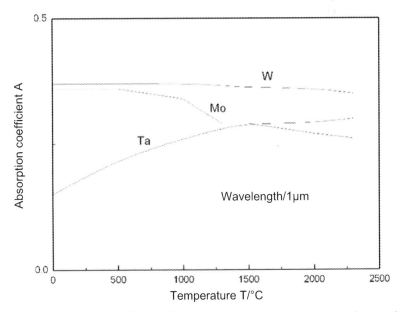

FIGURE 5.7 Absorption coefficient of three metals to 1 μm wavelength laser changes with temperature.

the experimental results of the absorption coefficient of the three metals with respect to the wavelength of the laser changing with the temperature are shown in Fig. 5.7. This indicates that the absorption rate of metal for shorter wavelength lasers increases with temperature.

5.2.2.4 Effects of metal surface roughness

The surface roughness of the metal has a significant effect on the absorption rate of the laser, and the absorption of the laser by the rough surface can be doubled compared to the polished surface.

5.2.2.5 Effects of laser polarization

When the laser beam is incident perpendicularly, the absorption rate is independent of the polarization of the laser beam. However, when the laser is incident obliquely, the effect of polarization on absorption is very important.

The light wave is a transverse electromagnetic wave, and it consists of electrical and magnetic vibrations that are perpendicular to each other and perpendicular to the direction of propagation. The orientation of the electrical vector (E) in the electromagnetic field determines the polarization direction of the laser beam. During laser transmission, if the electric field vector vibrates in the same plane, it becomes plane polarized light (or linearly polarized light). The linearly polarized light of the two polarization planes is superimposed, and when the phase is fixed, elliptically polarized light is obtained. When the two polarized lights have the same intensity and the phase difference is $\pi/2$ or $3\pi/2$, circularly polarized light is obtained. At any fixed point, when the orientation of the instantaneous electric field vector is irregularly and randomly changed, the beam is not polarized.

According to the normal measurement of the plane, at an incident angle θ, $\theta \leq 90$ degrees is assumed, n is the refractive index, k is the absorption coefficient, and $n^2 + k^2 \geq 1$, the linearly polarized light (P-polarized light) parallel to the incident surface and the linearly polarized light (S-polarized light) perpendicular to the incident surface in polarization direction have the refractive index on the surface as below:

$$\rho_{R_P}(\theta) = \frac{(n^2 + k^2)\cos^2\theta - 2n\cos\theta + 1}{(n^2 + k^2)\cos^2\theta + 2n\cos\theta + 1} \tag{5.11}$$

$$\rho_{R_S}(\theta) = \frac{(n^2 + k^2) - 2n\cos\theta + \cos^2\theta}{(n^2 + k^2) + 2n\cos\theta + \cos^2\theta} \tag{5.12}$$

While the metal is nontransparent to the laser, the dependence of the absorption rate on the polarization and incident angle is:

$$\alpha_{A_P}(\theta) = 1 - \rho_{R_P}(\theta) = \frac{4n\cos\theta}{(n^2 + k^2)\cos^2\theta + 2n\cos\theta + 1} \tag{5.13}$$

$$\alpha_{A_S}(\theta) = 1 - \rho_{R_S}(\theta) = \frac{4n\cos\theta}{(n^2 + k^2)\cos^2\theta + 2n\cos\theta + \cos^2\theta} \tag{5.14}$$

Fig. 5.8 shows the relationship between the absorption rate of polarized light and the incident angle by metal Fe, that is, diagram to the Eqs. (5.13 and 5.14). For vertically incident S-polarized light, A_S decreases slowly with increasing incident angle; for parallel incident P-polarized light, A_P first

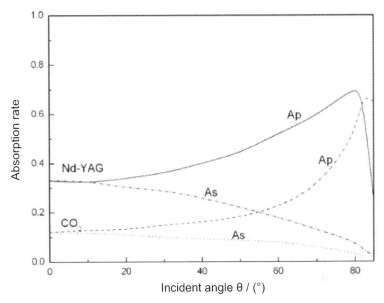

FIGURE 5.8 Relationship between the absorption rate of polarized light and the incident angle by iron (Fe).

increases with increasing incident angle, and at a large angle of incidence, A_P reaches the maximum value, and then decreases sharply with the further increase of the incident angle. The incident angle when the A_P reaches the maximum value is called the Brewster angle, and the absorption rate has a minimum value when the incident angle is 0–90 degrees.

5.2.3 Absorption of laser by metal powder

The absorptivity of powder to laser is an important characteristic of powder and is the ratio of absorbed laser energy to incident laser energy. Generally, the absorptivity of powder to laser is affected by coupling mode, transmission depth, laser wavelength λ, powder composition, particle size of powder, melting process and other factors.

5.2.3.1 Coupling mode

There are two action mechanisms between laser and powder: coupling of laser and block and coupling of laser and powder. When the laser pulse irradiates the metal powder, the laser energy transmission time is short, the heated layer of the spherical particle powder is instantaneously at a high temperature (surface temperature T_s), and is far higher than the temperature of the rest of the spherical particle powder and the average temperature T_{av} of the surrounding powder; when the laser continuous wave irradiates the powder, the action time is long, and the generated liquid phase causes higher average temperature and more liquid quantity.

5.2.3.2 Transmission depth of laser

In addition to transparent and dense materials, the absorption distance of bulk metals to laser is very short, generally within the range of 10 nm−1 μm. For powder, only part of the incident laser light is absorbed by the outer surface of the loose powder particles, while another part of the laser light passes through the interparticle pores and interacts with the deep powder particles. Metal powder composed of particle size corresponding to laser wavelength size has different absorption characteristics to laser from compact metal, mainly reflected in different laser transmission paths: laser has multiple reflections in powder, while laser does not have multiple reflections in bulk metal. The powder can be regarded as a black body due to the influence of the powder morphology (rough surface and pores between particles) on the absorption characteristics of incident laser. When the laser intensity is reduced to $1/e$ (36.8%), the depth at which the laser reaches the powder is the transmission depth D_p. During the transmission process of laser in the powder, it is reflected many times (see Fig. 5.9), so its transmission depth D_p in the powder is far greater than its transmission depth in dense material. For example, the laser transmission depth D_p of titanium powder is about 65 μm.

5.2.3.3 Influence of composition on laser absorption rate

The absorptivity of elemental powder to laser with different wavelengths is shown in Table 5.2. The test conditions are: laser power density of $1-10^4$ W/cm^2 and high purity argon atmosphere. Table 5.2 shows that the absorptivity of powder varies with laser wavelength. The absorptivity of metal and carbide to laser decrease with the increase of laser wavelength.

The absorption mechanism of powder to laser with different wavelengths is different, resulting in different absorption rates. The absorption of laser by metal depends on the electronic state close to Fermi level, which mainly occurs in the surface range of 10 nm level. In the absence of excitation, the insulator only

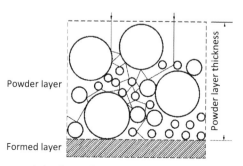

FIGURE 5.9 Laser transmission in powder.

TABLE 5.2 Absorption rate of different simple powder to laser with different wavelengths.

Powder	Absorptivity	
	$\lambda = 1064$ nm	$\lambda = 10640$ nm
Cu	0.59	0.26
Fe	0.64	0.45
Sn	0.66	0.23
Ti	0.77	0.59
Pb	0.79	—
Co-alloy(1%C; 28%Cr; 4%W)	0.58	0.25
Cu-alloy(10%Al)	0.63	0.32
Ni-alloy I(13%Cr; 3%B; 4%Si; 0.6%C)	0.64	0.42
Ni-alloy11(15%Cr; 3.1%Si; 0.8%C)	0.72	0.51
ZnO	0.02	0.94
Al_2O_3	0.03	0.96
SiO_2	0.04	0.96
BaO	0.04	0.92
SnO	0.05	0.95
CuO	0.11	0.76
SiC	0.78	0.66
Cr_3C_2	0.81	0.70
TiC	0.82	0.46
WC	0.82	0.48
$NaNO_3$	0.16	0.80
NaCl	0.17	0.60
Polytetrafluoroethylene	0.05	0.73
Poly(methylacrylate)	0.06	0.75
Epoxypolyether-based polymer	0.09	0.94

binds electrons, and the absorption of laser is mainly through lattice vibration, while the polymer absorbs laser through molecular vibration. Generally, the absorption mechanism of the same material to laser is very complex, and its mechanism will also change with the change of laser spectral region.

TABLE 5.3 Absorption rate of laser for mixed powder of Fe-alloy and TiC.

Powder	Absorptivity	
	$\lambda = 1064$ nm	$\lambda = 10640$ nm
Fe-alloy(3%C; 3%Cr; 12%V) + 10%TiC	0.65	0.39
Fe-alloy(0.6%C; 4%Cr; 2%Mo; 1%Si) + 15%TiC	0.71	0.42
Fe-alloy(1%C; 14%Cr; 10%Mn; 6%Ti) + 66%TiC	0.79	0.44

TABLE 5.4 Absorption rate of binary mixed powder (Ni-alloy I + epoxy polyether-based polymer) to laser.

Ni-alloy I mass fraction/(%)	Absorptivity	
	$\lambda = 1064$ nm	$\lambda = 10640$ nm
100	0.72	0.51
95	0.68	0.54
50	0.38	0.69
25	0.23	0.76
0	0.09	0.94

5.2.3.3.1 Absorption rate of mixed powder to laser

The absorptivity of mixed powder of iron-based alloy and titanium carbide to laser with different wavelengths is shown in Table 5.3. The results show that the greater the volume fraction of titanium carbide in the mixed powder, the greater the absorption rate of the mixed powder to laser. Because titanium carbide has a high absorptivity to laser, the absorptivity of mixed powder to laser is also high.

The absorptivity of the mixed powder of Ni-alloy I and polymer to laser is shown in Table 5.4. Table 5.4 shows that its absorptivity satisfies the relational expression $\alpha_A = \alpha_{A1}\gamma_1 + \alpha_{A2}\gamma_2$, where γ_1 is the volume fraction of powder 1, α_{A1} is the absorptivity of powder 1 to laser, γ_2 is the volume fraction of powder 2, and α_{A2} is the absorptivity of powder 2 to laser. However, this absorptivity formula is suitable for only the mixed powder of polymer and metal.

5.2.3.4 Influence of powder particle size on absorption rate

The laser absorptivity of Ni-alloy I powder with different particle sizes is shown in Table 5.5. The table shows that the particle size distribution of powder particles has little effect on the laser absorptivity.

TABLE 5.5 Influence of particle size of Ni-alloy I powder on absorption rate of laser (1064 nm).

D/μ.m	<50	50–63	63–100	100–160
α_A	0.64	0.65	0.62	0.62

5.2.3.5 Influence of power melting process on absorption rate

The absorptivity of powder to laser is affected by the thermophysical properties, particle recombination, phase change, residual oxygen, etc. of powder. The melting process of powder by laser depends on the laser power density. If the laser power density is low, the sintering process will stop due to thermal balance. If the laser power density is high, the powder will melt and the porosity will decrease sharply until it is completely melted, thus affecting the absorption of laser by the powder.

To sum up, the absorptivity of metal powder to laser is not only affected by carbon, nickel, chromium, manganese, copper and other components, but also affected by powder accumulation characteristics, laser wavelength, particle recombination, phase change, residual oxygen, temperature and so on. The absorptivity of metal powder to laser with a wavelength of nearly 1000 nm is more than 60%.

5.2.4 Temperature, stress and strain fields in selective laser melting forming process

Since the melting point of most liquid metals exceeds 1000°C, the temperature of SLM forming process changes greatly, which is easy to generate thermal strain and thermal stress, resulting in residual stress in SLM parts. If SLM forming process is not well controlled, it is easy to produce poor interlayer bonding, leading to failure of the forming process.

In the SLM forming process, the laser beam melts the metal powder in a very short time (0.5–25 ms), and the metal melting point is high, which makes the temperature dynamic change larger. Therefore, the temperature changing in the forming process will bring about corresponding real-time changes in thermal stress and thermal strain. These real-time changes in temperature, stress and strain will bring difficulties to direct measurement.

Using theoretical method and finite element numerical simulation method to study SLM forming process can reveal the distribution of temperature field, thermal stress field and thermal strain field, which provides important technical means for optimizing process parameters and scanning strategy and theoretical support for SLM forming technology.

5.2.4.1 Temperature field

5.2.4.1.1 Basic equation of three-dimensional transient temperature field

According to the principle of heating balance, the heat required for microbody heating is balanced with the heat transferred from X, Y, and Z directions to microbody and the heat generated by heat sources in microbody. Therefore, the field variables of transient temperature field should be satisfied in rectangular coordinate system

$$\rho c \frac{\partial \varphi}{\partial t} - \frac{\partial}{\partial x}\left(k_x \frac{\partial \varphi}{\partial x}\right) - \frac{\partial}{\partial y}\left(k_y \frac{\partial \varphi}{\partial y}\right) - \frac{\partial}{\partial z}\left(k_z \frac{\partial \varphi}{\partial z}\right) - \rho Q = 0 \quad \text{(within domain D)} \tag{5.15}$$

The temperature field distribution in D domain satisfies the following three kinds of boundary conditions.

The first kind of boundary conditions: given the temperature $\Gamma_1 + \Gamma_2 + \Gamma_3 = \Gamma$ on the Γ_1 boundary

The second kind of boundary condition: given the heat flux $q = q = q$ (Γ γ, t) on the Γ_2 boundary, when $q = 0$, it is the adiabatic boundary condition.

The third kind of boundary conditions: given convective heat transfer conditions on Γ_3 boundary.

The second and third kinds of boundary conditions are natural boundary conditions. All boundaries r in the Ω domain shall satisfy

$$\Gamma_1 + \Gamma_2 + \Gamma_3 = \Gamma \tag{5.16}$$

Then, the boundary conditions that Eq. (5.15) should satisfy are

$$\varphi = \overline{\varphi} \quad \text{(on border A)} \tag{5.17}$$

$$k_x \frac{\partial \varphi}{\partial x} n_x + k_y \frac{\partial \varphi}{\partial y} n_y + k_z \frac{\partial \varphi}{\partial z} n_z = q \quad \text{(on the } \Gamma_2 \text{ boundary)} \tag{5.18}$$

$$k_x \frac{\partial \varphi}{\partial x} n_x + k_y \frac{\partial \varphi}{\partial y} n_y + k_z \frac{\partial \varphi}{\partial z} n_z = h(\varphi_a - \varphi) \quad \text{(on } \Gamma_3 \text{ boundary)} \tag{5.19}$$

In the above formulas, ρ is the material density (kg/m^3); c is the specific heat capacity of the material (J/(kg·K)); t is time (s); k_x, k_y and k_z are the thermal conductivity coefficients (W/(m·K)) of the material along the X, Y and Z directions respectively; $Q = Q(x, y, x, t)$ is the heat source density inside the object (W/kg); n_x, n_y, n_z are the direction cosine of the normal outside the boundary; $\overline{\varphi} = \overline{\varphi}(\Gamma, t)$ is the given temperature on the Γ_1 boundary; $q = q = q(\Gamma, t)$ is the given heat flux on the Γ_2 boundary; h is the heat release coefficient (W/(m^2·K)); In the formula $\overline{\varphi} = \overline{\varphi}(\Gamma, t)$, under the condition of natural convection, φ_a is the ambient temperature; under forced convection, φ_a is the adiabatic boundary temperature of the boundary layer.

Solving the 3D transient temperature field is to solve the field function φ satisfying the transient heat conduction equation and boundary conditions under the initial conditions ($t = 0$, $\varphi = \varphi_0$), φ is a function of coordinates and time.

The problem of transient heat conduction temperature field depends on time variation. After the discretization of space domain finite element, the first-order ordinary differential equations are obtained, which cannot be directly solved, but can be solved by modal superposition method or direct integration method. In practical application, direct integration method is mainly used to solve the problem.

5.2.4.1.2 Finite element method for transient heat conduction

The temperature field function of transient temperature field is not only a function in space domain Ω, but also a function in time domain t. However, the time domain and the space domain are not coupled, so the finite element scheme is established by partial discrete method.

The general format of 2D transient temperature field finite element: firstly, the space domain Ω is discretized into finite element bodies, and the temperature φ in a typical element can be approximately obtained by interpolating temperature φ_i, which is a function of time, that is,

$$\varphi = \tilde{\varphi} = \sum_{i=1}^{n_e} N_i(x, y)\varphi_i(t) \tag{5.20}$$

The interpolation function is used to solve the heat balance differential Eq. (5.15) to generate the margin:

$$R_\Omega = \frac{\partial}{\partial x}\left(k_x \frac{\partial \tilde{\varphi}}{\partial x}\right) + \frac{\partial}{\partial y}\left(k_y \frac{\partial \tilde{\varphi}}{\partial y}\right) + \rho Q - \rho c \frac{\partial \tilde{\varphi}}{\partial t} \tag{5.21}$$

$$R_{\Gamma_2} = k_x \frac{\partial \tilde{\varphi}}{\partial x} n_x + k_y \frac{\partial \tilde{\varphi}}{\partial y} n_y - q \tag{5.22}$$

$$R_{\Gamma_3} = k_x \frac{\partial \tilde{\varphi}}{\partial x} n_x + k_y \frac{\partial \tilde{\varphi}}{\partial y} n_y - h(\varphi_a - \tilde{\varphi}) \tag{5.23}$$

Let the weighted integral of the margin be zero, that is,

$$\int_\Omega R_\Omega w_1 d\Omega + \int_{\Gamma_2} R_{\Gamma_2} w_2 d\Gamma + \int_{\Gamma_3} R_{\Gamma_3} w_3 d\Gamma = 0 \tag{5.24}$$

According to Galerkin method, the option function is

$$\begin{array}{c} w_1 = N_j \quad (j = 1, 2, \ldots, n_\ell) \\ w_1 = w_3 = -w_1 \end{array} \tag{5.25}$$

After step-by-step integration, the matrix equation for determining the temperature φ_i of n nodes can be obtained as follows

$$C\dot{\varphi} + K\varphi = P \tag{5.26}$$

where C is the specific heat capacity matrix; K is the heat conduction matrix; both C and K are symmetric positive definite matrices; P is the temperature load column vector; φ is the node temperature column vector and $\dot{\varphi}$ is reciprocal column vector of node temperature to time, $\dot{\varphi} = d\varphi/dt$. Elements of matrices K, C, and P are integrated by matrix elements corresponding to the units:

$$\begin{cases} K_{ij} = \sum_{\ell} K_{ij}^e + \sum_{e} H_{ij}^e \\ C_{ij} = \sum_{e} C_{ij}^e \\ P_i = \sum_{e} P_{Qi}^e + \sum_{e} P_{\varphi}^e + \sum_{e} P_{Hi}^e \end{cases} \tag{5.27}$$

where $K_{ij}^e = \int_{\Omega^e}(k_x \frac{\partial N_i}{\partial x}\frac{\partial N_j}{\partial x} + k_y \frac{\partial N_i}{\partial y}\frac{\partial N_j}{\partial y})d\Omega$, is the contribution of the unit to the heat conduction matrix; $H_{ij}^e = \int_{\Gamma_3^e} hN_iN_j d\Gamma$, is the correction of heat conduction matrix for unit heat exchange boundary; $C_{ij}^e = \int_{\Omega^e} \rho c N_i N_j d\Omega$, is the unit's contribution to the specific heat capacity matrix. $P_{Qi}^e = \int_{\Omega^e} \rho Q N_i d\Omega$, is the temperature load generated by unit heat source; $P_q^e = \int_{\Gamma_2^e} q N_i d\Gamma$, is the temperature load of the heat flow boundary given by the unit; $P_{Hi}^e = \int_{\Gamma_3^e} h\varphi_a N_i d\Gamma$, is the temperature load at the boundary of unit convection heat transfer.

For n_1 nodes on the boundary Γ_1 of a given temperature value, the following conditions should be introduced into the corresponding formula in Eq. (5.26):

$$\varphi_i = \tilde{\varphi}_i \quad (i = 1, 2, \ldots n_1)$$

where i is the number of n_1 nodes on Γ_1.

5.2.4.1.3 Mathematical model of heat transfer in selective laser melting forming process

If the SLM forming equipment does not use a preheating device, the ambient temperature of its forming chamber is room temperature (298K). Since the molten pool is constantly moving during SLM forming, the temperature field is also changing in real time. Changes in temperature will cause changes in the thermophysical parameters (specific heat capacity, thermal conductivity and density) of the corresponding metal powder. The changes of thermophysical parameters of common metal powders with temperature changes are shown in Tables 5.6 and 5.7. The thermophysical parameters of metal powder are also affected by components, and can be simplified according to the content of each component.

TABLE 5.6 Thermophysical parameters of pure iron.

Temperature T/K	273	573	873	1173	1473	1773
Thermal conductivity λ/[W/(m·K)]	74.7	55.5	38.2	28.2	32.2	32.2
Specific heat capacity c/[J/(kg·K)]	435.1	552.3	753.1	656.9	640.2	836.8
Density ρ(kg/m³)	7870	7770	7700	7620	7630	7640

TABLE 5.7 Thermophysical parameters of pure nickel.

Temperature T/K	373	573	873	1073	1273	1473
Thermal conductivity λ/[W/(m·K)]	82.8	90.5	65.5	67.4	71.8	76.1
Specific heat capacity c/[J/(kg·K)]	444	447	589	523	548	577
Density ρ(kg/m³)	8900	8899	8782	8696	8608	8517

The specific heat capacity of the alloy material is:

$$c = \sum_{i=1}^{2} x_i c_i \tag{5.28}$$

where c_i is the specific heat capacity of component i; x_i is the mass fraction of component i.

The thermal conductivity of alloy material is:

$$\lambda_e = \lambda_1 \left[\frac{\lambda_2 + 2\lambda_1 - 2\varphi(\lambda_1 - \lambda_2)}{\lambda_2 + 2\lambda_1 + \varphi(\lambda_1 - \lambda_2)} \right] \tag{5.29}$$

where λ_1 represents the thermal conductivity of components with large content; λ_2 represents the thermal conductivity of components with less content; φ represents the volume fraction of the components with less content in the components with large content.

The thermal conductivity of the powder is

$$\lambda = 6\rho\lambda_e/\pi \tag{5.30}$$

where ρ is the relative density of powder spreading, and its value is equal to the ratio of density of compacted powder bed to density of alloy material.

The molten pool exists for a very short time (0.5–25 ms) during SLM forming. After laser scanning, the molten pool rapidly cools and solidifies.

Therefore, the change of temperature is also related to the latent heat of metal phase change and the distribution of laser energy. The fundamental mode laser energy is Gaussian distribution, that is,

$$I = \frac{2\alpha_A P}{\pi r_b^2} \exp\left(\frac{-2r^2}{r_b^2}\right) \tag{5.31}$$

Therefore, the average heat flux of laser is:

$$I_m = \frac{1}{\pi r_b^2} \int_0^{r_b} I(2\pi r)dr = \frac{2\pi}{\pi r_b^2} \int_0^{r_b} \frac{2\alpha_A P}{\pi r_b^2} \exp\left(\frac{-2r^2}{r_b^2}\right) rdr = \frac{0.865\alpha_A P}{\pi r_b^2} \tag{5.32}$$

where αA is the absorption rate of the material; P is laser power; and r_b is the effective laser radius.

Latent heat of phase change is calculated by enthalpy at different temperatures. Enthalpy can be defined as the product of density and specific heat capacity and the temperature is integrated, that is,

$$H = \int \rho c(\varphi) d\varphi \tag{5.33}$$

In any period of time, the heat absorbed by the laser per unit area on the surface of the metal powder shall be equal to the sum of the heat transferred from the surface to the inside of the powder, convection heat transfer and the heat dissipated by surface radiation. Powder undergoes a phase change process from melting to solidification under the action of laser, so latent heat of phase change is also a form of laser energy. There is heat exchange between the molten pool and the surrounding air and powder bed. Let the heat dissipated by the material to the air be q_m and the laser power density be q_{laser}. Therefore, the heat conduction equation of the 3D transient temperature field in the SLM forming process of metal powder can be established according to the principle of heat balance. The heat required for microbody heating is balanced with the heat transferred to microbody in X, Y and Z directions and the heat generated by heat source in microbody. Therefore, the field variables of the transient temperature field satisfy the differential Eq. (5.15) in the rectangular coordinate system.

For SLM forming process, the thermal conductivity coefficient $k_x = k_y = k_z = \lambda_e$ of powder along X, Y and Z directions, the heat dissipated by air is q_m, and the laser power density is q_{laser}. Therefore, the heat source density $\rho Q = q_m + q_{laser}$ in the powder. The boundary conditions are

$$\lambda_e \frac{\partial \varphi}{\partial n} + h(\varphi - \varphi_0) + \sigma\varepsilon\left(\varphi^l - \varphi_0^l\right) - q \tag{5.34}$$

where λ_e is the effective thermal conductivity of metal powder; φ is the surface temperature of the material at time t; φ_0 is the initial temperature (295K); h is the convective heat transfer coefficient; ε is the coefficient of thermal radiation; σ is Stefan-Boltzmann constant with a value of 5.670×10^{-8} W/(m² · K⁴).

5.2.4.1.4 Temperature field analysis of molten pool in selective laser melting forming process

In order to simulate SLM forming process, the CAD modeling function of ANSYS software is used to build the model. The size of the base plate is 3 mm × 1.2 mm × 0.2 mm, and the size of the formed metal powder layer is 2.2 mm × 1.2 mm × 0.2 mm. Constraints are added to the bottom of the model base plate to fix it. Solid70 hexahedral solid units are selected to divide the metal powder layer, and the unit size is 0.1 mm × 0.1 mm × 0.1 mm. Solid90 tetrahedral solid units are used to divide the substrate solid units. The finite element model after meshing is shown in Fig. 5.10. The scanning strategy for simulating the forming process is shown in Fig. 5.10B.

Fig. 5.11 shows the temperature field when the process parameter is laser power 98 W, scanning speed is 90 mm/s, layer thickness is 0.1 mm, and forming to 0.231 s (1.0 mm from the starting point). The temperature field shows that when the SLM forming process has a large temperature gradient, the temperature of the molten pool exceeds the melting point of the metal, while the ambient temperature of the molten pool is relatively low. In order to study the influence of process parameters on the temperature field of the molten pool during SLM forming, the temperature distribution of the molten pool in the X-direction, Y-direction, and Z-direction was investigated respectively.

The simulation results at laser power of 98 W and different scanning speeds are shown in Fig. 5.12A, indicating that the width of Y-direction molten pool (temperature: 1600°C) is about 0.150 mm. The width of molten pool in SLM forming process determines the width of forming track.

The width of single-pass scanning forming track with the same process parameters is affected by powder, as shown in Fig. 5.13. The width of −250 mesh water atomization 304 L powder SLM forming track (see Fig. 5.13A) is 0.3520−420 mm. The width of −500 mesh water atomization 316 L powder forming track (see Fig. 5.13B) is 0.144−0.162 mm. However, the width of the-800 mesh atomization 316 L powder forming track (see Fig. 5.13C) is 0.231−0.240 mm. However, the width of the forming track of substrate scanning by the same process parameters is 0.090 mm (see Fig. 5.13D). Therefore, the simulation results are somewhat different from the actual ones

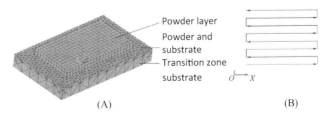

FIGURE 5.10 Finite element model and scanning strategy of metal powder forming (A) finite element model; (B) scanning strategy.

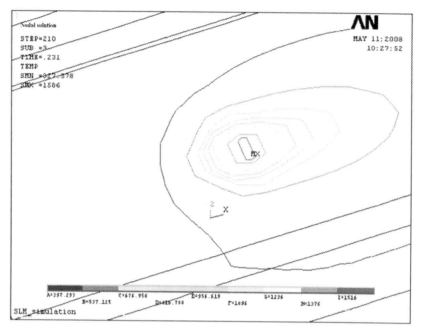

FIGURE 5.11 Temperature field when scanning 0.231 s with the scanning laser power of 98 W and scanning speed of 90 mm/s.

because of the phenomena of powder splashing and particles being dragged by the surface tension of liquid metal during SLM forming.

The single track 30 scan forming trajectory width of the same process parameters is affected by the powder [1]. The SLM forming track width of −250 mesh water mist 304 L powder (Fig. 5.13A) is 0.352−0.420 mm; the forming track width of −500 mesh water mist 316 L powder (Fig. 5.13B) is 0.144−0.162 mm; and the forming track width of −800 mesh aerial fog 316 L powder (Fig. 5.14C) is 0.231−0.240 mm. However, the scan of substrate with the same process parameters has a forming trajectory of 0.090 mm (Fig. 5.13D). Therefore, the simulation results are somewhat different from the actual ones, because powder splashes and particles are dragged by the surface tension of the liquid metal during the SLM process.

The simulation results of the laser power of 98 W and the scanning speed of 30 mm/s (Fig. 5.12A) indicate that the Y-direction melting pool (temperature: 1600°C) has a width of about 0.200 mm. The single-pass scanning forming track width of the same process parameters (Fig. 5.14) is affected by the powder. The forming track width of −250 mesh water mist 304 L powder (Fig. 5.14A) is 0.425−0.466 mm; the forming track width of −500 mesh water mist 316 L powder (Fig. 5.14B) is 0.146−0.204 mm; and the forming track width of −800 mesh aerial fog 316 L powder (Fig. 5.14C) is

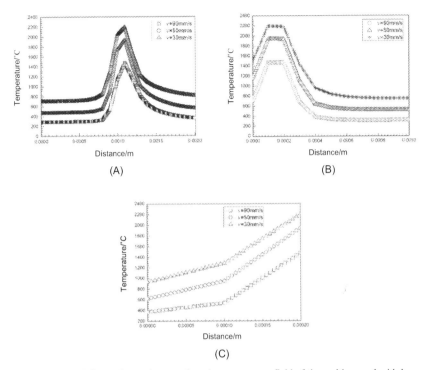

FIGURE 5.12 Effects of scanning speed on the temperature field of the melting pool with laser power of 98 W. (A) Y direction. (B) X direction. (C) z direction.

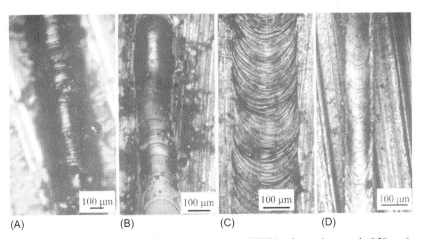

FIGURE 5.13 Forming track width with laser power of 98 W and scanning speed of 50 mm/s. (A) −250 mesh water mist 304 L (B) −500 mesh water mist 316 L (C) −800 mesh aerial fog 316 L (D) 316 L substrate.

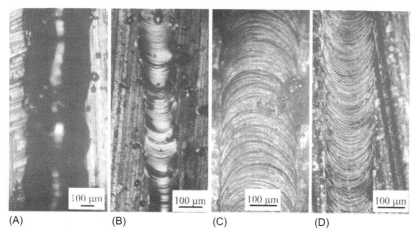

FIGURE 5.14 Forming track width with laser power of 98 W and scanning speed of 30 mm/s. (A) −250 mesh water mist 304 L (B) −500 mesh water mist 316 L (C) −800 mesh aerial fog 316 L (D) 316 L substrate.

0.275–0.280 mm. However, the scan of substrate with the same process parameters has a forming trajectory of 0.170 mm (Fig. 5.14D).

The simulation results of the laser power of 98 W and the scanning speed of 90 mm/s (Fig. 5.12) indicate that the width of the Y-direction melting pool (temperature: 1600°C) is zero. Although this result is quite different from the experimental results, the trend of the whole temperature field changing with the scanning speed is consistent with the actual situation.

The same process parameters have great influence on the Y-direction and X-direction width of the melting pool during the SLM forming process (Fig. 5.12A,B). When the scanning speed is slow, the laser and the powder act for a long time, resulting in a wide width of the melting pool. According to the principle of conservation of energy, when the scanning speed is slow, the accumulated energy is high, and the substrate temperature is also high; when the scanning speed is fast, the accumulated energy is small, and the substrate temperature is also low. The multilayer heat accumulation effect during the SLM forming process reduces the temperature gradient. Fig. 5.12B shows that the temperature in the X direction is higher near the melting pool and lower far away from the melting pool.

Fig. 5.12C shows the effects of scanning speed on the Z-direction depth of the melting pool. With the same laser power, the bath depth decreases as the scanning speed increases, and even the melting pool cannot be formed. The simulation results show that the depth of melting pool can reach 65 μm at a scanning speed of 30 mm/s, and the depth of melting pool is only 35 μm at 50 mm/s, that is, the depth of melting pool is less than 100 μm (0.1 mm). In order to ensure that the previous layer is remelted during SLM forming process, the thickness of the selected forming layer is required to be lower than the simulation result of the depth of melting pool.

5.2.4.2 Thermal stress and thermal strain field

Problems such as cracks, warpage, delamination, etc. occur during the SLM forming process, causing failure of the forming process or degradation of the mechanical properties of the formed part. This is mainly due to the large temperature gradient, thermal stress and thermal strain during SLM forming. Through the obtained temperature field of the SLM forming process, the temperature stress is calculated, and the stress and strain changes during the SLM forming process due to the temperature gradient are revealed. Since the thermal expansion in the SLM forming process only produces linear strain (initial strain), the shear strain is zero. Therefore, the basic idea of the thermal stress finite element method for solving the temperature (thermal) stress is to first calculate the initial strain caused by the temperature gradient, and then solve the equivalent node load (temperature load) caused by the corresponding initial strain, and then solve the node displacement caused by the temperature load, finally the thermal stress is obtained by the node displacement.

5.2.4.2.1 Basic equations of stress-strain relationship

5.2.4.2.1.1 Stress and strain of thermal elastoplastic
The incremental constitutive equation $d\sigma = \boldsymbol{D}_T(\varepsilon)d\varepsilon$ of the material is used when applying Newton-Raphson and incremental loading method. $\boldsymbol{D}_T(\varepsilon)$ is elastic matrix.

Under the high temperature conditions, the yield limit σ_y of the material is reduced, the strengthening property is also reduced, and the ideal plasticity is gradually approached as the temperature rises. The linear elastic constant also varies with temperature. Therefore, the material strain rate $\dot{\varepsilon}$ under the action of external force and temperature should be composed of four parts: elastic strain rate $\dot{\varepsilon}^e$, plastic strain rate $\dot{\varepsilon}^p$, creep strain rate $\dot{\varepsilon}^c$ and the strain rate caused by temperature change $\dot{\varepsilon}^T$, $\dot{\varepsilon} = \dot{\varepsilon}^e + \dot{\varepsilon}^p + \dot{\varepsilon}^c + \dot{\varepsilon}^T$.

Because the elastic constant changes with temperature, so

$$\dot{\varepsilon}^e = \frac{d(D_e^{-1}\sigma)}{dt} = D_e^{-1}\dot{\sigma} + \frac{d}{dt}\left(D_e^{-1}\right)\sigma \tag{5.35}$$

where D_e is elastic matrix.

According to flow theory and creep theory:

$$\dot{\varepsilon}^p = \dot{\lambda}\frac{\partial F}{\partial \sigma} \tag{5.36}$$

where F is the yield function.

$$\dot{\varepsilon}^c = \frac{3}{2}\frac{\overline{\dot{\varepsilon}}^c}{\overline{\sigma}}\sigma' \tag{5.37}$$

where,

$$\tilde{\dot{\varepsilon}}^c = \frac{d\tilde{\varepsilon}^c}{dt} = \frac{\sqrt{2}}{3}\left[(\dot{\varepsilon}^c_{11} - \dot{\varepsilon}^c_{22})^2 + (\dot{\varepsilon}^c_{22} - \dot{\varepsilon}^c_{23})^2 + (\dot{\varepsilon}^c_{33} - \dot{\varepsilon}^c_{11})^2 + 6(\dot{\varepsilon}^{c2}_{12} + \dot{\varepsilon}^{c2}_{23} + \dot{\varepsilon}^{c2}_{31})^2\right]^{1/2}$$
(5.38)

Temperature strain rate is

$$\dot{\varepsilon}^T = \dot{T}A \tag{5.39}$$

where $A = a\{1, 1, 1, 0, 0, 0\}^T$, a is the rate of change of temperature with time, then

$$\dot{\varepsilon} = D_c^{-1}\dot{\sigma} + \left(\frac{d}{dt}D_c^{-1}\right)\sigma + \dot{\lambda}\frac{\partial F}{\partial \sigma} + \dot{\varepsilon}^c + \dot{\varepsilon}^T \tag{5.40}$$

Multiply the two sides by the modulus of elasticity matrix, then

$$\dot{\sigma} = D_c\dot{\varepsilon} - \dot{\lambda}D_c\frac{\partial F}{\partial \sigma} - D_c^{-1}(\dot{\varepsilon}^c + \dot{\varepsilon}^T) + \left(\frac{d}{dt}D_c\right)\varepsilon^c \tag{5.41}$$

Let the yield condition be

$$F(\sigma_{ij}, \varepsilon^p_{ij}, T) = 0 \tag{5.42}$$

It can be solved:

$$\dot{\lambda} = \frac{q^T D_c\dot{\varepsilon} - q^T D_c(\dot{\varepsilon}^c + \dot{\varepsilon}^T) + q^T\left(\frac{d}{dt}D_c\right)\varepsilon^c + \frac{\partial F}{\partial T}\dot{T}}{p^T q + q^T D_c q} \tag{5.43}$$

Therefore, the stress-strain relationship of the incremental type thermal elastoplastic is

$$\dot{\sigma} = \left[\frac{D_c - D_c q(D_c q)^T}{W}\right](\dot{\varepsilon} - \dot{\varepsilon}^c - \dot{\varepsilon}^T) + \left[\frac{\dot{D}_c - D_c q(\dot{D}_c q)^T}{W}\right]\varepsilon^c - D_c\frac{\partial F}{\partial T}\frac{\dot{T}}{W}$$
(5.44)

where $W = p^T q + q^T D_c q$, $\dot{D}_c = \frac{d}{dt}D_c q$.

5.2.4.2.1.2 Thermal elastoplastic finite element method

The finite element method is used to deal with the thermal elastoplastic problem. In essence, the nonlinear stress-strain relationship is gradually transformed into a linear problem according to the loading process. Since there is no external force in the SLM forming process, the load term is actually caused by the temperature change ΔT. The temperature change ΔT is divided into several incremental loads and gradually added to the structure to solve.

Considering a certain unit constituting the whole object has the temperature of T at time t, the external force of the node is F^e, the displacement of the node is δ, the strain is ε, and the stress is σ; at time $t + dt$, it becomes

$T + dT$, $\{F + dF\}^e$, $\delta + d\delta$, $\varepsilon + d\varepsilon$ and $\sigma + d\sigma$. Apply the principle of virtual displacement, the following equation can be obtained:

$$\{d\delta\}^T \{F + dF\}^e = \iint_{\Delta V} \{d\delta\}^T [B]^T (\{\sigma\} + [D]\{d\varepsilon\} - \{C\}dT) dV$$
$$= \{d\delta\}^T \iint_{\Delta V} [B]^T (\{\sigma\} + [D]\{d\varepsilon\} - \{C\}dT) dV \qquad (5.45)$$

where $[B]$ is the geometric matrix related to the unit geometry.

Since the object is in equilibrium at time t, so

$$\{dF\}^e = \iint_{\Delta V} [B]^T \{\sigma\} dV \qquad (5.46)$$

$$\{dF\}^e = \iint_{\Delta V} [B]^T ([D]\{d\varepsilon\} - \{C\}dT) dV \qquad (5.47)$$

$$\{dF\}^e + \{dR\}^e = [K]^e \{d\delta\} \qquad (5.48)$$

Here the equivalent nodal force of the initial strain is

$$\{dR\}^e = \iint_{\Delta V} [B]^T \{C\} dT dV \qquad (5.49)$$

The element stiffness matrix is

$$[K]^e = \iint_{\Delta V} [B]^T [D][B] dV \qquad (5.50)$$

According to the elastic state or plastic state of the unit, the equivalent node load and stiffness matrix of the unit formed by Eqs. (5.49 and 5.50) respectively are substituted into the total stiffness matrix and the total load column vector to obtain the algebraic equations of the node displacement.

$$[K]\{d\delta\} = \{dF\} \qquad (5.51)$$

where,

$$[K] = \sum [K]^e \qquad (5.52)$$

$$\{dF\} = \sum (\{dF\}^e + \{dR\}^e) \qquad (5.53)$$

The incremental tangent stiffness method is used to solve the thermal elastoplastic problem. The incremental tangent stiffness method is to obtain an approximate solution by adjusting the stiffness matrix according to the stress state of the unit during each loading process. For the purpose of linearization, a method of gradually increasing the load is used: a load is added at a certain stress and strain level. Then

$$[K]\{\Delta\delta\}_i = \{\Delta F\}_i \tag{5.54}$$

where $\{\Delta\delta\}_i$ is the displacement increment obtained by the ith load; $\{\Delta F\}_i$ is the load of the ith load, $\{\Delta F\}_i = \frac{1}{n}\{F\}$ and n are positive integers.

Since the differential of stress and strain is replaced by increment $[K]$ in Eq. (5.54) is only related to the stress level before loading, so the load and displacement increment are linear. It is not difficult to find the increment of displacement, strain and stress, and then superimpose the total displacement, total strain and total stress after the $(i - 1)$th loading to obtain the total displacement, strain and stress after the first loading, and the stress is used to calculate the next load.

After each incremental solution is solved, the effect of the above incremental method is to adjust the stiffness matrix to reflect the nonlinear variation of structural stiffness before adding next incremental load. However, pure increments inevitably accumulate errors with each load increment, resulting in a loss of balance eventually. To this end, the Newton-Raphson iteration method is used. Before each solution, the Newton-Raphson method is used to estimate the residual vector (the difference between the load of the unit stress and the applied load), then the unbalanced load is used for linear solution and check of the convergence. If the convergence criterion is not met, the unbalanced load should be reestimated, and the stiffness matrix should be modified to find a new solution. This iterative process continues until the problem converges.

5.2.4.2.2 Analysis of thermal stress and strain field

In order to study the interlaminar thermal stress during the forming process, a single-channel multilayer scanning process in the X direction is simulated to facilitate observation of the interlaminar stress distribution due to the phase transformation process in which the powder is melted to solidify. The model size is 20 mm × 8 mm and the layer thickness is 0.1 mm. The Solid70 eight-node solid element is selected. The finite element model after meshing is shown in Fig. 5.15.

During the SLM forming process, the laser-irradiated powder undergoes solid-liquid and liquid−solid phase transitions within a few milliseconds. Since the powder around the liquid metal has low thermal conductivity, the temperature gradient between the liquid metal and the powder and the substrate is large. The temperature change causes the line strain $a \cdot \Delta T$ of the material. When the positive temperature changes, the linear strain of temperature is restrained by the surrounding material, and the material will be subjected to compressive stress; in the case of negative temperature change (cooling process), the material will be subjected to tensile stress.

The residual stress distribution of the single-pass scanning forming trajectory is shown in Fig. 5.16, indicating that the residual stress distribution in

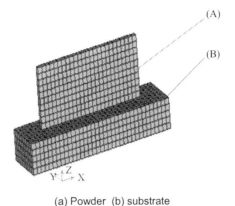

(a) Powder (b) substrate

FIGURE 5.15 Thermal stress-strain model. (A) Powder (B) substrate.

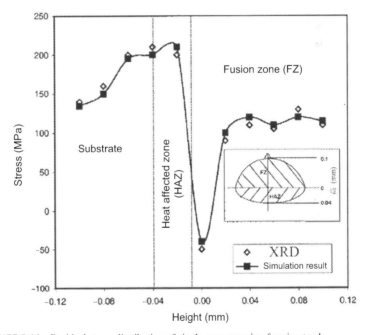

FIGURE 5.16 Residual stress distribution of single-pass scanning forming track.

the SLM forming layer changes greatly. The residual stress near the heat affected zone (HAZ) of the forming layer is compressive stress, and the tensile stress is on both sides. The substrate immediately adjacent to the HAZ is subjected to tensile residual stress. Therefore, as the forming layer is accumulated, the compressive stress of the HAZ is restored to the tensile stress state. The heat treatment by laser remelting facilitates the release of residual stress. The results of thermal stress finite element analysis (Fig. 5.17) show that the residual stress of the forming layer is tensile stress. Therefore, the finite element analysis results of the thermal stress in the SLM forming process can prove that the compressive stress in the HAZ returns to the tensile stress state during the multilayer forming process.

In Fig. 5.17A, there is an order of magnitude difference between the maximum tensile stress (0.107×10^{10} MPa) and the maximum compressive stress (0.479×10^9 MPa). Therefore, the main residual stress in the SLM forming process is tensile stress. The maximum tensile stress value in Fig. 5.17B is also greater than the maximum compressive stress. The area around the melting pool produces large thermal strain under the action of large temperature gradient, and also produces large thermal strain after cooling, thus exhibiting large residual tensile stress. Fig. 5.18 shows cracks perpendicular to the scanning direction during the forming process due to the accumulation of residual stress. The simulation results (Fig. 5.17) show that the residual tensile stress at the contact between the molded part and the substrate is large, that is, there is stress concentration. Fig. 5.19 shows that the SLM-formed parts of Steel 45 and Powder 316 L are prone to cracking or warping deformation due to stress concentration at the point of contact with the substrate.

During the SLM forming process, the temperature of the formed parts rises due to the increase of the layer thickness and the accumulation of heat, and the temperature gradient is reduced during the forming process, thereby

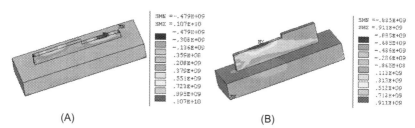

FIGURE 5.17 Stress (σ_x) distribution of SLM forming parts. (A) σ_x distribution at 0.16 s, (B) σ_x distribution at 0.34 s.

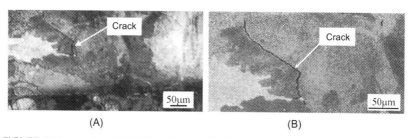

FIGURE 5.18 Cracks of SLM-formed parts of -250 mesh water mist 304 L powder under stress action.

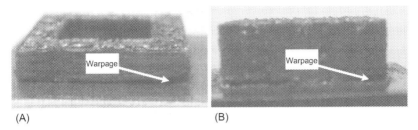

FIGURE 5.19 Cracks or warpage of SLM-formed parts under stress concentration. (A) −800 mesh aerial fog 45 steel powder formed part, (B) −500 mesh water mist 316 L powder formed part.

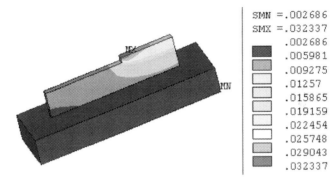

FIGURE 5.20 Thermal strain distribution of SLM-formed parts at 0.34 s.

contributing to the reduction of thermal strain and thermal stress. Therefore, the stress distribution of the latter (maximum tensile stress in Fig. 5.17B is 0.911×10^9 MPa) is lower than that of the former (maximum tensile stress in Fig. 5.17A is 0.107×10^{10} MPa).

The thermal strain field distribution is shown in Fig. 5.20. The high temperature of the melting pool forms a large temperature gradient with the surrounding temperature, so the thermal deformation near the melting pool is large. The temperature gradient away from the melting pool is small and the thermal strain generated is small.

5.2.4.2.3 Method of reducing thermal stress and strain

Residual thermal stress and thermal strain during SLM forming can be reduced by preheating, scanning strategy and postprocessing.

1. Preheating method

   Preheating reduces the temperature gradient around the melting pool and produces less thermal strain, thus reducing residual thermal stress caused by thermal strain. By preheating to 400°C, the residual tensile stress of the SLM-formed part can be reduced to 400 MPa.

2. Scanning strategy method
   The temperature gradient during the forming process can be changed by the scanning strategy. The residual stress produced due to the temperature changing in the SLM forming process can be attributed to the following two reasons: first, the cross-section of the formed part is scanned and formed by parallel scanning of vector. If the area of the scanned zone is small, a shorter scan vector appears. When adjacent scanning paths are sequentially swept, there is no long time left for cooling between adjacent scanning paths, resulting in high forming temperature. When the scanning area is large, the laser scanning time is long, and the adjacent scanning path has a long time to cool, resulting in a low forming temperature in the scanning area. High forming temperature produces better wetting conditions, which is beneficial to increase the density of the material; while low forming temperature produces poor wetting conditions and reduces the density of the material. Second, the temperature changes with time during the forming process due to the difference in heat transfer coefficient between the powder and the solid. The density of the material increases (e.g., increasing from 40% to 95%) when the metal powder melts. When scanning a small area, less heat is lost due to the isolation of the surrounding powder, while a large scanning area produces relatively more heat loss, especially at the corners of the formed part. As a result of the high temperature, better wetting conditions and higher forming densities are produced. To this end, the scanning strategy of variable scanning vector length grouping is adopted to reduce the temperature gradient during the forming process by setting a short scanning vector length to improve the wettability of the material and reduce the residual stress.
3. Postannealing method
   SLM-formed parts are postannealed to reduce residual stress. The SLM-formed part is treated after annealing at 600°C for about 1 hour, and the residual tensile stress can be reduced to only 200 MPa. Annealing treatment of the forming trajectory can also be realized by immediate postannealing method in the forming process, such as using a dual laser to achieve this solution for eliminating residual stress.

5.2.5 Dynamics and stability of melting pool

The laser beam, as a heat source for heating and melting the metal powder material, undergoes a process of heating (temperature rise) and change by the applied material and when it acts on the surface of the solid material. During the action of the laser beam and the material, the heat transfer from the laser beam to the metal surface layer is achieved by the inverse bremsstrahlung effect, and the high energy density laser beam (10^5–10^7 W·cm^{-2}) interacts with the material in a short time (10^{-4}–10^{-2} s). Such high energy is sufficient to cause the local surface of the material to quickly heat up to thousand degrees, thus melting or even vaporizing, and then the very thin surface of the molten layer rapid cures when the laser beam leaves by heat exchange of the substrate which is still in cold state, with the cooling rate of 10^5-10^9 K/s. Once the laser power density acting on the material is above the ionization temperature, plasma will be generated on the surface of the material. The plasma gas forms "vapor plume" or plasma on the metal surface of the laser action zone. When the laser power density is very high, due to the large amount of metal gasification, a shape similar to "crater" can be seen in the laser action area, and there is a circle of high protrusions around the "crater." At this time, the metal liquid is severely impacted by the formed plasma stream and is sputtered around. After stopping the laser irradiation, traces of metal sublimation can be seen on the metal surface [2].

5.2.5.1 Dynamics of melting pool

In the SLM process, the high-energy laser beam continuously melts the metal powder to form a melting pool. The hydrodynamic state and heat and mass transfer state in the melting pool are the main factors affecting the process stability and the quality of the parts. In the SLM process, the material melts due to the absorption of the energy of the laser. Due to the distribution of the intensity of the Gaussian beam—the light intensity at the center of the beam is the largest, there is a temperature gradient in the radial direction on the surface of the melting pool, that is, the temperature in the center of the melting pool is higher than the temperature of the edge region. Due to the uneven temperature distribution on the surface of the melting pool, uneven distribution of surface tension is caused, so that there is a surface tension gradient on the surface of the melting pool. For liquid metal, the higher the temperature, the smaller the surface tension, that is, the surface tension temperature coefficient is negative. The surface tension gradient is one of the main driving forces of fluid flow in the melting pool, which causes the fluid to flow from a portion with low surface tension to a portion with high surface tension. For the melting pool formed by the SLM process, the temperature at the center of the melting pool is high and the surface tension is small; while the temperature at the edge of the melting pool is low and the surface tension is large. Therefore, under this surface tension gradient, the liquid metal in the melting pool flows radially from the center to the edge, and flows from the bottom to the top at the center of the melting pool (Fig. 5.21). At the same time, the shear force causes the material at the edge to flow along the solidus line. At the bottom center of the melting pool, the melt flow meets and then rises to the surface, thus forming two characteristic melt vortices in the melting pool, which are called Marangoni convection. During this process, the outward flow of the melt causes deformation of the melting pool, which leads to the surface of the melting pool to exhibit fish scales.

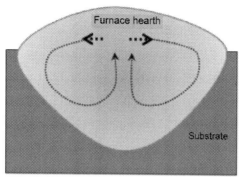

FIGURE 5.21 Schematic diagram of melt flow inside the melting pool.

5.2.5.2 Stability of the melting pool

The SLM process is an AM process from line to surface and then from surface to volume. Whether the metal cladding pass formed under the action of high-energy laser beam can exist stably and continuously determines the quality of the final part. According to the pinch instability theory, the smaller the volume of the liquid metal, the better the stability; at the same time, the sphere has lower free energy than the cylinder. The volume of the liquid metal is mainly determined by the size and energy of the laser spot. The spot of large size is more likely to form a melting pool of large size. The more powder that enters the melting pool, the more unstable the melting pool is. At the same time, too large spot will significantly reduce the laser power density, which is easy to produce a series of defects such as sticky powder, holes, and reduced bonding strength. When the size of the spot is too small, and the laser-irradiated metal powder absorbs much energy to vaporize, significantly increasing the impact of the plasma jet on the melting pool. Therefore, it is necessary to control the size of the spot to ensure the stability of the melting pool. At the same time, because the sphere has lower free energy than the cylinder, the liquid column-shaped melting pool tends to shrink and form small droplets, causing the surface to fluctuate. When a certain condition is met, the pressure difference between the two points on the liquid column promotes the conversion of the liquid column into a sphere. This requires proper matching of laser power and scanning speed. In the SLM process, as the laser power increases, the amount of molten metal in the melting pool increases, and the stability of the liquid column formed by the melting pool is weakened. On the one hand, the larger the laser power, the larger the area of the melting pool formed, the more powder will enter the melting pool, resulting in an increase in the instability of the melting pool; on the other hand, when the laser power is too large, the depth of penetration is increased. When the surface tension of the liquid metal cannot be balanced with its gravity, it will flow down both sides until the melting pool becomes wider and shallower, so that the surface tension and gravity can reach equilibrium again.

5.3 Metal powder for selective laser melting

Forming material is one of the key links in the development of SLM technology. It plays a decisive role in the physical and mechanical properties, chemical properties, precision, and application fields of the parts, directly affecting the use of SLM parts and the competitiveness of SLM technology and other metal AM technologies. This section focuses on the effects of powder material properties on SLM formability.

5.3.1 Effects of powder particle size on formability

5.3.1.1 Analysis of powder particle size and apparent density

The average particle size and particle size distribution characteristics of the powder are analyzed by laser diffraction method. The average particle size is obtained by using the volume average method in this experiment. The experimental results are shown in Fig. 5.22 [3,4].

It can be seen from Fig. 5.22 that the volume distribution of the powders in various particle size ranges is normally distributed, and the average particle size of each group of powders is obtained according to the theory of volume average particle size: Fig. 5.22A (No. 1 powder) characterizes the average particle size of the powder of 50.81 μm; Fig. 5.22B (No. 2 powder) characterizes the average particle size of 26.36 μm; Fig. 5.22C (No. 3 powder) characterizes the average particle size of 13.36 μm; and Fig. 5.22D shows the particle size analysis chart of the mixed powder of No. 1 and No. 3 powder (No. 4 powder), because it is the presence of two particle sizes, the powder particle size exhibits "bimodal" distribution characteristic with an average particle diameter of 47.15 μm. The apparent densities of the powders of various average particle sizes are tested, as shown in Table 5.8:

The No. 1 powder with an average particle size of 50.81 μm has an apparent density of only 54.98%, which is the lowest in the four powders. Since the powder particle size distribution range is the narrowest, similar to the accumulation of sphere with single particle size. According to the bulk density theory of sphere, the bulk density of single particle size sphere is the smallest, with an average value of 53.3%. When the particle size of the powder is decreased, the average bulk particle diameter decreases from 50.81 to 13.36 μm, and the powder bulk density gradually increases. It can be seen from Fig. 5.21 that the curves of b and c are wider, indicating that the particle size distribution is wider, and the spheres of different particle sizes are mixed together to reduce the porosity of the powder. According to the theory of sphere accumulation, in the accumulation of spheres having only two particle sizes, the accumulation reaches the maximum when the particle

440 Materials for Additive Manufacturing

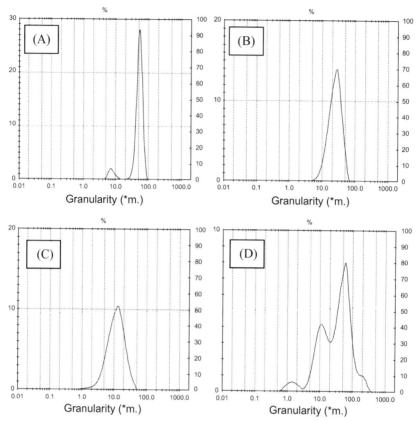

FIGURE 5.22 Test results of four powders of different particle sizes. (A) An average particle diameter of 50.81 m; (B) An average particle diameter of 26.36 m; (C) An average particle diameter of 13.36 m; (D) An average particle diameter of 47.15 m. In the Figure, the left ordinate represents the particle size distribution, and the right ordinate represents the cumulative size distribution.

TABLE 5.8 Apparent density of various particle sizes.

Powder no.	1	2	3	4
Average particle size (μm)	50.81	26.36	13.36	47.15
Apparent density (%)	54.98	55.79	56.13	59.83

diameter ratio of small and large sphere is 0.31. No. 4 powder is made by mixing No. 1 and No. 3 powders. We assume that both No. 1 and No. 3 powders are composed of spheres with particle sizes of 50.81 and 13.36 μm, then the particle size ratio of small and large spheres in mixed powder is

0.26, and the ratio is close to the ideal ratio of 0.31. Finally, the measured apparent density reaches 59.83% of the highest ratio.

5.3.1.2 Analysis of powder particle size on single-pass forming shape and spheroidization

Comparing the single-pass scanning trajectory shape of 316 L stainless steel powder with three different particle sizes, the powder with a particle size of 50.81 μm has the worst single-pass formability, as shown in Fig. 5.23, which shows uneven scanning line width, rough surface and phenomenon of large particles spheroidization. The main reason for this phenomenon is that the powder is prone to uneven distribution due to the large particle size of the powder, and the powder has small specific surface area, less energy absorption, and the energy of the edge of laser scanning line is low in the scanning process since the laser beam energy is Gaussian distribution mode, so that some of the powder particles are not completely melted, resulting in uneven scanning lines, and the spheroidization occurs due to solidification of the molten metal liquid adhered to the unmelted large particle powder. As shown in Fig. 5.23B and C, as the laser power is increased, the energy absorbed by the powder is increased, the powder is sufficiently melted, and the quality of the melt channel is improved. In Fig. 5.23D, due to the increase of the scanning speed, the impact of the laser on the powder and the melting pool is strengthened, so that the surface of the melting channel formed after the melting appears coarse fish scale-like lines, and the splashed metal liquid solidifies besides, forming some small spheroidization phenomena; Fig. 5.24 shows the melting track shape of stainless steel powder with a particle size of 26.36 μm. Stainless steel powder with an average particle size of 13.36 μm has the best formability, and has good formability under various process parameters. As shown in Fig. 5.25, the single-pass scanning line is continuous under each parameter, the solidification shape is uniform, the spheroidization phenomenon is suppressed, and the forming quality is good. The forming properties of powder with particle size of 26.36 μm are

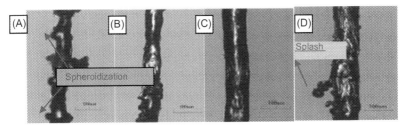

FIGURE 5.23 Surface morphology of single-pass scanning track of stainless steel powder with particle size of 50.81 μm under different parameters. (A) $P = 120$ W, $v = 650$ mm/s; (B) $P = 130$ W, $v = 650$ mm/s; (C) $P = 150$ W, $v = 650$ mm/s; (D) $P = 150$ W, $v = 700$ mm/s.

FIGURE 5.24 Surface morphology of single-pass scanning track of stainless steel powder with particle size of 26.36 μm under different parameters. (A) $P = 120$ W, $v = 650$ mm/s; (B) $P = 130$ W, $v = 650$ mm/s; (C) $P = 150$ W, $v = 650$ mm/s; (D) $P = 150$ W, $v = 700$ mm/s.

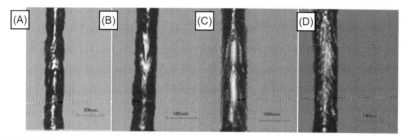

FIGURE 5.25 Surface morphology of single-pass scanning track of stainless steel powder with particle size of 13.36 μm under different parameters. (A) $P = 120$ W, $v = 650$ mm/s; (B) $P = 130$ W, $v = 650$ mm/s; (C) $P = 150$ W, $v = 650$ mm/s; (D) $P = 150$ W, $v = 700$ mm/s.

between 50.81 μm and 13.36 μm. When the power is 130 W, the powder melting effect is not good and there is spheroidization. However, with the appropriate laser power and scanning speed, a better scanning trajectory can be obtained.

In the SLM process, if the metal powder is not melted evenly after being melted by the laser, and a large number of metal balls isolated from each other are formed, this is called spheroidization of the SLM process. Spheroidization is a ubiquitous forming defect for SLM technology, which seriously affects the forming quality of SLM. Its damage is mainly manifested in the spheroidization leads to the formation of pores inside the metal parts: after spheroidization, the metal balls are separated from each other, and there are a large number of pores between the isolated metal balls, which greatly reduces the mechanical properties of the formed parts and increases the surface roughness.

The phenomenon of spheroidization comes down to the problem of wetting of liquid metal and solid surfaces. Fig. 5.26 shows the wetting state of the melting pool and the substrate. Where, θ is the angle between the surface tension $\sigma_{L/V}$ between the gas and liquid and the surface tension $\sigma_{L/S}$

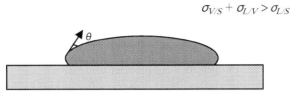

FIGURE 5.26 Schematic diagram of the wetting of the melting pool and the substrate.

between the liquid and solid. When the three stress contact points reach equilibrium, the resultant force is zero, namely:

$$\sigma_{V/S} = \sigma_{L/V}\cos\theta + \sigma_{L/S} \qquad (5.55)$$

when $\theta < 90$ degrees, the SLM melting pool can be spread evenly on the previous layer without spheroidization. Conversely, when $\theta > 90$ degrees, the SLM melting pool will solidify into a metal ball and adhere to the previous layer. At this time, $-1 < \cos\theta < 0$, it can be concluded that the relationship between the interfacial tensions during spheroidization is

$$\sigma_{V/S} + \sigma_{L/V} > + \sigma_{L/S} \qquad (5.56)$$

It can be seen that for the laser-melted metal powder, the surface energy after wetting of the liquid metal is smaller than the surface energy before wetting, so from the thermodynamic point of view, the wetting of the SLM is a process of reducing the free energy. The reason for the spheroidization is mainly the energy minimum principle of Gibbs free energy. During the solidification of the molten metal pool, under the action of surface tension, the melting pool forms a spherical shape to reduce its surface energy.

There are three main spheroidizing factors in the SLM forming process: (1) the oxygen content of the powder; (2) the particle size of the powder; and (3) the forming atmosphere. Due to the presence of oxygen, the metal oxide formed after the powders melting floats on the surface of the molten metal, and the wetting effect is reduced and the spheroidization is formed when the connection between the melting channel and the melting layer is achieved. However, the effect of the particle size distribution on spheroidization is also very important. In this experiment, the powder with the largest average particle size and the powder with the two particle sizes mixed are easy to form spheroidization. The principle is as shown in the Fig. 5.27.

5.3.1.3 Effects of powder particle size on single-pass scanning line width

As can be seen from Figs. 5.23–5.25, the powder with different particle sizes tends to increase width the increase of laser power, and the width of scanning line also increases. At the same time, it can be found that the variation of the width of the melting pool of the stainless steel powder with the

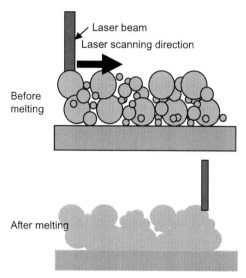

FIGURE 5.27 Schematic diagram of SLM forming spheroidization.

particle size of 13.36 μm is the most uniform, because the forming process of the powder with smaller particle size is more stable, and the trajectory is continuous, so that the width of the melting pool is very uniform, and the size measured conforms to the ideal rule to a large extent; while the powder with relatively large particle size is unstable, it is easy to form fluctuations, resulting in uneven size of the melting pool.

It is found that the single-pass scanning line width (w) stainless steel powders of different particle sizes has a certain relationship with the linear energy density (φ) of laser, that is, the influence of the laser power (P) and the scanning speed (v) on the melting pool. According to the measured data, an exponential relationship between the width of the melting pool and the linear energy density of the laser can be fitted.

The relationship between the width of the SLM single-pass forming track and the linear energy density of laser is shown in Fig. 5.28. The relationship between the width (w) of the single-pass scanning trajectory of the 316 L stainless steel powder with a powder particle size of 50.81 μm and the linear energy density of laser (φ) is obtained:

$$w = 0.08797 - 0.25505\exp(-14.39072\varphi) \qquad (5.57)$$

The relationship between the width w (mm) of the single-pass scanning trajectory of the 316 L stainless steel powder with a particle size of 26.36 μm and the linear energy density of laser (φ) is satisfied:

$$w = 0.09885 - 0.13232\exp(-7.35164\varphi) \qquad (5.58)$$

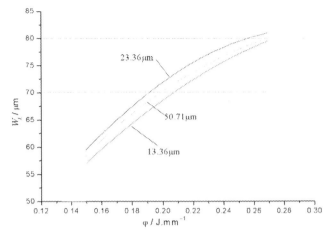

FIGURE 5.28 Relationship between the width of the melting path of different particle sizes and the linear energy density.

The width (w) of the SLM single-pass scanning forming track of the 316 L stainless steel powder with a particle size of 13.36 μm and the linear energy density of laser (φ) satisfy the equation:

$$w = 0.08965 - 0.1545\exp(-10.33076\varphi) \tag{5.59}$$

According to the above analysis, the SLM single-pass forming track width of 316 L stainless steel powder increases with the increase of linear energy density of laser, and there is a function relationship with the linear energy density of laser as an independent variable, and its form is as follows:

$$w = w_0 - A\exp(-K\varphi) \tag{5.60}$$

It can be seen from the aforementioned data function relationship that the powder parameters have a great influence on the shape and size of the melting pool of SLM single-pass forming. Where the constant (w_0) is the maximum value of the scanning line width of each powder and the width of the single-pass forming track of the stainless steel powder having an average particle diameter of 50.81 μm is the smallest. According to the law that the powder has a certain volume after melting, the width of the scanning line is small, indicating that the height of the melting channel is high, which is not conducive to a good continuous lap between the melting pools during surface scanning, and also affects the flatness of the surface scanning surface, and eventually lead to voids due to the scanning line, which seriously affects the density and mechanical properties of the final formed part.

In contrast, it can be found that the width of the single-pass scanning track of the stainless steel powder having a particle diameter of 26.36 μm under the same conditions is the largest, which indicates that the powder of the particle size can be melted and flattened evenly on the substrate under the same powder layer thickness, this helps the good overlap between the scanning lines. Meanwhile, the output power of the laser is also lower, and the quality of the formed surface is better, and the resulting formed part has higher density and better mechanical properties. The single-pass forming trajectory of stainless steel powder with an average particle size of 13.36 μm is generally linearly continuous, and the trajectory is relatively flat, the surface quality is good, and there are few defects such as spheroidization. The forming properties of the powder are between No. 1 powder and No. 2 powder.

5.3.1.4 Effects of powder particle size on surface scanning

According to the single-pass scanning measurement and analysis results, the melting pool width of various particle size powders under different process parameters can be calculated, providing the basis for the calculation of the overlapping ratio of scanning line of powder having different particle sizes scanned under different process parameters in the next step and the calculation of scanning pitch required in the physical manufacturing. Under certain process parameters, the setting of the scanning pitch (S) is guided by the lap ratio (η). For example, for stainless steel powder with an average particle diameter of 50.81 μm, when the laser power is 140 W, the scanning speed is 550 mm/s, and the scanning pitch is 0.08 mm, the scanning lines are overlapped, the spheroidization phenomenon is more serious, and the surface is rough; when the scanning pitch is reduced to 0.06 mm (overlapping ratio is 21.6%), the structure is basically free of pores, the spheroidization phenomenon is reduced, and the surface is flatter, as shown in Fig. 5.29A; when the scanning pitch is reduced to 0.04 mm (47.7%), only a small amount of metal balls splash and the structure is denser.

The stainless steel powder with an average particle diameter of 26.36 μm has a relatively continuous pore gully between adjacent scanning lines and almost no bonding at a laser power of 140 W, a scanning speed of 550 mm/s,

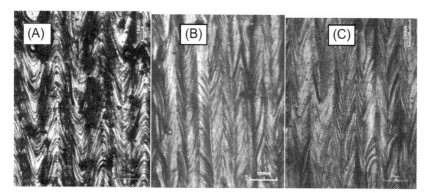

FIGURE 5.29 Surface scanning of metal surface morphology (laser power: 140 w, scanning speed: 550 mm/s, scanning pitch: 0.06 mm).

and a scanning pitch of 0.08 mm, so the surface is extremely uneven; when the scanning pitch is reduced to 0.06 mm (the overlapping ratio is 25%), the scanning lines are well overlapped and the surface is even and flat, as shown in Fig. 5.29B. When the scanning pitch is reduced to 0.04 mm (the overlap ratio is 39.4%), the surface quality is poor, and the spheroidization phenomenon is somewhat serious.

Stainless steel powder with an average particle size of 13.36 μm has good overlapping quality of scanning line but low overlapping ratio when the laser power is 140 W, the scanning speed is 550 mm/s, and the scanning pitch is 0.08 mm (the overlapping ratio is 12%), which may affect the compactness of the formation. As shown in Fig. 5.29C, when the scanning pitch is 0.06 mm, the overlapping ratio is 16%, and the forming surface quality is better, but it is worse than the results when the scanning pitch is 0.08 mm; when the scanning pitch continues to decrease the surface quality is worse, the spheroidization phenomenon is more serious, and the roughness is higher.

5.3.1.5 Density and mechanical properties of formed entities

The powder parameters and process parameters optimized by single-pass scanning and surface scanning experiments are combined to select four kinds of particle size powders were selected with a laser power of 140 w, a scanning speed of 650 mm/s, and a layer thickness of 0.02 mm cubic blocks. Fig. 5.30 shows the curve of the apparent density of the powder with four different particle sizes and the density of the formed parts. The relationship between process parameters and relative density of parts is shown in Fig. 5.31. The densities of formed parts No. 1, No. 2, and No. 3 powder are sequentially increased, and the density of No. 3 powder is the highest. The density of No. 4 powder is slightly higher than that of No. 1 powder and lower than that of No. 2 powder. It can be seen from the densification results of the No. 1 powder, No. 2 powder, and No. 3 powder that as the apparent density of the powder increases, the density of the formed parts also increases. The No. 4 powder has the highest apparent density but its relative density is relatively lower than that of No.3 powder. This is because the two powder particle sizes in the No. 4 powder are quite different. During the melting process, the powder of small particles preferentially melts, and some powder of large particles is not melted to form spheroid phenomenon, resulting in uneven distribution of the next layer of powder, eventually appearing pores. A fault occurs in the sample during the manufacturing process, and the sample is broken from the fault. The scanning electron microscopy (SEM) photo taken is shown in Fig. 5.32A. A large amount of unmelted metal powder and some pores formed can be seen from the break point.

Previous studies have shown that the powder having an average particle size of 26.36 μm has the best forming properties, so the stainless steel powder is used as forming material to make eight groups of formed parts for density testing, and then wire-cut into tensile test strips for mechanical properties testing. The tensile curve is shown in Fig. 5.33.

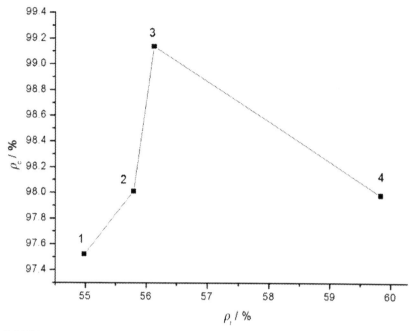

FIGURE 5.30 Relationship between apparent density of powder and density of formed parts.

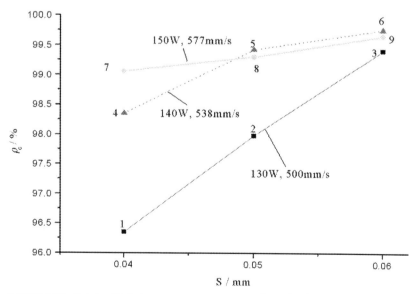

FIGURE 5.31 Relationship between process parameters and part density.

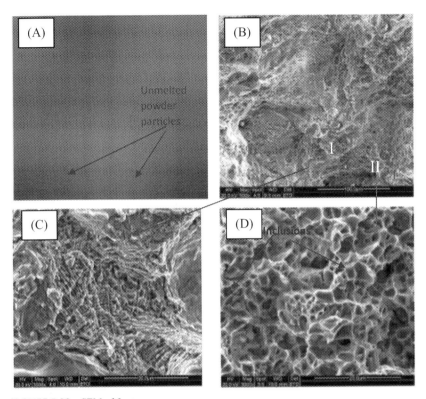

FIGURE 5.32 SEM of fracture.

Fig. 5.31 shows the effects of the change in scanning pitch on the density at the same linear energy density. Previous studies have shown that each process parameter corresponds to a certain melting pool width at the same linear energy density, and the scanning pitch is determined according to the width of the melting pool and the reasonable overlapping ratio. Density test shows that when the linear energy density is 0.26 J/mm, the scanning pitch is 0.06 mm. At this time, the overlapping ratio between the melting pools is 25%, and the density is the largest, which is consistent with the surface scanning results of the previous study. It can also be seen from the Fig. 5.31 that as the laser power increases, the density increases, but relatively, the density is the highest at a power of 140 W. This is because the power is increased when the linear energy density is the same, and the scanning speed is also increased, the scanning speed is too fast, and the impact of the laser on the powder is increased, thereby causing an increase in powder splash, the surface smoothness of the melting pool is deteriorated, and the roughness is increased, thereby increasing the probability of formation of pores and decreasing the density.

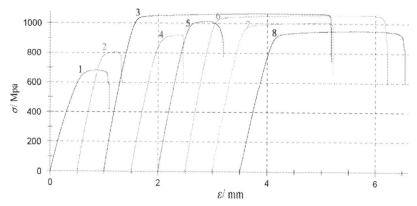

FIGURE 5.33 Tensile curve of the formed part at room temperature.

Fig. 5.33 shows the tensile strength curve. It can be seen from the Figure that the tensile strength of samples No. 1, No. 2 and No. 3 are increased sequentially, and the tensile strength of sample No. 3 is up to more than 1000 MPa. It can be known from the density curve of Fig. 5.31 that the tensile strength is consistent with the relative density, that is, the tensile strength of the formed part with high density is high; the tensile strength of samples No. 4, No. 5, and No. 6 is sequentially increased, and the relative density is also increased. It can be seen from the tensile curve that the sample with high tensile strength also has long tensile elongation, mainly because the part with relatively high density has a longer crack formation and expansion time during the deformation process. Its tensile elongation is longer. Fig. 5.32B−D are SEM photos of the tensile fracture of the sample No. 3, wherein C and D are enlarged views of B. It can be seen from the Fig. 5.32 that it belongs to the mixed fracture mode, some areas have ductile fracture, and some areas have brittle fracture. The fracture morphology shown in Fig. 5.32C is intergranular fracture, which is brittle fracture characteristic. Fig. 5.32D shows the dimples that occur during the fracture process. This fracture first forms microscopic voids (micropores) where the plastic deformation is severe. Inclusions are the locations where microscopic voids nucleate. Under the action of tension, a large amount of plastic deformation breaks the brittle inclusions or disengages the inclusions from the matrix to form voids. Once the void is formed, it begins to grow and gather, eventually forming a crack and finally causing the fracture.

5.3.2 Effects of powder sphericity on formability

The test is carried out by using two kinds of stainless steel powders of the same size of 400 mesh gas atomization and water atomization [5]. The particle size distributions of the two powders are basically the same, but due to the different powder atomization methods, the particle shape has big difference. Fig. 5.34 shows the microscopic morphology of two stainless steel powders. The particles of the water-atomized stainless steel powder have an irregular shape, and the particles of gas atomized stainless steel powder have a relatively regular spherical shape.

The process parameters are set to the laser power of 98 W and the scanning speed of 90 mm/s. Single-pass single-layer scanning test is carried out using this process condition, and the single-pass single-layer melting result is obtained as shown in Fig. 5.35. The results show that there is no significant difference in the single-pass scanning trajectory of the two stainless steel powders. Both scanning lines are continuously linear, with no fracture in the middle and no spheroidization effect. That is, when the laser energy density is sufficiently large, the particle shape of the powder does not affect the quality of the single-pass single-layer scanning.

The shape of the powder particles affects the fluidity of the powder, which in turn affects the uniformity of the powder leveling. In the multilayer forming process, uneven distribution of the powder leveling will result in uneven metal melting in various parts of the scanning area, so that the

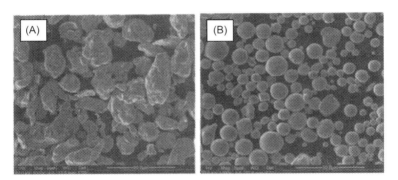

FIGURE 5.34 Microscopic morphology of stainless steel powder particles. (A) Water atomization (B) Gas atomization.

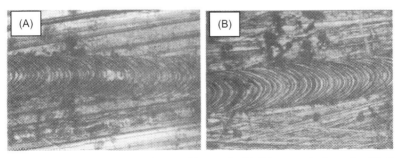

FIGURE 5.35 Single-pass single-layer melting pool. (A) Water-atomized powder (B) Gas atomized powder.

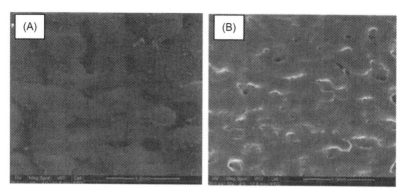

FIGURE 5.36 Surface microscopic morphology of the block parts. (A) Water-atomized powder (B) Gas atomized powder.

internal structure of the formed part is uneven. It is possible that part of the area is densely structured, while other areas have more voids. In order to verify the inference, the above-mentioned single-pass single-layer process parameters and forming powder are used to test and manufacture multi-pass multilayer block parts, and obtain the surface topography. Fig. 5.36 shows that the surface morphology of the two powder parts has obvious difference. At the same magnification, the surface of the water-atomized powder parts is rough, and there are a large number of voids on the surface. The surface of the gas atomized powder parts is relatively flat, and there are few voids with small volume. The density of the water-atomized powder part is measured to be about 90% in the test, and the density of the gas atomized powder part is above 90%. The results show that under the same process parameters, the shape of the powder particles directly affects the density and surface quality of the SLM parts. Spherical particle powder is more suitable for SLM forming than relatively irregular particle powder.

5.3.3 Effects of powder oxygen content on formability

The bulk forming test is carried out using the three powders in Table 5.9. The process parameters are set to the laser power of 98 W, scanning speed of 90 mm/s and powder layer thickness of 0.1 mm. The density and tensile strength of each part are measured, as shown in Table 5.10. As the oxygen content of the powder increases, the density and tensile strength of the formed part decrease significantly. When the mass fraction of oxygen exceeds a certain limit, its performance deteriorates drastically. The reason is that during the forming process, the metal powder absorbs high-density laser energy in a short time under the action of the laser, and the temperature rises sharply. If the oxygen is present, the part is easily oxidized. On the other hand, the residual oxide in the powder will also cause oxidation of the liquid

TABLE 5.9 Characteristics of test powder.

Grade	Mesh	Mass fraction of oxygen /%	Powder manufacturing technology	Particle size distribution			
				D10/μm	D50/μm	D90/μm	Average particle size/μm
316 L	400	1.46	Aerial fog	5.13	15.23	30.02	21.34
316 L	400	1.81	Water mist	9.68	24.21	49.43	27.31
316 L	200	2.11	Water mist	9.76	29.81	62.92	33.67

TABLE 5.10 Characteristics of test powder.

Powder type	Mass fraction of oxygen/%	Density/%	Tensile strength/MPa
400 mesh aerial fog 316 L	1.46	95	626.69
400 mesh water mist 316 L	1.81	90	331.91
200 mesh water mist 316 L	2.11	74	3.84

phase metal under the action of high temperature, thereby increasing the surface tension of the liquid phase pool, increasing the spheroidization effect and directly reducing the density of the formed part, so that the internal organization of the part is affected.

The electron microscopic results of the three powder formed parts are shown in Fig. 5.37. It can be seen from the microscopic results that there are significant differences in the internal organization of the three powder parts. The microstructure of the 400 mesh gas atomized powder parts is fine and uniform, the shape is flat and the crystals are very tightly bonded. The 400 mesh water-atomized powder has a compact microstructure and uniform composition, and the shape is relatively flat. There are obvious pot holes; the crystal structure of the 200-mesh water-atomized powder is fibrous, and the bonding between them is not dense, there are a large number of voids.

In summary, the difference in powder oxygen content leads to a difference in the quality of the melting pool during the forming process. The shape and spheroidization effect of the melting pool ultimately determines the internal structure of the part, which in turn affects the performance of the final formed part.

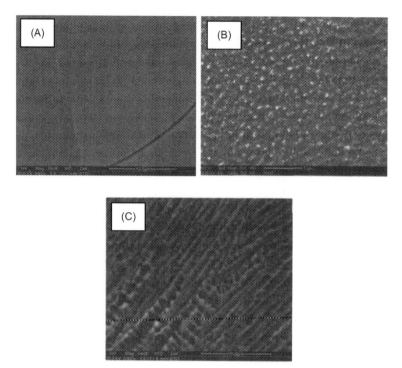

FIGURE 5.37 Surface microscopic morphology of the block parts. (A) 400 mesh gas atomized water (B) 400 mesh water-atomized powder (C) 200 mesh water-atomized powder.

5.3.4 Common metal and alloy powder materials for additive manufacturing

The SLM forming technique is characterized by complete melting and solidification of the material. Therefore, it is mainly suitable for the formation of metal materials, and one of its advantages is that it can form most of the metal materials, including pure metal materials, alloy materials, and metal matrix composite materials. The SLM forming materials are classified and discussed below.

Pure metals have been used in a variety of AM, such as SLS, SLM, etc., but compared with alloys, pure metal powder is not the main research object of AM, the main reasons are attributed to the following three aspects: First, the nature of the pure metal itself (relative to its alloy) is weak, it not only has low mechanical properties, but also weak properties such as oxidation resistance and corrosion resistance. This leads to the fact that the interest in researchers in AM has been greatly reduced. Second, before the advent of SLM, researchers repeatedly tried to use SLS to form pure metal powder but failed, which prompted researchers to discourage the sintering of pure metal powder. Third, the metal powder used in AM, especially in SLM, has very fine particle size (20–100 μm), which makes it difficult to manufacture spherical pure metal powder with good fluidity, this is also a factor hindering the research of forming pure metal powder by SLM. At present, research on SLM forming pure metals is mainly focused on non-ferrous pure metals such as Ti, Cu and Au.

Iron-based alloy is a type of alloy that is deeply carried out the research on SLM forming in the very early stage. The first is because iron-based alloy powders are easy to prepare, difficult to oxidize, and have good fluidity. Secondly, iron-based alloys are the most widely used in engineering technology and also the most important alloy. Among them, the main iron-based alloys studied include Fe−C, Fe−Cu, Fe−C−Cu−P, stainless steel and M-2 high-speed steel.

Titanium and titanium alloys are widely used in medical equipment, chemical equipment, aerospace and sports equipment due to their remarkable light weight, high strength, good toughness and corrosion resistance. Titanium is an allotrope with a melting point of 1668°C. It is a close-packed hexagonal lattice structure below 882°C, which is called α titanium, and is a body-centered cubic structure above 882°C, which is called β titanium. Titanium alloys of different microstructures can be obtained after gradual change of the phase transition temperature and phase content by adding appropriate alloying elements using different characteristics of the above two structures of titanium. At room temperature, titanium alloys have three matrix structures, α alloy, (α + β) alloy and β alloy. At present, the types of SLM forming titanium alloys are mainly concentrated on pure Ti, Ti6Al4V and Ti-6Al-7Nb alloy powders, which are mainly used in aerospace parts and artificial implants (such as bones, teeth, etc.).

Nickel-based alloy refers to a kind of alloy with high strength and certain anti-oxidation and corrosion resistance at high temperature of 650°C−1000°C. It is widely used in aerospace, petrochemical, shipbuilding, energy and other fields. For example, nickel-based superalloy can be used in turbine blades and turbine disks of aerospace engines. Commonly used SLM nickel-based alloys are mainly Inconel 625, Inconel 718 and Waspaloy alloys.

As a light metal material, aluminum alloy has been widely used in aerospace, high-speed trains and light-duty vehicles due to its excellent physical, chemical and mechanical properties. The commonly used SLM aluminum alloys are AlSi10Mg and Al12Si.

5.4 Properties and microstructure characteristics of metal powder for additive manufacturing

5.4.1 Metallurgical characteristics of selective laser melting metal powder

5.4.1.1 Spheroidization

A remarkable feature of SLM forming technology is that the melt is easily formed into a spherical shape after solidification, which is called the "spheroidization" of SLM. According to the principle of minimum surface energy, in the SLM forming process, in order to reduce the surface energy, the liquid melting channel tends to solidify and shrink into a ball under the surface tension. However, the occurrence of spheroidization will have an adverse effect on SLM forming, which seriously hinders the progress of SLM technology at home and abroad and the expansion of application fields. According to the current progress of research on spheroidization (research progress in SLM spheroidization at home and abroad), and combined with the actual problems met by Binhu Electromechanical Technology Co., Ltd., Huazhong University of Science and Technology, in the research and development of HRPM-II and HRPM-IIA laser melting forming system, the harm of spheroidization is summarized as follows:

1. The spheroidization phenomenon greatly increases the surface roughness of the formed parts, so that the SLM parts need to be polished for use, but it is difficult to ensure the dimensional accuracy of the metal parts after polishing;
2. When spheroidizing occurs, the forming trajectory is discontinuous and is accompanied by a large number of pores, thereby greatly reducing the mechanical properties of the SLM-formed parts;
3. After the spheroidization is formed, the metal ball will form a convex. When the next layer is powder leveled, the raised metal ball is easy to rub the powder leveling roller, hindering the movement of the powder leveling roller, and in severe cases, it can damage the powder leveling roller or the part, or directly cause the forming process to stop.

It can be seen from the above that the spheroidization phenomenon has seriously hindered the progress of SLM technology and the promotion of application fields. How to eliminate the spheroidization effect and reduce the surface roughness of metal parts has become an international problem in the field of SLM.

5.4.1.1.1 Analysis of spheroidization

Fig. 5.38 shows the microscopic morphology of the surface of the SLM-formed specimen, which can reflect the spheroidization characteristics of the metal when solidified. It can be seen that the forming surface is discontinuous and is separated by large-sized metal balls (Fig. 5.38A). Further enlargement of the Figure can reveal a large number of small metal balls (Fig. 5.38B); Fig. 5.38C has a relatively flat surface without large-size spheroidization. However, further enlargement of the microscopic region can reveal a large number of fine metal balls; in general, the spheroidization of SLM has different shapes and sizes. SLM

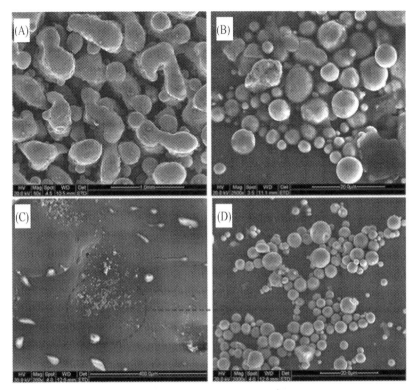

FIGURE 5.38 Microscopic scanning electron microscopy (SEM) photo of spheroidization. (A) Large-size spheroidization, about 500 μm; (B) small size spheroidization, about 10 μm; (C) (D) flat surface, no large-size spheroidization.

spheroidization can be roughly divided into two categories in terms of shape and size:

1. Ellipsoidal shape with non-flat solidified agglomerate features, large size, about 500 μm;
2. Spherical shape with high degree of sphericity, smaller size, about 10 μm.

The classification of the above spheroidization and the analysis from the macroscopic and microscopic perspectives help to further explore its formation mechanism and propose a reasonable process to slow the formation of spheroidization. Professor Gu has similar research on the spheroidization of stainless steel, which divides the spheroidization into big balls and small balls. However, the study uses CO_2 laser with laser mode, spheroid size and shape similar to the spheroidal characteristics formed by the fiber laser.

In general, the formation of spheroidization is due to the fact that the sphere has the lowest surface energy. According to the principle of the minimum

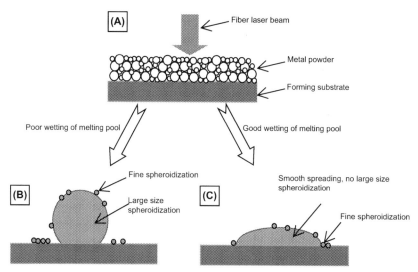

FIGURE 5.39 Schematic diagram of the spheroidization mechanism in SLM forming. (A) The rapidly moving laser beam scans the metal powder; (B) The wettability of the melting pool and the substrate is poor; (C) The melting pool and the substrate are well wetted.

surface energy, the metal droplets can reach a stable state after forming a spherical shape. However, the specific mechanism of SLM spheroidization remains to be further studied. The characteristics of the big ball and the small ball are described below and the formation mechanism is analyzed.

Fig. 5.39 shows the schematic diagram of two types of spheroidization. When the rapidly moving laser beam scans the metal powder (Fig. 5.39A), the time required for the metal powder to melt and solidify is extremely short, only a few milliseconds. As shown in Fig. 5.39B, when the wettability of the melting pool and the substrate is poor, it is difficult to spread flat on the substrate, and tends to form large-sized ball. At this time, the spheroid size is equivalent to the size of the laser melting pool. At the same time, a large amount of fine spheroidization will also occur, which is caused by the impact of the laser beam on the melting pool that converts the kinetic energy of the laser beam into the surface energy of fine spheroidization, thereby forming a large number of small metal spheres. As shown in Fig. 5.39C, when the melting pool and the substrate are well wetted and spread flat, no large-size spheroidization occurs (Fig. 5.38C). However, based on the same mechanism as described above, the melting pool is still subjected to the impact of the laser beam, which converts kinetic energy into surface energy and forms fine metal balls. In summary, the large-size spheroidization is caused by the poor wettability of the melting pool and the substrate, which can be avoided through process adjustment; and the fine spheroidization is generated by the conversion from laser beam kinetic energy to the surface energy of the microsphere, which is hard to avoid.

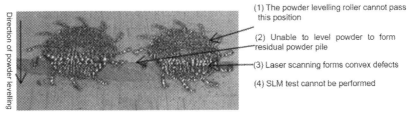

FIGURE 5.40 Example of SLM forming defects caused by spheroidization.

The micro-size ball in the SLM process is caused by the conversion of the impact kinetic energy of the laser into the surface energy of the sphere. It is commonly found in the SLM forming process and does not hinder the powder leveling. Therefore, the defects caused to the SLM are negligible. However, the large-size metal ball produced during the SLM process not only causes the surface of the formed body to be rough, but also hinders the movement of the powder leveling roller, which in turn leads to SLM defects, mainly reflected in:

1. The friction between the protruding large-size metal ball and the powder leveling roller is likely to cause the fracture of the part;
2. When the friction is severe, the forming cannot progress directly.

Fig. 5.40 shows examples of SLM forming defects caused by spheroidization.

In combination with the test of the HRPM-IIA metal fusion rapid prototyping equipment used by the research group, the following is an analysis of the defect that the spheroidization hinders the movement of the powder leveling roller and forms a projection, as shown in Fig. 5.41.

- The first layer of powder leveling (Fig. 5.41A): the powder leveling roller spreads out the metal powder of thickness H on the forming substrate;
- After laser scanning (Fig. 5.41B): the laser scans the powder layer with thickness H, resulting in spheroidization, the spheroidized height is greater than 2H thickness;
- The second layer of powder leveling (Fig. 5.41C): When the powder leveling roller is used for spreading, since the spheroidizing height exceeds the bottom of the powder leveling roller, the forward movement of the powder leveling roller will be hindered. At this time, the powder leveling roller will jump under the driving of the motor, spreading the powder layer with a thickness exceeding 2H. At this time, the upper surface of the powder layer is not flat and has fluctuation of the thickness;
- After the laser is scanned again, the powder leveling is blocked (Fig. 5.41D): due to the uneven thickness of the previous powder leveling, the powder layer is too thick at some locations. After the laser is scanned again, the thicker layer intensifies the spheroidization, producing a metal sphere with a larger ball diameter. At this time, the height of the uppermost end of the metal ball

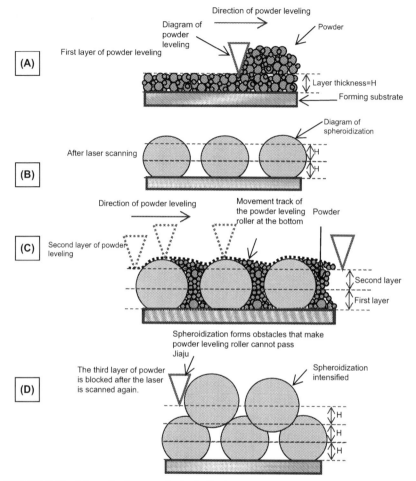

FIGURE 5.41 Schematic diagram of sphericalization hindering the movement of the powder leveling roller.

is far more than 3H. When the powder layer of 3H thickness is spread again, the powder leveling roller is subjected to greater resistance, and it may be difficult to pass the higher metal ball, and the powder leveling motion stops.

It can be seen that the generation of metal spheroidization (mainly large-size spheroidization) is a serious forming defect for SLM technology, hindering the development of SLM technology. If the spheroidization phenomenon can be minimized, it will be beneficial to improve the quality of the SLM-formed metal parts and effectively reduce the forming defects. Therefore, it is especially necessary to study the influencing factors and mechanisms of spheroidization and to reduce the spheroidization effect by using a reasonable process. An introduction to the factors affecting spheroidization will be detailed below.

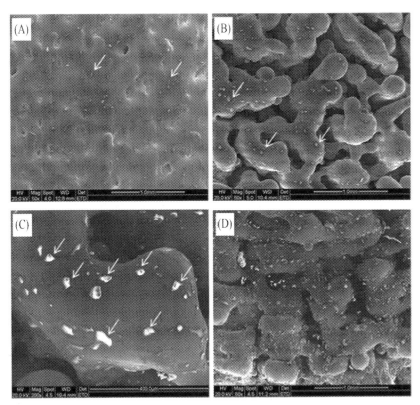

FIGURE 5.42 Effects of 316 L stainless steel powder with different oxygen content (molar fraction) on large-size spheroidization. (A) Gas atomized powder having an oxygen content of 4.52%; (B) and (C) water-atomized powder having an oxygen content of 5.44%; (D) water-atomized powder having an oxygen content of 5.9%.

5.4.1.1.2 Influencing factors of spheroidization

5.4.1.1.2.1 Powder properties Fig. 5.42 shows the results of surface spheroidization after SLM forming of different powders. It can be seen that the gas atomized powder has a relatively flat forming surface and is free from spheroidization; while the water-atomized powder exhibits poor wetting ability on the forming surface and severe spheroidization. It can be seen that the influence of powder characteristics on spheroidization is obvious, which can be analyzed from two aspects: one is the chemical properties of the powder, such as oxygen content; the other is the physical properties of the powder, such as powder shape and bulk density.

1. Oxygen content of metal powder
   The analysis of Fig. 5.42 shows that the oxygen content of powder has obvious effects on spheroidization. As the oxygen content increases, large-size spheroidization will occur. When the laser is melted by 316 L stainless steel

powder (gas atomized) of low oxygen content, the forming surface is relatively flat, the single-melting channel is continuous, and the melting channels are well overlapped, and there is still no large-size spheroidization after overlapping (Fig. 5.42A). When the SLM test is carried out with 316 L stainless steel powder of high oxygen content (water-atomized), it's found that the single-melting channel of laser is discontinuous and split into multiple metal spheres, with multi-channel and multilayer scanning, the forming surface eventually deteriorates, forming a large number of isolated large-sized metal balls (Fig. 5.42B−D), which seriously affects the forming quality of the SLM. By partially magnifying Fig. 5.42B, as shown in Fig. 5.42C, it can be found that a large amount of white matter is distributed on the surface of the large-sized metal ball (this substance universally exists on the surface of the SLM forming sample). However, this white matter differs in the amount and distribution on the surface of the sample formed by the powder of different oxygen contents. The EDX spectrum analysis of the ubiquitous white matter is performed below, and an area is randomly selected on the forming surface for analysis, as shown in Fig. 5.43.

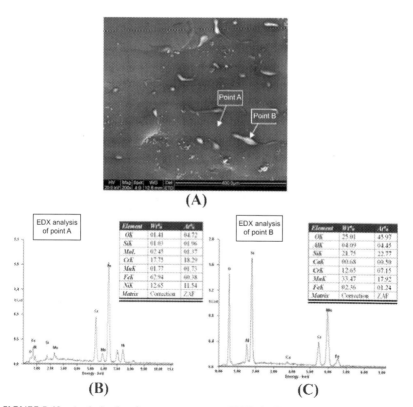

FIGURE 5.43 Analysis of surface energy spectrum (EDX) for SLM-formed samples. (A) SEM photograph of surface; (B) EDX analysis of matrix; (C) EDX analysis of white matter.

The black region is the main component of AISI316L, in which the oxygen content is 4.72% (molar fraction, the same below); and the white region mainly contains oxygen, silicon, manganese and aluminum, and the oxygen content is 45.97%. Thus, the white substance ubiquitous on the surface of the SLM-formed sample is oxide. The following rules can be found by combining low-magnification images of the surface of the metal powder SLM-formed sample with different oxygen contents in Fig. 5.42. First, in the SLM forming, after the melting pool is partially oxidized, a large amount of oxides escapes and float on the surface; secondly, if the oxygen content of the formed powder is high, the surface oxide of the formed sample also increases. As can be seen from Fig. 5.42A, a small amount of oxide is distributed on the surface of the sample. As the oxygen content of the powder increases, it can be seen from Fig. 5.42B and D that the amount of oxide on the surface of the sample also increases.

The effects of oxygen content on SLM forming spheroidization can be attributed to the interfacial wetting problem of oxides. When the SLM is carried out under low oxygen content, the stainless steel melting pool can be spread on the surface of the relatively clean solid stainless steel. The wetting interface is mostly liquid stainless steel/solid phase stainless steel; however, if the SLM is carried out under high oxygen content, the stainless steel melting pool will be spread on the oxide surface, and its wetting interface is mainly liquid stainless steel/oxide. The former is wetted by homogenous material of stainless steel, while the latter is wetted by heterogeneous material of stainless steel and oxide (poor wettability). It can be seen that when the metal powder has a high oxygen content, its surface shape It can be seen that when the oxygen content in the metal powder is high, a large amount of oxide formed on the surface is disadvantageous to the wetting and spreading of the liquid melting pool, thereby forming a large number of metal balls, which reduces the forming quality of the SLM. It can be seen that the gas atomized 316L stainless steel prepared by using the protective atmosphere has less spheroidization, the scanning line is continuous and the lap joint is good, the forming surface is relatively flat, and the SLM formability is good.

Fig. 5.44 shows the micro-size spheroidization analysis of the surface of SLM samples at different oxygen contents. It can be seen that the oxygen content of the metal powder has little effect on the spheroidization of the fine size. That is to say, the metal powder prepared by gas atomization and water atomization is prone to micro-size spheroidization during the SLM forming process. This is because, as analyzed in the previous section, micro-size spheroidization is caused by laser shock waves, which impinge liquid metal into a large number of small metal spheres to form a splash. That is to say, the kinetic energy of the laser beam is partially converted into the surface energy of the splash metal ball, thereby forming a large number of fine metal balls as shown in the Figure. Thus, the

464 Materials for Additive Manufacturing

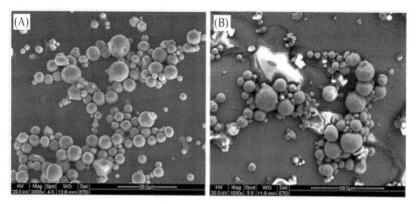

FIGURE 5.44 Micro-sized spheroidized SEM image of the surface of the SLM-formed sample. (A) Oxygen content: 4.52%; (B) oxygen content: 5.9%.

micro-size spheroidization is related to the impact energy of the laser beam, regardless of the oxygen content of the metal powder. Large-size spheroidization is caused by poor wettability due to high oxygen content of the metal powder. Therefore, SLM forming materials should use low-oxygen gas atomized powder to ensure the quality of SLM forming.

2. Grain shape of metal powder

It should be noted that the gas atomized powder is sphere that has relatively high apparent density, and has small volume shrinkage during the melting process. In the SLM process, since the powder tends to shrink, the actual powder layer thickness is larger than the slice thickness, which has an "actual powder layer thickness." Due to the irregular shape of the water-atomized powder, the actual powder layer thickness is also lower than the actual powder layer thickness of the gas atomized powder, and the thicker powder layer thickness is disadvantageous for its wettability. Therefore, spherical powder (gas atomized) is more conducive to the spreading and wetting properties of the melt.

5.4.1.1.2.2 Forming atmosphere In the SLM process, the oxygen content is mainly introduced from the oxygen content of the metal powder itself and the oxygen content in the forming atmosphere. Therefore, this section discusses the effects of oxygen content on spheroidization from two aspects of metal powder and forming atmosphere.

Fig. 5.45 shows the spheroidization of the formed surface under different oxygen content atmospheres. The scanning speed is 50 mm/s, the laser power is 190 W, the scanning pitch is 0.2 mm, and the layer thickness is 0.05 mm. It can be seen that the atmospheric oxygen content has a great influence on the spheroidization condition of the formed surface. When the atmosphere oxygen content is 0.1%, the scanning channels are continuous and well

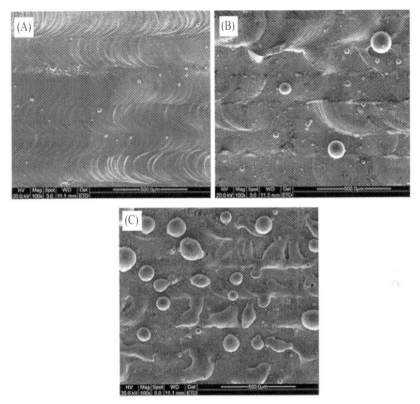

FIGURE 5.45 Spheroidization characteristics (SEM) under different oxygen content atmospheres. (A) 0.1%; (B) 2%; (C) 10%.

overlapped with each other, the fish scales are clearly visible without large-size spheroidization, and there's only a small number of fine droplets caused by splashing (Fig. 5.45A); when the atmospheric oxygen content is increased to 1%, the morphology of scanning channel becomes blurred and spheroidization occurs (Fig. 5.45B). As the atmospheric oxygen content continues to increase to 10%, the scanning channel is discontinuous, a large number of metal tumors and metal spheres are formed, and the quality of the forming surface is deteriorated (Fig. 5.45C). It can be seen that the oxygen content of the forming atmosphere also has a large effect on the spheroidization. The oxygen in the atmosphere easily reacts with the molten metal pool to form an oxide film, which is very unfavorable for the soldering and wetting of the metal melting pool and the previous layer, resulting in a more serious spheroidization phenomenon. Table 5.11 shows the relationship between oxygen content in atmosphere and oxygen content on the surface of shaped samples. It can be seen that oxygen content on the surface of shaped samples increases with the rising oxygen content in atmosphere.

TABLE 5.11 Relationship between oxygen content in atmosphere and oxygen content on the surface of shaped samples.

Oxygen content in atmosphere (%)	0.1	2	10
Oxygen mass fraction of formed sample (%)	2.11	5.81	7.69
Oxygen atom fraction of formed sample (%)	6.01	14.34	19.5

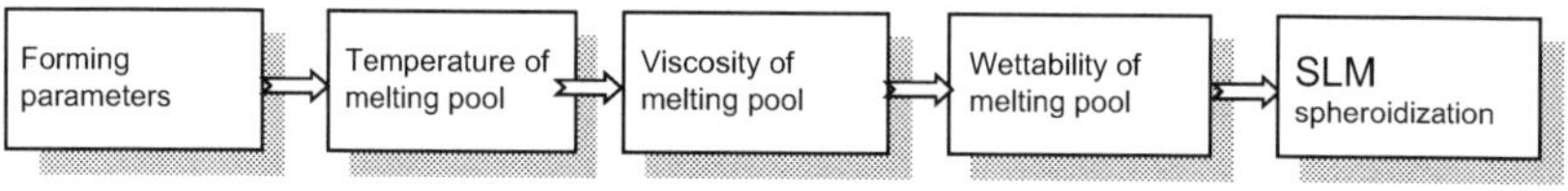

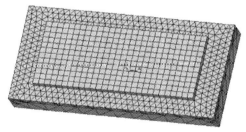

FIGURE 5.46 Correlation between SLM melting temperature and factors such as spheroidization.

FIGURE 5.47 SLM 3D model.

5.4.1.1.2.3 Forming process parameters Previous studies by international scholars have shown that the temperature of the liquid metal determines its viscosity. The better the viscosity, the better the fluidity of the melting pool, which in turn affects its fluidity, spreading and wetting properties, as shown in Fig. 5.46. Therefore, the SLM forming temperature has a decisive influence on the wetting and spheroidization of the melting pool. This section firstly analyzes the influence of laser forming parameters on the temperature field of the melting pool in theory. On this basis, the SLM forming test is carried out to analyze the influence law of forming parameters on spheroidization.

1. Analysis of temperature field of melting pool
   This section focuses on the effects of process parameters such as scanning speed, laser power, and scanning pitch of SLM process on the melting temperature field of 316 L stainless steel powder. Based on ANSYS environment, the finite element model of SLM 3D temperature field is established, shown in Fig. 5.47. The laser heat source is approximately Gaussian, and the moving loading of laser Gaussian heat source is realized by AYSYS parameterized language (APDL). The nonlinear conditions of thermal

properties of 316 L stainless steel changing with temperature and phase are considered.

a. Finite element model of powder and substrate

The SLM 3D finite element model is shown in Fig. 5.50. The model consists of lower substrate and a long layer of powder. The substrate size is 3 mm × 1.8 mm × 0.4 mm, and the powder size is 2.2 mm × 1.2 mm × 0.1 mm. In order to obtain accurate calculation, the powder portion is divided into small cell grids of 0.1 mm × 0.1 mm × 0.1 mm. the upper part where the substrate is directly in contact with the powder is divided into free mesh by Solid70 thermal unit. Considering that too many grids may lengthen the calculation time, the lower part of the substrate is divided into larger grids by Solid90 thermal unit.

b. Determination of boundary conditions and initial conditions

During the melting of the metal powder, there is heat exchange inside the melting pool with the surrounding air and powder bed. This process can be described by the classical Fourier heat conduction equation:

$$\lambda_e \left(\frac{\partial^2 T}{\partial x^2} + \frac{\partial^2 T}{\partial y^2} + \frac{\partial^2 T}{\partial z^2} \right) + q_c + q_g = \rho C \frac{\partial T}{\partial t} \tag{5.61}$$

where λ_e is the effective thermal conductivity of the powder; T is the temperature; t is the time; ρ is the powder compaction density; C is the specific heat capacity of the material; q_c is the heat lost by the material to the air, and q_g is the laser power density.

Assume that the metal powder has an initial temperature of T_0, and initial conditions of the Eq. (5.61) can be obtained:

$$T(x, y, z, 0) = T_0, \quad \text{where } (x, y, z) \in D \tag{5.62}$$

For the determination of the boundary conditions, it can be assumed that the powder bed temperature is an initial value at time $t = 0$, that is, the temperature of the bottom layer of the metal powder is constant at $t = 0$; when $t > 0$, the heat of laser absorbed by the surface unit area of the metal powder in any time period should be equal to the sum of the heat transferred from the surface into the powder, the convective heat transfer and the heat dissipated by the surface radiation. Therefore, the natural boundary conditions can be expressed as:

$$\lambda_e \frac{\partial T}{\partial n} - q + h(T - T_0) + \sigma \varepsilon (T^4 - T_0^4) = 0 \tag{5.63}$$

where λ_e is the effective thermal conductivity of the metal powder; T is the surface temperature of the material at time t; T_0 is the initial temperature (or ambient temperature); h is the convective heat transfer coefficient; σ is the Stefan-Boltzmann constant, and its value is 5.607×10^{-8} W/m$^2 \cdot$ K^4; and ε is the thermal emissivity.

The thermophysical parameters changing with the temperature and the radiation in the boundary conditions cause the finite element analysis to be highly nonlinear. Radiation will greatly increase the solution time, so the Vinokurov empirical relationship is used to solve:

$$H = 2.4 \times 10^{-3} eT^{1.61} \tag{5.64}$$

c. Loading of moving Gaussian heat source
During the SLM forming process, the laser energy is input into the powder bed at a heat flux density, which is subject to Gaussian distribution, that is:

$$q = \frac{2AP}{\pi w^2} \exp\left(-\frac{2r^2}{w^2}\right) \tag{5.65}$$

where w is the spot radius, that is, the distance from the heat flux density at $1/e^2$ of the spot center to the spot center; A is the absorption rate of the laser beam by the powder bed; P is the laser power; r is the distance from a point on the surface of the powder bed to the center of the spot:

$$r^2 = (x-x_0)^2 + (z-z_0-vt)^2 \tag{5.66}$$

d. Thermal properties of metal powder
The thermal property parameters of metal powder materials are one of the important factors affecting the calculation accuracy in numerical simulation calculation. Since the temperature field distribution is related to the effective thermal conductivity, specific heat capacity, heat absorption rate and melting point of the powder material, setting accurate thermal property parameters has an important influence on the simulation results. Considering the relationship between the thermal property parameters of the powder and the temperature, the thermophysical parameters corresponding to different temperatures are set in the general finite element analysis software, ANSYS, and ANSYS will automatically calculate the unset values by linear interpolation as needed. The temperature set by the materials used in the simulation and the corresponding thermal properties are shown in Table 5.12.

TABLE 5.12 Thermal properties of 316 L stainless steel powder.

Temperature (T/K)	200	500	800	1100	1500	2000
Thermal conductivity (λ/w/m/K)	37	26	17	19	700	900
Specific heat capacity (C/J/kg/K)	435.1	602.5	832.6	606.7	836.8	836.8

Among the thermophysical parameters that affect the powder forming properties, the effective thermal conductivity of the powder bed is the most important, and its conduction mechanism is complicated. Assuming all the powders are spherical and there is no contact deformation, the effective thermal conductivity of the powder bed can be estimated by the following equation:

$$\frac{\lambda_e}{\lambda_g} = \left(1 - \sqrt{1-\varphi}\right)\left(1 + \frac{\varphi k_r}{\lambda_g}\right) + \sqrt{1-\varphi}\left(\frac{2}{1-\frac{\lambda_g}{\lambda_s}}\left(\frac{1}{1-\frac{\lambda_g}{\lambda_s}}\ln\left(\frac{\lambda_s}{\lambda_g}\right) - 1\right) + \frac{k_r}{\lambda_g}\right) \quad (5.67)$$

where λ_g and λ_s are the thermal conductivity coefficients of the ambient gas and the solid material, respectively; Φ is the porosity, which is about 0.477; k_r is the heat transfer coefficient caused by the radiation in the powder.

During the SLM process, the stainless steel powder undergoes the transition from solid to liquid to solid phase. ANSYS considers the latent heat of phase change by defining the enthalpy of the material changing with the temperature, which can be expressed as:

$$H = \int \rho C(T) dT \quad (5.68)$$

e. Effects of scanning speed

Fig. 5.48 shows the relationship between the maximum temperature of the powder bed surface and the scanning speed during laser melting, in

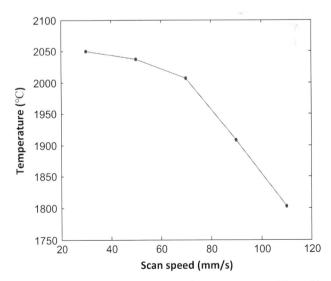

FIGURE 5.48 Effects of scanning speed on the maximum temperature of the melting pool.

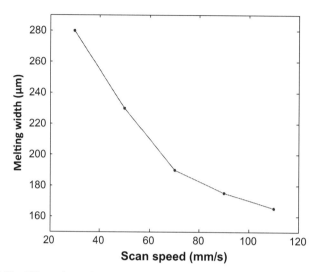

FIGURE 5.49 Effects of scanning speed on the width of the melting channel.

which the laser output power is 95 W. It can be seen that the temperature generated by laser melting at various scanning speeds exceeds the melting point of 316 L stainless steel powder (about 1370°C), the metal powder is completely melted, and the maximum temperature at which the powder melts gradually decreases with the increase of the scanning speed, showing a nonlinear change. This is because the increase in scanning speed causes the laser energy density to decrease, thereby reducing the temperature field of the SLM. According to the melting point and temperature nephogram of 316 L stainless steel powder, the melting width at different laser scanning speeds can also be obtained, as shown in Fig. 5.49. It can be seen that as the scanning speed increases, the melting width decreases continuously, which coincides with the melting temperature variation in Fig. 5.48.

f. Influence of laser power

Fig. 5.50 shows the relationship between the maximum temperature of the powder bed surface and the laser output power during laser melting, in which the scanning speed is 50 mm/s. Fig. 5.51 shows the relationship between laser melting width and laser power.

It can be seen that as the laser power increases, the maximum temperature of the melting pool also increases and exceeds the melting point of the stainless steel. The increasing laser power leads to a corresponding increase in energy density, which in turn increases the surface temperature of the powder bed. An increase in laser power results in an increase in the melting width, which is greater than the laser scanning pitch, thereby making the scanning lines overlap closely.

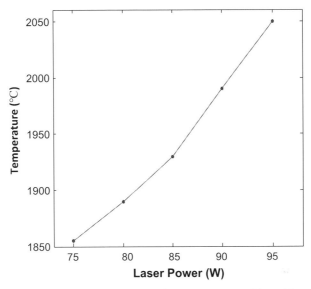

FIGURE 5.50 Effects of laser power on the maximum temperature of the melting pool.

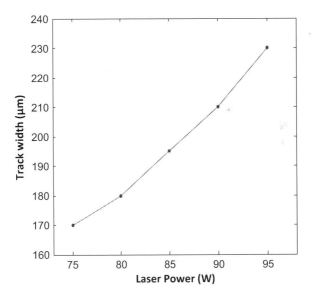

FIGURE 5.51 Effects of laser power on the melting width.

g. Impact of scanning pitch

Considering that the unit size of the finite element model of the powder bed is 0.1 mm, the laser scanning pitch is selected to be 0.1, 0.2, and 0.3 mm, as shown in Fig. 5.52. The laser power is 95 W and the

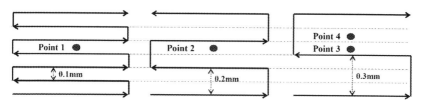

FIGURE 5.52 Schematic diagram of different scanning pitches.

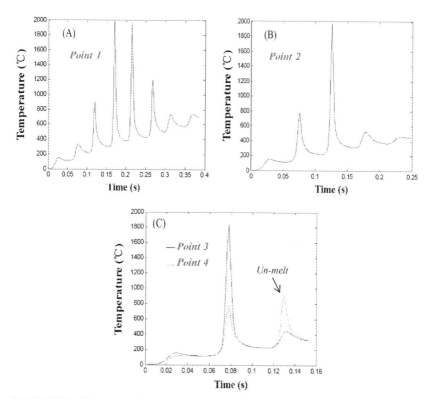

FIGURE 5.53 Temperature-time relationship of the center position of the powder bed under different laser scanning pitches. (A) 0.1 mm; (B) 0.2 mm;(C) 0.3 mm.

scanning speed is 50 mm/s. The influence of laser scanning pitch on the temperature field is studied.

Fig. 5.53 shows the temperature in the center position (606th node) of the powder bed changing with time at different scanning pitches. It can be seen that as the scanning pitch continues to increase, the maximum temperature at the center of the powder bed reduced. When the scanning pitch is 0.1 mm, the stainless steel powder first undergoes a preheating process under about 900°C, then two melting processes, and finally a high temperature heat treatment process under 1200°C; when the scanning pitch is 0.2 mm, the stainless steel powder first undergoes a preheating process under about 800°C, then one melting, and the finally gradual decrease in temperature, which does not undergo a high temperature heat treatment process when the scanning pitch is 0.1 mm; when the scanning pitch is 0.3 mm, the stainless steel powder at point 3 undergoes only one melting, and no high temperature heat treatment process before and after the melting, and the powder temperature at point 4 does reach the melting point. The above results can be explained as follows: the energy density of the laser multi-channel scanning is related to the laser power and the scanning speed, and is also affected by the scanning line overlap, that is, due to the scanning pitch, the finer the scanning pitch, the higher the energy density of the multi-line scanning, and the higher the final temperature.

h. Test verification

The test used the SLM rapid prototyping machine developed by the Rapid Prototyping Center of Huazhong University of Science and Technology. A single-line scanning test is performed on the 316 L stainless steel powder with a laser power of 95 W and different scanning speeds. Fig. 5.54 shows the morphology of the scanning line at different scanning speeds. It can be seen that the laser melting line is fish scale, and the line width becomes narrower as the scanning speed increases. Fig. 5.58 compares measured value of the laser melting line width with the predicted value. It can be seen that the actual melting width of the laser is basically consistent with the simulation results. The experimental results verify the numerical simulation results. The prediction results show that the isotherm shape on the surface of the powder bed shows the change from dense to sparse, and the maximum melting temperature, the laser power and scanning speed change nonlinearly. Fig. 5.55 shows that the low scanning speed, high laser power and small scanning pitch are beneficial to improve the temperature of melting pool, increase the width of the laser melting channel. The above parameters are also closely related to the continuity of the scanning line and the spheroidization. The following is a detailed study on how the process factors influence spheroidization through process experiments.

FIGURE 5.54 Melting channels at different laser scanning speeds.

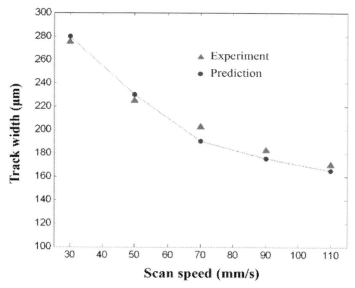

FIGURE 5.55 Relationship between the width of the melting channel and the scanning speed.

2. Linear energy density

   Since the SLM technology is based on the line to surface and then surface to volume forming methods, it is important to study the laser single-melting trajectory characteristics to understand the surface and volume forming laws and to optimize the forming parameters. The metal powder absorbs the laser energy and undergoes a melting and solidification process to form a single scanning track. From an energy point of view, power and speed can be summarized as the concept of "linear energy density (E_l)":

   $$E_l = \frac{P}{v} \qquad (5.69)$$

   where P is the laser power and v is the scanning speed of the laser beam. Since the laser single-melting channel is determined by the linear energy density. Based on this, this section focuses on the spheroidization characteristics of single-pass melting trajectories at different powers (P) and different scanning speeds (v).

   a. Longitudinal section characteristics of single-melting channel

      The scanning speed is fixed at $v = 50$ mm/s, and the single-melting channel is prepared under different laser powers. The wire-electrode cutting method is used to cut along the longitudinal plane and embed into a metallographic section. Fig. 5.56 is the SEM photo of longitudinal section of single-melting channel at different laser powers. Fig. 5.67 shows the wetting angle of the melting pool at different powers.

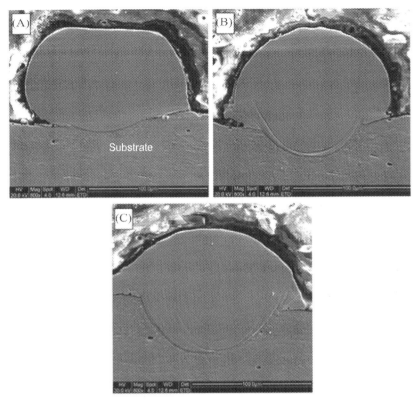

FIGURE 5.56 SEM photo of the longitudinal section of single-melting channel under different laser powers. (A) $P = 60$ W; (B) $P = 80$ W; (C) $P = 100$ W.

In general, the cross-section of single-melting channel at different powers shows the spreading and solidification characteristics of the melting pool. However, its shape, depth of fusion, wetting angle, and spreading performance show significant differences. When the laser power is 60 W (Fig. 5.56A), the melting pool has a large wetting angle (wettability and spreading ability are poor) and tends to form "spheroidization". It can also be seen that at this time, the laser penetration depth is shallow, only a small part of the melting pool is melted into the substrate, and the soldering ability to the substrate is poor, and whether the spheroidization or the shallow soldering depth is disadvantageous to the multilayer SLM forming; as the laser power increases to 80 W (Fig. 5.56B), the wettability of the melting pool is improved, the wetting angle is reduced, and the depth of soldering to the substrate is also increased; When the power increased to 100 W (Fig. 5.56C), the melting pool can be well spread on the substrate, the wetting angle is further reduced, and the soldering depth with the substrate is further increased.

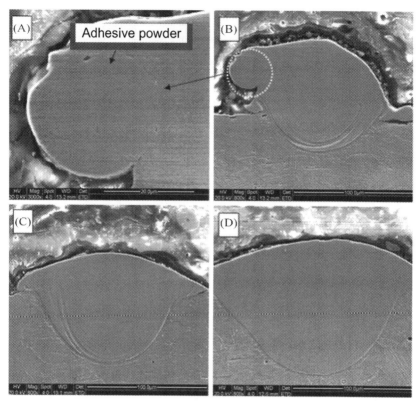

FIGURE 5.57 SEM photograph of the longitudinal section of the single-melting channel at different scanning speeds. (B) $v = 150$ mm/s; (A) enlarged view of B; (C) $v = 60$ mm/s; (D) $v = 20$ mm/s.

The laser power is fixed at $P = 100$ W, a single-melting channel is prepared at different scanning speeds, and the sample is prepared according to the same method and the longitudinal section features are observed. Fig. 5.57 is a SEM photo of the longitudinal section of single-melting channel at different scanning speeds. Fig. 5.58 shows the wetting angle of the melting pool to the substrate at different scanning speeds. At a scanning speed of 150 mm/s, the upper part of the melting pool is shown as a "bulge" shape (Fig. 5.57A), accompanied by bonded powder (Fig. 5.57B), the melting pool shows poor wetting and spreading ability; as the scanning speed reduced to 60 mm/s (Fig. 5.57C), the upper part of the melting pool is flat, showing better wetting ability and fluidity. At the same time, as the scanning speed increases, the penetration depth and melting width of the melting pool also increase; at a scanning speed of 20 mm/s (Fig. 5.57D), the size of the melting pool is further

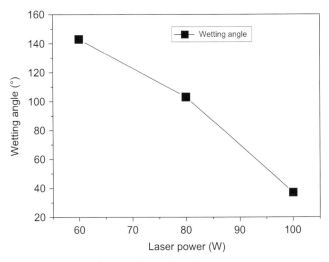

FIGURE 5.58 Wetting angle of the melting pool at different powers.

increased, and the penetration depth and melting width of the melting pool further increased. On the other hand, it can be found from the change in the wetting angle that as the scanning speed decreases, the wetting angle also decreases and the degree of spheroidization decreases.

When the rapidly moving laser beam is applied to the metal powder, the energy is quickly absorbed by the metal powder, and a series of processes such as melting, wetting, flowing, spreading, solidification of the metal powder occurs, which is extremely short, about several milliseconds. The higher linear energy density (high power, low scanning speed) facilitates the increase of the SLM forming temperature field, resulting in a wider HAZ, while reducing the viscosity of the melting pool and increasing the fluidity. Therefore, as the linear energy density of laser increases, the size of the melting pool increases, and its wettability is also improved, as shown in Fig. 5.59 [6].

b. Solidification morphology of single-melting channel

Fig. 5.60 shows the morphology of single-melting channel at different scanning speeds, where the laser power is fixed at 190 W and the forming equipment is HRPM-IIA. It can be seen that as the scanning speed increases, the melting width decreases, and the continuity of the scanning line deteriorates, gradually showing spheroidization. An increase in the scanning speed may result in a decrease in the melting temperature and deterioration in the penetration depth of the laser. On the one hand, the viscosity and fluidity of the melting pool are related to temperature. The lower temperature is not conducive to the spreading and wetting of the melting pool, and eventually leads to spheroidization. On the other hand, under shallow depth of laser penetration,

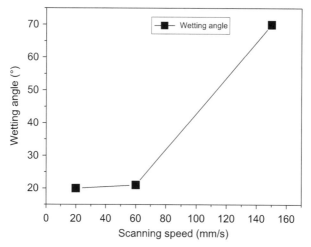

FIGURE 5.59 Wetting angle of the melting pool at different scanning speeds.

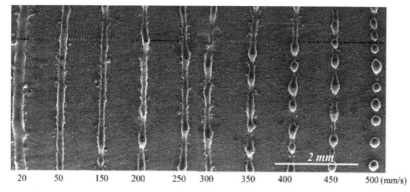

FIGURE 5.60 Forming condition of single-melting channel at different scanning speeds (SEM).

the liquid melting pool has a small contact area with the substrate, and the fluidity of the melting pool is hindered by the surrounding powder, so that the wettability deteriorates and spheroidization occurs.

Fig. 5.61 shows the morphology of single-melting channel under different laser powers, in which the scanning speed is fixed at 200 mm/s and the forming equipment is HRPM-IIA. It can be seen from the figure that at low power (70−110 W), the scanning line is discontinuous, resulting in spheroidization; as the power increases (150−190 W), the scanning line gradually becomes continuous, and the melting width also increases. Thus, high laser power favors wetting and spreading of the melting pool, which is similar to the results reflected in Fig. 5.55.

Metal materials for additive manufacturing Chapter | 5 **479**

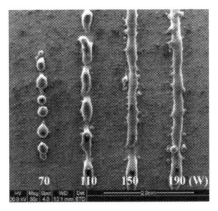

FIGURE 5.61 Single-melting channel at different laser powers.

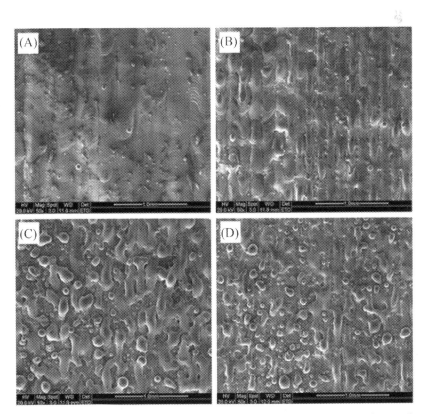

FIGURE 5.62 Single-layer scanning surface morphology (SEM) at different scanning speeds. (A) 50 mm/s; (B) 400 mm/s; (C) 600 mm/s; (D) 800 mm/s.

c. Morphology of single-layer scanning surface

Fig. 5.62 shows the SEM photo of the single-layer scanning surface at different scanning speeds. The power P = 190 W and the scanning pitch S = 0.15 mm are fixed values, and the forming equipment is HRPM-IIA.

At a low scanning speed of 50 mm/s, the scanning lines are continuous and well overlapped, the scanning surface is flat, and the entire scanning surface is not significantly spheroidized; as the scanning speed increases to 400 mm/s, the scanning lines become discontinuous with intermittent phenomenon, the scanning surface is also rough; when the scanning speed increased to 600 mm/s, the scanning line is unrecognizable, a large number of metal balls appear; at a high scanning speed of 800 mm/s, the scanning surface continues to appear a large amount of metal balls (small size). The above test results show that the high scanning speed is not conducive to the wetting and spreading of the melting pool, forming serious spheroidization phenomenon. Therefore, higher laser energy (low scan speed and high laser power) should be used to avoid spheroidization.

1. Scanning line interval

Fig. 5.63 shows the morphology (SEM photo) of the surface of the formed sample block at different scanning pitches. The power P = 190 W

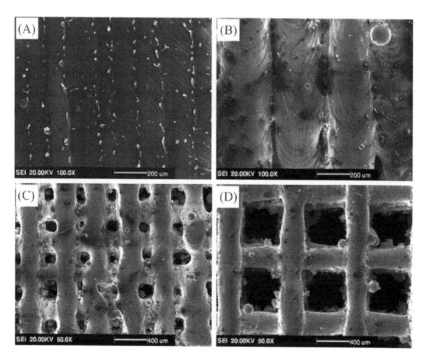

FIGURE 5.63 Spheroidization at different scanning pitches. (A) 0.15 mm; (B) 0.3 mm; (C) 0.4 mm; (D) 0.8 mm.

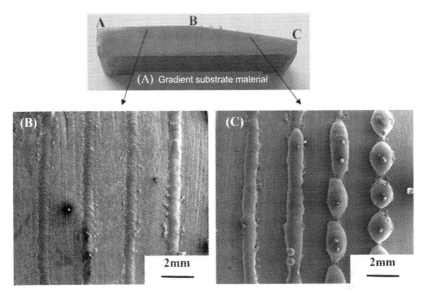

FIGURE 5.64 Single scanning line of 316 L stainless steel powder with different layer thickness.

and the scanning speed $v = 100$ mm/s are fixed values, and the forming equipment is HRPM-IIA. Since both power and speed are constant, the corresponding laser linear energy input (P/v) is also a fixed value.

As can be seen from the analysis of Figs. 5.63 and 5.64, the scanning line is continuous without spheroidization at a power of 190 W and a speed of 100 mm/s. From the Fig. 5.66, it can be analyzed that at lower scanning pitch (Fig. 5.63A and B), the scanning lines are overlapped to each other; while at higher scanning pitch (Fig. 5.63C and D), although the scanning lines cannot be overlapped to each other and pore structure is formed, the scanning lines are continuous and have no spheroidization. Therefore, if the scanning line is continuous (no spheroidization), the scanning surface will not be spheroidized at any scanning pitch, but the overlapping ratio of the scanning lines is different; if the scanning line is discontinuous (spheroidized), spheroidization occurs at any scanning pitch. Therefore, under the condition that the laser line energy density is fixed, the scanning pitch does not affect the SLM spheroidization.

2. Slice thickness

   In order to study the influence of the thickness of the powder leveling on the spheroidization of the SLM process, it is investigated from the aspects of line scanning and volume scanning.

   a. Stainless steel powder line scanning

      In order to reveal the influence of the thickness of the powder leveling on the single-line scanning spheroidization, a substrate with a certain slope is fabricated, and its side is shown in Fig. 5.63. Above the substrate is a layer of powder spread flat. It can be seen that the thickness of the powder layer changes gradient at the slope of the substrate (the layer thickness gradually increases from left to right). Furthermore, the variation law of spheroidization conditions under different powder layer thicknesses can be investigated.

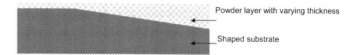

A single-line scanning test ($P = 190$ W, $v = 50$ mm/s are fixed) is performed on the substrate, and the forming material is 316 L stainless steel powder, and the characteristics of the scanning line are observed by SEM, as shown in Fig. 5.64. It can be seen that the thickness of the powder leveling layer has a great influence on the spheroidization of the scanning line. Fig. 5.64A is a physical picture of the substrate, the substrate is divided into two parts:

- Section AB: The substrate in this area is horizontal, and after powdering, the powder above is the same in thickness;
- Section BC: The substrate in this area has a wedge shape (having a certain slope). After powdering, the thickness of the powder above is gradually thickened from left to right.

As shown in Fig. 5.64B, the melting channel of section AB powder is continuously scanned and wetted well with the substrate after being scanned by the laser line. This is because the thickness of the powder in this area is the same, and the melting-wetting-solidification conditions are the same under the same laser parameters, so the scanning lines are basically indistinguishable.

After the powder in the section BC is scanned by laser, the melting channel characteristics of different parts are different: the melting channel becomes discontinuous, and spheroidization gradually appears from left to right, as shown in Fig. 5.64C. It can be seen that the thickness of the powder layer has large influence on the spheroidization condition: the low layer thickness of the powder layer is beneficial to the wetting and spreading of the melting pool, and the high thickness of the layer causes spheroidization easily, which affects the forming quality of the SLM. At the same time, it can be noticed that as the thickness of the powder leveling layer increases, the width of the melting channel also increases. This is because the more powder can be supplied for laser melting under the thick powder, and the melting volume is correspondingly larger, but it contributes to the phenomenon of spheroidization. Therefore, in order to effectively suppress the spheroidization of the SLM, precise control of the fine layer thickness is an important step.

b. Three-dimensional body forming of stainless steel powder
The influence of the thickness of the powder leveling layer on the line scanning is examined in the foregoing. The influence law of the forming on the body is examined below. Two kinds of powder layer thickness (0.05, 0.1 mm) are selected for SLM formation ($P = 190$ W, $v = 100$ mm/s, $h = 0.15$ mm are fixed), and the surface of the sample block is observed by SEM, as shown in Fig. 5.65. Under the low powder thickness of 0.05 mm, the forming surface is relatively flat, without spheroidization, and has good wettability; under the thick layer of 0.1 mm, the forming surface has more serious nodules, which is caused by poor wetting. Therefore, whether it is line scanning or body forming, the thickness of the powder layer will have a great influence on the surface quality, wetting and spheroidization of the SLM. In order to obtain SLM samples with good wettability and surface condition, a thinner powder layer thickness is required.

Based on the above-mentioned results of line scanning and bulk forming under different powder layer thicknesses, it is found that the thickness of the powder leveling is a key part of controlling the spheroidization and quality of the SLM. The effect of different powder thickness on spheroidization is shown in Figs. 5.66 and 5.77. The

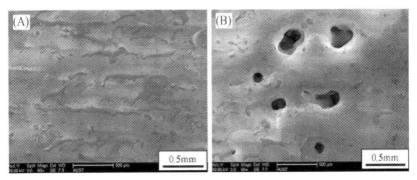

FIGURE 5.65 Surface wetting of the body forming under different layer thicknesses. (A) 0.05 mm; (B) 0.1 mm.

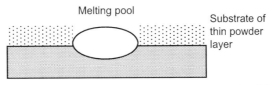

FIGURE 5.66 Effect of thinner powder layer (large wetting area) on spheroidization.

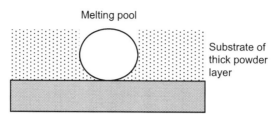

FIGURE 5.67 Effect of thicker powder layer (small wetting area) on spheroidization.

thicker paving layer results in poor wettability and is prone to cause spheroidization. The main reasons are as follows:

- Under thick powder leveling layer, the input of laser energy per unit volume will be correspondingly reduced, and the temperature field generated will be lower, so the fluidity and wetting of the melting pool will also be deteriorated to form spheroidization;
- Under thick powder layer, although the powder is melted to a great extent, it is difficult for the laser energy to completely penetrate the powder layer to bond with the substrate. At this time, as shown in Fig. 5.67, the contact surface (or effective wetting surface) where the melting pool wets the substrate is small. It is difficult for a large volume melting pool to wet the substrate through a small contact surface, thus forming spheroidization.

The above analysis shows that thick powder layer is liable to cause spheroidization. However, the excessively thin powder layer thickness greatly increases the forming time of the SLM on one hand; on the other hand, the thickness of the powder layer is also limited by the particle size of the powder layer, and the thickness of the powder layer is slightly larger than the particle size of the powder.

c. Pure nickel powder line scanning

The above spheroidization theory is based on the AISI316L material. In order to reveal the universality of the above spheroidization theory, this section uses pure nickel powder to test the variation of the laser single-channel scanning line under different layer thicknesses ($P = 190$ W, $v = 50$ mm/s are fixed), as shown in Fig. 5.68.

It can be seen that from (A)–(D), as the thickness of the powder leveling layer increases in turn, the scanning line becomes thicker, but the continuity of the scanning line deteriorates. Under thicker powder conditions, significant spheroidization occurs (Fig. 5.68D). The mechanism of its formation is similar to that of the above-mentioned stainless steel powder spheroidization. At the same time, for thicker powder layers, a large number of holes appear in the laser scanning line, and the formation mechanism is still unclear, which may be caused by the laser shock wave resulting in partial melt to splash. Based on the above, the theory of

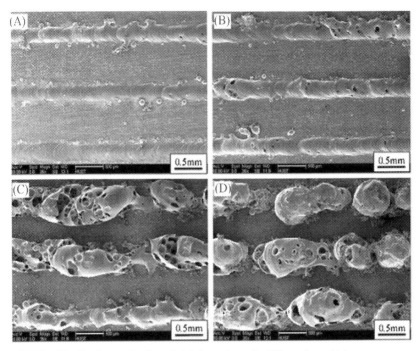

FIGURE 5.68 Single scanning line of nickel powder under different layer thicknesses.

stainless steel SLM spheroidization is also applicable to SLM forming of nickel powder, so the spheroidization theory has certain universality.
3. Remelting scanning
   To investigate the effects of laser remelting scanning on SLM-formed spheroidization, this section uses primary scan and secondary scan to prepare the block samples (the same laser process parameters) and observe the surface spheroidization condition of the samples by SEM (forming equipment is HRPM-IIA). Fig. 5.69 shows the spheroidization of the sample surface before and after primary scan and secondary remelting.

   Fig. 5.69A shows the surface topography of the sample before remelting (primary scan). It can be seen that the laser beam scanning lines are densely overlapped and the surface is generally flat, but some metal balls appear that are not conducive to the powder leveling. Fig. 5.69B shows the surface morphology of the samples after remelting (secondary scanning). It can be seen that the laser beam scanning lines are overlapped clearly and the surface is not spheroidized. It can be inferred from the above analysis that the laser secondary remelting can re-melt the surface of the metal sphere, making it into a melt and rewetting with the substrate. Therefore, the secondary remelting of the laser is beneficial to alleviate the spheroidization of the surface to a certain extent.

486 Materials for Additive Manufacturing

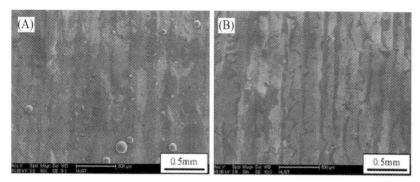

FIGURE 5.69 Effects of scanning method on spheroidization of forming surface. (A) Primary scanning; (B) secondary remelting.

5.4.1.2 Pores

Although the SLM technology has been greatly developed in recent years, since the SLM is a near-net forming technology based on powder deposition forming, pores are easily formed inside the formed part. The existence of pores drastically reduces the mechanical properties of SLM parts, seriously affecting the practicality of SLM technology and hindering the advancement of this technology. Therefore, it is necessary to optimize the process to promote densification; on the other hand, the application of some metal parts requires porous structure, for example, biological materials, filter materials, etc., thus controlling porosity to form porous materials has become another important issue. The pores formed in SLM are caused by many factors, however, there are few studies on the classification of SLM pores and the corresponding formation mechanism. Therefore, it is of great significance to study the pore formation law of SLM forming and its control method.

5.4.1.2.1 Factors affecting porosity

5.4.1.2.1.1 Powder characteristics The SLM forming test is carried out using three different 316 L stainless steel powders, and the low-power polished section of the SLM sample is observed by SEM, as shown in Fig. 5.70. The relative density is tested by the drainage method, as shown in Fig. 5.71. It can be seen that there are a large number of pores inside the SLM sample formed by water-atomized powder (Fig. 5.70A and B), while the part formed by gas atomized powder has few internal pores (Fig. 5.70C), its forming density is significantly higher than the density of water-atomized powder (Fig. 5.71). The difference between gas atomized and water-atomized powder SLM forming densification can be considered from the aspects of powder oxygen content and bulk density.

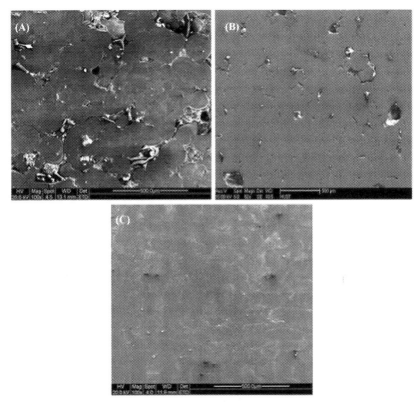

FIGURE 5.70 SEM photo of low-magnification polishing cross-section of SLM-formed parts. (A) Water-atomized powder 1#; (B) Water-atomized powder 2#; (C) Gas atomized powder 3#.

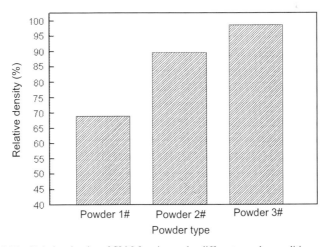

FIGURE 5.71 Relative density of SLM forming under different powder conditions.

From the law of the influence of powder characteristics on spheroidization, it is known that the water-atomized powder has high oxygen content, which leads to the formation of more oxide film on the forming surface that is not conducive to the wetting and spreading of the melting pool, so it is easy to cause serious spheroidization. In SLM forming, each single scanning line cannot be well wetted with the previous layer, so that a continuous straight line cannot be formed, thereby splitting into a large number of metal balls, and a large number of pores exist between the large number of split metal balls. As the SLM multilayer stacking proceeds, the pores between the metal balls remain in the formed part to form the pore structure as shown in Fig. 5.70A and 70B.

The water-atomized powder is mostly irregular in shape, while the gas atomized powder is generally spherical. The irregular water-atomized powder has poor fluidity, and the accumulation coordination between the powder particles is poor, so the bulk density during the powder leveling is low; while the gas atomized powder is spherical, the powder has good fluidity and the bulk density is higher. The influence of the bulk density of metal powder on its formation densification has been reported, and the results show that the powder material with high apparent density facilitates the densification of the final form. Therefore, the spherical gas atomized powder used in this experiment has high apparent density, and its relative density of forming is higher than that of the water-atomized powder.

Since the powder is a non-dense body, there is shrinkage during layer-by-layer melting and stacking of laser, so that the height of the upper surface of the solidified layer is lower than the upper surface height of the powder layer before melting after the completion of each layer of SLM (assuming the melt solidifies into a dense body); when the upper surface of the solidified layer is powder leveled, the thickness of the powder layer is greater than the actual thickness of the powder layer (software slice thickness), so that each layer of the SLM has an actual layer thickness. The published literature deduces the change of the actual shrinkage of the powder during the SLM multilayer forming process. The calculation results show that the final shrinkage of the powder tends to a fixed value. The calculation process is as follows:

$$a_n = h\left[(1-k) + (1-k)^2 + (1-k)^3 + \cdots + (1-k)^n\right] = \frac{h(1-k)[1-(1-k)^n]}{k}$$

(5.64)

where a_n is the amount of shrinkage of the nth layer of powder, h is the set layer thickness (slice thickness), and k is the actual relative density of the powder of apparent density, and the limit value is taken to a_n to obtain:

$$\lim_{n \to \infty} a_n = \frac{h(1-k)}{k}$$

(5.65)

It can be seen that as the SLM forming continues (n value increases), a_n tends to a certain value, the actual powder layer thickness of the SLM is:

$$H_{\max} = h + a_n = \frac{h + h(1-k)}{k} = \frac{h}{k} = \frac{h\rho_m}{\rho_l} \tag{5.66}$$

where ρ_m is the standard density of the metal body and ρ_l is the apparent density of the corresponding powder. It can be seen that the actual layer thickness of the powder is always greater than the set layer thickness, and the thicker the set layer thickness, the thicker the actual layer thickness. The gas atomized powder has higher apparent density, which is beneficial to reduce the thickness of the actual powder layer, so that the forming process results in relatively good wettability (the actual powder layer is thin, which is beneficial to the wetting), and it is easy to form a dense parts.

In summary, since the gas atomized powder has relatively high sphericity and low oxygen content, on the one hand, the powder layer after the powdering has relatively high apparent density; on the other hand, the wetted surface is relatively clean (less oxide), and the melting pool has good wettability. The above two factors together contribute to the excellent densification performance of the gas atomized AISI 316 L powder.

5.4.1.2.1.2 Forming process

1. Laser power

   Two kinds of block samples are prepared by using two kinds of laser powers (100, 80 W), and other parameters are fixed values (scanning pitch: 0.1 mm, scanning speed: 90 mm/s, layer thickness: 0.05 m, forming equipment is HRPM-II system)). The pore distribution and morphology on the surface of the formed body are observed by SEM, as shown in Fig. 5.72.

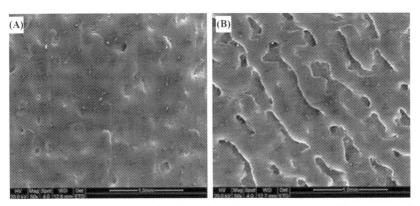

FIGURE 5.72 Surface porosity of the formed samples under two powers. (A) P = 98 W; (B) P = 80 W.

It can be seen that under the action of 98 W laser power, the surface of the formed ample is relatively flat, the scanning channel itself has no splitting phenomenon, the scanning channels are arranged neatly, the lap joint effect is good, and basically no pores are generated (Fig. 5.72A); at low power (80 W), the surface of the formed sample is rough, the scanning track is discontinuous, the wettability of the melt is poor, and a large number of pores appear between the discontinuous melts (Fig. 5.72B). The above phenomenon is similar to the effect of power on spheroidization: as the power is reduced, the temperature of the melt is lowered, the wettability is deteriorated, the melt spreadability is deteriorated, and a large number of pores not filled with the melt are formed. Thus, high laser power facilitates densification of the SLM process.

2. Scanning speed

The effects of several sets of scanning speeds on the relative density and low-magnification microstructure of 316 L stainless steel forming samples are investigated. Other parameters are fixed values (scanning pitch $S = 0.15$ mm,

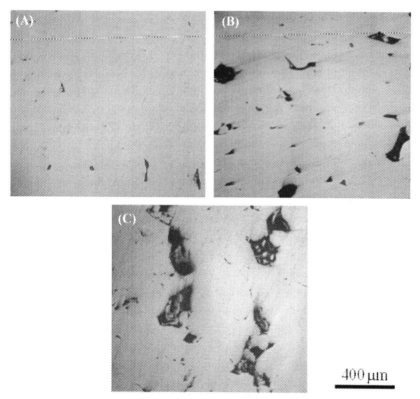

FIGURE 5.73 Light microscopic photos of formed samples after polishing at different scanning speeds (not corroded). (A) $v = 100$ mm/s; (B) $v = 200$ mm/s; (C) $v = 300$ mm/s.

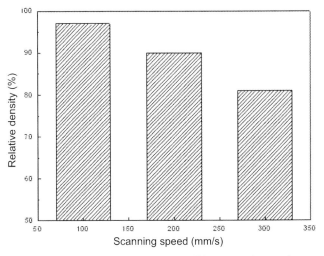

FIGURE 5.74 Relative density of formed samples at different scanning speeds.

layer thickness $d = 0.05$ mm, power $P = 190$ W, forming powder is 1# water-atomized powder, forming equipment is HRPM- IIA). Fig. 5.73 shows the low-magnification light microscopic photos of polished section of the SLM parts, which are not corroded. Fig. 5.74 is showing the relationship between the scanning speed and the relative density of the formed samples.

It is obvious that the laser scanning speed has great influence on the porosity of the SLM sample. When the scanning speed is 100 mm/s, although there are individual small pores in the polished cross-section of the formed sample, no macropores are generated and the relative density is high (Fig. 5.73A); when the scanning speed increased to 200 mm/s, the number and size of the pores increase, and the relative density decreases gradually (Fig. 5.73B); as the scanning speed continues to increase to 300 mm/s, the melting state of powder further deteriorates, forming a pore with large pore size, and resulting in a large amount of pores in the part (Fig. 5.73C). The effects of scanning speed on relative density in this section is similar to the effects of scanning speed on spheroidization in the previous chapter, that is, high scanning speeds tend to result in poor wetting-spreading ability and also increased porosity.

1. Scanning pitch

   SLM block samples are prepared with different scanning pitches, and other parameters are fixed (laser power $P = 190$ W, scanning speed $v = 100$ mm/s, layer thickness $d = 0.05$ mm, forming powder is 1# powder, and forming equipment is HRPM- IIA). The formed sample is cut and polished, and then observed under optical microscope for low-magnification (not corroded) microstructure, as shown in Fig. 5.75. The metallograph of cross-section is consistent with the surface morphology results, as shown in

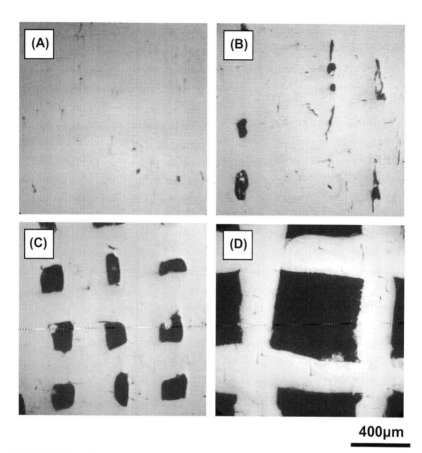

FIGURE 5.75 Light microscopic photos of formed samples after polishing at different scanning pitchs(not corroded). (A) 0.15 mm; (B) 0.3 mm; (C) 4 mm; (D) 0.8 mm.

Fig. 5.63. It can be seen that the overlapping ratio of the scanning channels under different scanning pitches is different, and the scanning line spacing has a significant influence on the pore size, distribution, shape and porosity of the sample. At a low scanning pitch (0.15 mm), the formed sample is denser in structure, the number and size of the pores are small, and the porosity is also extremely small (Fig. 5.75A); when the scanning pitch increased to 0.3 mm, the number and size of the pores increase, and the pores exhibit an irregular shape. At this time, although the scanning lines are just overlapped, the accidental fluctuation of the scanning line width caused by the low overlapping ratio between the scanning lines generates pores (Fig. 5.75B); as the scanning pitch further increased to 0.4 mm, the scanning lines cannot be overlapped to each other due to the excessive scanning pitch, thereby forming a large number of pores (Fig. 5.75C). At this time, the pores are characterized by large pore size and relatively regular pore shape, which is close to square hole. The arrangement between the holes is relatively regular. When the scanning pitch increased to 0.8 mm, there is not only overlap between the scanning lines, the distance between the lines is large, at which point the hole is a very obvious square hole, the size of the hole is large (close to 0.5 mm), forming a regular hole visible to the naked eye (Fig. 5.75D). It can be known from the above phenomena: (1) For the SLM parameters used in this experiment (the higher the linear energy density of the laser, the thinner the powder layer), the scanning line will not be spheroidized (see the research results on the spheroidization in the previous chapter); (2) By adjusting the overlapping ratio or spacing of the scanning lines, the dimension, size, distribution and porosity of the pores can be flexibly controlled [7].

2. Slice thickness
   SLM samples are formed using two powder thicknesses (0.05 and 0.1 mm), other parameters are fixed (laser power $P = 100$ W, scanning speed $v = 90$ mm/s, scanning pitch $S = 0.1$ mm, forming material is 3# powder, forming equipment model is HRPM-IIA). The sample is cut, polished, and observed for low magnification by SEM, as shown in Fig. 5.76. At a layer thickness of 0.05 mm, the sample structure is extremely dense and substantially free of pores; while at a layer thickness of 0.1 mm, a large number of flat pores appear in the formed sample.

   Similarly, HRPM-II equipment is used to study the effect of different layer thickness on densification. Other parameters are fixed (laser power $P = 190$ W, scanning speed $v = 100$ mm/s, $h = 0.15$ mm, $d = 0.05$ mm, gas atomized AISI316L powder). The surface porosity of the two-layer thickness is observed by SEM, as shown in Fig. 5.77.

   Similar to the results in Fig. 5.76, at a layer thickness of 0.05 mm, the forming surface is flat and no pores are visible. When the layer thickness is increased to 0.1 mm, the forming surface has poor wettability and the

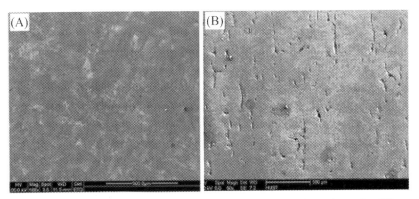

FIGURE 5.76 Low-magnification microstructure (SEM) of polished samples under different thicknesses. (A) 0.05 mm; (B) 0.1 mm.

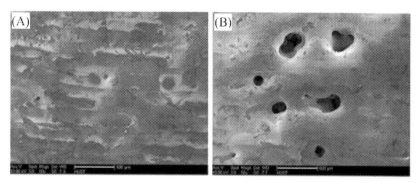

FIGURE 5.77 Surface morphology of the formed sample under different layer thickness (SEM). (A) 0.05 mm; (B) 0.1 mm.

spreading ability of the melt is difficult to fill the pores. The effects of layer thickness on SLM forming densification can be considered from the following perspectives: (1) There is shrinkage in the original SLM forming process, resulting in the actual powder layer thickness always greater than the setting thickness. When the setting thickness is increased, a thicker layer thickness is generated, resulting in spheroidization (the scanning line is not easily continuous and split into a large number of solidified bodies); (2) the excessively thick powder layer is not conducive to the laser penetration, resulting in poor bonding between the melt and the substrate; (3) when the powder layer is thick, the time for the bubbles to escape is increased, and many bubbles are not allowed to escape from the melt and are solidified in the formed body. Therefore, SLM forming requires thinner powder layer to form relatively dense metal parts. In general, the thickness of the powder layer should be less than 0.06 mm, but it needs to be larger than the powder particle size.

5.4.1.2.2 The Classification of pores and their formation mechanism

The process of laser melting and forming 316 L stainless steel involves multiple laying by the powder leveling roller, repeated scanning of the laser beam in each layer, and layer-to-layer welding (the basic principle of rapid prototyping), so the densification mechanism of the SLM process is extremely complicated. When the laser beam scans the metal powder bed, the energy of the laser beam is quickly absorbed by the metal powder, causing the temperature of the metal powder to rapidly increase, melting beyond the melting point. At this point, the interior of the melt is accompanied by high temperature gradient and surface tension. Driven by surface tension, the overlapping of the scanning lines and the wettability of melt determine the surface morphology; and the surface morphology of each layer will affect the spreading of the next layer of powder, ultimately affecting the SLM densification. However, the pore formation of the SLM process involves many factors and mechanisms, and there is a corresponding formation mechanism behind the formation of each pore. The types of pores in the SLM process and the formation mechanism of each pore are discussed in detail below, and methods of controlling densification are proposed.

5.4.1.2.2.1 Melt spheroidization One of the most important reasons for pore formation is the spheroidization of the SLM process. When the melting channel of the SLM has high viscosity and high wetting angle, the melt spontaneously splits into a large number of metal spheres or metal tumors under the action of capillary force to reduce the surface energy of the melt itself. In this case, the scanning surface formed will be very rough and accompanied by a large number of metal spheres or metal tumors with poor wettability, in which case a large number of pores exist between the metal spheres. As shown in Fig. 5.69A, the pores inside the tissue are mainly caused by spheroidization. Since the SLM technology is based on a rapid prototyping technique in which the powder is leveled on the forming surface and melted and solidified layer by layer, the formation of a rough surface is not conducive to the spreading and pore filling of the next layer of powder.

The schematic diagram of powder leveling on a rough surface is shown in Fig. 5.78: after paving the first powder layer, the surface of the powder bed is scanned using a laser beam, which produces many metal balls and a rough surface; after paving the second powder layer, although the upper surface of the metal powder layer can form a flat surface, it is still difficult to achieve densification, which can be attributed to two main reasons:

1. Since the gap between the metal balls of the previous layer is difficult to be filled with powder, when the second layer of powder is laser-scanned, the gap unfilled by the powder in the previous layer of will become the pores inside the SLM part.
2. Even if some of the deeper metal ball gaps can be filled with powder (equivalent to increasing the thickness of the powder layer), the penetration depth of the laser beam energy is limited, and it is difficult for the

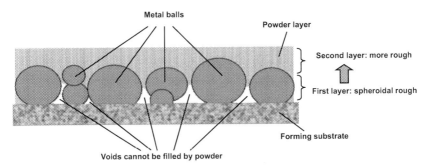

FIGURE 5.78 Schematic diagram of pores caused by spheroidization.

laser energy to contact the powder of the deep gap, which is difficult to achieve sufficient melting or to be melted into spherical shape, so that it is still difficult to achieve densification.

In summary, if a layer forms a rough surface (a large amount of spheroidization), it will inevitably cause the surface of the lower layer to be rougher (including spheroidization and pores), forming a "chain effect," which eventually leads to a large number of pores in the formed part; meanwhile, if the spheroidization is serious, it will directly damage the metal parts or make the powder leveling process impossible. It can be seen that the generation of spheroidization is the main reason for the pores of the SLM parts, and the shape of the formed pores is irregular. Therefore, in order to increase the relative density of the SLM, it is necessary to suppress the SLM spheroidization.

5.4.1.2.2.2 The scanning line is not overlapped Even if the laser scanning line is continuous and does not produce spheroidization, voids may still be produced, as shown in Fig. 5.79. It can be seen that the scanning lines are continuous, but do not overlap each other. Although there is no spheroidization as a whole, a large number of square holes are arranged neatly, and the pore size is large, close to 0.5 mm. The formation of such pores is due to the fact that the scanning lines are not tightly overlapped. Thus, in order to form a dense part, the following two main conditions should be met: (1) First, the scanning line itself is continuously non-spheroidized; (2) Second, the continuous scanning line needs to be closely overlapped.

It should be pointed out that the pores caused by not overlapping the scanning lines are formed by artificial design. Therefore, many metal parts require porous structure. The porous metal has the following excellent characteristics: small density, large specific surface area, good heat dissipation, good sound absorption, excellent permeability and wave absorption. Thus,

FIGURE 5.79 Porosity caused by not overlapping the scanning lines.

porous metal materials have been used in various industries as complex parts required for filtration, silencing, energetic, electrochemical, biomedical, and heat exchange. When the above porous metal parts require complex shapes, SLMtechnology can be used to produce the required porous complex metal parts without being limited by the material.

5.4.1.2.2.3 Crack Cracks are easily generated during the SLM forming process, and cracks are also the cause of pore formation in the SLM formed part, as shown in Fig. 5.80. This is because SLM is a rapid melt-solidification process with a high temperature gradient and cooling rate of melt. This process occurs in a short period of time and will produce the following stresses:

Thermal stress: In general, the thermal stress of SLM is caused by the inconsistency between the thermal expansion and shrinkage deformation of various parts when the laser heat source acts on the metal. Specifically, as shown in Fig. 5.81A, during the melting process, since the SLM melting pool instantaneously rises to a very high temperature, the melting pool and the region with a relatively high temperature around the melting pool tends to swell, while the region farther away from the melting pool is cooler and has no tendency to swell. Since the two parts are mutually restrained, the position of the melting

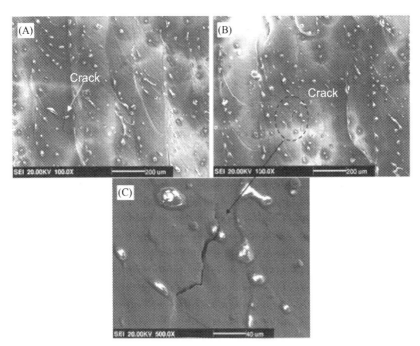

FIGURE 5.80 Surface crack of SLM forming sample. (A) (B) The cracks (C) The larger version of (B).

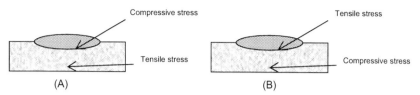

FIGURE 5.81 Schematic diagram of thermal stress generation. (A) Solid-liquid transition (B) Liquid-solid transition.

pool will be pinched by the compressive stress, and the part away from the melting pool is subjected to tensile stress; during the cooling process of the melt, as shown in Fig. 5.81B, the melt gradually contracts, and conversely, the solidified portion of the melt is subjected to tensile stress, while the portion away from the melt is subjected to compressive stress. It can be seen that the uneven heating of the SLM process is the main cause of thermal stress.

Structural stress: For some metals, in the solid-state phase transformation, such as transforming from phase A to phase B, the specific volume of the two phases is inconsistent, causing volume expansion or volume shrinkage, which will pinch the surrounding metal and generate stress. When austenite is transformed into ferrite or martensite, volume expansion occurs to generate tissue stress.

Thermal stress is the most common and is the main cause of cracking in the SLM process. When the internal stress of the SLM part exceeds the yield strength of the material, cracks will be generated to release the thermal stress. The presence of microcracks reduces the mechanical properties of the part, damaging the quality of the part and limiting the practical application. Currently, the method for eliminating internal cracks in SLM parts is hot isostatic pressing (HIP). Professors F. Wang and X. Wu of the University of Birmingham in the United Kingdom successfully formed Hastelloy X nickel-based superalloy of complex shape using the SLM method. However, since the nickel-based alloy has higher coefficient of thermal expansion than other metals, the thermal stress inside the nickel-based alloy is higher, thereby forming cracks. After the HIP treatment to the SLM parts of Hastelloy X nickel-based alloy, the internal cracks are closed and the mechanical properties are greatly improved. From the fracture morphology, after the HIP treatment, the pores are closed.

5.4.1.2.2.4 Hole Since the powder bed is a porous structure, inert gas is intrinsic inside (the forming chamber is filled with argon gas in this experiment), and the gas remains in the melt during the powder forming melt. Since the SLM is completed in an instant, some of the gas has not yet escaped and has solidified in the sample, eventually forming pores. It should be noted that the SLM pores caused by the air hole are spherical because the bubbles always exist in the minimum surface energy (only the spheres satisfy the minimum surface energy). As shown in Fig. 5.82, the spherical pores are caused by the air hole in the melt.

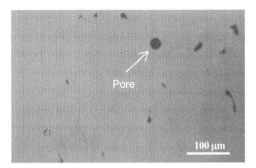

FIGURE 5.82 SLM pores caused by air hole (OM, not corroded).

Baroody E.M. et al. studied the formation mechanism of bubble and divided the formation of bubble pores into two main processes: (1) the movement of bubbles in the melt; and (2) the combination of bubbles. Finally, the expression of the number of bubbles in a unit volume after bubble combination is given:

$$\frac{N}{N_0} = \left[0.87 \left(\frac{N_0 mkT}{\sigma} \right)^{1/2} Ct + 1 \right]^{-2} \tag{5.67}$$

where N is the number of bubbles in the unit volume after the combination, N_0 is the number of bubbles in the initial unit volume, m is the number of gas atoms in the bubble per unit volume, t is the melting time, T is the temperature, σ is the surface tension, and C is the specific heat capacity. It can be seen that the reasonable control of the SLM process can regulate the number of bubbles. Decreasing the scanning speed increases the melting time t and the melting temperature T, and lowers the surface tension σ, all of which reduce the number of pores N. Therefore, in the SLM forming experiment, the scanning speed can be appropriately reduced to promote the escape of gas, thereby increasing the relative density of the SLM.

5.4.1.3 Organization characteristics

Metal materials formed by SLM technology have microstructures that are distinct from other forming methods, so it is particularly necessary to study the unique microstructure formation in the SLM mode. Fig. 5.83 shows the front and side metallographies of the SLM-formed samples. The microstructure of each layer can be seen. In general, in the low-magnification metallography, it can be seen that the formed sample has few holes and dense structure. Metal materials with high relative density tend to be superior to low-density materials, so the density of materials is also an important indicator for measuring the quality of SLM parts. Fig. 5.83A depicts the metallography of the formed sample in the front view. It can be seen that many of the scanning lines still show

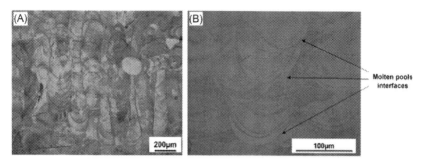

FIGURE 5.83 Metallography of polished section of SLM sample. (A) Front view; (B) side view.

fish scale features after polishing, very similar to laser single scanning track. In the production process of each layer of the SLM, the two scanning lines need to overlap each other, and the previously scanned scanning track needs to be remelted again by the new scanning track, so that a strong overlap is formed between the scanning tracks. Fig. 5.83B shows the metallography of the SLM-formed sample in the side view direction, reflecting the condition of soldering and overlapping between the melting pools. The interface of the melting pool is clearly visible, and the overlapping is densely connected.

Fig. 5.84 shows the high-magnification microstructure of the SLM-formed sample. Overall, it can be seen that the grains are very fine and the crystal growth direction is intricate in all directions. According to the Hall-Petch formula of metallurgical principle, for dense metal materials, when the crystal grain is finer, it is more favorable for the mechanical properties of the metal materials. The extremely fine grains are caused by the rapid cooling rate of the melting pool during solidification and the large degree of subcooling. In summary, the microstructure in Fig. 5.84 is mainly characterized by fine grain and multiple directions of crystal growth. The finer grains formed by the SLM can be considered from the nucleation and growth angle of the metal solidification process.

For liquid metals, the nucleation rate (number of nucleation) per unit volume and unit time is related to the degree of subcooling ΔT of the melt below the melting point of T_m. According to the Johnson-Mer equation, the relationship between the numbers of crystal nuclei P(t) formed in time t and the nucleation rate N and the growth rate v is as follows:

$$P(t) = k\left(\frac{N}{v}\right)^{3/4} \tag{5.68}$$

In the equation, k is a constant, $N \propto \exp(-\frac{1}{\Delta T^2})$, and $v \propto \Delta T$ (continuously grow) or (screw dislocation grows), so as the degree of supercooling increases, the speed that N becomes large is faster than that of v, the increase in the degree of supercooling can increase the number of crystal nuclei to refine the grains. In

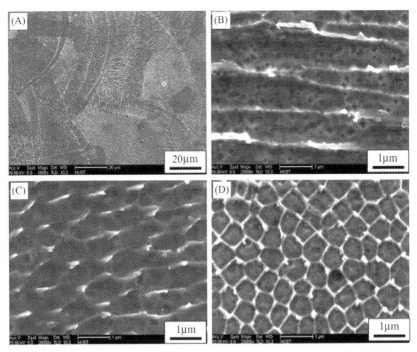

FIGURE 5.84 High-magnification microstructure (SEM) of SLM-formed samples.

the SLM metallurgical mode (Gaussian heat source and fast scanning), it has the characteristics of rapid cooling. As mentioned above, the degree of supercooling is large. From the theory of metal solidification, it is known that the liquid melting pool is likely to form a large amount of crystal nucleus under the condition of large degree of subcooling. The rapid solidification can inhibit the growth of crystal grains, so that a large number of spontaneous nucleation is difficult to grow in a short time, thereby forming fine-grained structure.

Since AISI316L is an austenitic alloy with Fe-Cr-Ni as the main element, not only thermal subcooling but also constitutional supercooling exists in the SLM metallurgical mode. When the alloy starts to precipitate solid phase from pure liquid state, the fraction of the solute in the solid phase precipitated first is so low that the remaining solute is released from the solid phase into the liquid phase, thereby changing the temperature of the liquid phase solidification, so that the degree of supercooling of liquid phase appears to be large. It should be noted that this supercooling due to compositional changes is different from thermal subcooling. As shown in Fig. 5.85, at the front of the solidification interface, due to insufficient diffusion of the solute in the liquid phase, there is a sharp increase in the solute fraction at the interface front (Fig. 5.85A). The liquidus temperature is shown in Fig. 5.85B and C. At this time, there is a temperature gradient at the solid-liquid interface. When the temperature

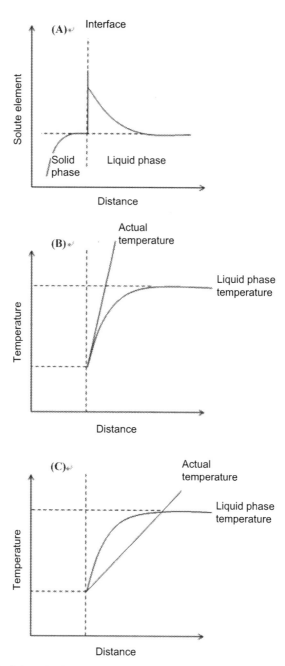

FIGURE 5.85 Schematic diagram of constitutional supercooling when the alloy is solidified. (A) Solid phase and liquid phase composition at the solidification interface; (B) stable interface; (C) interface is unstable and the composition is supercooled to produce cell crystal grains.

gradient is large, the liquidus temperature is lower than the actual temperature. At this time, there is no constitutional supercooling; and when the temperature gradient so small that the liquidus temperature is higher than the actual temperature, the supercooling will occur and the criteria are as follows:

$$\frac{G_L}{R} < -\frac{M_L C_0(1-K)}{KD_L} \tag{5.69}$$

Or

$$\frac{G_L}{R} < \frac{\Delta T}{D_L} \tag{5.70}$$

where G_L represents the temperature gradient of the solidification interface; R is the solidification rate of the melt; M_L is the slope of the liquidus; C_0 is the overall composition of the alloy; K is the equilibrium partition coefficient; D_L is the diffusion coefficient of the solute in the liquid phase; ΔT_0 is the solidification temperature range at the solute concentration of C_0.

If $G_L/R > \Delta T/D_L$, the interface front will tend to be stable without constitutional supercooling, and the metal at the front of the solidification interface will grow forward in a planar shape. Under the SLM condition, the temperature gradient is small, and the solidification speed is fast, which is extremely prone to s constitutional supercooling. Under the condition that the constitutions are undercooled, the solid-liquid interface will grow as cell crystal grains, thereby forming hexagonal cell crystal grains as shown in Fig. 5.84D.

On the other hand, crystals tend to grow in a certain optimal direction during solidification of the liquid melting pool. For cubic crystals, the preferred growth direction is <100>. When one of the directions most likely to grow coincides with the direction of heat flow, crystal growth is promoted. From this, it is understood that in any crystal growth direction, growth in the direction closest to the heat flow direction in the <100> crystal orientation is promoted, and other growth is suppressed. Moreover, since the SLM technology is a rapid prototyping technique based on multi-line scanning, the moving laser heat source controlled by the scanning path and the computer system causes a complicated heat transfer process, which in turn produces a complicated change in heat transfer direction. During solidification, crystal growth will follow a plurality of heat dissipation directions, thereby forming crystal structures grown in various directions to coordinate the direction of heat conduction. Thus, the microstructure and crystal orientation grown in various directions as shown in Fig. 5.84 are formed (Fig. 5.84A, C and D). It can also be seen that the cell crystal grains in the Figure are hexagonal (Fig. 5.84D), which can be analyzed from the characteristics of the intersection of three grain boundaries, as shown in Fig. 5.86. Let γ represent the grain boundary energy between two adjacent grains, and θ represents the

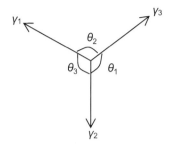

FIGURE 5.86 Schematic diagram of the intersection of three grain boundaries.

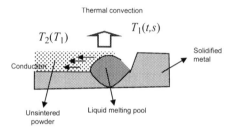

FIGURE 5.87 Solidification heat transfer of the melting channel.

angle between the grain boundaries. In order to achieve the condition of energy balance, the following equation must be satisfied.

$$\frac{\gamma_1}{\sin\theta_1} = \frac{\gamma_2}{\sin\theta_2} = \frac{\gamma_3}{\sin\theta_3} \tag{5.71}$$

Under this condition, when $\theta = 120$ degrees, the crystal grains are in a relatively stable equilibrium state. In this experiment, the formed grains are mostly close to the hexagon (Fig. 5.84D), and the angle is 120 degrees, which has the smallest interfacial energy. The formation of the above microstructure is due to the heat transfer process of the SLM process.

Fig. 5.87 is a schematic diagram of heat transfer during melting and solidification of the melting pool. It can be seen that after the metal powder absorbs the laser energy, the heat can be transmitted in the following ways. First, part of the heat can be lost to the air by convection; second, another part of the heat can be transferred to the loose metal powder by heat conduction and dissipated; third, the rest of the heat can be conducted through the solidified metal solids and eventually dissipated through the substrate. In the SLM process, the temperature of the surface of each processing layer is changing, and the repeated changes in temperature also contribute to the diversity of SLM organization. As can be seen from the SEM photo (Fig. 5.88), there are several typical microstructures of the SLM-formed body. It can be seen that the material has a non-uniform tissue distribution and has a cell shape or a columnar shape (the cell structure is a cross-section of the

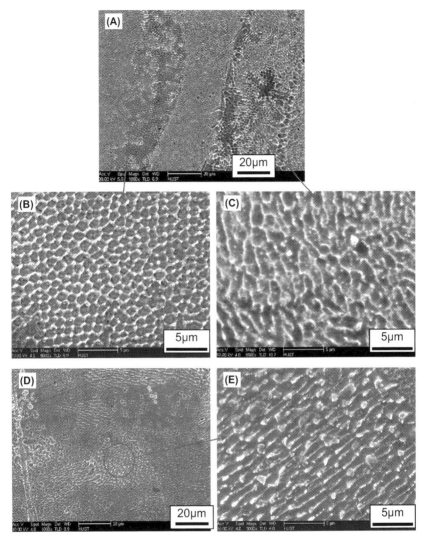

FIGURE 5.88 Typical microstructure photo of SLM. (A) and (E) The structure at the welding interface. (B) and (C) The larger version of (A). (E) The larger version of (D).

columnar structure). The tissue in the Figure can also be seen as a thin honeycomb and a dot. Moreover, in Fig. 5.88A, the structure at the welding interface is thicker than that in the interface, because part of the structure undergoes "heat treatment" of the post-solidification liquid phase, causing the growth of the crystal grains.

Fig. 5.89 shows the results of surface energy spectrum analysis of the microstructure of SLM-formed samples, which mainly characterize the distribution of various elements in the SLM forming material. It can be seen that

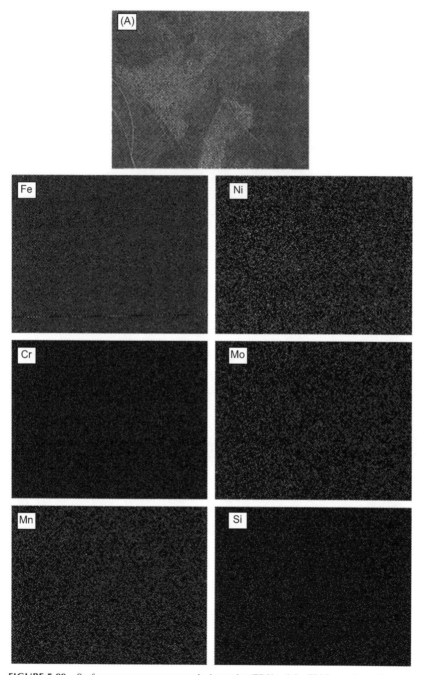

FIGURE 5.89 Surface energy spectrum analysis results (EDX) of the SLM sample section.

the elements inside the material are evenly distributed, without segregation and aggregation, which is different from the typical casting structure (serious segregation), indicating that the microstructure for SLM forming by the pre-alloyed powder is relatively uniform.

5.4.2 Surface roughness and dimensional accuracy of formed parts by the selective laser melting technology

5.4.2.1 Surface roughness

The SLM technology is one of the most viable and complex direct metal manufacturing techniques. This is because the laser has a small spot to form minute details, and the high-precision galvanometer can precisely control the scanning area of the laser. Compared with traditional processing methods, directly manufacturing metal parts reduces many intermediate links and saves a lot of time and manpower. Without special requirements, metal parts manufactured by SLM technology can be applied directly without any post-treatment or with simple surface treatment. This requires a higher surface quality of the manufactured part. The surface quality of the part is the key factor determining the performance of the part. Many practical parts require a surface roughness less than $Ra = 0.8$ m to avoid premature failure of the part due to cracks on the surface during use.

The surface roughness of parts manufactured by SLM technology is mainly divided into upper surface roughness and side surface roughness. The upper surface roughness is mainly determined by the surface quality of the melting channel formed by the melting of single-pass laser melting, the undulating surface formed by the overlapping between the melting channels, and the step formed by the connection between the molten layers, as shown in Fig. 5.90. When the angle between the side of the part and the Z-axis direction is greater than 0 degree, the step on Fig. 5.90B appears on the side of the part. When the angle between a surface and the Z-axis is acute angle, when the next layer of laser is melted, there is a base that can be used for support on the surface, and the powder solidifies on the base after melting; when the angle between a surface and the Z-axis is obtuse angle, it will always have a part of the powder not supported by the metal matrix after melting at the time of manufacture of the next layer. The powder is equivalent to sintering directly on the powder bed. After the powder is melted, the powder at the bottom is adsorbed on the lower surface. In addition to the step effect, this surface also adheres to a large amount of unmelted metal powder on the surface, resulting in deterioration of surface quality.

The upper surface roughness is determined by the surface roughness of the laser-scanned melting channel. However, due to the particularity of the SLM technology, the upper surface and the lower surface and the side surface are affected by many factors, and the surface quality laws are different.

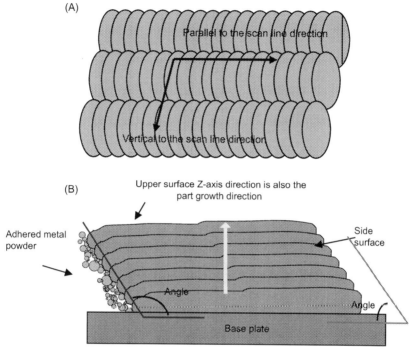

FIGURE 5.90 Surface roughness of the part. (A) Upper surface of the part; (B) Side surface of the part.

5.4.2.1.1 Surface roughness of a single pass

As shown in Fig. 5.91, as the scanning speed increases, the surface roughness shows a trend of decreasing first and then increasing. When the laser power is 120 W, the scanning speed is increased from 500 to 900 mm/s, and the surface roughness is reduced from $Ra = 11-7$ μm. However, when the scanning speed is further increased ($v = 1200$ mm/s), the surface roughness starts to increase. Since the energy of the laser cannot sufficiently melt the metal powder, the spheroidization of the molten metal causes the undulation or even fracture of the melting channel, which improves the unevenness of surface to cause an increase in surface roughness. As can be seen from the figure, when the scanning speed is increased, the width of the melting channel becomes smaller, and due to the impact of the laser energy, more metal droplets splash at the edge of the melting channel, and become small balls after cooling, so that when manufacturing the solid part, splashing balls also reduce the surface quality. This is because when the laser energy is sufficient to melt the powder within the laser scan width, the scanning speed is increased to reduce the time the laser stays in the molten pool. The low scanning speed increases the temperature and width of the molten pool, and the

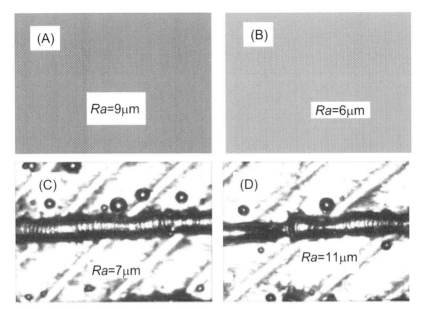

FIGURE 5.91 Surface roughness of the melt channel when the laser power is 120 W. (A) Scanning speed: 500 mm/s; (B) Scanning speed: 700 mm/s; (C) Scanning speed: 900 mm/s; (D) Scanning speed: 1200 mm/s.

molten pool absorbs more powder particles at the edge of the scanning line, increasing the volume of liquid in the molten pool. The powder particles solidify together with the liquid in the molten pool after being absorbed in the molten pool, causing the surface to be uneven. As the scanning speed increases, the moving speed of the laser beam in the molten pool increases and the cooling time of the metal liquid in the molten pool decreases, thus reducing the poor surface tension caused by the heat influence in the molten pool and decreasing the small droplets formed from the liquid in the action of surface tension, so that the surface of the melting channel becomes flat. As the scanning speed is increased to a larger value, the energy of the laser is insufficient to melt the powder within the width of the spot. The ability to melt the powder of fine particles is weak, and the powder of large particles is not melted, so that the molten metal liquid adheres to the powder of the large particles to form spheroidization. During the SLM process, the movement of the laser spot forms a constantly moving molten pool. The connection between the molten pools forms a continuous melting channel. As the moving speed is too fast, the cooling rate of the metal liquid in the molten pool is also increased. After the molten metal of the previous molten pool has begun to solidify, the molten metal of the next molten pool has just formed, causing a temperature difference between the molten pools, and a large surface tension difference is formed between the molten metal. In this surface tension,

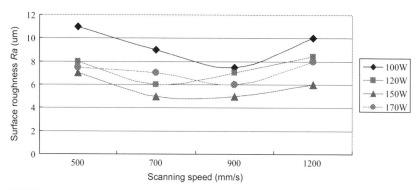

FIGURE 5.92 Effects of selective laser melting process parameters on the surface roughness of single pass.

the molten pool disconnects. The melting channel formed after cooling shows unevenness or even disconnection, as shown in Fig. 5.91D.

As shown in Fig. 5.92, with the increase of laser power, the surface roughness of single-melting channel undergoes a process of first falling and then rising. When the power is increased from 100 to 150 W, the surface roughness has an improved surface quality. This is because during the process of increasing the laser energy, the temperature in the molten pool formed by the melting of the metal powder is gradually increased, and a result of the rise in temperature is a decrease in the surface tension of the metal, and the molten metal can be spread flat on the base plate. The surface tension becomes smaller, which also reduces the occurrence of spheroidization and the unevenness of the spherical surface of the melted surface due to spheroidization. When the laser power is 170 W, the surface roughness of the melting channel becomes larger than that when the power is low. When the laser power is 150 W, the surface roughness of the melting channel is relatively lower. When the laser power is constant, the surface roughness of the melting channel shows the trend of decreasing first and then increasing as the scanning speed increases. The roughness is the lowest at a scanning speed of 900 mm/s.

5.4.2.1.2 Surface roughness of a single layer

For single-layer scanning, the factors affecting the surface roughness are mainly the surface quality of the melting channel along the scanning line direction and the surface quality perpendicular to the scanning line. The surface roughness perpendicular to the scanning line is generated because the scanning lines overlap each other to form a surface, while the cross-section of the scanning line is not a completely flat plane, and the cross-section of solidified melting channel is approximately a circle, so that the edges of various melting channels are overlapped to form a gully-like surface between the melting channels. From the single-channel scanning experiment, the better parameters are obtained for the surface scanning experiment. A single-pass scanning forms a melting channel, and adjacent scanning lines of the same layer are in contact to complete the layer manufacturing. The scanning pitch is set to 0.06 mm, the laser power is 120–180 W, and the scanning speed is 450–900 mm/s.

5.4.2.1.2.1 Impact of power and speed Fig. 5.93 shows the roughness of single-layer scanning surfaces measured along the direction of scanning line changing with the power and velocity (Fig. 5.93A). The surface roughness measured in the direction along the scanning line is mainly determined by the surface roughness of single-melting channel. Of course, when performing layer scanning, there is effect between the melting channels, which is unlike the single-channel scanning. During layer scanning, due to the overlap between the melting channels, at the edge of the melting channel, the illumination of the laser causes the previously solidified melting channel to change. It can be seen from the figure that the change of surface roughness is a process of decreasing first and then increasing with the increase of scanning speed. The surface roughness is the lowest at a scanning speed of 600 mm/s. When the scanning speed is increased from 600 to 900 mm/s, the surface roughness increases, but the surface roughness value is the largest at a scanning speed of 450 mm/s. The value of surface roughness decreases with increasing laser power, which is consistent with the results of surface roughness in single-channel scanning in Fig. 5.92.

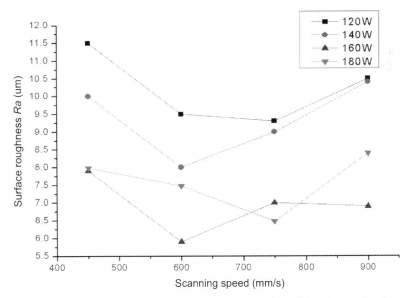

FIGURE 5.93 Curve of effects of power and speed measured parallel to the scanning line on surface roughness when the scanning pitch is 0.06 mm.

Fig. 5.94 shows the surface roughness perpendicular to the scanning direction changing with the process parameters. It can be seen that the scanning speed has less influence on the surface roughness perpendicular to the scanning direction, and the surface roughness is slightly lower when the scanning speed is low. The surface roughness decreases as the laser power increases. The surface roughness is the lowest when the laser power is 180 W and the scanning speed is 750 mm/s. The low scanning speed increases the time during which the molten metal is present in the molten pool, thereby providing more time to level the peaks of the previously formed melting channels and reduce the height difference between the peaks and valleys of the melting channels. In fact, the generation of surface roughness perpendicular to the direction of the scanning line is largely caused by the peaks and valleys formed between the melting channels. The higher laser power forms flatter molten pool, and the height difference between the peaks and valleys formed by the interconnection of the flat melting channels is relatively small. The high-energy density can increase the wettability of the melt, and the improvement of the wettability can effectively reduce the surface tension and the spheroidization tendency, forming a smooth surface of the melting channel. However, if the laser power is too large, the temperature in the molten pool will be too high, which will cause vaporization of the metal. The back-flushing pressure caused by vaporization will destroy the surface of the molten pool and increase the surface roughness.

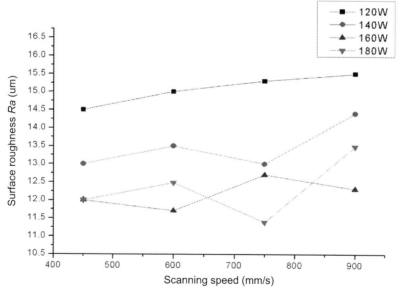

FIGURE 5.94 Fig. 5.93 Surface roughness measured vertical to the scanning line direction when the scanning pitch is 0.06 mm.

5.4.2.1.2.2 Impact of the scanning strategy The measurement direction of surface roughness is perpendicular to the scanning line direction, and the surface roughness measurement takes its *Ra* value as a reference.

Fig. 5.95 shows the surface morphology and surface profile of the parts in different scanning modes. The forming method of Fig. 5.95A is a single scanning method on the same direction, that is, the laser is scanned on one layer along a single direction; the forming method of Fig. 5.95B is spiral scanning method from the outside to the inside. The forming method of Fig. 5.95C is a variable direction scanning mode, that is, after the first scanning is completed in one direction, then the powder is leveled, and the second scanning is carried out along the vertical direction of the last scanning direction. The forming method of Fig. 5.95D is a one-way two-filling mode, as shown in Fig. 5.96, the scanning pitch of this scanning mode is set to be twice of the normal scanning pitch in the first scanning, and the working plane remains stationary after scanning is completed, with the powder leveling roller performing powdering once layering. The scanning pitch is set to a normal value during the second scanning, and the scanning direction is unchanged.

Where *Ra* is the arithmetic mean of the absolute value of the contour offset over the length of the sample. Its calculation method is:

$$Ra = \frac{1}{n}\sum_{i=1}^{n} |yi| \qquad (5.78)$$

According to the optical micrograph of Fig. 5.95, the surface of the single scanning of the same direction and the spiral scanning method is rough, while the surface quality of single scanning of variable directions and two scanning of single-layer with variable scanning pitches is better, in which the surface quality of two scanning of single-layer with variable scanning pitches is the best, the surface of the melting channel is smooth, and the connection between the melting channels is flat. From the surface roughness value *Ra* measured in Table 5.13, the surface roughness of the single scanning of the same direction is up to 12.66 μm, and the surface roughness of two scanning of single-layer with variable scanning pitches is as low as 6.75 μm. After analysis, the main reason is that in single scanning of the same direction, the laser starts and ends the scanning at the same position. When the scanning of the upper layer is completed, the position of the laser scanning is at the highest position of the melting channel of the last melt forming in the lower layer scanning, and the layers are accumulated, the position which is already high becomes higher, and the uneven surface formed by the multilayer forming is more obvious, increasing the surface roughness. The spiral scanning method is similar to the single scanning of the same direction. For single scanning of variable scanning directions, since the scanning direction is perpendicular to the first scanning during the second layer scanning, the convex-concave phenomenon between the melting channels formed by the first scanning is not further enlarged, so the surface quality is

514 Materials for Additive Manufacturing

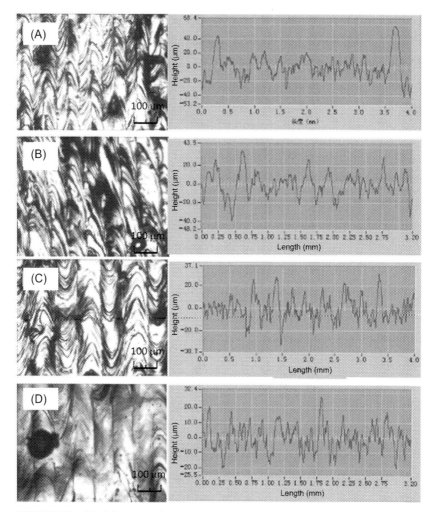

FIGURE 5.95 Single-layer scanning surface morphology and surface contour of different scanning modes observed by optical microscope. (A) Single scan in the same direction; (B) Spiral scan; (C) Single scan with variable directions; (D) Two scans with variable scanning pitches.

good. The surface quality of two scanning of the single layer with variable scanning pitches is the best. In the single-layer variable-pitch scanning, the scanning pitch in the first scanning is wider, which is twice of the normal scanning pitch. After the scanning is completed, the powder leveling is repeated, so that the powder is filled in the depression between the melting channels. At the secondary scan, when the laser is scanned to the top of the previous melting channel, the surface of the melting channel is heated by the laser since there is no powder, the burr and micro balls are melted and become smooth. When the scanning pitch is half of the previous one, the laser beam will illuminate the depression

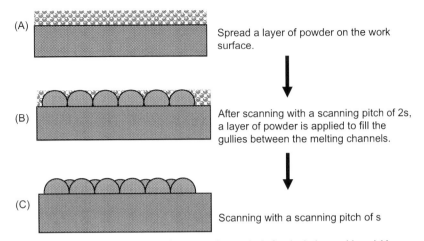

FIGURE 5.96 Schematic diagram of two scanning methods for single-layer with variable scanning pitches.

TABLE 5.13 Surface roughness data of different scanning methods.

Scanning method	Arithmetical mean deviation of the profile (μm)
Single scan in the same direction	12.66
Spiral scan	9.53
Single scan with variable directions	7.52
Two scans with variable scanning pitches	6.75

between the two melting channels, and the molten powder compensates for the depression between the melting channels, reducing the height difference between the melting channels and the surface roughness.

5.4.2.1.2.3 Impact of the overlapping ratio The scanning pitch S refers to the distance of the center of the laser spot of two adjacent scanning lines. The main purpose of the surface scanning experiment is to determine a good scanning pitch. One concept related to scanning pitch is the "overlapping ratio." The overlapping ratio refers to the percentage of the overlap width D_w and the scanning pitch S between two adjacent melting channels. As shown in Fig. 5.97, W is the width of the molten pool, S is the scanning pitch, and D_w is the overlap amount. "Overlap" refers to the area where two single-pass pools overlap, and "overlapping

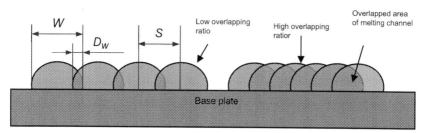

FIGURE 5.97 Schematic diagram of overlapped melting channel.

ratio" is defined as the ratio of the width of the overlap area to the width of the corresponding molten pool. Different scanning pitches result in different overlapping ratios of the melt channels and different forming qualities.

The overlapping ratio is defined as:

$$\eta = \frac{D_w}{W} \times 100\% = \frac{W-S}{W} \times 100\% = \left(1 - \frac{S}{W}\right) \times 100\% \quad (5.79)$$

The scanning pitch S should be set in actual SLM manufacturing. According to the above equation, we can obtain:

$$S = (1 - \eta) \times W \quad (5.80)$$

For example, when the overlapping ratio is smaller, it means that the overlapped area is smaller, liquid metal mixes less, so that the bond between the molten pools is weaker. When the overlapping ratio is 0, that is, the scanning pitch S is equal to the melting width W, it means that the single-channel molten pools are in a critical separation state, and there is almost no bonding at this time. When the overlapping ratio is 50%, it means that the distance between adjacent laser scanning routes is just half of the melting width. When the overlapping ratio is about 30%, the bond between the single-channel molten pools is better. Therefore, the overlapping ratio η is designed to be 10%, 20%, 30%, 40%, 50%, 60% and 70%. The scanning pitch is set to obtain an accurate overlapping ratio from the previously determined width data of the melting channel under certain process parameters, shown in Tables 5.14 and 5.15.

It can be seen from Fig. 5.98 that the roughness value increases with the increase of the scanning pitch, and the reason is that the overlapping ratio is the main factor affecting the surface roughness of the horizontal plane. When the overlapping ratio is 40%, the surface roughness value is the smallest. As the overlapping ratio increases, the surface roughness value increases and the surface quality decreases. The overlapping relationship between the two cladding channels is shown in Fig. 5.97. Due to the effect of remelting expansion, the height of the remelted parts of the two cladding channels must

TABLE 5.14 Scanning pitch at different overlapping ratios.

Laser power (W)	Scanning speed (mm/s)	Melting channel width (μm)	Overlapping ratio (%)	Scanning pitch (μm)
160	750	97.2	10	87.48
			20	77.76
			30	68.04
			40	58.32
			50	48.6
			60	38.88
			70	29.16

TABLE 5.15 Theoretical values of the best overlapping ratio with different process parameters.

Laser power (W)	Scanning speed (mm/s)	Melting channel width (μm)	Melting channel height (μm)	Theoretical optimal scanning pitch (μm)	Overlapping ratio (%)
120	450	106.75	35.14	75.66	12.68
	600	97.21	38	62.77	19.26
	750	69.93	45.9	44.97	29.17
	900	63.53	54	42.46	29.88
140	450	110.00	31	84.46	4.89
	600	94.35	36	61.78	18.45
	750	93.65	43.95	57.53	24.28
	900	65.04	53.38	43.66	29.92
160	450	115.51	32.5	87.99	4.80
	600	111.43	31	81.97	4.15
	750	97.20	38.52	62.76	19.71
	900	89.39	40.52	54.82	23.47
180	450	125.51	30.7	105.22	3.03
	600	116.92	34.84	99.53	7.90
	750	105.37	31	75.39	7.24
	900	101.84	46.66	63.21	23.73

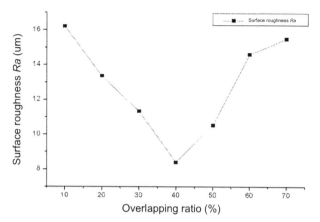

FIGURE 5.98 Relationship between the overlapping ratio and surface roughness when the laser power is 160 W and the scanning speed is 750 mm/s.

increase. If the height is increased, it is equal to the height of the single cladding channel, and the effect is the best; if it is larger or smaller than the height of the single cladding channel, it will inevitably affect its surface roughness.

It can be seen from the optical micrograph that when the scanning pitch is large, the corresponding overlapping ratio is small, such as hen the overlapping ratio is 10% or 20%, and the "groove" formed between the scanning lines is obvious, and the scanning surface is uneven—flatness influences the bonding between layers, affecting the powder leveling, and thus affecting the quality; when the scanning distance is moderate, the corresponding overlapping ratio is 30% or 40%, and part of the "groove" is filled and the surface quality is good; when the scanning pitch is small, that is, the overlapping ratio is 50%, the alignment between the scanning lines becomes tight, but the "groove" is not further weakened.

Fig. 5.99 shows the morphology of the melt-formed surface when the overlapping ratio is set to 50%, 30%, and 0. As can be seen from the figure, when the laser power is 150 W, the scanning speed is 500 mm/s, and the overlapping ratio is 30%, the surface is the more flat. When the overlapping ratio is 50%, the surface is excessively overlapped, and the raised ridges appear at the overlapping portions of the melting channels, which reduces the flatness of the surface. When the overlapping ratio is 0, there is almost no connection between the melting channels, and a few places are connected, but most of them are separated from each other.

As can be seen from Fig. 5.97, the overlapping ratio between the melting channels directly determines the surface quality of the molten metal part. On the surface, the greater the overlapping ratio between the melting channels, the smaller the gully between the melting channels and the smoother the

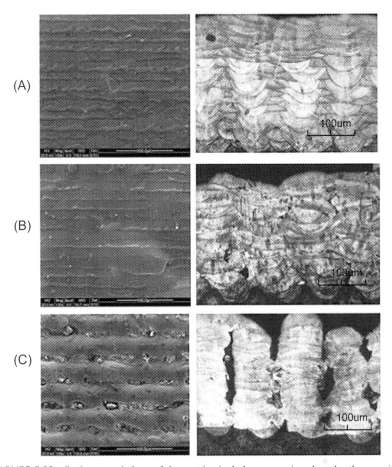

FIGURE 5.99 Surface morphology of the part in single-layer scanning shoot by the scanning electron microscope and profile morphology of the part in the height direction observed by the optical microscope (laser power: 150 W, scanning speed: 500 mm/s). (A) Overlapping ratio = 50%; (B) overlapping ratio = 30%; (C) overlapping ratio = 0.

surface. But in reality, the overlapped area is not as big as possible. When the overlapping ratio is 30%−40%, the gully phenomenon can be weakened, so that the surface of the scanned layer is scanned. From this point of view, if the overlapping ratio is too small, it will obviously form an undulating wave shape, that is, a wave peak formed on the top of the melting channel, and a valley formed at the junction of the melting channel. If the overlapping ratio is too large, although the peak and valley phenomenon can be avoided, the surface quality is not improved due to excessive overlapped area, and on the contrary, it is decreased, and the manufacturing process is time consuming due to an excessive overlapping ratio.

The surface quality is not simply determined by a single parameter. For the melting channels with the same width, the most reasonable overlapping ratio is not necessarily the same, and it also depends on the morphology of the melting channel, mainly including the width and height of the melting channel. We can theoretically analyze the width and height parameters of melting channel to determine the reasonable overlapping ratio, and finally determine the scanning pitch by the overlapping ratio and the width of the melting channel. In the melting channel, the laser melts the metal powder to form liquid that solidifies on the formed substrate under the action of gravity. Due to the surface tension, the molten metal tends to form a spherical shape, and the morphology of the melting channel should be close to part of the cylinder. For the convenience of calculation and analysis, we consider the surface of the cross section of melting channel as part of a circular arc, the shape of which is shown in Fig. 5.100.

Let the radius of the circle be R, then according to the Pythagorean theorem, the following can be obtained:

$$R^2 = \left(\frac{W}{2}\right)^2 + (R-H)^2$$
$$R = \frac{(W^2 + 4H^2)}{8H} \tag{5.81}$$

In the figure, zone I is the overlapping zone of two melting channels, and zone II is the groove of two melting channels. It is envisaged that if the area of zone I and the area of zone II are exactly equal, then the metal of the overlapping zone can just fill the groove of the melting channel, so that the gullies between the melting channels disappear, and the surface formed will be smooth and flat. When the area of zone I = the area of zone II, the area of the rectangular, abcd, is exactly equal to the area of an arch.

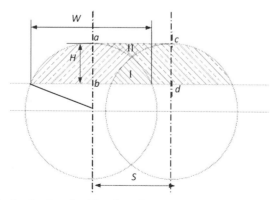

FIGURE 5.100 Overlapping of melting channel.

The area of the arch is obtained by simplifying the Simpson formula:

$$A = \frac{H(a + 4b + c)}{6} \tag{5.82}$$

where a is 2R,

$$b = 2\sqrt{R^2 - \frac{h^2}{4}}, \quad c = 0 \tag{5.83}$$

Then

$$H \cdot S = \frac{H\left(2R + 8\sqrt{R^2 - \frac{H^2}{4}} + 0\right)}{6} \tag{5.84}$$

$$S = \frac{\left(2R + 2\sqrt{R^2 - \frac{H^2}{4}}\right)}{6} \tag{5.85}$$

$$R = (W^2 + 4H^2)/8H \tag{5.86}$$

Comparing the data obtained from the table with the experimental results, it is found that the overlapping ratio of melting channel calculated theoretically is smaller than the best overlapping ratio measured in the experiment. According to theoretical calculations, when the laser power is 160 W and the scanning speed is 750 mm/s, the optimal overlapping ratio is 20%. In Fig. 5.98, it can be seen that the surface roughness is minimal when the overlapping ratio is about 40%. However, when the overlapping ratio is 20%, the surface roughness is slightly larger than that the overlapping ratio is 40%, and the difference is not obvious. Therefore, the optimum overlapping ratio calculated by this method has certain reference value. The error mainly comes from that we consider the surface of the melting channel as a part of the cylinder, but in fact its shape is closer to half of the ellipse. When calculating the radius of the circle, there is a certain error, which leads to a difference in the calculation results. On the other hand, since the molten pool is formed after laser melting of the powder, the molten pool absorbs the powder at the edge portion, as shown in Fig. 5.95. Fig. 5.101 shows that when the laser melts the powder, the edge powder is absorbed, and the adjacent melting channel will lead to less metal liquid formed due to few powders melted. In practice, when the overlapping ratio is small, the powder melted by the laser becomes small, and it is not enough to fill the gully between the melt channels, and the overlapping ratio rate must be increased to fill in the gully between the melting channels, which is one reason why the actual overlapping ratio is higher than the theoretical calculation.

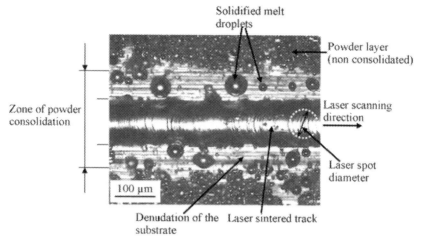

FIGURE 5.101 Powder at the edge of the melting channel being absorbed.

5.4.2.1.3 Side surface roughness of the part

For a part of simple geometric shape, its surface can be simply divided into two types: one is the surface horizontal to the working plane (XY plane), that is, the angle formed by the plane and the work surface is 0, which we call the upper surface, the other is a surface that is not parallel to the working plane, and the angle between the surface and the work surface is between 0 and 90 degrees, which we call the side surface. The surface quality of the upper surface has been studied in the previous section. This section mainly studies the influence law of surface quality when the angle between the surface of the part and the work surface is in the range of 0–90 degrees. The surface on which the surface of the part is above the powder is called the A-side, and the surface on which the surface of the part is below the powder is called the B side. When the inclination of the part is less than 30 degrees without support, the part is difficult to form, so this experiment sets the inclination from 40 to 90 degrees.

5.4.2.1.3.1 *Impact of the tilt angle*
Fig. 5.102 shows the morphology of the A-side surface of the part below the powder layer at different angles. It can be seen from the figure that the surface is rough when the inclination angle is 40°. The surface roughness is mainly formed by two aspects: one is a powder having bonding surface, the other is the step effect formed between the layers when the slices are layered manufactured, and the formation of the steps increases the surface roughness of the parts. As the tilt angle increases, the surface quality of the part becomes better, mainly because the step effect becomes weaker as the tilt angle increases, and the factors affecting the surface quality are mainly the powder adhered to the surface. It can be seen in

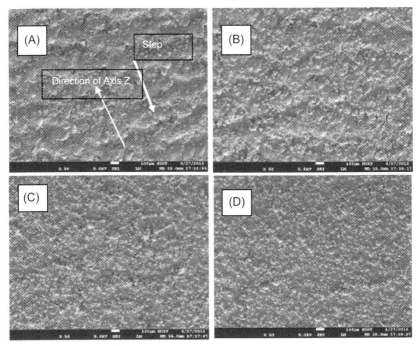

FIGURE 5.102 A-side morphology of different inclinations of parts. (A) Tilt angle: 40 degrees; (B) tilt angle: 60 degrees; (C) tilt angle: 80 degrees; (D) tilt angle: 90 degrees.

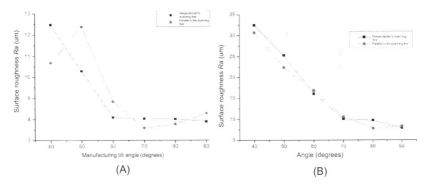

FIGURE 5.103 Surface roughness of parts with different tilt angle when the laser power is 120 W and the scanning speed is 750 mm/s. (A) Roughness of A surface (B) Roughness of B surface.

Fig. 5.102D that there are no visible uneven steps except some bonded powders on the surface.

Fig. 5.103 shows the curve which shows how the surface roughness and forming angle for both surfaces of the part change. The roughness of the sample with different tilt angles parallel to the scanning line direction and

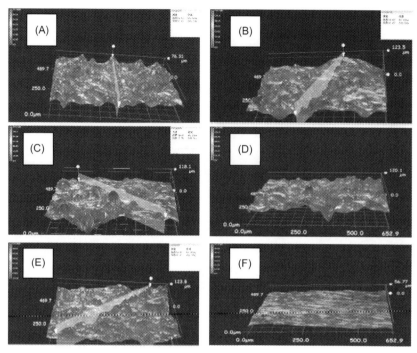

FIGURE 5.104 Morphology of A-side surface of the part taken by the 3D super depth of field optical microscope when the laser power is 120 w and the scanning speed is 750 mm/s. (A) Tilt angle: 40 degrees; (B) Tilt angle: 50 degrees; (C) Tilt angle: 60 degrees; (D) Tilt angle: 70 degrees; (E) Tilt angle: 80 degrees; (F) Tilt angle: 90 degrees.

perpendicular to the scanning direction is measured. As can be seen from the figure, the roughness of A and B side surfaces decreases as the inclination angle increases. The roughness perpendicular to the scanning line is slightly larger than the roughness parallel to the scanning line. Especially when the tilt angle is small, the roughness perpendicular to the scanning line is larger than that of the parallel to the scanning line, which indicates that the step effect has great influence on the surface quality when the tilt angle is small; when the tilt angle is between 70–90 degrees, the roughness of A and B side surfaces has few difference; when the tilt angle is smaller, the roughness of the B side surface is increased more than the roughness of the A side surface (Figs. 5.104 and 5.105). This is because the B side surface is located in the upper part of the powder layer, when the laser scans, melting is carried out directly on the powder layer without the metal substrate. Since the heat transfer performance of the powder layer is inferior to that of the metal body, the molten metal cooling time is greater than the cooling time of the substrate, and a large amount of metal powder is adhered below during the cooling of the molten metal, resulting in serious deterioration of the

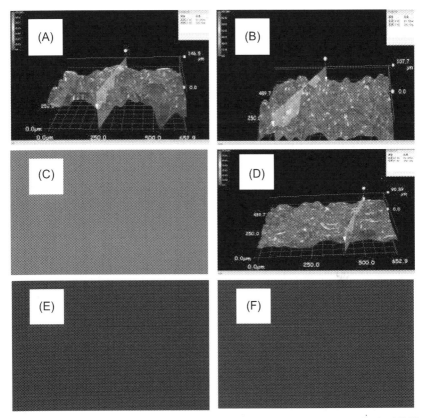

FIGURE 5.105 Morphology of B side surface of the part taken by the 3D super depth of field optical microscope when the laser power is 120 w and the scanning speed is 750 mm/s. (A) Tilt angle: 40 degrees; (B) Tilt angle: 50 degrees; (C) Tilt angle: 60 degrees; (D) Tilt angle: 70 degrees; (E) Tilt angle: 80 degrees; (F) Tilt angle: 90 degrees.

surface quality. The smaller the tilt angle is, the more obvious this trend is. When the tilt angle is 70–90 degrees, the area of direct sintering and melting on the powder layer of the B side surface is small, even close to being melt-formed on the metal substrate, and the surface roughness is only affected by the step effect, which is similar to the A side surface. Therefore, there is only little difference between the roughness of Side A and that of Side B.

5.4.2.1.3.2 Impact of laser power In the previous study of single-channel scanning and single-layer scanning, it is found that different laser powers have an effect on the surface quality. The main reason is that the temperature and cooling time of the molten pool formed by the laser melting of the metal powder change under different laser powers. The numerical value of the surface roughness is simplified, and the roughness data parallel to the scanning line and perpendicular to the scanning is averaged to obtain the integrated surface roughness.

Fig. 5.106 shows the change of roughness of the two surfaces of the part when the scanning speed is 750 mm/s and the laser power is set to 120 and 140 W, respectively. It can be seen from the measurement results that when the laser power is 140 W, the surface roughness of the surface under the lower powder layer, that is, the A-side surface, is better than the surface roughness when the laser power is 120 W, and when the laser power is increased, the surface quality of the A-side surface is improved. Because on the A-side surface, the surface roughness is mainly affected by the step effect. When the laser power is high, the morphology of the melting channel tends to be flat. As shown in Fig. 5.107, the melting channel structure is flat, and the edge curves of the melting channel are connected to be flatter, thus the side surface formed has small undulation and small surface roughness. On the B side surface, when the laser power is 140 W, the surface roughness

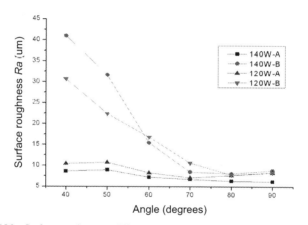

FIGURE 5.106 Surface roughness at different powers when the scanning speed is 750 mm/s.

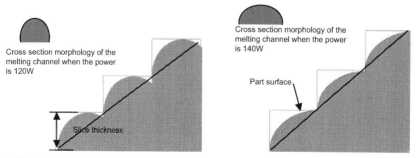

FIGURE 5.107 Schematic diagram of the influence of different melting channel geometry on the side surface morphology of the part when the tilt angle is 45 degrees.

value is larger than that when the power is 120 W, especially when the inclination is less than 60 degrees. Because when the inclination is small, the surface roughness of the B side surface is mainly affected by the factor that after the laser melts the powder, the molten metal liquid bonds the powder on the powder bed to the bottom. When the power is larger, the penetration depth is greater, and the more powder is bonded, so that the roughness value of the formed surface.

5.4.2.1.3.3 Impact of processed layer thickness In the SLM forming process, in addition to the laser process parameters and surface tilt angle have an impact on the surface quality of the part, thickness level of the powder will also affect the surface roughness of the part. In theory, the thinner the powder leveling, the smaller the step formed on the side. When the thickness of the powder leveling is close to infinity, the step on the side will disappear. No matter how large the tilt angle of the side, the surface will be a smooth curve. However, in actual production, depending on the production efficiency, the powder characteristics and the interaction between the parameters, there is a most reasonable value for the thickness of the powder leveling. It not only ensures production efficiency, but also guarantees the surface quality of the parts. To this end, this experiment set the processing parameters of different layer thicknesses to study the surface quality changes of parts with different layer thickness and strive to obtain the most suitable process parameters.

The parts shown in Fig. 5.108 are designed. In this part, four different manufacturing angles are designed, namely 45, 60, 75, and 90 degrees, to study the influence of the manufacturing angle on the surface quality of the formed part. In forming, the scanning direction is parallel to the short side of the cross-section, and the thickness of the powder layer is also divided into three types, namely 0.02, 0.04, and 0.06 mm, respectively, and the position of the red line in the figure is about the position where the thickness of the layer is changed. The thickness of the powder leveling increases from bottom

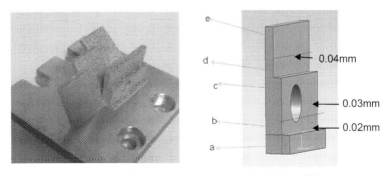

FIGURE 5.108 Setting different layer thicknesses for manufacturing at different angles.

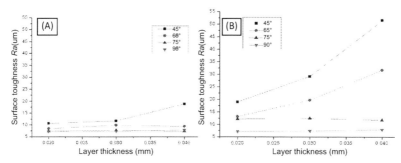

FIGURE 5.109 Effects of different powder leveling thickness on the roughness of the upper surface.

to top, and the design can be used to study the law about how the thickness of the powder leveling affects the surface quality of the formed part.

The change in the thickness of the powder leveling does not have much influence on the roughness of the upper surface, and the surface roughness increases only slightly when the thickness of the powder leveling is increased. This is mainly because the increase of the thickness leads to an increase in the molten metal formed in the molten pool formed by single melting, and the height of the melting channel formed after solidification of the molten metal pool is increased, and the gully formed after the overlapping of the formed melting channels becomes deep, so that the roughness is increased. However, this effect is not very obvious.

Fig. 5.109 shows the effects of different thickness of powder leveling on the roughness of the two surfaces of the part. As can be seen from the figure, the increase in the thickness of the powder leveling will result in an increase in the surface roughness, and the greater the tilt angle, the more the surface roughness increases. Especially for the surface above the powder layer (the angle formed by the substrate is an acute angle), the increase in the thickness of the powder leveling is more pronounced on the surface roughness. When the tilt angle of the part is 45 degrees, the thickness of the layer is increased from 0.02 to 0.04 mm, and the surface roughness is increased from 18.9 to 51.5 μm. The change is very obvious. While on the surface under the powder layer (the surface formed with the substrate is an obtuse angle), when the title angle changes from 60 to 90 degrees, the change of the roughness is small, which is below 5 μm. The change in roughness when the tilt angle is 45 degrees is slightly more obvious. When the thickness of the powder leveling is increased from 0.02 to 0.04 mm, the surface roughness is only increased by about 8 μm. The reason is similar to the effect of the front tilt angle on the surface roughness, which is mainly due to the step effect produced by the slicing. When the slice is thicker, the step effect is more obvious. Especially for the surface at an acute angle to the substrate, the increase of the step causes the increase in the area where the laser directly scans on the powder bed, and more powder is adhered, resulting in a decrease in surface roughness.

5.4.2.2 Dimensional accuracy

Compared to other traditional manufacturing methods, SLM technology has the advantage of being able to quickly manufacture complex and delicate parts, so dimensional accuracy is very important for SLM-formed parts. The dimensional accuracy of SLM-formed parts is affected by powder characteristics, process parameters, and equipment hardware conditions. The characteristics of the powder include composition, particle shape and particle size distribution. Process parameters include parameters such as laser power and scanning speed. Equipment hardware conditions include the mechanical motion accuracy of the equipment, the size of the laser spot, and so on. The hardware conditions are difficult to be optimized and improved since they have been fixed, we only study conditions that can be changed and optimized: powder characteristics and process parameters. According to the process characteristics of the SLM technology, the parts of the SLM process are actually accumulated by numerous melting channels. In the XY plane, its accuracy is mainly affected by the change of edge powder absorption and the width of the melting channel. In the scanning path planning software, the scanning path of the most edge has been appropriately expanded to facilitate post processing. In the Z-axis direction, it is mainly affected by the cumulative error between layers. Since there is an error in the set layer thickness and the actual layer thickness formed after melting, this error causes the accuracy of the z-axis direction of the part to decrease. In this chapter, the influence law of process parameters and powder characteristics on the accuracy of parts is studied to propose the methods for improving the machining accuracy of parts.

5.4.2.2.1 Machining error of XY plane

5.4.2.2.1.1 Influence of process parameters on XY plane accuracy
Since the size and morphology of the melting channels formed during the manufacturing process vary with different process parameters, in this test, the blocks with a side length of 10 mm are fabricated using nine process parameters to study the dimensional change of XY plane under different process parameters, shown in Fig. 5.110. It can be seen from the previous analysis that the change of the laser process parameters causes the width of the melting channel to change, resulting in a change in the size of the formed entity, and the influence is mainly manifested in the change of the width of the melting channel at the most edge.

As can be seen from Table 5.16, there are slight differences in the size of the blocks produced by different process parameters. This difference is mainly caused by the influence of laser process parameters on the width of the melting

FIGURE 5.110 Block samples manufactured at nine process parameters.

TABLE 5.16 Process parameters and block size.

Item	Laser power (W)	Scanning speed (mm/s)	Pitch(mm)	Measured size (mm)
1	140	600	0.06	10.010
2		750		10.007
3		900		10.006
4	160	600		10.014
5		750		10.013
6		900		10.011
7	180	600		10.016
8		750		10.014
9		900		10.014

channel. Because the manufacturing process software calculates the scanning path for the part surface in the same manner for the parts with the same size and shape, the change in size can only be caused by the width of the most edge scanning pass. When the laser energy density is low, the width of the melting channel is small, and the contour size of the manufactured part is small. When the laser energy density is large, the width of the melting channel is large, and the contour of the formed part is slightly larger. In fact, the influence of the width of the melting channel on the contour size of the part is relatively small. In addition, when the energy density is large, the edge sticking powder is more serious, resulting in a larger measurement size.

5.4.2.2.1.2 Machining error of thin-walled parts The dimensional accuracy of the manufactured part mainly includes the dimensional accuracy on the XY plane and the accuracy of the Z-axis height direction, which mainly affect the dimensional accuracy of the part. Thin walls having wall

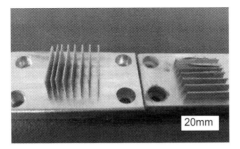

FIGURE 5.111 Thin wall of different thicknesses processed.

thicknesses of 0.1, 0.15, 0.2, 0.25, 0.3, 0.35, 0.4, and 0.45 mm are produced in this test. Since the spot size of the laser is close to 100 μm, when processing a thin wall of 0.1 mm thick, the laser is scanned a pass and melted to form a thin wall. Two process parameters are used for processing with the laser power of 140 and 160 W, respectively, and the scanning pitch of 0.06 mm. When the laser power is 140 W, the strength of the machined parts is not enough. During the manufacturing process, the powder leveling roller is encountered to cause scratch, resulting in the parts of smaller thickness being scratched and deformed, deviating from the original position, and defects appear in subsequent manufacturing, as shown in Fig. 5.111.

A cross-sectional optical microscope image of thin-walled part machined with a laser power of 160 W and a scanning speed of 800 mm/s is shown in Fig. 5.112. The thickness of the thin-walled parts is measured and analyzed using the dimensional measurement software of optical microscope as shown in Fig. 5.113.

As can be seen from Fig. 5.113, when the set wall thickness is 0.1 mm, the wall thickness of the actual formed part is 0.199 mm, which exceeds the set size by nearly 100%. However, as the set wall thickness increases, the difference between the actual formed wall thickness and the set wall thickness decreases. When the wall thickness is 0.45 mm, the error between actual formed size and the set size is only 2%. As shown in Fig. 5.114, when the wall thickness of the set part is larger, the dimensional error of the final machined part is smaller, and the machining error is large when the part with a size smaller than 0.2 mm is processed. This is because the wall thickness is close to the laser spot size, but larger than the laser spot diameter. When scanning, the set scanning line trajectory has an area slightly larger than the set size after overlapping. The scanning pitch is set to 0.06 mm, and when processing the thin wall with a size of 0.1 mm, since the laser spot is smaller than the set size, in order to compensate, the laser actually scans two lines. From the data of the single-pass scan analysis performed above, it can be seen that when the laser power is 160 W, the

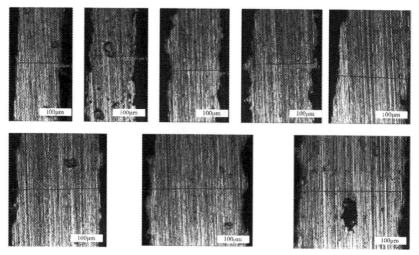

FIGURE 5.112 Optical microscopy measurement of wall thickness: laser power: 160 W, scanning speed: 800 mm/s, scanning pitch: 0.06 mm.

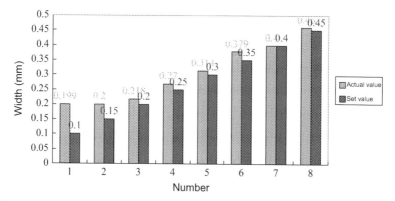

FIGURE 5.113 Comparison of the set wall thickness and the wall thickness with of the actual formed part.

scanning speed is 750 mm/s, and the width of the melting channel is about 95 μm, the interval between the melting channels is 0.06 mm, and the size of the actual machined parts should be 155 μm, which is 0.155 mm. Besides, since the powder particles bonded to the edge of the melting channel has influence on the measured value, it is close to the actual measured result of 0.199 mm. When machining a part with large wall thickness, the influence brought by the edge error generated by the scanning path planning and the powder bonding is smaller.

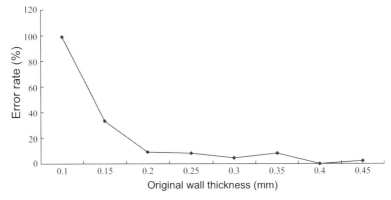

FIGURE 5.114 Relative error of different wall thickness processed.

Model	PowerScanner-STD (standard type)
Single amplitude measurement range	400x300 nun²
Single amplitude measurement point	1,300,000
Measurement accuracy	0.05nmi
Point spacing	0.3 5 nun
Single amplitude measurement time	1.5 seconds
Output file format	ASC, PLY

FIGURE 5.115 3D contour measurement software.

5.4.2.2.1.3 3D scanning measurement of manufactured parts

The 3D measuring device in Fig. 5.115 is a high-precision 3D measuring system, PowerScanner-STD, developed by the Rapid Prototyping Center of Huazhong University of Science and Technology. The system pre-generates structured light with sinusoidal distribution of light intensity in the horizontal direction in the computer, and then projects the grating onto the measured object to be synchronously collected by the camera. The phase demodulation, expansion and 3D reconstruction are carried out to the captured images to obtain 3D morphology of the measured object. The measurement speed of the system is fast and tens of thousands to millions of 3D points can be obtained in seconds. It is featured

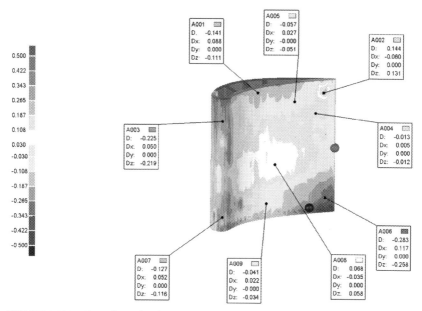

FIGURE 5.116 Three-dimensional measurement of SLM-formed turbine blades.

by high measurement accuracy, with single amplitude measurement accuracy of 0.05 mm, and has good portability. The device is simple in structure and suitable for complex on-site measurements. It can be widely used in industrial inspection, reverse design, cultural relics reproduction and biomedical fields under conventional sizes (several millimeters — several meters).

Place the manufactured turbine blades on a workbench, scan the 3D contour data by using the scanning system, and then compare the scanning result with the original 3D data. The comparison result is shown in Fig. 5.116. From the measurement results, the maximum size error of the turbine blade is within 0.4 mm, the maximum size of the turbine blade is 110 mm, and the relative dimensional error is within 0.4%, thus the result is ideal. The maximum deviation occurs at the sharp corners of the blade, where the part is slightly deformed due to heat concentration, and the dimensional error of other parts is relatively uniform.

5.4.2.2.2 Dimensional changes at different processing positions

Since the SLM forming technology uses a scanning galvanometer to scan the focused laser beam according to a programmed scanning trajectory. The position of the galvanometer is fixed and the beam moves within the forming area as the lens is reflected. At the center of the forming area, the beam is perpendicular to the working plane and has the shortest length. At the edge of the forming area, the beam is not perpendicular to the working plane, the

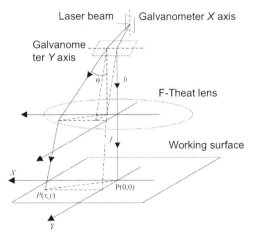

FIGURE 5.117 Schematic diagram of scanning galvanometer.

beam becomes longer, and the focal plane of the laser may change. The forming system in this test scans in front of galvanometer laser objective lens of the F-theta lens, which requires the incident laser beam to be parallel light, and the focal plane is at an ideal focal length. In practice, the laser beam undergoes optical changes and transmission over long distances, it is difficult to ensure that the incident laser beam is parallel, resulting in an indeterminate focal plane. As shown in the figure, after the incident laser beam is reflected by the X-axis and Y-axis of the galvanometer, it is focused on the working surface by the F-Theta lens. Ideally, the distance (L) from the focus to the center of the work plane satisfies the following relationship: $L = f \times \theta$, as shown in Fig. 5.117.

In order to test the machining accuracy at various positions in the machining plane, the test of Fig. 5.118 is designed. The size of the formed block is set to 11 mm, the diameter of inner circle is 10 mm, and the projected cylindrical diameter is also 10 mm. The parts are distributed along the diagonal line in the machining plane. The center coordinate of the machining plane is (0,0), and the value in the coordinate XY represents its distance from the center in mm. The coordinates of the test samples from No. 1 to No. 6 on the machining plane are (−80, 80), (−50, 50), (−20, 20), (20, −20), (50, −50), and (80, −80). The dimensions of each part are measured after the processing. The size results are shown in Table 5.17.

As can be seen from Table 5.17, the size of the formed block is slightly larger than the set size, the inner circle size is smaller than the set size, and the cylindrical size is larger than the set size. This is because the forming process software compensates for the size of the edge, resulting in the actual forming size slightly larger than the set size, which facilitates subsequent processing, leaving a little machining allowance. On the other hand, the

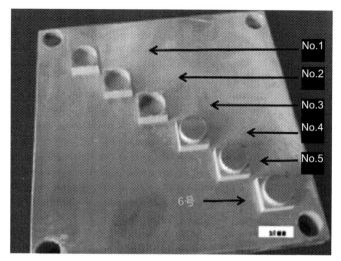

FIGURE 5.118 Collaboration diagram of machining parts at different positions.

TABLE 5.17 Dimensions of each part.

No.	Item	Set size (mm)	Measured size (mm)
1	Rectangular side length	11	11.014
	Diameter of inner circle	10	9.950
2	Rectangular side length	11	11.102
	Diameter of inner circle	10	9.912
3	Rectangular side length	11	11.124
	Diameter of inner circle	10	9.906
4	Rectangular side length	11	11.143
	Diameter of inner circle	10	10.106
5	Rectangular side length	11	11.126
	Diameter of inner circle	10	10.012
6	Rectangular side length	11	11.122
	Diameter of inner circle	10	10.014

sticky powder on the edges also brings some errors to the actual measurement. It can also be seen from the above that the size of the part near the center is slightly larger than the size of the part far from the center. During the forming process, at the center position, the laser travels the nearest route, the energy is slightly higher, while at the position away from the center, the energy is slightly low, but the difference is relatively small.

5.4.2.2.3 Z-axis accuracy

As can be seen from Fig. 5.119, the height of the melting channel formed after the powder is melted is not consistent with the thickness of the powder, as shown in Fig. 5.119A and B. When the laser energy is low, the height of the melting channel is higher than that of the powder leveling thickness, at this time, it is easy to form spheroidization. After melting, the metal and the base have no overlapping region, and the melting channel and the substrate have low bonding strength. When the powder leveling roller levels the powder again, since the height of the melting channel is higher than the thickness of the powder layer, it is easily scratched by the powder leveling roller and separated from the base to form manufacturing defects. When the laser power meets the manufacturing needs, the melting channel becomes wide and flat, and has coincident region with the base. Since the density of the powder after melting is greater than the apparent density of the powder, so the height of the melting channel is lower than the thickness of the powder layer. After the scan of one layer is completed, the overall height of the formed surface will be lower than the powder layer thickness. The thickness of the powder leveling layer is set to a fixed value, so that the thickness of the next layer is higher than that of the upper layer.

Fig. 5.120 shows the schematic diagram of the thickness variation of the powder. It can be seen from the figure that when a layer of powder melts and solidifies, the volume shrinks, and the height after solidification is lower than the height of the powder layer. Since the decreased height of the working cylinder is constant, so the powder leveling thickness of the next layer will be higher than that of the previous layer. The change of the powder layer thickness will affect the forming precision and performance, but the final powder thickness will become a fixed value. This value can be determined by the

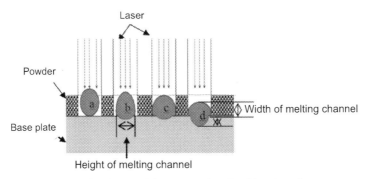

FIGURE 5.119 Schematic diagram of the cross-section of melting channel.

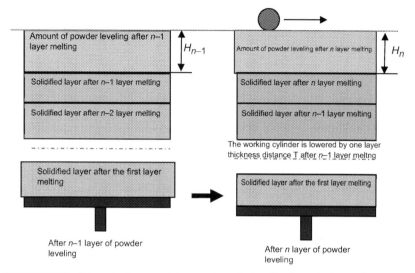

FIGURE 5.120 Schematic diagram of the actual powder leveling layer thickness after powder melting.

calculation method below: the length of the working cylinder is set to L, the width is W, the layer thickness is T, the apparent density is set to ρ_0, and the melting density is ρ. As shown in Fig. 5.120, after the melting of the $n-1$ layer, the distance from the plane on the solid to the powder leveling plane is H_{n-1}, and the amount of powder leveling is H_n after the melting of the n layer. According to the principle of conservation of mass, the mass of the n layer of powder is equal before and after melting, and the mass after melting is:

$$(T + H_{n-1})WL\rho_0 = (T + H_{n-1} - H_n)WL\rho \quad (5.81)$$

It can be derived that:

$$H_n = \left(1 - \frac{\rho_0}{\rho}\right)(T + H_{n-1}) \quad (5.82)$$

Set

$$\left(1 - \frac{\rho_0}{\rho}\right) = K, \quad K > 1, \quad K > 0 \quad (5.83)$$

Then

$$H_n = T + KH_{n-1} \quad (5.84)$$

It can be derived from recursive formula that:

$$H_n = T\left(1 + K + K^2 + K^3 + \cdots + K^{n-1}\right) \quad (5.85)$$

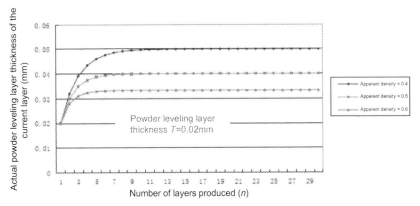

FIGURE 5.121 Relationship between the actual powder leveling layer thickness and number of layers produced for powers with different apparent densities when the powder leveling layer thickness is 0.02 mm.

It can be further concluded that:

$$H_n = \frac{(1-K^n)}{1-K}T \tag{5.86}$$

Assuming that the density after melting is 100%, the relationship between H_n and the number of layers is shown in the Eq. (5.86): Note: $K = (1 - \frac{\rho_0}{\rho})$, and T is the layer thickness.

Fig. 5.121 shows the relationship between the actual thickness of the powder and the number of layers produced after each powder leveling when the working cylinder is lowered a height of 0.02 mm each time. The apparent density is a ratio of the mass to the volume of certain volume of powder compared to the metal density of the powder material, with a maximum value of 1, representing the apparent density of 100%. It can be seen from the figure that the powder with higher apparent density has smaller change in the final layer thickness than the initial set layer thickness. Since the actual powder apparent density is between 0.4 and 0.6, we have studied only the situation with powder apparent density in this range.

As can be seen from the foregoing analysis, the thickness of the powder gradually increases as the number of layers produced increases due to volume shrinkage after melting of the powder. However, the powder thickness will tend to a fixed value after manufacturing several layers, as can be seen from Table 5.18. When the thickness of the powder leveling is set to 0.02 mm, the thickness of the powder layer is fixed to a value after 20 layers of manufacturing. For powder of different apparent densities, the final powder layer thickness is not the same, and the higher the apparent density of the powder, the smaller the final powder leveling thickness increases. When the apparent density is 0.4, the theoretical final thickness of the powder layer

TABLE 5.18 Final powder layer thickness of powder with different apparent densities.

Apparent density of powder	Set layer thickness (mm)	Final powder layer thickness (mm)	Layer thickness increase ratio (%)
0.4	0.02	0.050	150
0.5	0.02	0.040	100
0.6	0.02	0.033	65

is 0.05 mm, the limit is reached and the thickness no longer increases. At this time, the final powder leveling thickness is 2.5 times of the initial layer thickness, which is increased by 150%. For powder with apparent density of 0.6, the final powder layer thickness is 0.033 mm, an increase of 65% over the initial set powder thickness. It can be seen that the apparent density of the powder has great influence on the layer thickness and manufacturing performance of the manufactured parts. The larger the layer thickness of the powder, the greater the laser energy required and the higher the probability of defects such as spheroidization and porosity. Since the laser processing parameters are configured according to the set powder layer thickness, when the final layer thickness of the powder differs greatly from the set value, the previously set processing parameters deviate from the optimal value, and the process parameters need to be modified to adapt to the actual situation.

Fig. 5.122 shows the final stable powder leveling thickness of the powder at different apparent densities when the thickness of the layer is different. It can be seen from the figure that the powder with lower apparent density has greater powder leveling thickness finally, and the smaller the thickness of the powder leveling, the smaller the thickness of the final powder leveling is. As mentioned above, too thick powder leveling is liable to cause the laser to fail to penetrate the powder layer, the powder cannot be completely melted, and since there is a large of amount of melted powder, the molten metal generated is large, thus increasing the tendency to form spheroidization. From the forming results of the two tower parts shown in Fig. 5.123, the parts with powder of large apparent density have rough surface and obvious spheroidized particles after being manufactured with large layer thickness; while the parts with powder of large apparent density have flat surface and spheroidized particles that are not visible to the naked eye after being manufactured with a small layer thickness.

It can be seen from the previous theoretical analysis that the metal powder becomes dense metal body after melting due to the elimination of the gap between the powders, and the final surface height is lower than that of the powder leveling, which leads to the gradual increase of lower layer

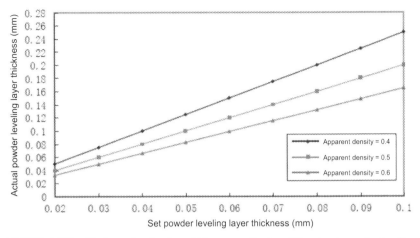

FIGURE 5.122 Final powder leveling thickness results for different set powder leveling thickness.

FIGURE 5.123 Tower parts made with two processing parameters. (A) Apparent density of the powder: 0.45, layer thickness processed: 0.05 mm; (B) apparent density of the powder: 0.59, layer thickness processed: 0.025 mm.

thickness, finally reaching a limit value, and the final thickness of the powder is more than 2.5 times the thickness of the actually set powder layer. Such a thickness causes difficulty in laser melting and a tendency to increase spheroidization. However, in the actual manufacturing process, the molten metal surface is not a completely flat surface as envisioned in the theory, as shown in Fig. 5.124.

The surface formed by melting the metal powder is formed by overlapping a plurality of semi-cylindrical surfaces. The surface morphology is composed of a plurality of raised ridges and recessed grooves. Therefore, in fact, the surface height formed by the melting of the metal powder is higher than the theoretically calculated height, and there is a certain error in the theoretical

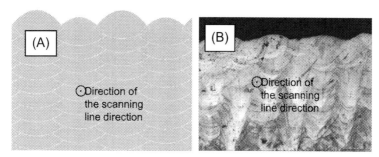

FIGURE 5.124 Profile of SLM-formed part. (A) Formed profile for CAD modeling; (B) Actual formed profile.

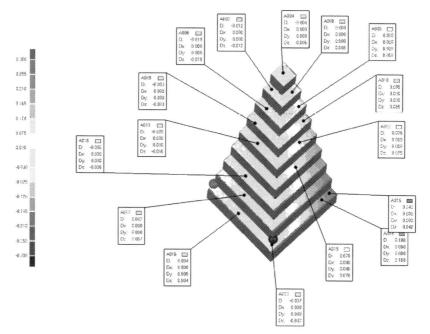

FIGURE 5.125 Scanning results of tower parts.

calculation. Therefore, a test for the height direction accuracy is designed. A tower part with a height of 45 mm is designed on the computer using CAD software. The height of each layer is 5 mm, with a total of 9 layers, as shown in Fig. 5.125. After the parts are manufactured, the parts are scanned and measured using the PowerScanner-STD 3D measurement system to obtain the dimensional data of the parts, as shown in Fig. 5.125.

The height data of the nine step faces in the scan data is extracted, as shown in Fig. 5.126. Using the software measurement function, the distance from the surface to the reference plane is measured, and the height of each

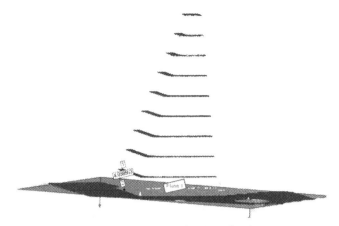

FIGURE 5.126 Surface data of the extracted part in height direction.

TABLE 5.19 Height measurement values of the tower part with a powder apparent density of 0.45 and a layer thickness of 0.08 mm.

Number of layers	Model height (mm)	Measurement of actual height (mm)(mm)	Relative difference of each layer (mm)	Cumulative height difference (mm)
Layer 1	5	4.90	0.1	0.10
Layer 2	10	10.33	−0.43	−0.33
Layer 3	15	15.14	0.19	−0.14
Layer 4	20	20.06	0.08	−0.06
Layer 5	25	24.97	0.09	0.03
Layer 6	30	29.81	0.16	0.19
Layer 7	35	34.90	−0.09	0.10
Layer 8	40	39.80	0.10	0.20
Layer 9	45	44.89	−0.09	0.11

step from the reference plane can be obtained. In fact, the surface data extracted by software is not a single surface. In fact, it is a sheet with a certain thickness composed of many data. It is measured in multiple points for many times and then averaged to obtain the height values of each step, as shown in Table 5.19.

It can be seen from Table 5.19 that in the case of using powder with low apparent density, the parts are manufactured with high manufacturing layer thickness, and the fluctuation of the height direction has a certain fluctuation law. In the first step of 5 mm height, there is a certain contraction in the height direction of the part, and the height direction is 0.1 mm lower than the model. However, the height of the second step is higher than the height of the model by 0.43 mm. It indicates that when the number of layers produced is increased, the powder leveling thickness of the lower layer is increased due to the shrinkage of the front layer after being melted. The powder leveling height is set to a high value in this test, when the stable powder leveling thickness is reached during the manufacturing process, the powder thickness is as high as 0.16 mm. The powder of this thickness cannot be completely melted by the laser energy, and the partially melted powder bonds the unmelted powder together, equivalent to one sintering. Since the powder is not completely melted, the plane height after laser scanning does not decrease, which is consistent with the height of the model. At the next powder leveling, the thickness of the powder is just set to a thickness of 0.08 mm. At this time, the laser can fully melt the metal powder to show height shrinkage, thus entering the same cycle, so that there are many unmelted parts in the height direction, as shown in Fig. 5.127 and Table 5.20.

Due to the excessive thickness of the layer, the thickness of the stable powder leveling is more than the limit of laser penetration. We optimized the process parameters, and selected the powder with higher apparent density and lower layer thickness. As can be seen from Fig. 5.128, the height of the 5 mm part in the Z-axis direction is reduced by about 0.04 mm, the total height is reduced by 0.435 mm, and the total error in the height direction is 0.9%. From the previous theoretical formula of the height direction contraction, although there is certain shrinkage after each layer of powder is melted, since the descending height of the working cylinder is constant, the amount of powder will be added to the lower layer of powder leveling to compensate for the previous shrinkage. Theoretically, the difference between the final height of the formed part and the height of the model is due to the shrinkage of the last layer of powder, and the difference should be less than the maximum powder leveling thickness, that is, 0.04 mm. In fact, there is a certain error due to the movement accuracy of the working cylinder, and the height of the part sliced by the process software of the forming system may

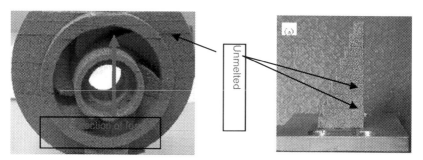

FIGURE 5.127 SLM-formed parts.

TABLE 5.20 Height measurement values of optimized tower parts.

Number of layers	Model height (mm)	Measurement of actual height (mm)(mm)	Relative difference of each layer (mm)	Cumulative height difference (mm)
Layer 1	5	4.937	0.063	0.063
Layer 2	10	9.851	0.036	0.149
Layer 3	15	14.812	0.039	0.188
Layer 4	20	19.793	0.019	0.207
Layer 5	25	24.756	0.037	0.244
Layer 6	30	29.708	0.047	0.292
Layer 7	35	34.655	0.053	0.345
Layer 8	40	39.598	0.058	0.402
Layer 9	45	44.565	0.032	0.435

Note: Apparent density of powder: 0.59, powder layer thickness: 0.025.

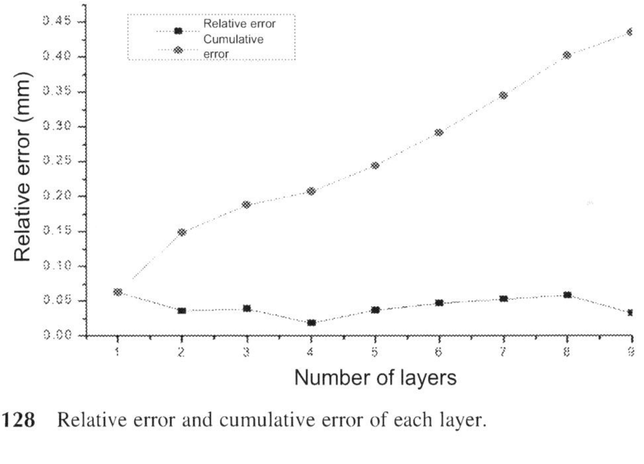

FIGURE 5.128 Relative error and cumulative error of each layer.

not reach one layer thick at the top level of the layered channel, and the software automatically removes the layer, that is, the final layer may not be processed and melted. Thus, due to these two reasons, the final height difference of the entire part is 0.435 mm, and the relative error is within 1%.

5.4.3 Microstructure characteristics and mechanical properties of typical metal materials for additive manufacturing

SLM forming technology is characterized by complete melting and solidification of the material. Therefore, it is mainly suitable for the formation of metal materials, and one of its advantages is that it can form most of the metal materials, including pure metal materials, alloy materials, and metal matrix composite materials. The microstructure characteristics and mechanical properties of typical metal materials formed by SLM are analyzed below.

5.4.3.1 316 L stainless steel
5.4.3.1.1 Microstructure

Fig. 5.129 shows the microstructure of SLM-formed 316 L stainless steel at low magnification (× 100). It can be seen that there are no obvious macroscopic defects such as pores and cracks inside the sample, indicating that the density of the formed parts is high. At the same time, the fish-scale crystal boundary formed by the overlapping trajectories between the laser cladding channels can be clearly seen. In the production process of each layer of the SLM, the two scanning lines need to overlap each other, and the previously melted scanning track needs to be remelted again by the new scanning track, so that strong overlap is formed between the scanning tracks. It can be seen that under the optimized forming process conditions, 316 L stainless steel parts with a density nearly 100% can be prepared by the SLM technology.

Since SLM parts are gradually formed based on the AM process from line to surface and from surface to volume, the microstructure inside the part tends to have obvious orientation. Different microstructures can often be observed in different viewing directions. In order to better study the internal microstructure, we mainly study the microstructure of three regions with typical characteristics in the part: (1) the microstructure in the cladding channel; (2) the microstructure between the horizontal channels; (3) the microstructure between the vertical channels (i.e., between layers).

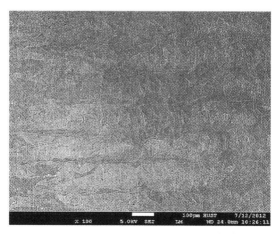

FIGURE 5.129   SEM of the sample at low magnification (× 100).

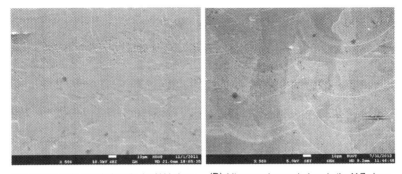

(A) Microscopic morphology in the X-Y plane   (B) Microscopic morphology in the Y-Z plane

FIGURE 5.130 Typical morphology of SLM-formed 316 L stainless steel (SEM). (A) Microscopic morphology in the X-Y plane (B) Microscopic morphology in the Y-Z plane.

Fig. 5.130 shows a typical microscopic morphology of SLM-formed 316 L stainless steel in different viewing planes. It can be seen from the figure that the sample has obvious "channel-channel" overlapping crystal boundary after slight corrosion. Fig. 5.130A shows the microscopic morphology in the XY plane. The overlapping crystal boundary between the n cladding channel and the $n + 1$ cladding channel in the same layer can be clearly seen from the figure. At the same time, it can be seen that the columnar crystals which are mainly grown in the vertical cladding channel in the cladding track have obvious turning dendritic crystal in the overlapping region of the cladding track. Fig. 5.130B shows the microscopic morphology in the Y-Z plane. The overlapping crystal boundary between the n cladding track and the $n + 1$ cladding track, and the overlapping crystal boundary between the cladding tracks of the n layer and the $n + 1$ layer can be seen in the figure. The grains mainly exhibit epitaxial growth characteristics. At the same time, it is also possible to clearly observe the fish scales formed by the superposition of the upper and lower cladding channels. One fish scale is one melting channel, and numerous cladding channels are repeatedly cycled to eventually form dense SLM parts. It can be seen that the SLM-formed 316 L stainless steel has a significant regular crystal boundary inside.

Fig. 5.131 shows the microstructure in the cladding channel in the Y-Z plane. It can be seen that the fine columnar crystals in the cladding channel have significant epitaxial growth characteristics. Fig. 5.131A shows the microscopic morphology of the center of the cladding track parallel to the direction of the cladding track. The fish scales formed by the superposition of the cladding track can be seen. Fig. 5.131B shows the dendrites of epitaxial growth inside. One fish scale is one cladding channel, that is, in the SLM process, the dendrites produced by the solidification of the molten pool formed by the former spot continue to grow during the heating and cooling of the latter spot, and the cycle

548 Materials for Additive Manufacturing

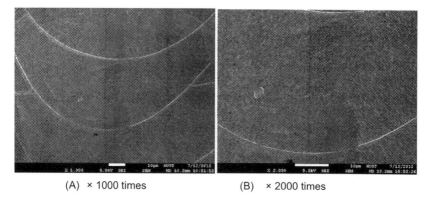

(A) × 1000 times   (B) × 2000 times

FIGURE 5.131 Microstructure of the cross-section of the cladding track in the Y-Z plane (SEM). (A) ×1000 times (B) ×2000 times.

is repeated to make the dendrite grow continuously. This is due to the fact that during the SLM process, heat transfer is dependent on the adjacent solidified molten pool and is transmitted backward (opposite to the laser moving direction). Thus, both in the cladding track and between the cladding tracks, the crystal growth direction is along the largest heat dissipation direction (opposite to the heat dissipation direction), and is a nonuniform nucleation based on the melting boundary. In the cladding track, the heat in the molten pool mainly diffuses downward through the solidified cladding track and the bottom of the base, and has a large degree of subcooling in a direction parallel to the scanning direction, that is, perpendicular to the plane of the molten pool. When the liquid metal in the molten pool solidifies, the grains grow preferentially along the direction of the larger temperature gradient, thus forming crystal grains having distinct orientation as shown in Fig. 5.131.

Figs. 5.132 and 5.133 show the microstructure of the different sides of the center of the cladding track. It can be seen that the microstructures show completely different morphology in different viewing directions. Fig. 5.132A shows the morphology of the microstructure parallel to the direction of the cladding track. It can be seen that the morphology is mainly cell structure with a diameter of about 0.3 μm. Fig. 5.133 shows the microstructure morphology perpendicular to the direction of the cladding track. It can be found that the morphology is mainly columnar crystal with a diameter of about 0.3 μm, and the orientation of the columnar crystals is highly directional. Obviously, the cell crystal in Fig. 5.132A is the cross-section of the columnar crystal in Fig. 5.133. It can be seen that the grain orientation inside the molten pool is highly directional.

Fig. 5.134 shows the microscopic morphology between the "layer-layer" overlapping perpendicular to the direction of the cladding track in the Y-Z plane, apparently in the region where the overlapping is the turning of the

Metal materials for additive manufacturing **Chapter | 5** 549

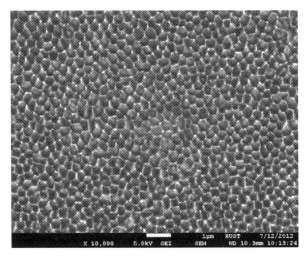

FIGURE 5.132 Microstructure (SEM) of the side at the center of the cladding channel (parallel to the direction of the cladding channel).

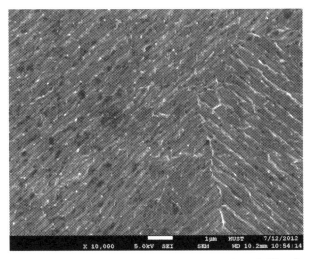

FIGURE 5.133 Microstructure (SEM) of the side at the center of the cladding channel (vertical to the direction of the cladding channel).

dendrites. Fig. 5.134A shows the overlapped region of the center of the molten pool. It can be seen that the growth of the grains has significant epitaxial features. This is mainly due to the fact that during the SLM forming process, there is a large positive temperature gradient at the front edge of the solid-liquid interface in the molten pool, and the nucleation phenomenon generally

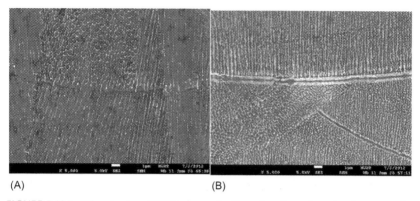

FIGURE 5.134 Microstructure (SEM) between the "layer-layer" overlapped region perpendicular to the scanning direction in the X-Z plane. (A) Morphology of the center of molten pool in the overlapped region (B) Boundary area of the molten pool in the overlapped region.

does not occur at the interface front, so the solidified structure is mostly formed by epitaxial growth. In the overlapped region, heat is transferred from the current melting channel to the previous melting channel, so the grain growth is opposite to the heat transfer direction; at the same time, the overlapping boundary provides heterogeneous nucleation base for crystal growth, promoting nucleation and growth of crystal, and eventually developing into columnar crystals. Fig. 5.134B shows the overlapped region of the boundary of the cladding track. It can be seen that there are some epitaxial features in the grain in this region, and the orientation of the crystal is significantly different in different molten pools. This is mainly because the boundary area of the molten pool is generally the intersection of different molten pools, and the direction of heat transfer in each molten pool is not exactly the same. The growth direction of dendrites is mainly determined by the heat transfer mechanism in the molten pool. Therefore, it shows the characteristics of multi-orientation growth, and it can be clearly seen that the growth of the crystal grains exhibits the characteristics of "mixed but not confused."

It can also be seen from Fig. 5.134 that the direction of crystal growth is not completely perpendicular but close to be perpendicular to the overlapping boundary. The formation of this microstructure depends on the heat transfer mechanism during SLM laser scanning. For a single molten pool, the heat of the molten pool is first transmitted to the solidified portion through the boundary of the molten pool, and further transmitted through the substrate; in the heat conduction process of the molten pool, heat is easily conducted outward along the boundary of the molten pool, thereby forming the direction of crystallization perpendicular to the boundary of the molten pool. Therefore, the nucleation at the boundary of the molten pool can be considered as a non-uniform nucleation, and the solidified portion provides "base" of heterogeneous nucleation for the subsequently solidified liquid metal. Under this condition, the spindle with fast growth rate, that is, the crystal orientation perpendicular to the boundary of the molten pool can preferentially grow; while the other crystal orientations are inhibited by heat transfer. Usually, the columnar crystal region portion can be regarded as "crystalline texture." At the same time, it can be seen that some of the columnar crystals are characterized by cell crystals. The nucleation is mostly spontaneous nucleation. Due to the Marangoni convection inside the molten pool, the columnar crystals growing from the boundary to the inside are easily broken by the convection cell to form many small grains.

In fact, the SLM process is a process of rapid melting/solidification of crystal with large temperature subcooling, which is a forced crystal growth process. For forced dendrite growth, if the maximum local subcooling of the interface front is higher than the nucleation supercooling, it may lead to the growth of equiaxed grain with consistent neuclation. However, if the volume fraction of growth of equiaxed grain is too small, the equiaxed grain will be entrapped by the growing columnar crystal. Only when the volume of the equiaxed grain reaches a certain value, it is possible to block the growth of the columnar crystal, showing fully equiaxed grain growth. It is generally believed that when the volume fraction φ of the equiaxed grain is less than 0.66%, the solidified structure is columnar crystal; when φ is greater than 49%, the solidified structure is an equiaxed grain structure; when φ is between 0.66% and 49%, the solidified structure is in a growth pattern combined with columnar crystal and equiaxed grain.

During the solidification of the SLM molten pool, there is a close relationship between the crystal growth direction and the growth rate of the crystal spindle and the laser scanning speed. There is a great temperature gradient in the SLM process. Under this forced condition, the velocity (v_s) at the leading edge of the solidified interface of the molten pool is mainly determined by the temperature field. The heat flow and fluid convection in the molten pool determine the shape of the molten pool. Under the steady state processing, the velocity (v_s) of the solidified interface can be determined by the shape of the molten pool. Normally, v_s basically follows the direction with the largest temperature gradient in the molten pool, that is, the normal direction of the solid-liquid interface, so its scanning speed (v_b) with the laser beam satisfies:

$$v_s = v_b \cdot \vec{n} = v_b \cos\theta \qquad (5.94)$$

where $\vec{n}$ is the normal vector of the solid-liquid interface, parallel to the v_s direction; θ is the angle between the normal vector and the scanning speed. However, the above equation ignores influence of the preferred orientation of dendrite growth on the growth rate of the solidified interface caused by the anisotropy of crystallographic orientation. For the cell crystal/eutectic crystal, the growth direction inversely parallel to the direction of the heat flow, so the growth direction is close to the normal direction of the leading edge of the solidified interface, so the growth rate is approximately equal to the velocity (v_s) at the leading edge of the solidified interface. That is to say, it can be determined by the above equation.

Further, the crystal orientation, the crystal orientation, and the laser scanning direction also have certain orientation relationship. For the microscopic crystallographic orientation [hkl] of the crystal, the crystal growth rate in this direction is V_{hkl}, then the relationship between V_{hkl} and R_v is as follows:

$$v = V_{hkl} \times \vec{n} = |V_{hkl}|\cos\varphi \tag{5.95}$$

where φ is the angle between the velocity of leading edge of the solidification and the fastest growing crystal orientation of the crystal. Then, the relationship between the growth rate V_{hkl} in the crystal [hkl] direction and the scanning speed v is:

$$\begin{aligned}|V_{hkl}| = V \times \cos\theta/\cos\Psi = V/\{&h\cos\alpha_{100} + k\cos\alpha_{010} + l\cos\alpha_{001} \\ &+ \tan\theta[\cos\varphi(h\cos\beta_{100} + k\cos\beta_{100} + l\cos\beta_{001}) \\ &+ \sin\varphi(h\cos\gamma_{100} + k\cos\gamma_{010} + l\cos\gamma_{100})]\}\end{aligned} \tag{5.96}$$

For the body-centered cubic 316 L stainless steel material, there are three <100> crystal orientations in the crystallographic preferential growth direction. In the laser metallurgy solidification process, the laser beam is generally scanned along the [100] direction on the (001) plane. The following equations can be obtained through Eq. (5.96):

$$V_{100} = V \tag{5.97}$$

$$V_{010} = \frac{V}{(\tan\theta\cos\varphi)} \tag{5.98}$$

$$V_{001} = \frac{V}{(\tan\theta\sin\varphi)} \tag{5.99}$$

It can be seen from the above analysis that in the liquid molten pool of SLM-formed 316 L, the nucleus at the bottom end and in the middle grow mainly along the [001] crystal orientation, while the two sides and the upper side are along [010] or [01(−)0] directions.

5.4.3.1.2 Mechanical properties

The room temperature tensile properties of the sample in the X-axis direction are tested in the experiment. The average tensile strength of the SLM-formed 316 L stainless steel is 795 MPa and the average elongation is 18.6%. The tensile strength of the 316 L stainless steel medical parts manufactured by the traditional method is 540−620 MPa, the yield strength is 200−250 MPa, the elongation is 50%−60%, and the hardness is 170−200 HRV. It can be seen that the tensile strength of SLM-formed 316 L stainless steel parts is significantly higher than that of stainless steel parts manufactured by traditional

Metal materials for additive manufacturing **Chapter | 5** 553

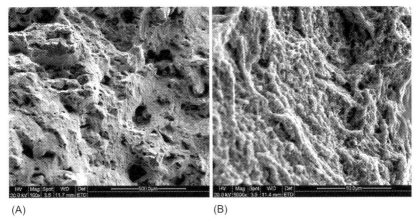

FIGURE 5.135 Typical morphology (SEM) of room temperature tensile fracture of SLM-formed 316 L stainless steel. (A) Macroscopic fracture morphology (B) Local typical features.

manufacturing methods, but its elongation is significantly reduced, about 1/3 of the traditional manufacturing method. This is mainly due to the fact that the microstructure of the SLM-formed 316 L part is formed at an extremely fast solidification rate, and its grain size is small. We can know that the size is generally less than 1 μm by the analysis of the previous microstructure. Through the Hall-Petch equation of metallurgical principle, it can be seen that for dense metal materials, the finer the crystal grains, the more favorable the mechanical properties of the metal materials. Therefore, the tensile strength of the part will be significantly enhanced, while the plasticity, that is, the elongation will be significantly reduced.

Fig. 5.135 shows the typical fracture morphology of 316 L stainless steel tensile specimen formed by SLM at room temperature. Fig. 5.135A shows the macroscopic morphology of the tensile fracture. It can be seen that there are pores inside the part, and the presence of these pores will reduce the mechanical properties of the part. A further magnified observation of the microscopic area of the section (Fig. 5.135B) reveals that the dimples with a shallow size of about 0.5 μm are distributed on the section, indicating that the fracture mode is intergranular fracture. Based on this, we can conclude that the fracture mechanism of SLM-formed 316 L stainless steel is intergranular fracture.

5.4.3.2 Ti6Al4V alloy
5.4.3.2.1 Microstructure

Fig. 5.136 shows the microstructure morphology of Ti6Al4V alloy parallel to the direction of the cladding channel in the XY plane. Fig. 5.136A shows the low-magnification microstructure morphology and Fig. 5.136B

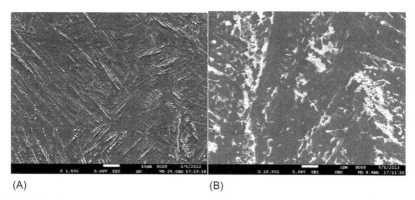

(A)   (B)

FIGURE 5.136 Microstructure morphology of Ti6Al4V alloy parallel to the cladding channel in X-Y plane (SEM). (A) Low magnification morphology ($\times$ 1000 times) (B) High magnification morphology ($\times$ 1000 times).

shows the high-magnification microstructure morphology. It can be seen from Fig. 5.136A that there is no obvious overlapping boundary between the cladding channels of the SLM-formed Ti6Al4V alloy, indicating that the phase change dominates the SLM forming Ti6Al4V alloy. At the same time, it can be seen that the internal microstructure is mainly needle-shaped martensite structure. This is mainly because during the SLM process, the process of transforming from β phase to α phase in the process of rapid cooling of Ti6Al4V alloy is too late, and the β phase will be transformed into supersaturated solid solution with the same composition as parent phase but different crystal structure, that is, martensite. The martensitic transformation is a nondiffusion phase transition, and atomic diffusion does not occur during the phase transition, and only lattice remodeling occurs. The martensitic transformation of Ti6Al4V alloy is a typical shear phase transition, and its lattice reconstruction speed changes at a speed close to the speed of sound. At this time, the atoms in the β phase of the body-centered cubic structure are regularly and closely migrated in a concentrated manner, and when the migration distance is large, the hexagonal α'' is formed; when the migration distance is small, the oblique square α'' is formed. Obviously, there's a large number of needle-like α'' clusters in the microstructure with V alloy of SLM-formed Ti6Al4 (Fig. 5.136A). At the same time, there is also α'' formed due to only short-range migration of atoms around the bundling α'' (Fig. 5.136B).

Fig. 5.137 shows the microstructure morphology of Ti6Al4V alloy perpendicular to the direction of the cladding channel in the XZ plane. The morphology is observed under low-magnification SEM (Fig. 5.137A). It is found that it still shows needle-like martensite structure in the vertical direction. A further magnified observation of the micro-area (shown in Fig. 5.137B) shows that the short-axis of the α bundling is perpendicular to the observation surface in this direction, and the α-phase and the β-phase are in a staggered phase

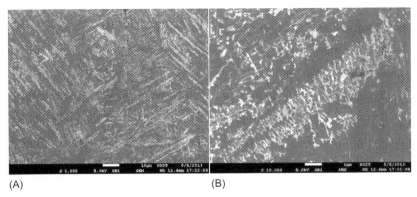

FIGURE 5.137 Microstructure morphology of Ti6Al4V alloy perpendicular to the cladding channel in X-Z plane (SEM). (A) Low magnification morphology (× 1000 times) (B) High magnification morphology (× 1000 times).

relationship. It shows that the growth of the internal α phase has certain orientation, and in the microstructure, the texture exists inside the part [8].

5.4.3.2.2 Mechanical properties

5.4.3.2.2.1 Density Fig. 5.138 shows the relative density of the test samples. It can be seen that the relative density of the test samples exceeds 96%, the lowest relative density is 96.2%, and the highest relative density is 98.6%. It is indicated that there are still many defects such as pores inside the test samples, which will reduce the mechanical properties of the test samples to some extent.

5.4.3.2.2.2 Mechanical properties The mechanical properties at room temperature are one of the key indicators for evaluating the quality of parts. The clinically available forged and annealed Ti6Al4V alloy has a yield strength of 830–860 MPa, a tensile strength of 900–950 MPa, and an elongation of 10%. Because the SLM process uses a "line-surface-volume" AM process, its mechanical properties tend to exhibit anisotropy, so we test the mechanical properties of the test samples from the direction parallel to the cladding track and the direction perpendicular to the cladding channel.

Table 5.21 shows the room temperature tensile properties of the samples in the X-axis direction. As can be seen from the table, the average yield strength of the samples in the X-axis direction is 1204 MPa, the tensile strength is 1346 MPa, and the elongation is 11.4%. Obviously, its tensile performance index is better than that of the clinically forged and annealed Ti6Al4V alloy, which meets the requirements for mechanical properties of medical Ti6Al4V alloy.

Fig. 5.139 shows the fracture morphology of the tensile specimen in the X-axis direction. From the low-magnification morphology of the fracture

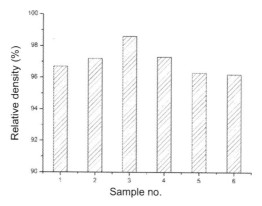

FIGURE 5.138 Relative density (%) of the sample.

TABLE 5.21 Tensile properties at room temperature in the X-axis direction.

No.	Yield strength (MPa)	Tensile strength (MPa)	Elongation (%)
1	1214.36	1387.47	11.461
2	1187.28	1263.98	11.782
3	1211.51	1387.17	11.057
Average value	1204.38	1346.21	11.433

side (Fig. 5.139A), it can be seen that there are many pore defects in the sample, which will result in a decrease in the mechanical properties of the sample compared with the dese SLM parts. The fracture of the sample is relatively flat, and the section has no obvious shrinkage, indicating that the sample appears sudden fracture, and the internal pore defect has little effect on uniaxial tensile properties. From the low magnification of the front of the fracture (Fig. 5.139B), it can be seen that there are some small cracks in the sample parallel to the direction of the cladding track. Further analysis shows that these cracks are mainly the overlapping gaps between the cladding channels, with original powder particles filled inside. Further analysis of the section morphology reveals that there are two typical morphologies in the fracture surface, one is the small step with relatively flat surface (as shown in Fig. 5.139C), and the other is the shallow dimple (Fig. 5.139D). From the small step with flat surface, we can see that the crack propagation is characterized by transgranular expansion, which is typical transgranular brittle fracture. During the stretching process, many cleavage crack nuclei generated at

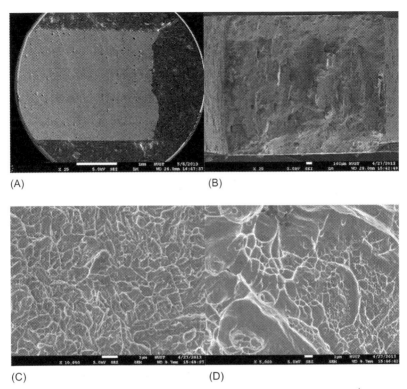

FIGURE 5.139 Fracture morphology of tensile specimens in the X-axis direction. (A) Low-magnification morphology on the side of the fracture (B) Low-magnification morphology on the front of the fracture (C) Typical morphology I of the fracture local area (D) Typical morphology II of the fracture local area.

different cleavage sites are then expanded into cleavage facets in cleavage manner, which is represented by small mesas (Fig. 5.139C); then it is torn in plastic manner and the adjacent lower facets are connected to finally form tearing ridge on the side, which is represented by shallow dimple (Fig. 5.139D). It can be seen that the X-axial tensile fracture mode parallel to the direction of the cladding track is a quasi-cleavage fracture mode, which is also consistent with the characteristics of the martensitic structure formed by the previous SLM process.

Table 5.22 shows the room temperature tensile properties of the sample in the Z-axis direction. As can be seen from the table, the average yield strength of the sample in the Z-axis direction is 1116 MPa, the tensile strength is 1201 MPa, and the elongation is 9.88%. Compared with the clinically forged and annealed Ti6Al4V alloy, the elongation is slightly smaller, and the yield strength and tensile strength are better than those of the forged and annealed Ti6Al4V alloy.

TABLE 5.22 Tensile properties at room temperature in the Z-axis direction.

No.	Yield strength (MPa)	Tensile strength (MPa)	Elongation (%)
1	1098.82	1140.42	10.355
2	1101.43	1175.06	9.422
3	1148.97	1290.30	9.864
Average value	1116.41	1201.93	9.880

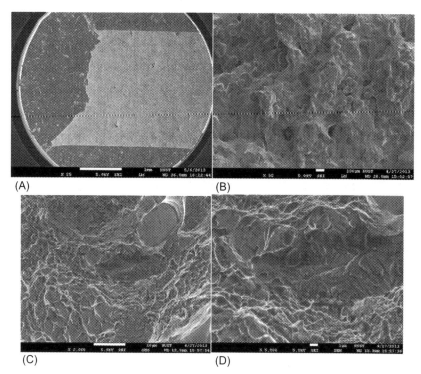

FIGURE 5.140 Fracture morphology of tensile specimens in the Z-axis direction. (A) Low-magnification morphology on the side of the fracture (B) Low-magnification morphology on the front of the fracture (C) Typical morphology I of the fracture local area (D) Typical morphology II of the fracture local area.

Fig. 5.140 shows the fracture morphology of the tensile specimen in the Z-axis direction. From the low-magnification morphology of the fracture side (Fig. 5.140A), it can be seen that there are fewer pores inside the specimen, and there is no obvious shrinkage in the section, but the fracture is uneven, indicating that the internal pore stretching process acts as crack source and has great influence on the uniaxial tensile properties. From the low magnification of the front of the fracture (Fig. 5.140B), it can be seen that there are a large number of raw unmelted metal powder particles on the surface of the fracture. It can be seen in the Z-axis direction of the powder layer accumulation that the overlapping between upper and lower layers is poor with obvious process defects, which directly leads to a significant decrease in mechanical properties in this direction. Further analysis of the section morphology reveals that there are two typical morphologies of the fracture surface, one is the tearing ridge (shown in Fig. 5.140C), and the other is cleavage with river converging pattern (shown in Fig. 5.140D). There is no obvious "dimple" of ductile fracture on the whole fracture surface, indicating that the fracture mode is a quasi-cleavage fracture closer to the cleavage fracture in the Z-axis direction perpendicular to the cladding channel, and its plasticity is poorer than that in the X-axis direction, which is consistent with the conclusion of the above mechanical properties.

5.4.3.3 Co-Cr alloy

5.4.3.3.1 Microstructure

Fig. 5.141 shows the microstructure of the original Co-Cr alloy powder. It is observed from the figure that the microstructure of the original powder material is mainly composed of a dendritic region (light color) and an interdendritic region (dark color). Both types are CoCrMo solid solutions, but the structure is different (fcc, hcp), and the light phase has an fcc structure, which is difficult to be etched; on the contrary, the dark phase has a hcp structure and is easily etched. Its microstructure consists mainly of austenite matrix and network and dendritic carbides.

Fig. 5.142 shows the microstructure of the SLM-formed Co-Cr alloy on different sides. It can be found that the Co-Cr alloy treated by the SLM

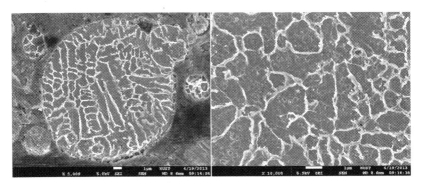

FIGURE 5.141 Microstructure (SEM) of Co-Cr alloy powder.

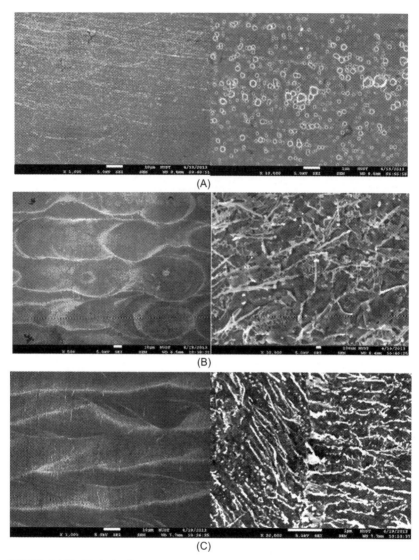

FIGURE 5.142 Typical Microscopic morphology of different sides of SLM-formed Co-Cr alloy. (A) Microstructure morphology (SEM) perpendicular to the scanning direction in the X-Z plane (B) Microstructural morphology (SEM) perpendicular to the direction of the additive material in the X-Y plane (C) Microstructure morphology (SEM) parallel to the scanning direction in the Y-Z plane.

process undergoes an extremely rapid melting/solidification process in which the solid solution reinforcement phase is remarkably refined. Fig. 5.142A shows the microstructure morphology perpendicular to the scanning direction in the X-Z plane. It can be seen that the cladding channels overlap each other to form fish scale crystal boundary, in which there are many white particles inside the molten pool. Observation of the molten pool under high-magnification SEM reveals that the white particles are mainly etch pits, and the carbide particles dispersed in the matrix are corroded by the corrosive liquid, but some carbide remains on the boundary and adheres around the pit. It is highly corrosive and macroscopically appears as white particles. Fig. 5.142B shows the microstructure morphology perpendicular to the direction of the additive in the X-Y plane. It can be seen that a large amount of white matter precipitates at the boundary of the cladding track. Further analysis reveals that the white material is mainly filamentous carbide. Fig. 5.142C shows the microstructure morphology parallel to the scanning direction in the Y-Z plane. It can still be found that the white carbides are enriched at the boundary of the molten pool. Further analysis shows that the carbide and matrix phases are arranged alternately, which is mainly determined by the heat transfer mechanism in the SLM process.

In summary, we can find that the SLM-formed Co-Cr alloy is mainly composed by austenite solid solution matrix formed by CoCrMo and filamentous carbides dispersed therein, and the carbides are enriched at the boundary of the molten pool.

5.4.3.3.2 Mechanical properties

5.4.3.3.2.1 Hardness The Vickers hardness of the SLM-formed Co-Cr alloy is measured to be 476 ± 6 HV, while the Vickers hardness of the conventional medical casted Co-Cr alloy is generally 300 HV, and the Vickers hardness of the medical forged Co-Cr alloy is 265–450. It can be seen that the hardness of the Co-Cr alloy formed by the SLM process is 58% higher than that of the conventional cast Co-Cr alloy, and is higher than the highest hardness of the conventionally forged Co-Cr alloy. This is mainly due to the fact that the CO-Cr alloy powder material undergoes rapid melting/solidification metallurgical process during the SLM forming process, in which the reinforced phase carbide is remarkably refined and dispersed in the matrix material, so that its hardness is significantly increased.

5.4.3.3.2.2 Tensile properties Fig. 5.143 shows the tensile curve of the Co-Cr alloy sample prepared by the SLM process. It can be seen that the curve has no obvious elastic deformation stage in both the horizontal direction and the vertical direction, and the strain before the failure of the sample is small, which is a typical tensile curve of brittle material. According to Table 5.23, the average yield strength of the sample in the horizontal direction of the X-axis is 1142 MPa, the tensile strength is 1465 MPa, and the elongation is 7.6%; the yield strength of the sample in the vertical direction of the Z-axis is 1002 MPa, the tensile strength is 1428 MPa and the

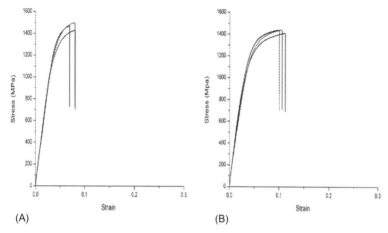

FIGURE 5.143 Tensile stress-strain curve of SLM-formed Co-Cr alloy at room temperature. (A) Horizontal direction (B) Vertical direction.

TABLE 5.23 Tensile mechanical properties of SLM-formed Co-Cr alloy at room temperature.

Tensile direction	Yield strength (MPa)	Tensile strength (MPa)	Elongation (%)
Horizontal direction	1142 ± 26.6	1465 ± 28.2	7.6 ± 0.45
Vertical direction	1002 ± 7.1	1428 ± 14.4	10.5 ± 0.49

elongation is 10.5%. It can be seen that the mechanical properties of SLM-formed Co-Cr alloy specimens have obvious anisotropy, and the tensile strength along the horizontal direction of X-axis is better than the tensile strength along the vertical direction in the Z-axis, but the elongation is worse on the contrary. Compared with the mechanical properties of traditional medical cast Co-Cr alloy (yield strength: 665 MPa, tensile strength: 860 MPa, elongation: 7.95–10.00%), its yield strength and tensile strength increase by about 50% in the horizontal direction and increase by about 70% in the vertical direction, and the elongation is roughly equivalent to the elongation of the cast Co-Cr alloy. Compared with the mechanical properties of traditional medical forged Co-Cr alloy (yield strength: 962 MPa, tensile strength: 1507 MPa, elongation: 28%), the yield strength and tensile strength of the two materials are very close, but the elongation of cast alloy is about 30% of Co-Cr alloy. It can be seen that the tensile strength of the SLM-formed

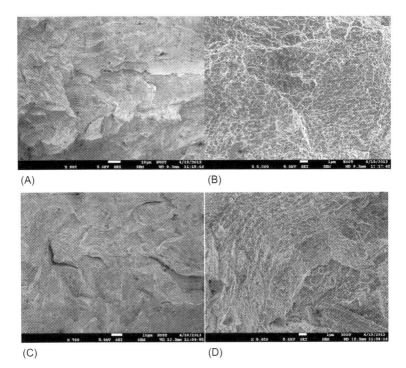

FIGURE 5.144 Tensile fracture morphology of SLM-formed Co-Cr alloy at room temperature. (A) Low-magnification morphology in the horizontal direction (B) High-magnification morphology in the horizontal direction. (C) Low- magnification morphology in the vertical direction (D) High-magnification morphology in the vertical direction.

Co-Cr alloy is mainly similar to that of the forging process, and the elongation is mainly similar to the elongation of the cast Co-Cr alloy.

Fig. 5.144 shows the fracture morphology of the SLM-formed Co-Cr alloy tensile specimen. Observed from the low-magnification morphology, it can be seen that the fracture morphology has clear cleavage area and cleavage step (Fig. 5.144A), as shown in Fig. 5.144C, it has no obvious dimple characteristics, indicating that it has poor plasticity. At the same time, we can see that there are wedge cracks on the section under the low-magnification shape. The wedge cracks are mainly parallel to the horizontal direction and perpendicular to the vertical direction. During the stretching process, the wedge crack acts as the initial source of expansion of cracking, this has significant effects on the performance of the part. This is one of the main reasons for the poor elongation in the horizontal direction. By observing its high-magnification microstructure, it is found that the size of the dimple-like structure in the horizontal direction (Fig. 5.144B) is significantly larger than that of the dimple-like structure in the vertical direction (Fig. 5.144D). At the same time, a significant stacking step can be observed under high-magnification morphology in the vertical direction, while in the horizontal direction, the stacking step can only be observed in the low-magnification morphology (Fig. 5.144A), indicating the fracture mode is mainly quasi-cleavage fracture. At the same time, it also shows that the SLM-formed Co-Cr alloy has obvious anisotropy in the growth direction, mainly in the vertical direction, which is mainly determined by the heat transfer mode in the SLM process [9−11].

5.4.3.4 AISI 420 stainless steel
5.4.3.4.1 Microstructure

The SLM point laser melts the fine metal powder channel by channel, which results in a very high cooling rate of the molten pool due to the small size of the molten pool and rapid movement. In addition, during the laser beam melting process channel by channel and layer by layer, the parts are subjected to complex thermal effects of continuous action and periodic changes, resulting in the great changes of the material with the as-cast and forged materials in the heat transfer and processes during the solidification process. Based on the above research, the effects of laser power on the microstructure, phase composition and properties of the formed parts are studied using four different process parameters. The specific process parameters are: laser power: 120−150 W, scanning speed: 550 mm/s, scanning pitch: 0.08 mm, and layer thickness: 0.02 mm, the scanning mode is progressive one-way scanning. Vacuuming and argon infusion are carried out twice during forming to reduce oxygen content. The size of the formed specimen is φ 10 mm × 40 mm, as shown in Fig. 5.145A. The relative densities of the samples measured by the drainage method are 99.04%, 99.94%, 99.95%, and 99.92%, respectively, and Fig. 5.145B shows the relative density results. The density of the samples produced in this process is higher than that of the process research in previous Section 5.3.3. On the one hand, the oxygen content

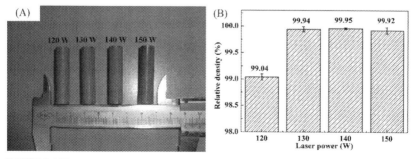

FIGURE 5.145 (A) Specimens made using different process parameters; (B) Relative density of the specimens.

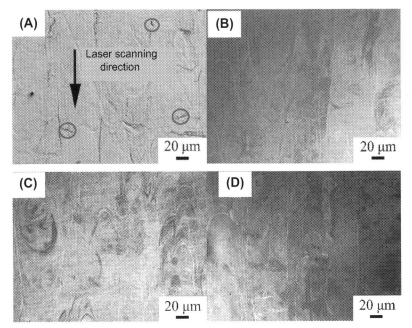

FIGURE 5.146 Microstructure of the formed specimen in the X-Y plane. (A) 120 W; (B) 130 W; (C) 140 W; (D) 150 W.

is reduced by the argon infusion twice, and it is also related to the stability and consistency of the current SLM equipment.

Fig. 5.146 shows the microstructure of the X-Y plane (scanning plane) of the samples formed by different processes. The black arrow indicates the scanning direction of the laser. The microstructures of the samples formed by the four process parameters are very similar. From the figure, we can see that the four samples are nearly fully dense, and no large-sized irregular pores caused by unmelted powder particles are observed. However, as shown in the red circle shown in Fig. 5.146A, there are some microscopic pores and microcracks in the specimen formed by 120 W laser power and the pore shape is irregular (<5 μm), which may be caused by insufficient local wetting. Microcracks are caused by thermal stresses in the SLM process. It can be seen from the figure that the overlapping between the melting channels after the multilayer processing is more disordered than the single-layer scanning. Because the laser energy density adopted this time is relatively close, the density of the formed parts exceeds 99%, and the influence of laser power on the microstructure is not obvious. Fig. 5.147 shows the microstructure of the X-Z plane (accumulation direction) of formed specimen with 120 W laser power. It can be seen from Figure 5.147A the typical scale-like morphology of the SLM-formed part, the semicircular molten pool

566 Materials for Additive Manufacturing

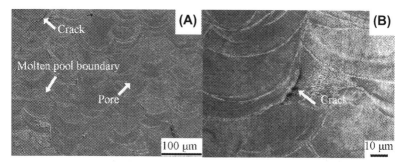

FIGURE 5.147 Microstructure of the X-Z plane using specimen formed by 120 W laser power.

(MPB) illustrates good remelting and overlapping between layers during the manufacturing process. At the same time, it can also be seen from the microstructure of the X-Z plane that there are microcracks and pore defects, and microcracks are likely to appear at the boundary of the molten pool (see Fig. 5.147B).

Since the influence of laser power on the microstructure of the formed part is not particularly obvious, the microstructure characteristics of the SLM-formed AISI 420 stainless steel are illustrated using the samples mainly formed with 140 W laser power. Fig. 5.148A shows the microstructure of 400-magnification. Due to the cooling and solidification layer by layer and channel by channel in the SLM process, the overlapping boundaries of the micro-melting pool can be clearly seen, and the molten pool is overlapped to form the appearance of fish scales, which are just in line with the characteristics of the Gaussian energy distribution of the laser beam. The laser remelts the upper layer at the current scanning layer, forming the molten pool boundary in Fig. 5.148A along the height direction. Fig. 5.148B shows the microstructure at the boundary of the molten pool in Fig. 5.148A, and Fig. 5.148C shows the microstructure at the center of the molten pool in Fig. 5.148A. Fig. 5.151D is the microstructure near the boundary of the molten pool. It can be seen from the figure that there is a big difference in the internal and boundary structure of the molten pool. Compared to the traditionally manufactured AISI 420 stainless steel, the SLM-formed parts have fine grains with the grain size less than 1 μm. The inner crystal grains of the molten pool are fine cell crystals, and the grain at the boundaries of the molten pool are smaller than the inner grain size of the molten pool, and show directional crystallization. The red arrow in Fig. 5.148B is the growth direction of crystal grains. It can be clearly seen from Fig. 5.148D that there are different regions at the boundary of the molten pool, including the HAZ, the boundary of the molten pool, the cell crystal region and the oriented crystal region. In the HAZ, the grain grows due to the periodic thermal action in the SLM process, and the oriented crystal region grows from the boundary of the molten

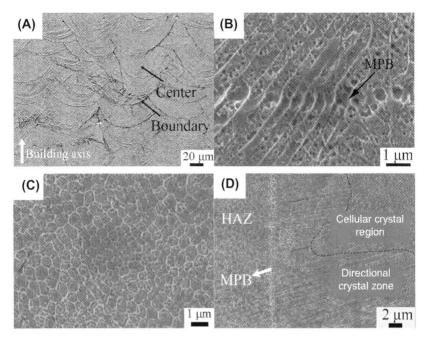

FIGURE 5.148 Microstructure of the X-Z plane of specimen formed using 140 W laser power. (A) 400-magnification morphology; (B) Tissue at the boundary of molten; (C) Tissue at the center of the molten pool; (D) tissue of the attachment at the boundary of molten pool.

pool in the direction of heat loss, and the cell crystal region is formed inside the molten pool due to rapid cooling.

Most metal materials undergo solidification and solid phase transitions during manufacturing, including nucleation, growth, and collision stop. However, classical theory cannot explain the non-equilibrium solidification process in the SLM process. In the SLM process, a higher degree of subcooling ΔT reduces the difference between the heterogeneous nucleation energy $\Delta G^{*\prime}$ and the homogeneous nucleation energy $\Delta G'$, and homogeneous nucleation is more likely to occur. The solid/liquid interface advances extremely rapidly during the SLM process, and the material undergoes supersaturated solid solution under non-equilibrium solidification. The grain growth mode near the boundary of the molten pool is mainly the growth mode driven by heat diffusion, and the rapid nucleation of the grains causes grain refinement, so no obvious secondary dendrites are observed in the formed parts.

5.4.3.4.2 Mechanical properties

The most materials prepared by SLM methods exhibit significant anisotropy, and many studies have focused on this issue. The placement direction of the part actually determines the angle between the manufacturing plane and the

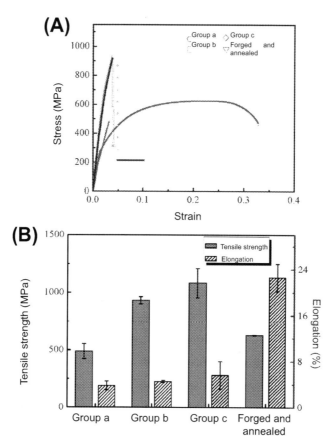

FIGURE 5.149 Tensile properties. (A) Stress-strain curves of four sets of samples; (B) Tensile strength and elongation of four sets of samples.

direction of load application. The research team proposed that the boundary of molten pool may be an important factor affecting the mechanical properties of SLM. This paper takes samples in different directions based on the previous research to study the influence of the special spatial topology of the molten pool on the mechanical properties. The stress-strain curves and tensile properties of the three sets of SLM formed samples and comparative samples are shown in Fig. 5.149. As shown in Fig. 5.149A, the SLM-formed parts and the forged and annealed samples exhibit completely different plastic deformation behavior. The forged and annealed 420 stainless steel material has obvious yield behavior, and the material plasticity greatly exceeds that of the SLM-formed part. While the SLM-formed part has no obvious yielding stage, the tensile specimen suddenly fractures during the continuous increase of the load. It can be seen that the anisotropy of the SLM-formed parts is very obvious, and the 420 stainless steel is more anisotropic than the materials such as 316 L and 304 stainless steel. It can be seen that the tensile curves of the samples in Group b and Group c are nearly identical, which indicates that the Young's modulus of the two sets of samples is relatively close, which is larger than the forged and annealed 420 stainless steel, while the Young's modulus of the samples in Group a is significantly smaller than the samples in other groups, which is mainly due to the high porosity in the samples of Group a, and the pores directly affect the elastic deformation behavior of the material and reduce the Young's modulus. The tensile strength and elongation of the four groups of samples are summarized in Fig. 5.149B. The tensile strengths of Group a, Group b and Group c formed by SLM are 488 ± 66, 932 ± 31, and 1082 ± 127 MPa, respectively, and the elongations are $3.8\% \pm 0.77\%$, $4.5\% \pm 0.2\%$, and $5.65\% \pm 2.4\%$, respectively. The tensile strength and elongation of the forged and annealed 420 stainless steel are 629 ± 4 MPa and $22.6\% \pm 2.4\%$, respectively. It can be seen from the above data that the tensile strength of other SLM samples except for those in Group a is much higher than that of the forged and annealed 420 stainless steel, but the tensile strength stability of SLM-formed parts is poor. The standard deviation of the tensile strength of three sets of samples is much larger than the standard deviation of the forged and annealed state. In addition, the elongation of the SLM-formed samples is below 6%, which is much lower than the elongation of 22.6% of the forged and annealed samples. Generally, the metal materials with tensile elongation less than 5% belong to brittle fracture, and with tensile elongation above 5% belong to plastic fracture. SLM-formed 420 stainless steel is largely characterized by low plasticity due to the composition of medium carbon and high Cr alloy.

The reasons of the special mechanical properties of 420 stainless steel SLM-formed parts can be discussed from the following aspects:

5.4.3.4.3 Effects of grain refinement on tensile strength

Usually the material is mainly strengthened by three mechanisms, and the linear superimposition of three mechanisms reinforces the material, just as described in the equation below:

$$\Delta\delta = \Delta\delta_{cr} + \Delta\delta_{wh} + \Delta\delta_{ps} \tag{5.100}$$

where $\Delta\delta_{cr}$ is the improvement of tensile strength brought by grain refinement, $\Delta\delta_{wh}$ is the improvement of tensile strength brought by work hardening, and $\Delta\delta_{ps}$ is the improvement of tensile strength brought by precipitation of second phase. Song studied the mechanical properties of the SLM-formed parts of pure Fe, and proposed that grain refinement is the main mechanism for the performance enhancement of SLM-formed parts. No significant martensitic transformation occurred during the SLM forming of 420 stainless steel, so the main reinforcing mechanism is still grain refinement. The effect of grain refinement can be expressed by the Hall-Petch relationship:

$$\Delta\delta_{cr} = kD^{-1/2} \qquad (5.101)$$

where D is the average size of the grains and k is the slope of Hall-Petch. When the average grain size is smaller, the strengthening effect of the material is more obvious. The grain size of 420 stainless steel SLM-formed parts is micron level and sub-micron level. Compared with the annealed material, the grain refining effect is obvious. Therefore, the tensile strength of the samples in Group b and Group c is significantly higher than that of forged and annealed 420 stainless steel.

5.4.3.4.4 Influences of microscopic defects on tensile strength

Although metal parts that are nearly fully dense can be formed by process control, some common microscopic defects, mainly microscopic pores (Fig. 5.150A) and cracks (Fig. 5.150B) in SLM-formed parts are difficult to completely eliminate. The pores are divided into regular spherical pores and irregular pores according to the morphology. Among them, the spherical pores are mostly caused by the entrapment of ambient gas into the molten pool during the forming process, or the inclusion of gas in the powder failing to overflow into the micro-melting pool in time. However, spherical pores have a low probability of occurrence, most of them are irregular pores, which are mainly caused by the instability of the molten pool morphology. Microcracks are mainly caused by microscopic defects such as maximal residual stress and element segregation generated during SLM forming. During the deformation of the material, stress concentration will be formed at these microscopic defects. As the load increases, the stress at the defects will first reach the yield strength and ultimate fracture strength of the material, which will cause the crack to expand and result in the material crack when far below the theoretical breaking load. The density of the sample in Group a is only 97.2%, with more pore defects than those of Group b and Group c. Generally, the better the plasticity of the material, the lower the sensitivity of the formed part to the initial pore

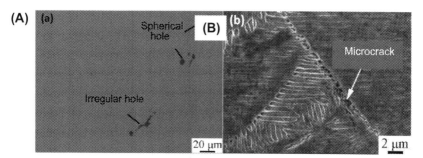

FIGURE 5.150 Forming defects of 420 stainless steel SLM parts: (A) Pores, (B) Cracks.

defect. Conversely, the greater the brittleness of the material, the more sensitive the formed part is to the initial defects such as pores and cracks. Because 420 stainless steel itself is poorly shaped, the initial defects in the stretching process will rapidly expand as crack sources. As a result, the tensile strength of samples in Group a is lower than that of the other two groups.

During the SLM forming process, the material undergoes rapid melting and solidification and periodic heat action. Different positions of the molten pool have undergone different thermal processes, and at the same time, physical and chemical effects such as decarburization and vaporization occur on the surface of the micro-melting pool, resulting in the non-uniformity of microstructure and chemical component, which directly affects the macroscopic and microscopic properties of the SLM-formed parts.

In order to study the difference in microscopic properties in the micro-melting pool, nano indentation tests are performed on different locations in the molten pool. Because the surface quality (flatness) of the sample has a significant effect on the nano indentation test results, the test data is not credible when the surface is uneven. Therefore, the polished sample is slightly wiped with alcohol to make the boundary of the molten pool appear. In Fig. 5.151A, as the dark lines where test points 1 and 2 are located, test points 3, 4, and 5 are randomly selected inside the molten pool. Fig. 5.151B shows a typical indentation 3D morphology. The edge of the indentation forms significant metal buildup due to plastic deformation of the material, and microcracks appear at the edge of the indentation. Other embossed positions to the indentation attachment are caused during the sample preparation process. Fig. 5.151C shows the load-depth curve, which is smooth and does not have displacement step. As shown in Fig. 5.151D, the Young's modulus and microhardness of test points 1 and 2 are lower than test points 3, 4, and 5. The Young's modulus and microhardness of the micro-melting pool boundary are 184.9 and 8.1 GPa, respectively, while the Young's modulus and microhardness inside the molten pool are 190.6 and 8.8 GPa, respectively. The C element is the main strengthening element of the Fe-based alloy. The decarburization phenomenon at the boundary of the molten pool may be the cause of the poor mechanical properties of the boundary. It can be inferred from the difference of the micromechanical properties that the boundary of the molten pool fractures earlier than the inside of the molten pool under the same stress.

Fig. 5.152 shows the topological structure of the boundary of the molten pool, and carbon depleted zone and carbon-rich zone caused by decarburization. The upper left corner is the coordinate system manufactured by SLM, d is the width of the molten pool (about 100 μm), and h is the depth of molten pool (approximately 15–50 μm). The loading directions of the samples in Group a and Group b are parallel to the Y axis, and the loading direction of the samples in Group c is parallel to the Z-axis. It can be seen that the boundary of molten pool is continuous in the Y-direction, forming the overlapping region within the black circle as shown in the figure. Since the

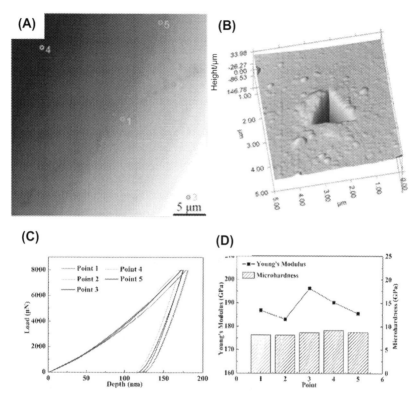

FIGURE 5.151 Nano indentation test. (A) Test point location; (B) Morphology of the indentation; (C) Load-depth curve; (D) Young's modulus and microhardness.

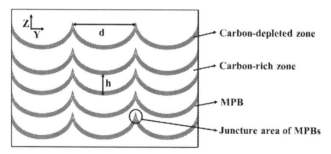

FIGURE 5.152 Topological structure of the molten pool boundary, and carbon depleted zone and carbon-rich zone caused by decarburization.

boundary of the molten pool is easy to adsorb some non-metallic elements, the overlapping strength of the molten pool boundary is lowered, and the overlapping area of the molten pool boundary may be the weak zone and the initial position of crack. When stretching in the Z-direction, the overlapping

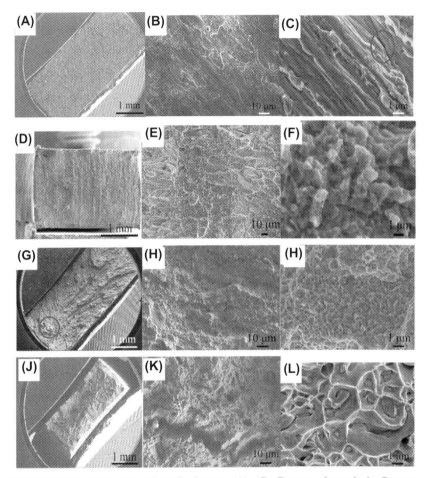

FIGURE 5.153 Morphology of tensile fracture. (A)–(C): Fracture of sample in Group a; (D)–(F): Fracture of sample in Group b; (G)–(I): Fracture of sample in Group c; (J)–(L) Fractures of forged and annealed samples.

area of the boundary of the molten pool in the same cross-sectional area is less than that in the Y-direction, and the gradient structure of carbon in the Z-direction stretching helps to reduce the residual stress, so the tensile strength of samples in Group c is higher than that of the group b.

As shown in Fig. 5.153, an observation is carried out for the fracture of the tensile specimen. Fig. 5.153A–C is the tensile fracture morphology of the sample in Group a. It can be seen from Fig. 5.153A that the fracture of the sample formed by the manufacturing method in Group a is very flat, the edge of the fracture has no obvious macroscopic deformation and necking, and the sample has brittle fracture. From the high-magnification SEM image in Fig. 5.153B and C, it can be seen that the fracture is composed of small faces, and at the same time, a parallel fibrous structure is observed, and microcracks present in the sample can be seen (as shown in the ellipse in Fig. 5.153C). Fig. 5.153D−F show the tensile fracture morphologies of the specimens in Group b. It can be seen from the low-magnification macro-fracture morphology that the tensile fracture is still relatively flat, but there are clear "channel-shaped" tear marks. As the magnified image shown in Fig. 5.153E, the fracture is a mixture of the dimple and the cleavage. The two types of fractures are uniformly and alternatively distributed. Under the high magnification, some tiny prototypes of the equiaxed dimple can be seen (Fig. 5.153F). It can be seen from the fracture morphology that the fracture mode of the samples in Group b is quasi-cleavage fracture, the plasticity and toughness of the material are enhanced, and the macroscopic elongation is also increased, which is consistent with the results obtained by the tensile specimen. Fig. 5.153G−I show the tensile fracture morphologies of the samples in Group C. Compared with the previous two sets of samples, the macro-fracture of the sample is more uneven. From Fig. 5.153G, it can be seen that the crack originates from the position indicated by the red circle, and the crack expands in the direction indicated by the red arrow, showing a typical river shape. It can be seen from Fig. 5.153H that there are deep pits in the fracture of sample in Group c, and the sample shows ductile fracture. However, the tensile results show that the macroscopic elongation of the sample does not increase greatly. It can be seen from Fig. 5.153I that the dimple size of the sample in Group c is very small (about 100 nm), which is consistent with the limited macroscopic elongation of the sample. Fig. 5.153J−L show the tensile fracture of the forged and annealed specimen. It can be seen that the specimen has obvious macroscopic deformation and necking occurs at the fracture. After the fracture is enlarged, many holes can be seen, as well as deep and long macroscopic cracks. Obvious dimples can be seen after further enlargement, which are close to the equiaxed and vary in size (a few microns to tens of microns). The particle of the second phase can be seen at the bottom of the dimple.

Fig. 5.154 shows the OM microscopic morphology of the fracture section after tensile fracture of the SLM specimen. It can be seen from Fig. 5.154A and B that the fractures of the sample in Group a and Group b are very flat. After the corrosion, it is observed that the fracture is near the juncture of the melting channel, due to the small deformation during the stretching process, the micro-melting pool does not deform significantly. The sample in Group c shows a fracture of nearly 45 degrees at the edge of the sample, while the fracture at the middle of the sample is flat, but not as flat as the fracture of the sample in Group a and Group b. As shown in the red circle in Fig. 5.154D, the fracture occurs at the juncture of the melting channel. As mentioned above, the juncture is a weak zone of mechanical properties of the formed part. The crack preferentially occurs here, and propagates in the direction of 45 degrees. Most of the cracks extend near the boundary of

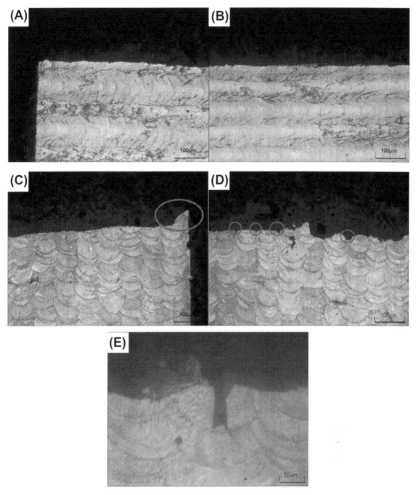

FIGURE 5.154 Fracture section of the SLM specimen after tensile fracture. (A) Sample in Group a; (B) Sample in Group b; (C) (D) (E) Samples in Group c.

the molten pool, and a few pass through the micro-melting pool along the Z-direction (Fig. 5.154E) [12−16].

5.5 Metal wire for arc fuse deposition forming

5.5.1 Design and preparation technology of wire materials

The traditional design of metal-based wire alloy system does not fully consider the metallurgical bonding of multi-material interfaces and particle precipitation during crystallization and stacking, which leads to the wire performance being difficult to meet the high-performance requirements of WAAM. The traditional preparation method of metal-based wire is mainly aimed at welding processes. The insufficient protection effect and unstable forming process when the wire is used in WAAM make it difficult to manufacture high-performance metal-based components [17−28]. In addition, metal-based wire materials are generally prepared through smelting, drawing and other processes. Due to the limitation of the process technology, it is very difficult to prepare some high-performance material or composite wire material with special alloying elements. Even some alloys can be smelted but cannot be manufactured into wires, which greatly limits the types and quantities of metal-based wire materials. For example, common metal-based wire materials are mainly iron-based alloys, and the types of materials are few, thus it is difficult to meet the actual application requirements.

(1) Design of high-performance flux-core wire with miniature slag pool

Materials play a decisive role in physical and chemical properties of the parts. For the metal-based parts made by arc AM, the bonding interfaces should be free of defects and have good bonding properties to ensure high property of formed parts. The traditional design of metal wire alloy system does not fully consider the metallurgical behaviors of multi-material bonding interfaces, the crystallization behaviors of the interfaces, and the particle precipitation during the stacking process, which results in the wire performance difficult to meet the high-performance requirements of WAAM. Therefore, a Ti-N alloy design system was established for high-performance wire materials. Based on the common properties of N and C, N is used partly instead of C for alloying, so as to overcome the defects such as cracks and CO pores which are easily generated in the manufacturing process of high C content wire materials. TiN particles are precipitated on-site during the stacking process, which hinder the grain growth in the HAZ and greatly improve the performance of the formed parts. Based on the metallurgical reaction law at the interface of multi-materials and the precipitation of second phase particles in metal materials, combined with Taguchi optimization design criteria, a series of special high-performance flux-core wire materials with miniature slag for WAAM are designed. The miniature slag wire adopts combined protection of gas and slag, thus overcoming the problems of easy high-temperature oxidation, low efficiency and poor moisture resistance of multi-slag wire. Miniature slag wire considers the advantages of gas protection and slag protection, so the WAAM is featured by good process performance and high performance of forming components.

(2) High-speed preparation of self-toughened flux-core wires

The preparation technology of metal-based wire plays an important role in the quality and formability of wire. Traditional metal-based wire is

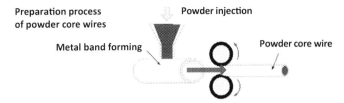

FIGURE 5.155 Preparation process of metal-based flux-core wire.

generally prepared by an integrated rolling and drawing method, and has the problems of easy wire breakage, easy burr on the surface, high cost, low efficiency, and low diameter precision. For this reason, based on the principle of dynamic recovery and recrystallization of metals, a method of high-speed preparation of flux-core wire by self-toughening is invented. The process of forming, powder adding, folding/sealing and diameter reducing of the flux core wire is integrated and shown in Fig. 5.155. Heat-generating agent is added into the drawing lubricant to regulate the temperature of the wire during drawing to reach the state of metal recovery and recrystallization. The method effectively overcomes the problems of low drawing speed, easy wire breakage and unstable arc caused by surface burrs existing in the traditional metal-based wire preparation method. Compared with the traditional preparation method, the wire drawing speed increased from 8 to 14 m/s, the diameter is 1.2 mm, the wire breakage rate is reduced from 5 times/10,000 m to 0.3 times/10,000 m, and the wire surface roughness (R_a) is reduced from 1.6 to 0.8 μm, which can realize efficient and high-precision preparation of small-diameter flux-core wire. The schematic and physical diagrams of the metal-based flux core wire material preparation equipment developed are shown in Figs. 5.156 and 5.157. Each drive roll of the wire rolling unit uses independent power and transmission system, which consists of pay-off machine, rough wire drawing machine, fine wire drawing machine, take-up machine, winding machine, layer winding machine and other units. The wire drawing unit uses AC frequency conversion motor as power and belt transmission to drive the winding drum to finish drawing.

Based on the above technology, a series of high-performance flux-core wires with miniature slag (MGY281, MGY381, MGY501, etc.) for WAAM have been developed. Among them, depositing metal via MGY281 miniature slag flux-core wire has a hardness of 28 HRC, good crack resistance and good yielding under high pressure. Depositing metal via MGY381 has a hardness of 38 HRC, which has higher strength, hardness and toughness. Depositing metal via MDY501 has a hardness of 54 HRC and has higher hardness, strength, thermal stability, fatigue resistance and wear resistance at high temperature, which is a good hard-facing material.

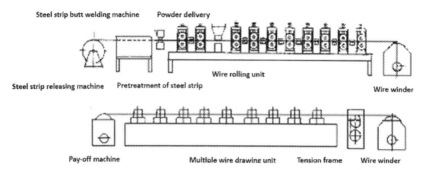

FIGURE 5.156 Schematic diagram of preparation equipment for metal-based flux-core wire.

FIGURE 5.157 Preparation equipment of high-performance metal-based flux-core wire.

5.5.2 Characterization of metal wire properties

A series of special flux-core wire (MGY281, MGY381, MDY501) to meet the requirements of "base layer + transition layer + hard-facing layer" of WAAM/remanufacturing for hot forging dies is introduced as an example in this section.

(1) Test standards

The test standards for flux-core wires are based on GB/T4336-2016, "Carbon and low alloy steel—Determination of multi-element contents—Spark discharge atomic emission spectrometric method (routine method)," GB/T20123-2006, "Steel and iron—Determination of total carbon and sulfur content—Infrared absorption method after combustion in an induction furnace (routine method)," and GB/T228.1−2010 "metallic materials—Tensile testing—Part 1: Method of test at room temperature."

(2) Test results

The MGY281 miniature slag flux-core wire specially used as the base layer has small spatter, good formability and good yield strength, and can effectively prevent the fracture failure of the forging die base material,

5CrNiMo, under high pressure. Its main performance indexes are shown in Table 5.24.

MGY381 miniature slag flux-core wire specially used as transition layer has small spatter, good formability, high strength, hardness and good toughness, and has good bonding with the base layer and the hard face layer. Its main performance indexes are shown in Table 5.25.

MDY501 miniature slag flux-core wire specially used as hard surfacing layer has small spatter and good formability, and has the highest hardness, strength, good thermal stability, fatigue resistance and wear resistance at high temperature, and good bonding with the transition layer. Its main performance indexes are shown in Table 5.26.

WA Group from British is currently a world-famous supplier of wire materials for welding and arc AM, the product performance of which represents the international advanced level. Table 5.27 compares MGY281 (base layer), MGY381 (transition layer) and MDY501 (cover layer) miniature slag flux core wires developed in this project with the similar products (Hard face) of British WA Company. It can be seen that the arc forming performance of wire materials in this project is not lower than that of similar products of British WA Group.

In general, the properties of miniature slag flux-core wires developed in this project can come up to the international advanced level, and some indexes are at the international leading level.

5.6 Microstructure and properties of wire and arc additive manufacture

5.6.1 Microstructure and properties of multiaxial pipe joint component

5.6.1.1 Scanning electron microscopy analysis of fracture surfaces

Impact and tensile tests were carried out on WAAM samples manufactured using self-designed low alloyed flux-core wire YHJ507JZ according to GB/T 229−2007 standard (Metallic materials—pendulum impact test method) and GB/T 228.1−2010 standard (Metallic materials-tensile testing), respectively. The chemical composition of the YHJ507JZ wire is listed in Table 5.28. SEM analysis were respectively performed on the fracture surfaces of tensile and impact samples. Fig. 5.158A shows the micro-morphology of the tensile fracture scanning under low-magnification microscope. It can be seen that most of the fracture area is relatively flat, and a few areas have uneven contours. Fig. 5.158B is the microscopic morphology of the tensile fracture under the high-magnification microscope, in which a large number of large and deep dimples are evenly distributed.

TABLE 5.24 MGY281 miniature slag flux-core wire.

Wire diameter (mm)	Chemical composition (wt. %)							Hardness (deposit 3 layers)	Crack sensitivity	Wear resistance	Impact resistance	Corrosion resistance
	C	Si	Mn	Cr	Ni	Mo	Others					
1.2	0.16	0.28	1.1	2.2	1.3	0.6	—	28 HRC	Excellent	Good	Good	Excellent

TABLE 5.25 MGY381 miniature slag flux-core wire.

Wire diameter (mm)	Chemical composition (wt%)							Hardness (deposit 3 layers)	Crack sensitivity	Wear resistance	Impact resistance	Corrosion resistance
	C	Si	Mn	Cr	Ni	Mo	Others					
1.2	0.22	0.32	0.9	5.0	1.1	1.0	V0.2 Ti0.1 N0.09	38 HRC	Good	Good	Excellent	Excellent

TABLE 5.26 MDY501 miniature slag flux-core wire.

Wire diameter (mm)	Chemical composition (wt. %)							Hardness (deposit 3 layers)	Crack sensitivity	Wear resistance	Impact resistance	Corrosion resistance
	C	Si	Mn	Cr	Ni	Mo	Others					
1.2	0.32	0.38	1.0	4.5	1.2	1.5	Ti0.12 N0.12	54 HRC	Excellent	Excellent	Good	Good

TABLE 5.27 Comparison between miniature slag flux-core wires in this project with similar products of international leading level.

Name	Wire diameter (Mm)	Hardness (deposit 3 layers)	Crack sensitivity	Wear resistance	Impact resistance	Corrosion resistance
MGY 281	1.2	28 HRC	Excellent	Good	Good	Excellent
MGY 381	1.2	38 HRC	Good	Good	Excellent	Excellent
MDY 501	1.2	54 HRC	Excellent	Excellent	Good	Good
HARD FACE WMOLC	1.2	52 HRC	General	General	General	Good

TABLE 5.28 Chemical composition of flux-core wire YHJ507JZ (wt.%).

Elements	C	Mn	Si	Cr	Cu	S	P
Weight percentage	0.06	1.40	0.45	≤0.20	≤0.50	≤0.030	≤0.030

Figure 5.158C shows the microscopic morphology of the impact tested sample fracture surface under low-magnification. It shows a distinct plastic deformation with a rugged surface and a stepped layer distribution. Fig. 5.158D shows the impact fracture surface under high magnification, similar to the tensile fracture scanning, the fracture morphology of impact sample shows a large number of uniform dimples, which is metal ductile fracture. The dimple is a microscopic cavity created by plastic deformation of the material in the micro-region, which is left on the surface of the fracture after nucleation, growth, aggregation and finally connected to each other [29,30].

5.6.1.2 Analysis of microstructure

It can be seen from the metallographic diagrams of YHJ507JZ sample in Fig. 5.159 that the microstructure of the sample consists of a large number of evenly distributed fine ferrite and a small amount of pearlite, the grain of which is fine, dispersedly distributed and finely divided. In the SEM micrographs in Fig. 5.160, the pearlite grains surrounded by complete ferrite grains can be clearly seen. When applied by external forces, the deformation will be dispersed into the fine grains, resulting in a decrease of stress concentration,

Metal materials for additive manufacturing **Chapter | 5** **583**

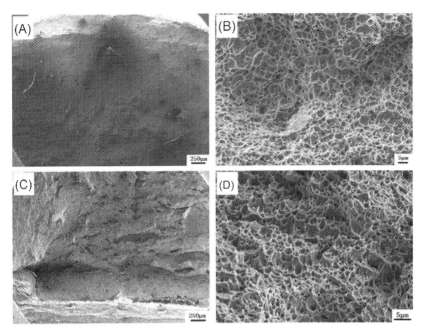

FIGURE 5.158 SEM of (A–B) tensile and (C–D) impact fracture.

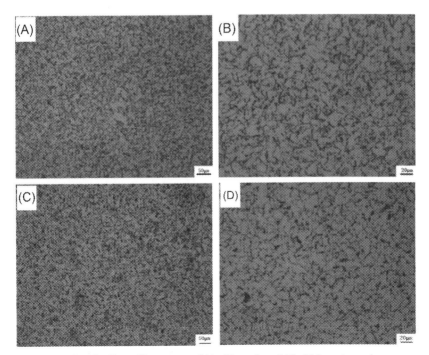

FIGURE 5.159 Metallographic structure of (A–B) tensile and (C–D) impact samples.

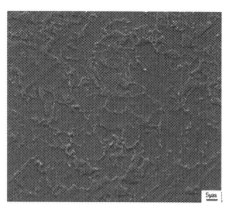

FIGURE 5.160 SEM micrographs of sample.

and thus, diminishing the cracking tendency. At the same time, the difference in strain between the grain and the grain boundary is low, so the material can be uniformly loaded. It indicates that the mechanical properties of the multiaxial pipe joint via WAAM technique are superior, mainly due to desired microstructure produced caused by quick heating and cooling processes in WAAM.

5.6.1.3 Analysis of mechanical properties of forming components

According to the relevant requirements, this section mainly introduces mechanical testing results including flattening test, tensile test, and impact test on the WAAM-manufactured multiaxial pipe joint. The metallographic analyses of the tested sample have been shown in the previous section.

5.6.1.3.1 Tensile and flattening tests

As shown in Fig. 5.161, an O-ring sample manufactured by WAAM was tested under GB T 246−2007 standard (Metallic materials_Tube_Flattening test). After the flattening test, there was no distinct cracks observed on the deformed area of the O-ring sample. The flattening test can check out the deformation resistance of the metal pipe under radial compressive stress to the specified level, and is one of the indicators to analyzing the mechanical properties of the multiaxial pipe joint. It can reflect the deformation resistance and applicable ability under different environments of the multiaxial pipe joints.

The strain-stress curves in Fig. 5.162 indicates that the tensile-tested sample mainly underwent three stages of deformation (two testing samples were used in tensile test in order to get a more precise result): the elastic deformation stage (AB), the plastic deformation stage (CD), and the necking stage (DE). In the elastic deformation stage, the load was relatively small,

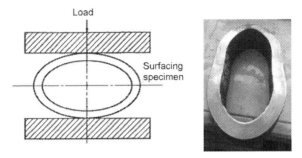

FIGURE 5.161 Schematic diagram of flattened specimen and deformation of specimen after test.

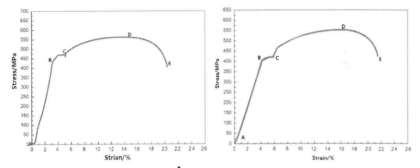

FIGURE 5.162 Stress-strain curves of two tensile samples.

and the stress was linear to the strain, as shown in curve AB of Fig. 5.162. As the load increased, the sample entered the yield stage, and point B is the yield point, as shown in the curve BC of Fig. 5.162. When the load continued to apply, the deformation resistance was increasing, and the sample entered the stage of plastic deformation, as shown in curve CD in Fig. 5.162. When it was increased to the point D, the deformation resistance reached the maximum value, that was, the tensile strength point. As the load continued to increase, it finally entered the necking stage, as shown in curve DE in Fig. 5.162.

Table 5.29 is the summarized data of the flattening and tensile tests. It can be seen that the measured values of the test equal to or greater than the standard values, which can ensure the good performance of the components.

Comparing the curves of the two tensile specimens and the tensile strength, yield strength, elongation and other parameters, it is found that the average tensile strength of the WAAM samples is 563 MPa, which is within the standard range; the average yield strength is 437 MPa, which is 46% higher than the standard value; the average elongation is 27.65%, which is

TABLE 5.29 Statistics of tensile and flattening test results.

Sample no.	Yield strength/MPa		Tensile strength/MPa		Elongation after break/%		Flattening test/H	
	Standard value	Measured value	Standard value	Measured value	Standard value	Measured value	Standard value	Measured value
1#	≥300	458	500–650	564	≥22	28.8	H = 120 mm	No crack
2#		417		562		26.5	No crack	

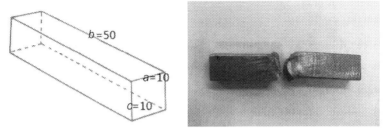

FIGURE 5.163 Dimensions of the impact specimen and the diagram after the fracture.

TABLE 5.30 Results of the impact test.

Testing temperature	Absorption work/J (Allow one of the values to be lower than the standard value, but no lower than 70% of the standard value)			Arithmetic mean value	
				Absorption work/J	
	1#	2#	3#	Standard value	Measured value
20°C	127	144	171	>60	147

25.7% higher than the standard value. Compared with the mechanical properties of national high-strength steel for construction (GB/Y 19879-2015 standard), the test results are obviously preferable, and there is no distinct crack in the flattening tested sample.

5.6.1.3.2 Impact test at room temperature

The impact test is to study the safety and reliability of WAAM component when subjecting to external impact forces, and the measured impact absorption work determines the toughness and impact resistance of the material. For the accuracy of the test results, three sets of impact test were performed. The size of the impact sample and the test results are shown in Fig. 5.163 and Table 5.30, respectively. The average impact absorption work is 147 J, and the impact absorption work of each sample is above 120 J, which is 145% higher than the standard value, indicating that the WAAM component has good toughness and impact resistance. The performance of the pipe joint under this process is superior, and the results are highly satisfactory.

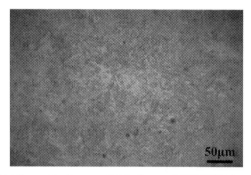

FIGURE 5.164 Metallographic diagram of high-alloy steel (MDY501) deposited layer.

5.6.2 Microstructure and performance of typical repaired components

5.6.2.1 Analysis of microstructure

From Fig. 5.164, the microstructure of the WAAM repaired component manufactured using the MDY501 flux-core wire (the composition is shown in Table 5.27) is fine lath martensite and retained austenite. In the high-alloyed steel flux-cored wire layers, due to the presence of a large amount of strong carbide forming element Cr, a large amount of Cr_7C_3 carbide were formed at the grain boundaries which could effectively increase the potential of grain boundaries and improve the corrosion resistance. At the same time, the addition of Cr significantly promoted the formation of twin martensite, which has high hardness and excellent wear resistance. Furthermore, the addition of strong carbide forming element V in the core wire could significantly improve the secondary tempering hardness of the metal, and ensure high hardness and stability at high temperature. Meanwhile, VC is an easily dispersed carbide, which hindered the migration of the austenite grain boundaries even in low equivalent carbon content, thereby preventing the growth of austenite grains, strengthening fine grains and improving the strength and toughness of the deposited metal, so that the wire has good formability and performance.

5.6.2.2 Heat treatment process and properties

The deposited metal of the flux-cored wire MDY501 is high-alloy steel with high strength and low toughness. The heat treatment method further enhanced the performance of the formed components, mainly improved the toughness. Meanwhile, the heating temperature caused the secondary hardening of carbides of V and Cr, so that the toughness could be greatly increased without sacrificing too much hardness, thereby satisfying the usage requirements.

Table 5.31 shows the Rockwell hardness values obtained after heat treatment of the post-deposited metal at a series temperature of 480°C, 500°C,

TABLE 5.31 Hardness values of as-deposited and heat treated of high-alloy steel samples (HRC).

Testing sample no.		1	2	3	4	5	Average value
1	As-deposited 480°C	49.6	48.6	48.5	48.3	50	49
	heat treated	43	41.2	43	41.5	43.2	42.4
2	As-deposited 500°C	48	48.6	48.4	47.9	49.1	48.4
	heat treated	45.5	46	44.5	47.2	46	45.8
3	As-deposited 520°C	49.9	50.5	49.4	49.8	49.5	49.82
	heat treated	43	41.5	43.5	41	42	42.2
4	As-deposited 560°C	49.2	49.1	49.1	49	49.4	49.16
	heat treated	39.1	39	38.9	38	38.5	38.7

520°C, and 560°C. It can be seen from the table that the average hardness of self-designed high-alloy steel flux-cored wire MDY501 depositing layer was 49.1 HRC, and its hardness reached the highest value after heat treated at 500°C, which was 45.8 HRC. The hardness value of the deposited metal decreased when heat treatment temperature increased further because of the atom migration and diffusion during the recovery process, and the dislocation density decreased, the number of internal defects of the metal and the distortion energy decreased, and the internal stress was eliminated, thereby causing a decrease in hardness.

The effect of heat treatment temperature on the hardness of the depositing layer is shown in Fig. 5.165. It can be seen from the figure that when the heat treatment temperature was in the range of 480°C–500°C, the hardness of the deposited metal showed an upward trend, mainly due to the precipitation of carbides in the martensite phase, increasing the non-uniform nucleation rate, refining the grains, and making the lath martensite transform into ferrite equiaxed grains. The hardness of the deposited metal reached a peak at 500°C, which was 45.8 HRC, indicating that the deposited metal had secondary tempering and hardening when heat treatment was performed under this temperature specification; when the heat treatment temperature was greater than 500°C, the hardness of the welding layer began to decrease, and the hardness value decreased to 38.7 HRC at 560°C. The hardness of the deposited metal was mainly determined by the carbon content in the martensite. When the temperature was greater than 560°C, the martensite decomposed and precipitated carbides, resulting in a continuous decrease in carbon content.

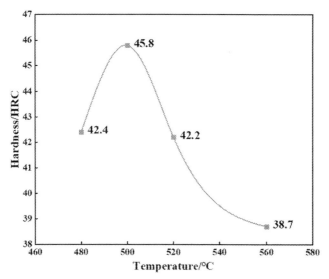

FIGURE 5.165 Effects of heat treatment temperature on hardness of depositing layer.

Fig. 5.166 shows the metallographic structure of the deposited metal after heat treatment at different temperatures (480°C, 500°C, 520°C and 560°C), respectively.

As shown in Fig. 5.167, when the heat treatment temperature was 480°C, the microstructure of the deposited metal was mainly tempered martensite and a small amount of retained austenite, the martensite was lath shaped and the carbide was an alloy carbide formed by Cr, V and C. At 500°C, the precipitation of carbides increased, and the particles were small and dispersedly distributed. As the heat treatment temperature rise to 520°C, the martensite in the welding layer metal decomposed, precipitated carbides and diffused along the grain boundaries, thus promoting nucleation and inhibiting grain nucleation, and the lath martensite transformed to ferrite equiaxed grains; when the heat treatment temperature reached 560°C, the cementite aggregated and the ferrite phase appeared recovery and recrystallization, but due to the decomposition of martensite, a large amount of carbide dispersed phase was precipitated, so that the ferrite in the metal of welding layer could not be sufficiently recrystallized to form large number of ferrite equiaxed grains.

It can be seen from the ESEM photos that when the heat treatment temperature was 500°C, the martensite was thin and uniform, the amount of the second phase particle was the largest, the particles were fine and finely distributed; lath martensite increased no matter whether the temperature was high or low, and the performance was not optimal. Through this series of comparative tests, the optimum heat treatment temperature of the flux-core wire can be obtained at

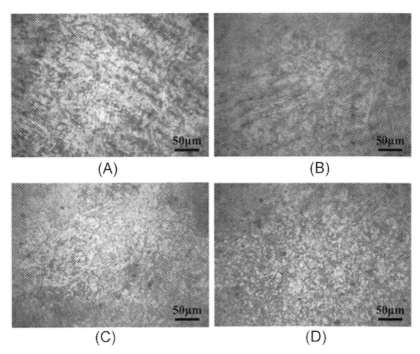

FIGURE 5.166 Metallographic structure of deposited metal after different heat treatment temperatures.

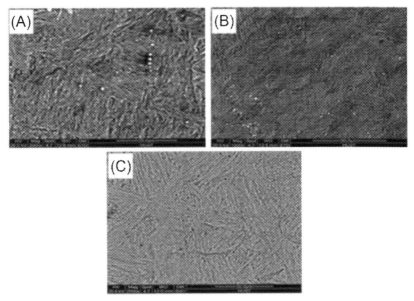

FIGURE 5.167 ESEM scan of WAAM samples heat treated at (A) 480°C, (B) 500°C and (C) 520°C.

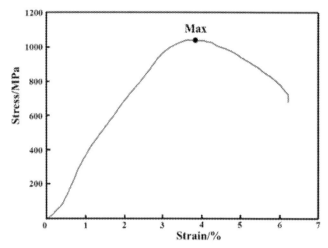

FIGURE 5.168 Tensile strength curve of high-alloy wire (MDY501) deposited sample.

about 500°C. In this temperature range, the hardness would reach a peak value due to "secondary hardening" effect, and the toughness was also enhanced, so that optimized microstructure and performance were obtained.

5.6.2.3 Tensile test

The Fig. 5.168 is the curve obtained by the tensile test on deposited sample manufactured via MDY501 wire. The test result shows that the strain linearly increased with the loading process in the elastic deformation range, and there was no yielding platform was shown in the test. It was because the high-alloyed steel contained a large amount of C, N, and alloy carbides, the "pinning effect" of the carbonitride particles caused disappearance of the yield platform. As the tension continues, the stress continued to increase up to 1039.4 MPa, followed by a "necking" phenomenon, and the tensile stress dropped rapidly until break down. The test results showed that the tensile strength of the high-alloy steel wire deposited layer was as high as 1039.4 MPa, and the material properties reach the required indexes.

5.6.2.4 Shearing test of wire arc addictive manufacturing-manufactured high-alloy steel

Three layers of self-designed flux-core wires were deposited on 5CrNiMo substrate, which were MGY281, MGY381 and MDY501 (Table 5.27). Fig. 5.169 shows the shearing test samples between two different materials. The shear samples used in the test had a 1 cm^2 bonding surface, which was used to detect the shear strength. The shearing strain versus stress between 5CrNiMo substrate and the MGY281 depositing layer is shown in

FIGURE 5.169 Photo of samples used in the shearing test.

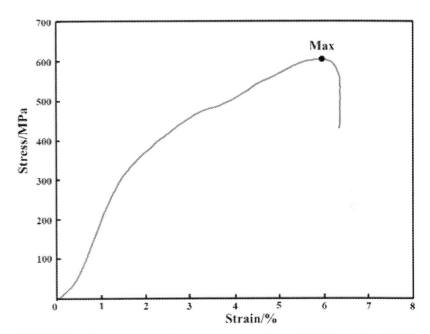

FIGURE 5.170 Shearing test results between depositing layer (MGY281) and the 5CrNiMo substrate.

Fig. 5.170. The stress-strain curve generated by the test indicated that the fracture limit of the testing sample was about 611.6 MPa, which was higher than the 542.1 MPa of the 5CrNiMo substrate. Therefore, it can be inferred that the bonding strength between the MGY281 layer and the substrate was high and reliable.

References

[1] R. Li, Y. Shi, J. Liu, et al., Densification and microstructure of selective laser melting forming of 304 L stainless steel powder, Appl. Laser 5 (2009) 369–373.

[2] W. Zhang, Y. Shi, J. Liu, Z. Lu, S. Huang, Complex structure micro-parts formed with selective laser melting method, in: P.J. Bartolo (Ed.), Virtual and Rapid Manufacturing-Advanced Research in Virtual and Rapid Prototyping. 11 New Fetter Lane, London EC4P 4EE, Taylor & Francis Ltd, England, 2007, pp. 537–540.

[3] L. Wang, Research on Properties of Selective Laser Melting Metal Parts, Huazhong University of Science and Technology, Wuhan, 2012.

[4] S. Zhang, Q.S. Wei, G.K. Lin, X. Zhao, Y. Shi, Effects of powder characteristics on selective laser melting of 316 L stainless steel powder, Adv. Mater. Res. 189 (193) (2011) 3664–3667.

[5] R. Li, Research on Key Basic Problems of Laser Powder Selective Laser Melting Forming, Huazhong University of Science and Technology, Wuhan, 2010.

[6] R.D. Li, Y.S. Shi, Z.G. Wang, et al., Densification behavior of gas and water atomized 316 L stainless steel powder during selective laser melting, Appl. Surf. Sci. 256 (13) (2010) 4350–4356.

[7] R.D. Li, J.H. Liu, Y.S. Shi, et al., 316 L stainless steel with gradient porosity fabricated by selective laser melting, J. Mater. Eng. Perform. 19 (5) (2010) 666–671.

[8] S. Zhang, Q.S. Wei, L. Cheng, S. Li, Y. Shi. Effects of scan line spacing on pore characteristics and mechanical properties of porous Ti6Al4V implants fabricated by selective laser melting. Mater. Design. 63 (2014) 185–193.

[9] S. Zhang, Study on Selective Laser Melting Process and Properties of Medical Alloy Powder, Huazhong University of Science and Technology, Wuhan, 2014.

[10] S. Zhang, Y. Li, L. Hao, T. Xu, Q. Wei, Y. Shi, Metal-ceramic bond mechanism of the Co-Cr alloy denture with original rough surface produced by selective laser melting, Chin. J. Mech. Eng. 27 (1) (2014) 69–78.

[11] S. Zhang, D. Yushu, B. Liu, B. Sun, C. Yan, L. Hao, et al. Effect of firing temperature on metal-ceramic bond strength of porcelain fused to metal restorations Co-Cr alloy fabricated by selective laser melting. Lasers Eng. 31 (2015).

[12] X. Zhao, Basic Research on Microstructure and Performance Evolution of Selective Laser Melting Formed Die Steel, Huazhong University of Science and Technology, Wuhan, 2015.

[13] X. Zhao, B. Song, W. Fan, Y. Zhang, Y. Shi, Selective laser melting of carbon/AlSi10Mg composites: microstructure, mechanical and electronical properties, J. Alloy. Compd. 665 (2016) 271–281.

[14] X. Zhao, B. Song, Y. Zhang, X. Zhu, Q. Wei, Y. Shi, Decarburization of stainless steel during selective laser melting and its influence on Young's Modulus, hardness and tensile strength, Mater. Sci. Eng. A 647 (2015) 58–61.

[15] X. Zhao, Q. Wei, B. Song, Y. Liu, X. Luo, S. Wen, et al., Fabrication and characterization of AISI 420 stainless steel using selective laser melting, Mater. Manuf. Process. 30 (2015) 1283–1289.

[16] B. Song, X. Zhao, S. Li, C. Han, Q. Wei, S. Wen, et al., Differences in microstructure and properties between selective laser melting and traditional manufacturing for fabrication of metal parts: A review, Front. Mech. Eng. 10 (2) (2015) 111–125.

[17] Y. Shike, Practical Basics of Welding Materials, second ed., Chemical Industry Press, 2015.

[18] G. Zhang, J. Zhang, Y. Pei, Process performance of CO_2 welding and MAG welding and development and application of welding wire, Welded Pipe Tube 29 (1) (2006) 11–15.

[19] J. Liu, S. Zhu, F. Yin, Influence law of process parameters on the limit dip angle of welding prototyping, J. Shenyang Univ. Technol. 34 (5) (2012) 515–519.

[20] B. Irving, How those million-dollar research projects are improving the state of the arc of welding, Weld. J. 72 (6) (1993) 41–45.

[21] S. Wu, Y. Liu, Principles of Material Forming, second ed., Mechanical Industry Press, 2008.

[22] Z. Cui, B. Liu, The Principle of Metallurgy and Heat Treatment, Harbin Institute of Technology Press, 2004.

[23] M. Terakubo, J. Oh, S. Kirihara, et al., Freeform fabrication of Ti-Ni and Ti-Fe intermetallic alloys by 3D micro welding, Intermetallics 15 (2) (2007) 133–138.

[24] A.F. Ribeiro, J. Norrish, R.S. McMaster, Practical case of rapid prototyping using gas metal arc welding, 1994.

[25] P. Kazanas, P. Deherkar, P. Almeida, et al., Fabrication of geometrical features using wire and arc additive manufacture, Proc. Inst. Mech. Eng. Part. B: J. Eng. Manuf. 226 (6) (2012) 1042–1051.

[26] J.D. Spencer, P.M. Dickens, C.M. Wykes, Rapid prototyping of metal parts by three-dimensional welding, Proc. Inst. Mech. Eng. Part. B: J. Eng. Manuf. 212 (3) (1998) 175–182.

[27] A.F. Ribeiro, AM with metals, Comput. Control. Eng. J. 9 (1) (1998) 31–38.

[28] A.F. Ribeiro, J. Norrish, Making components with controlled metal deposition, Int. Symp. Ind. Electron. 3 (1997) 831–835.

[29] Y.A. Song, S. Park, Experimental investigations into rapid prototyping of composites by novel hybrid deposition process, J. Mater. Process. Technol. 171 (1) (2006) 35–40.

[30] Y.A. Song, S. Park, D. Choi, et al., 3D welding and milling: Part I–A direct approach for freeform fabrication of metallic prototypes, Int. J. Mach. Tools Manuf. 45 (9) (2005) 1057–1062.

Chapter 6

Ceramic materials for additive manufacturing

As one of the three pillars of the global material industry, ceramic materials have outstanding characteristics, such as high strength, high hardness, high wear resistance, high-temperature resistance, and corrosion resistance. With the development of new ceramics with better mechanical, thermal, optical, electrical, chemical, and biological properties, it has drawn increasing application in the fields of mechanical, electronic, aerospace, military, bioengineering, and so on, which put forward new design requirements for the shape and structural complexity of ceramics. However, traditional ceramic forming methods such as dry pressing, isostatic pressing, tape casting, and extrusion forming have large restrictions by the geometry complexity of the fabricated parts, making it difficult to prepare parts of complex shapes. The mold manufacturing process is complicated, costly, and time-consuming, which is not enough to meet the needs of modern society for ceramic applications [1–4]. Therefore, the new forming process for realizing the preparation of ceramic parts with high-performance and complex structure without mold has gradually become one of the hot spots in ceramic research field. Additive manufacturing (AM) technology (also known as 3D printing) has changed the traditional "removal" or "equal volume" manufacturing to "additive" manufacturing and has a series of advantages such as no need for molds, short development cycle, and low cost. It has attracted wide attention of researchers around the world, and has become one of the most potential technical means for manufacturing complex ceramic parts. This chapter will systematically introduce several common AM technologies of ceramic materials and focus on the forming mechanism of ceramic materials prepared by selective laser sintering (SLS), the preparation method of ceramic powder used for SLS, and the properties of ceramic materials fabricated by SLS.

6.1 Additive manufacturing technology and principle of ceramic materials

Essentially, in the AM process of ceramic materials, a 3D modeling software is used to generate 3D models of parts in the computer, slice them based on the layering-superposition principle and import the information of each layer into the manufacturing equipment. Finally, any 3D solid parts with complex structures are manufactured by layer-by-layer stacking of the materials [5–6]. At present, there are several AM methods that can be applied to manufacture ceramic materials: stereolithography (SL) [7], laminated objected manufacturing (LOM) [8], fused deposition modeling (FDM) [9], three-dimensional printing (3DP) [10], selective laser melting (SLM) [11], and SLS [5]. These AM technologies have been widely applied and developed in the field of polymer and metal materials, but they are rarely used in the field of ceramic materials. Since the mid-1990s, many scholars have tried to form ceramic green parts by these AM technologies, and some even directly fabricate ceramic parts. Table 6.1 shows the main research work on the preparation of ceramic parts by AM methods.

TABLE 6.1 Main research work on the preparation of ceramic parts by additive manufacturing methods.

Additive manufacturing methods	Fabricating ceramic green parts	Sintering	Main research representatives
Stereolithography	Curing by UV light on a ceramic-photosensitive resin mixture	High-temperature furnace	Griffith et al. [12]
Laminated objected manufacturing	Bonding between layers of 3D parts of specific forming materials through laser beam.	High-temperature furnace	Griggin et al. [13]
Fused deposition modeling	Wire material is made by mixture of ceramic powder and organic binder to be used as raw material for ceramic green part.	High-temperature furnace	Agarwala et al. [14]
Three-dimensional printing	Adhesive selectively ejects from the nozzle and bonds the powder to form 3D part layer by layer.	High-temperature furnace	Yoo et al. [15]
Selective laser melting	Directly melt ceramic powder by laser beam, and the ceramic melted layer-by-layer solidifies into 3D part.	Forming machine	Shishkovsky et al. [16]

(*Continued*)

TABLE 6.1 (Continued)

Additive manufacturing methods	Fabricating ceramic green parts	Sintering	Main research representatives
Selective laser sintering	Ceramic powder and organic binder are mixed to be sintered by laser beam and compounded into powder to form 3D ceramic green part superimposed layer by layer.	High-temperature furnace	Subramanian et al. [17]

6.1.1 Stereolithography technology and principles of ceramic slurry

In SL technology, the ceramic slurry containing photosensitive resin is used as the raw material. Under the control of the computer, the ultraviolet light selectively scans the liquid photosensitive resin according to the cross-sectional information of the 3D solid model to realize single-layer curing, and then the workbench is lowered by the height of one layer of thickness, and the 3D solid part is finally obtained by repeating the above steps. The schematic diagram of SL forming process is shown in Fig. 6.1. SL is the most mature AM technology in present research and has been successfully applied in the field of biomedicine, microtechnologies and mechanical heat-resistant structures. Parts prepared by SL technology have high forming precision (± 0.1 mm) and high surface quality, which could be used to fabricate fine parts with complex structures.

In 1996, Griffin et al. [12] first produced ceramic parts through SL. They prepared ceramic-photosensitive resin slurry with a solid loading of 40%–55%, and formed three green parts of SiO_2, Al_2O_3, and Si_3N_4 by SL technology. The ceramic part is obtained after the binder removal and sintering of green part. In recent years, the Xi'an Jiaotong University, Tsinghua University, and other research institutes carried out preliminary research on the SL process of resin-based ceramic slurry. Zhou et al. [18] from Xi'an Jiaotong University found that the viscosity and solidification thickness of ceramic slurry played a decisive role in the forming process of ceramic green parts through experiments. The solid loading of the ceramic powder determines the shrinkage rate of the ceramic green part. Therefore, to reduce the shrinkage rate of the green part and improve the dimensional accuracy, it is necessary to increase the solid loading of the ceramic slurry, which is generally required to be higher than 40 vol.%.

Although SL technology has been widely used, it still faces the following problems for further development: (1) most of the currently available photosensitive

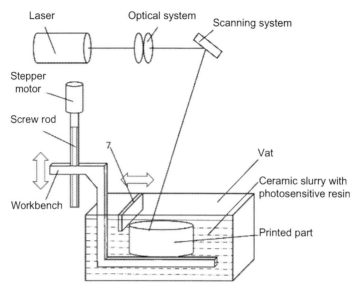

FIGURE 6.1 Schematic diagram of SL [12].

resins have a high price and are generally prone to cause environmental pollution because of their toxicity; (2) the price of the laser is high and the service life is limited, resulting in higher cost of the parts; and (3) to ensure the stability of the forming process, it is necessary to design a specific support structure according to the shape of the part, so that the predesign and postprocessing are complicated, causing a waste of resources.

6.1.2 Three-dimensional printing technology and principles of ceramic powders

Sachs and Cima et al. [15] from the Massachusetts Institute of Technology first proposed the concept of 3DP, and applied for a patent for this technology in 1991. The basic principle of 3DP is shown in Fig. 6.2. Under the control of the computer, the nozzle sprays the droplets in the working chamber to the specified position at a certain speed and frequency. As the droplets solidify layer by layer, the printed parts can be finally obtained [19]. The 3DP technology is featured by low equipment cost and wide range of forming materials, and complex structures such as inner cavities and cantilever beams without supporting structures.

In 1993, Yoo et al. [15] first applied 3DP technology to fabricate ceramic parts. Will et al. [20] obtained ceramic samples with different porosity (30%–64%) by 3DP using hydroxyapatite powder with adjusted particle size grading as raw material. Fierz et al. [21] used nanohydroxyapatite for 3DP forming after spray granulation to obtain biological scaffolds with macro- and micropores.

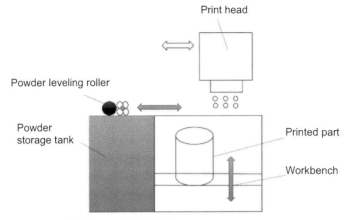

FIGURE 6.2 Schematic diagram of 3DP [15].

Fielding et al. [22] added ZnO and SiO_2 to the tricalcium phosphate powder, and successfully prepared ceramic parts with a relative density of 95%. Melcher et al. [23] prepared Al_2O_3 ceramics by high-temperature sintering of the infiltrated green parts, and not only the relative density of the obtained Al_2O_3 ceramics was greatly improved, but also the flexural strength and fracture toughness reached 236 MPa and 5.5 MPa m$^{1/2}$ respectively.

3DP technology can be used to prepare porous and dense ceramic parts, but the surface resolution of the parts is low, the precision is poor (about 0.2 mm), the relative density is not high, the nozzle is prone to blockage, and the maintenance is costly. These problems have severely limited the widespread use of 3DP technology.

6.1.3 Selective laser melting technology and principles of ceramic powders

The basic principle of SLM is shown in Fig. 6.3. Under certain powder preheating temperature conditions, a focused laser beam with a diameter of 30−50 μm is used to selectively melt the powder layer by layer, which accumulates into a printed part with metallurgical bonding and dense microstructure [24]. The biggest advantage of SLM is the short manufacturing cycle, good structure, performance of printed parts, and without postprocessing.

In 2007, Shishkovsky et al. [16] in France took the lead in the direct manufacturing of ceramic parts using SLM: the heating effect of the laser source made the powder preheating temperature up to 1450°C, and thus the YSZ-Al_2O_3 ceramic sample was fabricated. The surface is smooth, uniform and dense, but pores and cracks were still contained. Hagedorn et al. [25] in Germany directly produced Al_2O_3-ZrO_2 ceramic parts by SLM under the

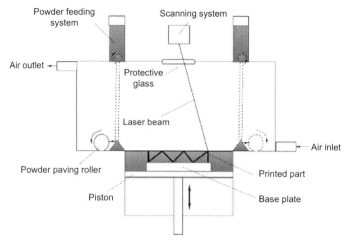

FIGURE 6.3 Schematic diagram of SLM [16].

powder preheating temperature up to 1700°C. Due to the limited heat-affected zone of the laser, the height of the printed parts was generally less than 3 mm. Wu et al. [26] used aerosol-assisted deposition to pave the powder, and dense Al_2O_3 parts were obtained after SLM process.

The relatively dense ceramic part could be prepared by SLM technology without subsequent sintering. However, due to the short action time between the laser and the powder, the physical and chemical changes during laser melting are complicated, the thermal shock resistance of the ceramic is poor, and the obtained parts often have defects such as pores and cracks. In addition, for equipment using high-temperature preheating system, the large molten pool that occurs during laser scanning often makes the rough surface of ceramics and the deteriorated accuracy.

6.1.4 Laminated objected manufacturing technology and principles of ceramic sheets

In 1984, Feygin proposed the idea of LOM technology and applied for a related patent in 1987 [27]. As shown in Fig. 6.4, the basic principle of LOM is a thin layer of material is coated with a layer of hot melt adhesive on one side, and the surface of the material reaches a certain temperature by hot-press arrangement to bond the thin layers together; according to the cross-sectional information of the 3D model, the CO_2 laser cuts the outline on the sheet coated with hot melt adhesive, cuts the noncontour area into a grid, and then the workbench is lowered by one layer thickness to lay a new layer of sheet. Under the rolling action of the hot pressing roller, the newly laid sheet is bonded to the cut layer, and the 3D solid part is finally obtained after the repetition of the steps above.

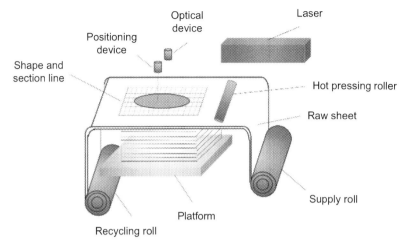

FIGURE 6.4 Schematic diagram of LOM [13].

In 1994, Griggin et al. [13] from Lone Peak used Al_2O_3 as raw material to prepare ceramic film by tape casting, and then they first used LOM technology to prepare ceramic parts with high purity and good performance that close to the properties of ceramics fabricated by the traditional hot pressing method. Rodrigues et al. [28] prepared Si_3N_4 ceramic film by tape casting and obtained Si_3N_4 ceramics with a relative density of 97% by LOM after sintering at the temperature of 1750°C, whose bending strength reached 918 MPa, and the fracture toughness reached 7.5 MPa m$^{1/2}$. Subsequently, in order to prepare parts with hollow structure and tilted surface, a "computer-aided manufacturing of laminated engineering materials" method [29] based on LOM technology has been rapidly developed in recent years. The basic principle of this method is similar to the traditional LOM, except that the scrap is removed before the new sheet is bonded, and when the slope or sphere is prepared, the cutting laser is tilted at an angle to prevent the staircase effect of the part surface. Cawley et al. [30] successfully prepared Al_2O_3 parts with internal pipe structure by this method.

However, the existing problem in fabricating ceramics by LOM is that the interlayer staircase effect on the surface of the green body, which is not smooth, and the boundary is required to be polished. The forming method of the ceramic material in the horizontal direction and the growth direction is different, as well as the relative density, resulting in the density of final ceramic component uneven, which is detrimental to the subsequent binder removal and sintering process and finally affect the properties of the ceramic parts.

6.1.5 Fused deposition modeling technology and principles of ceramic filaments

The principle of FDM technology is shown in Fig. 6.5. According to the path controlled by computer-aided design (CAD) layered data, the hot melt nozzle is used to squeeze the hot melt semiflow material so that the material is deposited in a specified position and solidified, and the solid part is finally obtained through deposition layer by layer. Because of its layer-by-layer stacking and solidification forming, this technology is also known as fused deposition method or melt extrusion forming [31].

In 1996, Agrarwala et al. [14] from the American Ceramic Research Center used Si_3N_4 with a small amount of acidic oxides as raw materials to prepare ceramic parts by FDM technology. Stuecker et al. [32] successfully prepared porous mullite green parts with a pore size of 100−1000 μm using mullite filaments in a diameter of 225−1000 μm. The support structure had a relative green density of 55%, which could reach 96% after sintering. To obtain a functionally graded material with a gradual transition of composition, Leu et al. [33] used an FDM apparatus equipped with three extrusion heads to form green parts, and the required parts were successfully obtained after freeze drying, binder removal, and high-temperature sintering.

At present, the FDM technology has the following problems: (1) the printing precision of parts is low, and the surface of the parts is rough; (2) compared with other AM technologies, the printing speed is slow, and it is not suitable for preparing large parts. Therefore, FDM is mainly used

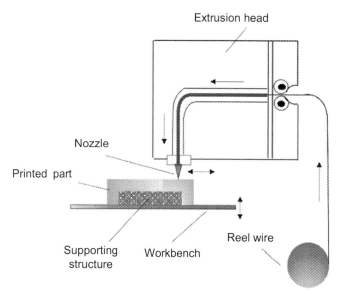

FIGURE 6.5 Schematic diagram of FDM [14].

commercially in the preparation of small-scale customized products, and there is still some distance away from largescale applications.

6.1.6 Selective laser sintering technology and principles of ceramic powders

In 1986, a graduate student at the University of Texas in the United States proposed the SLS, and then founded a company and proposed the SLS-based commercial forming system—Sinterstation [27]. The principle of SLS technology is shown in Fig. 6.6: a thin layer of powder is laid on the workbench, and the high-energy CO_2 laser beam selectively scans the powder layer under the control of the computer according to the CAD data; the powder in the scanned area is bonded together by sintering or melting, while the powder in the unscanned area is still loose and can be reused; after the processing is completed, the workbench is lowered by one layer thickness and the powder laying and scanning of the next layer continue, and the new layer will be bonded to the previous layer. The process is repeated until the entire part is formed, and then the part is finally removed from the working cylinder [34]. In 1995, Subramanian et al. [17] in the United States first applied SLS technology to the preparation of ceramic parts. Since the sintering temperature of the ceramics is high, it is difficult to be directly sintered using laser. At present, an indirect SLS forming strategy is generally used to coat or mix refractory high-melting ceramic particles with low-melting polymer binder, and the laser melts the binder to bond the ceramic powders to form ceramic green parts [6].

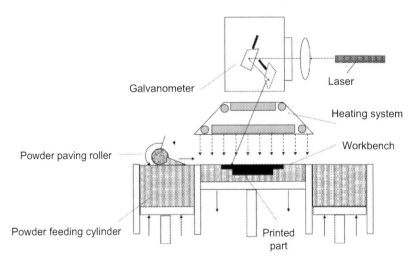

FIGURE 6.6 Schematic diagram of SLS [17].

The SLS technology of ceramic materials, integrated by CAD, computerized numerical control (CNC), laser processing technology, and material science and technology together, has many outstanding advantages:

1. Short production cycle and low manufacturing cost, it is suitable for the development of new products, especially for the production of complex-shaped parts, which has obvious advantages compared with other processes.
2. Combined with traditional process methods, functions such as rapid casting and rapid mold manufacturing can be realized, which injects new vitality into traditional manufacturing methods.
3. Wide application, SLS technology can be applied to many fields, such as automobiles, molds and home appliances.
4. Compared with other AM technologies, the most prominent advantage of SLS is that the raw materials are very extensive. Theoretically, any powder material that can be bonded by polymer material after heating can be used as the raw material for SLS.

Although the AM technology of aforementioned ceramic materials has been experimentally proved to be able to form ceramic parts with complex shapes, there are nonnegligible limitations in forming ceramic products, which seriously restricts the development and practical application of ceramic additive technology to some extent. However, the SLS technology of ceramic materials, as one of the most widely studied AM technologies to fabricate ceramics, has irreplaceable advantages in the manufacture of ceramic components of complex structures due to its wide variety of ceramic powder materials, wide range of sources, good surface quality of the parts, and high stability. Therefore, this chapter focuses on the forming mechanism of the ceramics prepared by SLS, the preparation of ceramic powder for SLS, and the properties of the prepared ceramic components.

6.2 Forming mechanism of the ceramics prepared by selective laser sintering

When the ceramic powder containing polymer binder (for example, epoxy resin) is formed by SLS, the energy of the laser mainly acts on the epoxy resin as binder, and the ceramic particles do not change. When the temperature gradually increases to the glass transition temperature (T_g), the epoxy resin powder absorbs the heat to change from glass state at normal temperature to a soft and high elastic state; when the temperature exceeds the melting temperature (T_m), the epoxy resin in high elastic state transforms into viscous flow state. When the temperature raises, the viscosity of the molten epoxy resin decreases and the fluidity increases. At this time, it is easy for resin to contact with the surrounding ceramic particles, and the ceramic powder can be bonded after cooling and solidification. The laser-formed green

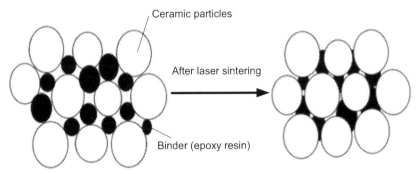

FIGURE 6.7 Principle of selective laser sintering of ceramic powder.

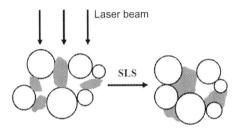

FIGURE 6.8 SLS process of mechanically mixed powders of polymer and ceramic powders.

part mainly bonds the powder together by the bonding force, which is determined by the cohesion and adhesion between the molecules of the polymer binder, namely the strength of the epoxy resin. Adhesion is the force between the epoxy resin and the ceramic particles, that is, the force by which the epoxy adheres to the surface of the ceramic particles. Fig. 6.7 shows the schematic diagram of SLS of ceramic powder, in which some polymer binder (for example, epoxy resin) has been added.

Fig. 6.8 shows the SLS process for mechanically mixed powders of polymer and ceramic powders, which can be divided into the following three processes:

1. The process of absorbing the laser energy by the mixed powder

    The absorption rate of the composite powder to the carbon dioxide laser can be expressed as:

    $$\alpha = \varphi_P \alpha_P + \varphi_M \alpha_M \tag{6.1}$$

    where α is the absorption rate of the mixed powder, α_P and α_M are the absorption rates of the polymer binder and the ceramic powder, respectively; φ_P and φ_M represent the volume fraction of polymer binder and ceramic powder, respectively.

2. Wetting process of ceramic powder by binder

The polymer binder absorbs laser energy, melts as the temperature increases, and wets and bonds the ceramic powder particles. The wetting rate is related to the viscosity of the binder, the contact angle and the surface tension. Considering the pores on the solid surface as capillaries, the time that the liquid with a viscosity of η flows through the capillary with a radius of R and a length of L can be calculated by Eq. (6.2) [35]:

$$t = \frac{2\eta L_2}{R\gamma \cos\theta} \quad (6.2)$$

Because the surface tension of various organic liquids is not much different, in the case of a certain capillary size, the wetting time is mainly determined by the viscosity of the liquid and the contact angle. The smaller the viscosity of the liquid and the contact angle are, the faster the wetting proceeds.

In addition, the surface roughness of ceramic particles also has an effect on the wetting rate. The thermodynamic analysis of this effect can be described by Eq. (6.3) [36]:

$$\Delta G = -\gamma_L \left[1 + \left(\frac{A_s}{A_L}\right)\cos\theta\right] A_L \quad (6.3)$$

where γ_L is the surface tension of the organic resin, A_L is the surface area of the wetted solid, A_S is the true area of the wetted solid and θ is the wetting angle. A_S/A_L is the roughness coefficient of the surface of the particle.

3. Sintering process of polymer binder particles

The sintering rate of the mixed powders depends on the wetting rate of the binder on the ceramic powders and the sintering rate between the binder particles, and the sintering rate between the binder particles is greater than the wetting rate of the ceramic powders by the binder.

Fig. 6.9 shows the SLS process of polymer-coated ceramic powders, which can be divided into the following two processes:

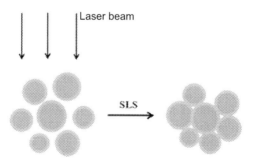

FIGURE 6.9 SLS process of polymer-coated ceramic powders.

1. The process of absorbing the laser energy by the powder coated with binder

    Since the nanoceramic aggregate is completely coated by the polymer binder in the coated powder, which substantially corresponds to the laser scanning of the polymer binder itself. Therefore, the laser absorption rate of the coated powder is the absorption rate of the polymer binder, which can be expressed as Eq. (6.4):

$$\alpha = \alpha_P \tag{6.4}$$

2. Sintering process of polymer binder layer

    The polymer binder layer on the surface of ceramic particles absorbs the laser energy, with the increase of temperature, the binder is sintered between the ceramic particles to form a sintering neck, and the sintering mechanism belongs to viscous flow mechanism.

6.3 Preparation of ceramic materials for selective laser sintering

6.3.1 Ceramic powders and binders for selective laser sintering

1. Ceramic powders for selective laser sintering

    The materials used for SLS technology are extensive, but the physical and chemical properties of the materials have great influence on the forming process. When selecting ceramic materials for SLS, the influence of material properties on the fabricated parts should be considered [37]. For example, the physical properties of powder material such as heat absorption, thermal conductivity, shrinkage, melting point, reaction curing temperature and time, crystallization temperature and rate, thermal decomposition temperature, flame retardancy and oxidation resistance, modulus, melt viscosity, melt surface tension, particle size distribution, particle shape, apparent density and fluidity of powder material have effects on the properties of the SLS-printed parts. At present, Al_2O_3, ZrO_2, cordierite and kaolinite are the major ceramic materials used to form ceramics by SLS at home and abroad.

    Al_2O_3 ceramics are one of the most commonly used ceramic materials. With the rapid development of medicine, electronics, machinery and other industries, the market has an increasing growth space for the demand of Al_2O_3 ceramics, and their output will continue to grow. Al_2O_3 ceramics are mainly divided into three types according to their phase: α-Al_2O_3, β-Al_2O_3 and γ-Al_2O_3. β-Al_2O_3 is an unstable transition phase between the formation from γ to α phase. It begins to change to α phase at a temperature of 1000°C, and its volume and density change. Therefore, this phase is not suitable for direct application in industrial production. γ-Al_2O_3 is industrial Al_2O_3 which is mainly used in the electrolytic aluminum industry to produce aluminum ingots. It has large burning shrinkage and is also an unstable phase. α-Al_2O_3 is a stable phase and the main component of corundum, it has stable structure, high apparent density and high strength, which is generally obtained by industrial calcination and phase transformation from β and γ phase. Therefore, α-Al_2O_3 is generally used ceramic products in the industrial production, and there are few cracks in the sintered products.

ZrO_2 ceramics are very important functional and structural ceramics, which have excellent physical and chemical properties such as good chemical stability, high-temperature resistance, corrosion resistance, good thermal stability and excellent mechanical properties. They are widely used in industrial production and are important raw materials for refractory materials, high-temperature structural materials, wear-resistant materials, and electronic materials [38]. ZrO_2 has three crystal forms: monoclinic, tetragonal, and cubic. ZrO_2 appears only in monoclinic phase at room temperature, but transforms into tetragonal phase when heated to about 1100°C, and further transforms into cubic phase at higher temperature. Due to the large volume change during the transition from the monoclinic phase to the tetragonal phase, and the opposite change appears when cooling, it is likely to cause product cracking, which limits the application of pure ZrO_2 in high-temperature field. However, after adding Y_2O_3 as stabilizer, the tetragonal phase can be stabilized at normal temperature, and no volume change occurs after heating, which greatly expands the application scope of ZrO_2 [39].

Cordierite, with the molecular formula of $2MgO \cdot 2Al_2O_3 \cdot 5SiO_2$, is a magnesium aluminosilicate mineral with a density of 2.53–2.78 g/cm^3, and has relatively low melting point of 1460°C among ceramic material. In addition, the most remarkable characteristic of cordierite is its low coefficient of thermal expansion, which is determined by the bond length and bond angle of its crystal chemical bond. In cordierite crystals, temperature has small influence on the internal Al-O and Si-O bond length of cordierite [40]. Cordierite is widely used in metallurgy, automobile, electronics, chemical industry, and environmental protection because of its low thermal expansion coefficient, good thermal shock resistance, and high hardness.

Kaolinite is a mixture of aluminosilicate minerals, with the main components of Al_2O_3 and SiO_2. Kaolinite can be divided into noncoal series kaolinite and coal-series kaolinite. The former is a kaolinite clay material commonly used in the fabrication of traditional ceramic products. It is prepared by inorganic acidification to remove Fe^{3+}, and so on, and then rinsed and dried by water. It has strong adsorption performance and can adsorb colored substances as well as organic substances. The latter is a kind of solid waste generated during coal mining and washing, which is mainly composed by kaolinite and carbonaceous, and is the main component of coal gangue [41].

Due to their high brittleness and high hardness, ceramic materials are prone to form defects in the manufacturing process, which is difficult to be dealt with through subsequent processing. Especially for ceramic parts with complex structures, they are more difficult to form and process [3]. Traditional ceramic forming methods mainly include dry pressing, isostatic pressing, slip casting, injection forming, and extrusion forming [4]. With the increasing demand for complex structural ceramic components in modern industry, and the increasingly fierce competition in various industries in the modern market, the aforementioned forming technologies not only require expensive molds, but also are difficult or unable to realize the manufacture of high-performance complex structural ceramic parts. As a typical AM technology that breaks through the deformation forming and dislodge forming of materials compared to conventional ceramic forming methods, SLS can form by "adding materials" without tooling or molds, which has unique advantages in the preparation of ceramics with complex shapes.

2. Polymer binders for selective laser sintering

The binder for SLS is required to have low melting point, good wettability, and low viscosity. The use of binder with low viscosity in liquid-phase conditions is beneficial to form parts by SLS. This is mainly because this kind of binder has good fluidity after melting at a high-temperature, which is favorable for the migration of substances, so that the parts tend to be uniform in structure and performance.

At present, there are mainly three types of binders used in SLS forming method: inorganic binders such as ammonium dihydrogen phosphate [42]; organic binders such as epoxy resin [43], phenolic resin [44], and nylon PA12 [45]; and metal binders such as aluminum powder [46]. Since the inorganic binder and the metal binder cannot be removed from parts in the posttreatment process easily, which might introduce impurities to destroy the phase structure of the raw material, and finally decrease the performance of the part, so the inorganic binder and the metal binder are mainly used for the preparation of some composite ceramic parts. For organic binders, epoxy resin, PMMA, and nylon are mostly studied. As shown in Table 6.2, cordierite ceramics are taken as an example. The performances of several typical epoxy resins are shown in Table 6.3.

When used as a binder, epoxy resin has the following characteristics: (1) the epoxy resin has strong bonding ability because the polarity of the hydroxyl group, and the ether group in the structure allows the epoxy resin molecule and the adjacent molecule to generate gravity. Therefore, the epoxy resin is suitable for bonding a variety of ceramic particles; (2) epoxy resin has low shrinkage rate and strong deformation resistance, which can reduce shrinkage and warpage during the thermal action of

TABLE 6.2 The comparison of green bodies obtained by adding different binders.

Binder type	Mass fraction	Forming effect	Strength
Epoxy resin E12	5%–20%	Good formability, no warpage, high precision.	High
PMMA	5%–20%	Less effective, PMMA particles are dispersed in the powder.	None
Nylon PA12	5%–20%	Requires high preheating temperature and poor formability.	None

TABLE 6.3 Performances of three kinds of epoxy resins.

Grade	Original grade	Color and shape	Softening temperature (°C)	Epoxy value
E03	609#	Yellow transparent solid	135–155	0.02–0.04
E06	607#	Yellow transparent solid	110–135	0.04–0.07
E12	604#	Yellow transparent solid	85–95	0.10–0.18

SLS; (3) epoxy resin has a low softening temperature, and is easy to achieve its fusion bonding.

Epoxy resin has good wettability and bonding ability that is suitable for SLS forming. Taking the cordierite as an example, three different epoxy resins are used as binders to make ceramic green parts, the quality of which is shown in Table 6.4. It can be seen from Table 6.4 that when E03 is used as the binder, the strength of the green part is low, and it is difficult to take out after posttreatment process. Both E06 and E12 can effectively bond ceramic powder to form green part, but E06 has higher softening temperature, which requires to heat the working chamber to higher temperature, while E12 only needs to be preheated to 45°C to be formed.

6.3.2 Preparation of composite ceramic powders for selective laser sintering

There are four main types of ceramic powder materials that have been extensively studied at this stage: ceramic powders with directly mixed binders,

TABLE 6.4 Quality of cordierite green parts prepared by SLS using three kinds of epoxy resins as binder.

	E03	E06	E12
Strength of green part	Relatively low	High	High
Required SLS laser power	High	Medium	Low
Warpage	Appears when the preheating temperature is low	Does not appear when the preheating temperature is low	Basically does not appear
Forming precision	High	High	High
Interlayer offset phenomenon	Mild	None	None

ceramic powders with surface-coated binders, surface-modified ceramic powders, and resin sands [6]. Among them, the composite ceramic powders suitable for SLS can be prepared by methods such as mechanical mixing, spray drying, and surface coating [47,48]. The ceramic/binder composite powders suitable for SLS forming are generally prepared by mechanical mixing or surface coating, and the polymer-coated ceramic particles are prepared by spray drying method and dissolution precipitation method.

6.3.2.1 Mechanical mixing method

In the mechanical mixing method, the ceramic powder and the appropriate amount of binder are mechanical ball milled in a planetary or a 3D powder mixer to achieve complete homogenization of the composite powder, and the prepared composite powders still maintain their respective forms and properties of original powder. This method is simple in operation, and has low requirements in equipment and short milling cycle, which can prepare composite ceramic powders suitable for the SLS forming via fully mixing. Therefore, mechanical mixing is most widely used.

6.3.2.2 Solvent evaporation method

Taking the stearic acid/nanoceramic composite powder as an example, the preparation of composite ceramic powder by solvent evaporation method is expounded. As shown in Fig. 6.10, first, the nanoceramic powder is mixed with absolute ethanol, and ZrO_2 grinding balls are added for ball milling to fully disperse the nanoceramic powders in the solvent; then the dispersed

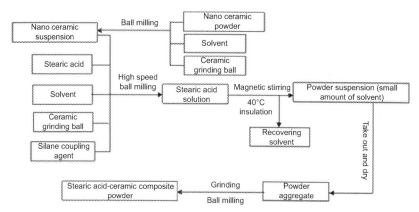

FIGURE 6.10 Preparation process of the stearic acid/nano-ZrO_2 composite powder.

nanoceramic powder mixture is taken out and added to the ball mill tank. The mass ratio of the ceramic powders, stearic acid, and ZrO_2 grinding balls is 4:1:10, and anhydrous ethanol is used as ball milling medium for ball milling at 300 rpm for 4 hours. After the ball milling, the mixture is poured into a flask connected to an ethanol recovery device and is placed on a thermostatic magnetic stirrer for stirring at a constant temperature of 40°C. When the solvent evaporates and remains small amount, the mixture is taken out and dried in an incubator, then the dried powder is slightly grinded or ball milled and passed through a 200-mesh sieve to obtain the nanoceramic−stearic acid composite powders.

6.3.2.3 Dissolution precipitation method

The principle of preparing ceramic composite powder by the dissolution precipitation method for SLS is like this: the polymer powder and the ceramic powder are put into organic solvent to dissolve the polymer in the solvent by heating, then the mixed solution is stirred vigorously; after the solution is cooled, a film formed by polymer crystallization is adhered on the surface of the ceramic powder particles, and then the aggregate of the film-coated powder is obtained by distillation drying. Finally, the polymer-coated-film composite ceramic powder with different particle diameters is obtained by ball milling.

Nylon is a kind of polymer with excellent solvent resistance. It is difficult to dissolve in common solvents under normal temperature conditions, but it is soluble in specific solvents at high-temperature. For example, ethanol is used as the solvent, nylon, coated powder, and antioxidant is added, the nylon is dissolved at high-temperature, and gradually cooled down by intense stirring. Since the coated ceramic powder has heterogeneous nucleation effect on the nylon crystallization, nylon will preferentially precipitate on the ceramic powder to form coated-film powder [49]. However, the nylon polymer powder material has a large specific surface area and is prone to oxidative degradation during the SLS forming process, resulting in poor performance. Therefore, it is necessary to add antioxidant to reduce the thermo-oxidative ageing during SLS and the use of molded parts.

In addition, the surface treatment of the ceramic powder with the silane-coupling agent improves the compatibility of the ceramic powder with the nylon and enhances the interfacial adhesion between the ceramic powder and the nylon, so it is favorable for uniform dispersion of the ceramic powder in the dissolution precipitation method [45]. Taking ZrO_2 as an example, in order to improve the interfacial adhesion between nylon 12 and nano-ZrO_2 matrix, the surface of nano-ZrO_2 particles is organically treated with silane-coupling agent, APTS. The reaction process of APTS with nano-ZrO_2 and nylon 12 is shown in Fig. 6.11.

First, APTS is hydrolyzed to hydrolysate containing alcohol group (−OH), and the reaction formula is shown in Fig. 6.11A. Second, the surface of the nano-ZrO_2 contains a large amount of -OH, which can undergo polycondensation reaction with the hydrolysate of APTS, forming siloxane, so

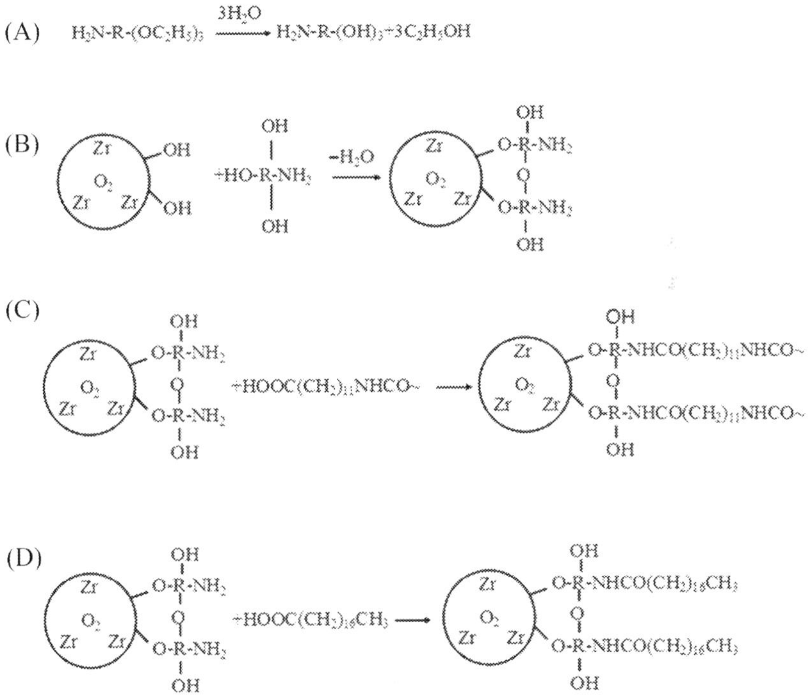

(A) $H_2N\text{-}R\text{-}(OC_2H_5)_3 \xrightarrow{3H_2O} H_2N\text{-}R\text{-}(OH)_3 + 3C_2H_5OH$

(B) [Reaction scheme showing ZrO_2 particle with surface -OH groups reacting with $HO\text{-}R\text{-}NH_2$, releasing H_2O, forming $O\text{-}R\text{-}NH_2$ groups on the surface]

(C) [Reaction scheme showing ZrO_2 particle with $O\text{-}R\text{-}NH_2$ groups reacting with $HOOC(CH_2)_{11}NHCO\sim$, forming $O\text{-}R\text{-}NHCO(CH_2)_{11}NHCO\sim$ groups]

(D) [Reaction scheme showing ZrO_2 particle with $O\text{-}R\text{-}NH_2$ groups reacting with $HOOC(CH_2)_{16}CH_3$, forming $O\text{-}R\text{-}NHCO(CH_2)_{16}CH_3$ groups]

FIGURE 6.11 Reaction formula of silane-coupling agent, APTS, with nano-ZrO_2 and nylon 12 resin.

that the amino group ($-NH_2$) is grafted onto the surface of the nano-ZrO_2 particles, and the reaction formula is shown in Fig. 6.11B; finally, the amino group grafted onto the surface of the nano-ZrO_2 can react with the carboxyl group in nylon 12 to form amide bond, and the reaction formula is as shown in Fig. 6.11C and D, and the interfacial adhesion of the nano-ZrO_2 to the nylon 12 and the stearic acid matrix is improved.

Taking nylon/nanoceramic composite powder as an example, as shown in Fig. 6.12: (1) Mixing a certain amount of nanoceramic powder with absolute ethanol, and adding ZrO_2 grinding ball for ball milling to make nanoceramic powder fully dispersed in the solvent; (2) taking out the nanoceramic mixture, and putting it into the jacketed stainless steel reaction kettle with nylon 12, solvent, antioxidant and silane-coupling agent in proportion, seal and pump the reaction kettle, then inject N_2 for protection (nylon 12 and nanoceramic powders are used to prepare the composite powder with two kinds of nylon contents at a mass ratio of 1:4 and 1:3), the antioxidant content is 0.1%−0.3% of nylon 12 mass, and the content of silane-coupling agent is 0.1%−0.5% of nylon 12 mass (PA12); (3) gradually heating to 140°C at a rate of 1−2°C/min, so that the nylon is completely dissolved in the solvent anhydrous ethanol, and keep the temperature and pressure for 1−2 hours; (4) gradually cooling to room temperature at a rate of 2°C−4°C under intense stirring, so that nylon is coated on the outer surface of ZrO_2 powder aggregate by gradually taking ZrO_2 as the powder aggregate as core, nylon coated-film nanoceramic powder suspension could be obtained; (5) taking out the coated-film nanoceramic powder suspension from the reaction kettle and stand for a few minutes, the coated-film nanoceramic powder in the suspension will precipitate, and the remaining anhydrous ethanol solvent can be

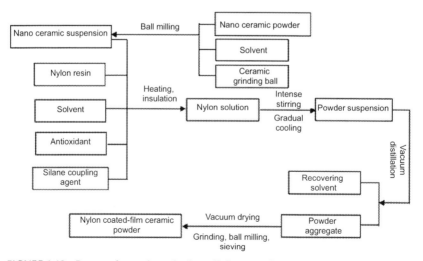

FIGURE 6.12 Process of preparing nylon/nano-ZrO_2 composite powder.

recycled; (6) vacuum drying the powder aggregates at 80°C for 24 hours to obtain the dried nylon coated-film nanoceramic composite powder, then slightly grind in the grinding bowl and ball milling at a speed of 200 rpm for 15 minutes, so that the nylon 12 coated-film nanoceramic powder is obtained after sieving through 200 meshes.

The composite ceramic powders for SLS prepared by the binder coating method can make the binder distribution uniform and reduce the phenomenon of powder segregation during laser sintering process. According to research by Vail [50], the ceramic green parts formed by the binder coating method have higher strength than those formed by the mechanical mixing, and the final parts have better forming precision and mechanical properties. This is because the green part obtained by the binder coating method has more uniform distribution of the binder and the ceramic particles inside, and the shrinkage and deformation of the green part are relatively small during the postprocessing, so the internal structure of the obtained part is also more even. However, compared with the powder prepared by the mechanical mixing method, the powder prepared by the binder coating method is more uniform, but the process is generally complicated, and it is easy to introduce impurities during the experiment, which requires more professional equipment. For example, to recycle binder-coated powder and the solvent, it needs a special vacuum filtration device, which shows low efficiency; the dried powder needs to be ground and sieved again. The process is complicated, time-consuming, and costly, and it is unfriendly to the environment. The SLS forming powder is usually prepared by the binder coating method for submicron ceramic powder having small particle size.

6.3.3 Properties of ceramic composite powders and ceramics prepared by selective laser sintering

When selecting ceramic materials for SLS, the influence of material properties on the prepared parts should be considered [37]. For example, the physical properties such as heat absorption, thermal conductivity, melting point, thermal decomposition temperature, melt viscosity, particle size distribution, particle shape, apparent density, and fluidity of the powder materials all have an effect on the properties of the prepared parts. At present, Al_2O_3, ZrO_2, cordierite and kaolinite are the major used ceramic materials for SLS studied at home and abroad [51−54].

6.3.3.1 Al_2O_3-polymer composite powder and properties of Al_2O_3 ceramics prepared by selective laser sintering

The Al_2O_3 powder used is a granulated α-Al_2O_3 powder of 140−180 mesh. It is prepared by spray drying of PVA with a mass fraction of 1.5% added to the submicron Al_2O_3 raw powder. The granulated α-Al_2O_3 ceramic powder and the binder pulverized to 50 μm are added to the ball mill with a mass ratio of 100:8, and the ball mixture having a volume ratio of 1:3 is mixed for 2 hours in ball mill to obtain the composite powder of uniform mixing.

Fig. 6.13A shows the SEM images of the Al_2O_3 raw powder. It can be seen that the Al_2O_3 particles are obviously irregular and the particles are fine. The particles are relatively uniform in size and slightly agglomerated. Most of the particles have an average particle size of 0.4 μm, while the agglomerated blocks generally have a diameter of no more than 1 μm. Fig. 6.13B shows the SEM morphology of binder epoxy resin, E06, and granulated Al_2O_3 homogeneous mixed powder. The particles or particle aggregates having irregular surfaces in the figure are epoxy resin E06, the particles are mostly polyhedrons with a water chestnut, and the particles having a smooth surface are granulated Al_2O_3 particles. Since the particle size of the powder is mostly in the range of 74−150 μm, the powder has high fluidity and thus has high formability.

Fig. 6.14 shows the fracture morphology of the SLS-fabricated sample using the granulated Al_2O_3-binder composite powder under laser power of 21 W, scanning speed of 1600 mm/s, scanning pitch of 100 μm, and single-layer thickness of 150 μm. It can be seen that after laser scanning, the PVA film-covered Al_2O_3 granulated particles are hardly affected, and still maintain the spherical morphology before SLS, but the particles are bonded by

FIGURE 6.13 SEM images of Al_2O_3 granulated powder and its composite powder formed with polymer: (A) Al_2O_3 granulated powder; (B) Al_2O_3-binder composite powder.

FIGURE 6.14 SEM images of fractured SLS sample of granulated Al_2O_3-binder composite powder: (A) low magnification; (B) high magnification.

the molten epoxy resin. It can be seen from Fig. 6.14B that there are many bonding necks between the particles. These bonding necks, formed by the melting and solidification of the epoxy resin after absorbing the laser heat, have high bonding strength since the epoxy resin wets on the surface of the PVA and both epoxy resin and PVA are polymers. However, there are still many pores inside the sample, which requires subsequent processing.

Fig. 6.15A shows the SEM images of fractured of the SLS-fabricated samples using granulated Al_2O_3-binder composite powder under low-magnification SEM when the cold isostatic pressing (CIP) holding pressure is 335 MPa. The composition distribution is relatively uniform, and there are no obvious pores, which is beneficial for further densification. It can be seen from Fig. 6.15B that the PVA-coated Al_2O_3 granulated particles are obviously compressed, the particles are closely arranged, and the epoxy resin bonded necks no longer exist, which are crushed and filled in the sample pores.

The cross section of the sintered sample is ground and polished, and heats at 1300°C for 2 hours. Fig. 6.16A and B are the SEM images of samples at the sintering holding temperature of 1510°C and 1650°C, respectively. As can be seen from the figure, the sample contains a lot of pores and the arrangement between the grains is not tight enough at 1510°C. When the sintering temperature increases to 1650°C, many pores are eliminated, and the grains are closely arranged, but some regions are still not dense.

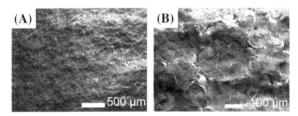

FIGURE 6.15 SEM images of fractured SLS sample of granulated Al_2O_3-binder composite powder after CIP: (A) SEM morphology of fractures of SLS/CIP specimens (low magnification); (B) SEM morphology of fractures of SLS/CIP specimens (high magnification).

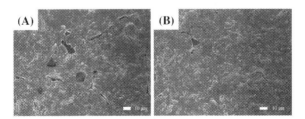

FIGURE 6.16 SEM images of fractured SLSed parts using granulated Al_2O_3-binder composite powder after sintering at different temperatures: (A) 1510°C; (B) 1650°C.

6.3.3.2 ZrO₂-polymer composite powder and properties of ZrO₂ ceramics prepared by selective laser sintering

6.3.3.2.1 ZrO₂-E06 composite powder and properties of ZrO₂ ceramics prepared by selective laser sintering

Fig. 6.17 shows the morphology of ZrO_2-10 wt.% epoxy resin E06 (EZ10) composite powder. The spherical particles with uneven size are ZrO_2 granulated powder with an average diameter of 45–60 μm. The irregular particles with a small particle size are epoxy resin E06, which have an average particle diameter of only 20–28 μm. Since the ZrO_2 granulated powder still maintains the spherical morphology before mixing and most of the particles are spherical and uniform in particle size, so the powder has good fluidity. However, the distribution of the epoxy resin particles in the composite powder is not uniform, because the density difference between the E06 powder and the ZrO_2 powder is large, and it is difficult to achieve complete uniformity in mechanical mixing. However, the uniformity of the distribution of the polymer binder in the ceramic powder will directly affect the pore and density distribution of the ceramic sample during the SLS process, and affect the subsequent CIP, thermal binder removal and the dimensional shrinkage and density distribution in high-temperature sintering.

Fig. 6.18A and B shows the particle size distribution diagrams of the granulated ZrO_2 powder and the epoxy resin E06 powder, respectively. It is found that the average particle size difference between the ZrO_2 powder and the E06 powder is large, which is 45–60 and 20–28 μm, respectively.

Fig. 6.19 shows the SEM images of fractured EZ10 specimens. The corresponding samples are prepared by optimized process with laser power = 5.5 W, scanning speed = 1400 mm/s, scanning space = 0.1 mm, laser energy density = 0.3929 J/mm², preheating temperature = 55°C. Fig. 6.19A and B shows the SEM images of fractured SLSed EZ10 specimens. The sample is formed mainly because the ZrO_2 granulated particles are bonded by the epoxy resin binder to form bonding neck, so the sample has high porosity. On the one hand, the distribution of these bonding necks

FIGURE 6.17 SEM images of ZrO₂-epoxy E06 composite powder (EZ10): (A) low magnification; (B) high magnification.

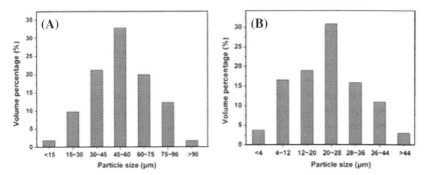

FIGURE 6.18 Particle size distribution of EZ10 powder: (A) granulated ZrO_2 powder; (B) epoxy resin E06 powder.

FIGURE 6.19 SEM images of fractured EZ10 specimens: (A, B) SLSed specimens; (C, D) SLS/CIPed green specimens; (E, F) SLS/CIPed sintered specimens.

in the SLS sample is not uniform; on the other hand, some of the ZrO_2 spherical particles are crushed by the action of the laser, which causes the ZrO_2 distributed with various morphologies in SLSed sample. These factors are not conducive for the sample to retain its original shape and density during subsequent CIP processing. Fig. 6.19C and D shows the SEM images of fractured SLS/CIPed green specimens. There are still many pores inside the sample. This is because the pores between the initial particles of ZrO_2 are large, and the high deformation resistance of ceramic particles will always hinder the reduction of these pore sizes during the CIP process, and the binder epoxy resin has poor fluidity in the CIP stage, only partially broken epoxy resin bonding neck is pressed into the pores, and there will be no largearea resin flow behavior, so there are still many pores. Fig. 6.19E and F shows the SEM images of fractured SLS/CIPed sintered EZ10 specimens. A large amount of pores remains inside the sintered sample. These pores mainly include two types: (1) pores existing in the SLS/CIP green part before binder removal, large pores change very little during binder removal and sintering, and finally remain in the sintered sample; (2) pores remaining in the original position after the degradation and volatilization of the binder during the binder removal process. Due to the low surface free energy of the granulated ZrO_2 and its low sintering activity, these pores will shrink to some extent in the subsequent high-temperature sintering, but most of the pores are difficult to eliminate, and these pores will remain in the sintered sample.

However, for the SLS/CIP/FS sample, due to the crushing of partial granulated ZrO_2 particles, the fine ZrO_2 powder exposed by different particles will be in close contact after CIP process, forming a "dense area" in the green part. The ZrO_2 fine powder in these areas has high sintering activity in the high-temperature sintering step and finally leads to high relative density. However, as shown in Fig. 6.20, there are also pores between the large particles in the "dense area."

FIGURE 6.20 SEM image of fractured SLS/CIPed sintered specimens with EZ10 powder.

6.3.3.2.2 ZrO$_2$-stearic acid composite powder and properties of ZrO$_2$ ceramics prepared by selective laser sintering

Fig. 6.21 shows the SEM images of ZrO$_2$−20 wt.% stearic acid (SZ20) composite powder prepared by solvent evaporation method. It is found that the particle size distribution of the SZ20 composite powder is non-uniform and wide (see Fig. 6.21A). It can be seen from Fig. 6.21B that the surface of SZ20 particles is very rough, which indicates that the crystallizing effect of stearic acid is not so good; the larger the particles are, the closer the shape is to the spherical shape, while the smaller particles take on irregular shape. The particle size distribution is shown in Fig. 6.22. The average particle size of SZ20 powder is 24−28 μm, but there are many SZ20 powder of 16−24 μm, which is almost similar to the powder content in the range of 24−28 μm. It indicates that the particle size distribution of SZ20 powder is too wide, which is not conducive to SLS process.

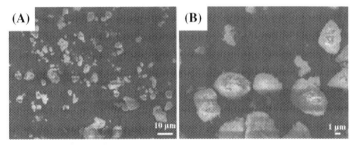

FIGURE 6.21 SEM images of ZrO$_2$-stearic acid composite powder: (A) low magnification; (B) high magnification.

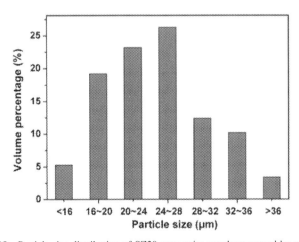

FIGURE 6.22 Particle size distribution of SZ20 composite powder prepared by solvent evaporation method.

Fig. 6.23 shows the SEM images of fractured SZ20 specimens. The observed samples are also prepared by optimized forming process with laser power = 5.5 W, scanning speed = 1400 mm/s, scanning space = 0.1 mm, laser energy density = 0.264 J/mm^2, preheating temperature = 40°C.

Fig. 6.23A and B shows the SEM images of fractured SLSed SZ20 specimens. It can be seen that the number and size of pores in the sample is smaller than that using EZ10. This is because the stearic acid has a lower melting point. When it is in a viscous state after being heated, it will wet and bind the nano-ZrO_2 powder, and the semi-solid stearic acid with good fluidity will fill the pores in the vicinity. Besides, the original nano-ZrO_2 powder also undergoes particle rearrangement at this stage. Therefore, the internal pore distribution of the SLS sample is more uniform and the size of pores is smaller.

Fig. 6.23C and D shows the SEM images of fractured SLS/CIPed SZ20 specimens. It can be seen that the pores inside the SLS/CIP sample are

FIGURE 6.23 SEM images of fractured SZ20 specimens: (A, B) SLSed specimens; (C, D) SLS/CIPed specimens; (E, F) SLS/CIP/FSed specimens.

FIGURE 6.24 High-magnification SEM images of fractured SZ20 specimens: (A) SLS/CIPed sample; (B) SLS/CIP/FSed sample.

further reduced and eliminated, and the relative density of the sample is high. Fig. 6.23E and F shows the SEM images of fractured SLS/CIP/FSed SZ20 specimens. It can be seen that the sintered sample has a dense fracture and a small amount of pores remained inside. The size is relatively small and the distribution is relatively uniform. This is because the nano-ZrO_2 powder has a large surface free energy and high sintering activity during the sintering process. Although a large amount of stearic acid binder leaves many pores after binder removal, but they can still be rapidly reduced and closed.

Fig. 6.24 shows the high-magnification SEM images of fractured SZ20 specimens. It can be seen from Fig. 6.24A that the morphology of the binder at different positions of the sample is different, which is related to the mechanism by which stearic acid is "melted-solidified" by heat treatment. The growth morphologies are different in different positions. Some are grown in needle-like shape, and some are grown in plush shape, which also leads to a certain difference in the distribution of the binder. Therefore, some pores in the sample have different shapes and sizes after binder removal and high-temperature sintering, as shown in Fig. 6.24B. These residual pores are detrimental to the mechanical properties of the sample, causing local defects.

6.3.3.2.3 ZrO_2-nylon 12 composite powder and properties of ZrO_2 ceramics prepared by selective laser sintering

Fig. 6.25 shows the SEM images of the film-covered powder ZrO_2−20 wt.% nylon 12 (PZ20) and ZrO_2−25 wt.% nylon 12 (PZ25) prepared by solvent precipitation method. Among them, the particle size distribution of PZ20 and PZ25 powders is more concentrated, and there is no obvious particle size unevenness. As shown in Fig. 6.25C and D, the PZ20 and PZ25 powder particles become more spherical in shape and the surface is smoother than the SZ20 powder, which is due to the better crystallization of the binder nylon 12, and the particle size of the PZ25 powder is larger than that of the PZ20 powder. Therefore, the powder is relatively concentrated in particle size

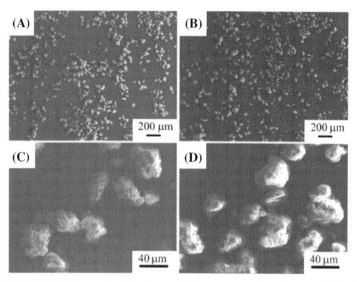

FIGURE 6.25 SEM images of nylon 12 coated-film nanoceramic powder: (A, C) PA-3 ZrO_2 powder with mass fraction of 20%; (B, D) PA-3 ZrO_2 powder with mass fraction of 25%.

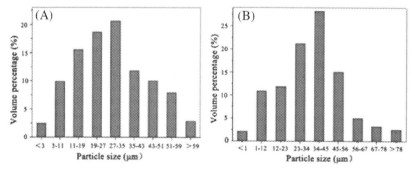

FIGURE 6.26 Particle size distribution of the composite powder prepared by the solvent precipitation method: (A) PZ20, ZrO_2−20% PA; (B) PZ25, ZrO_2−25% PA.

distribution and spherical with smooth surface, and the binder is uniformly distributed, which is suitable for SLS forming and subsequent processing.

Fig. 6.26 shows the particle size distribution of PZ20 and PZ25. Both powders have a particle size distribution of 1−80 μm and there is almost no nanosized powder, which further demonstrates that the nanopowder particles are coated with nylon 12 resin. Compared with that shown in Fig. 6.26A and B, the particle size of PZ20 powder is mainly distributed in the range of 3−59 μm, the average particle size is 27−35 μm, and the volume percentage of the powder in this particle size range is about 20%; while the particle size of PZ25 is mainly distributed in the range of 1−78 μm with an average of 34−45 μm, and the volume percentage of the powder in this particle size range is about 25%. From the analysis of laser particle size, the average particle diameters of the PZ20 powder and the PZ25 powder are 31.1 and 40.1 μm, respectively, and it can be seen that the average particle diameter of the PZ25 powder is larger than that of the PZ20 powder. Based on the above experimental results, it can be found that although both powders are prepared by solvent precipitation, the particle size of PZ20 is smaller than that of PZ25, which is mainly because nano-ZrO_2 acts as a nucleating agent in the crystallization of nylon 12. The content of nano-ZrO_2 in PZ20 is more than that in PZ25, so the nucleation center and the number of powder particles increase, leading to the decrease of the particle size of the powder. In addition, since the volume content of the PZ25 powder in the average particle size range is larger, the particle size distribution is more concentrated, which is advantageous for SLS forming.

Fig. 6.27 shows the SEM images of fractured PZ20 specimens. These samples are also prepared by optimized process with laser power = 6.6 W, scanning

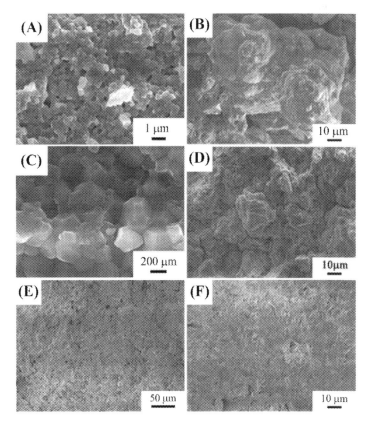

FIGURE 6.27 SEM images of fractured PZ20 specimens: (A, B) SLSed specimens; (C, D) SLS/CIPed specimens; (E, F) SLS/CIP/FSed specimens.

speed = 1600 mm/s, scanning space = 0.1 mm, laser energy density = 0.415 J/mm^2, preheating temperature = 155°C. Fig. 6.27A and B shows the SEM images of the fractured SLSed PZ20 specimens. It can be seen from Fig. 6.27A that although there are many pores in the sample, the pore distribution is very uniform, the PA12 material is still uniformly coated on the ZrO$_2$ material, and almost no internal ZrO$_2$ is exposed. Fig. 6.27B shows that PZ20 particles are bonded by nylon 12 material to form bonding neck, and the rest of the particles still maintain the shape before SLS. Fig. 6.27C and D shows the SEM images of the fractured SLS/CIPed sample. The large pores in the sample have been reduced, contracted and closed, and there is nearly no obvious porosity. Before the CIP is applied, the film-coated particles are only bonded by PA12, which has good plasticity and small deformation resistance. After the CIP process, the film-coated particles rapidly rearrange, plastic deform and creep, resulting in close and uniform arrangement of the particles in the samples as well as the high relative density. Due to the uniform solidification method of the binder, the morphology of the PA12 does not appear in various forms, and the distribution is also uniform. Fig. 6.27E and F shows the SEM images of the fractured SLS/CIP/FSed sample after binder removal and high-temperature sintering. Compared with the SLS/CIP/FSed SZ20 sample shown in Fig. 6.23A and F, the PZ20 sample is denser, and the pores are finer and more uniform. Although there are still a few pores, the pore size is very small, the shape is close to a circle, and there are no large and irregular pores in the sintered SZ20 sample.

Fig. 6.28 shows SEM images of different fractured sintered specimens. It can be seen from Fig. 6.28A and B that the internal pores of the PZ20

FIGURE 6.28 SEM images of different fractured sintered specimens: (A) SZ20 sintered specimen; (B) PZ20 sintered specimen; (C) SZ20 sintered specimen; (D) PZ20 sintered specimen.

sample are smaller than those of the SZ20 sample, and the ZrO_2 grain sizes inside the two samples are similar. It can be seen from Fig. 6.28C and D that the fracture of sintered SZ20 specimen is mainly the intergranular fracture, accompanied by a small amount of transgranular fracture; while the transgranular fracture of the sintered PZ20 specimen obviously increases. In addition, the grain shape of the PZ20 sintered sample is closer to a circle, and the crystallinity is better, which also indicates that the PZ20 sample may have better mechanical properties.

Fig. 6.29 shows the XRD patterns of PZ20, SZ20 sintered samples and yttria-doped ZrO_2 powder. The powder mainly contains m-ZrO_2 and yttria oxide. After high-temperature sintering, whether it is PZ20 sintered sample or SZ20 sintered sample, the phase completely converted into tetragonal phase ZrO_2, and a small amount of zirconium yttriate is formed. Zirconium yttriate is the liquid phase formed by the reaction of yttria and ZrO_2 at high-temperature, which could promote the sintering activity of the powder.

In addition, the diffraction spectrum of the tetragonal phase of the PZ20 sintered sample is stronger and sharper than that of SZ20, which indicates that the crystallinity of the sintered sample of PZ20 is better, its grains are slightly larger, while the crystallinity of the SZ20 sample is relatively poor, the grains are also finer, and there are defects such as dislocations. As shown in Fig. 6.28C and D, the grain size of the PZ20 sample is slightly larger, and the crystallinity is better. Nonetheless, the grain size of the SZ20 sample is relatively small and irregular, and the crystallinity is relatively poor, there are some large and irregular pores that could cause defects [55,56].

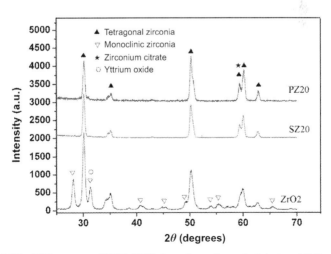

FIGURE 6.29 XRD patterns of PZ20, SZ20 sintered samples, and yttria-doped ZrO_2 powder.

6.3.3.3 Kaolinite-polymer composite powder and properties of kaolinite ceramics prepared by selective laser sintering

6.3.3.3.1 Noncoal series kaolinite-epoxy resin E06 composite powder and properties of kaolinite ceramics prepared by selective laser sintering

The raw ceramic material used in this experiment is noncoal series kaolinite, which is one of the traditional ceramics in China. These kaolinite powders are provided by Liling New Century Ceramics Co., Ltd., Hunan, which is a famous traditional ceramic pottery manufacturer in China. The main component of kaolinite is shown in Table 6.5.

Fig. 6.30 shows the SEM image of the kaolinite particles after granulation. The average particle size is about 100 μm. The powder has a good fluidity and is very suitable for SLS forming. In order to simplify the powder preparation for the traditional ceramic SLS technology, and considering the large average particle size and good fluidity of the kaolinite particles, the composite powder is prepared by the mechanical mixing method, and epoxy resin E06 is used as the binder. The kaolinite and epoxy resin powder are mixed with a mass ratio of 9:1 and uniformly mixed in a 3D mixer for 24 hours, and then it can be used for SLS forming. The method is simple and convenient, and it is suitable for the application and development of SLS forming in traditional ceramic products. Fig. 6.31 shows the SEM images of the fractured SLSed kaolinite specimen. Many bonding necks, which are generated after molten E06 binder wetting the ceramic powders, but there are still a lot of pores in the sample. Fig. 6.32 shows the SEM

TABLE 6.5 Main chemical components of kaolinite powder.

Composition	SiO$_2$	Al$_2$O$_3$	H$_2$O	Others
Content (%)	60	35	2	3

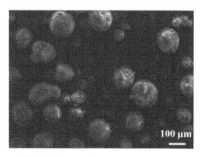

FIGURE 6.30 SEM image of kaolinite powders.

FIGURE 6.31 SEM images of the fractured SLSed kaolinite specimen. (A) High magnification; (B) Low magnification.

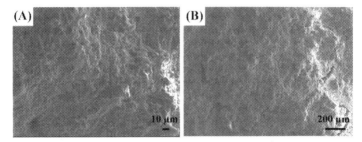

FIGURE 6.32 SEM images of the fractured SLS/CIPed kaolinite specimen. (A) High magnification; (B) Low magnification.

images of the fractured SLS/CIPed kaolinite specimen. After CIP, the porosity of the kaolinite green parts is greatly reduced, the particles are arranged more closely, and the binder is distributed uniformly.

Fig. 6.33 shows the SEM images of fractured kaolinite specimens sintered at different temperatures. As shown in Fig. 6.33A and B, when the sintering temperature is 1250°C, there are more pores in the sample, and the particles are still clearly visible, indicating that the sintering neck is initially formed among the ceramic particles at this temperature and many particles have not been bonded. As shown in Fig. 6.33C and D, as the sintering temperature increases to 1350°C, the number of pores is greatly reduced, mostly triangular small pores are not completely closed at the corners of the grains, and the grains are well bonded. Some areas have been completely sintered. As shown in Fig. 6.33E and F, with the further increase of sintering temperature to 1450°C, there is almost no pore in the sample, and the grains are completely sintered and fused, and the grains on the fractured kaolinite distribute in layers and grow up.

Fig. 6.34 shows the XRD patterns of kaolinite samples sintered at different temperatures. It can be seen that when the sintering temperature is 1250°C, the sample contains a large amount of SiO_2, which indicates that SiO_2 and Al_2O_3 have not started the chemical reaction in the kaolinite, and no sintering liquid phase is generated. When the temperature reaches

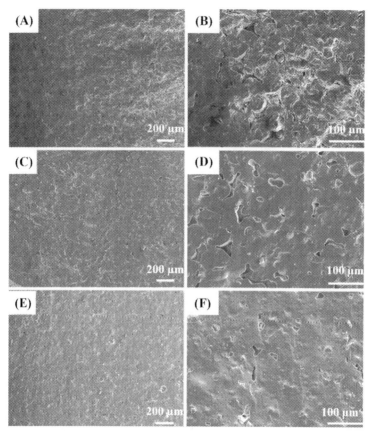

FIGURE 6.33 SEM images of fractured kaolinite specimens sintered at different temperatures: (A, B) 1250°C; (C, D) 1350°C; (E, F) 1450°C.

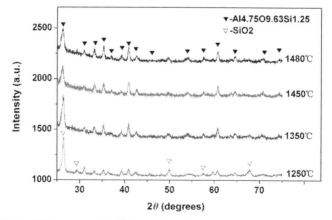

FIGURE 6.34 XRD patterns of kaolinite samples sintered at different temperatures.

1350°C, silicate aluminum is produced during the sintering process, and no SiO_2 is present. This shows that the sample completes the chemical reaction of SiO_2 and Al_2O_3 at 1250°C–1350°C. At 1350°C, the peak of silicate aluminum is strong and sharp, indicating that the crystallinity of kaolinite grains is good. As the sintering temperature increases to 1450°C and 1480°C, the peak is weaker and wider, indicating that the crystallinity of the kaolinite grain deteriorates and the mechanical properties of the sample will deteriorate. This is because the generated mullite grains will gradually grow up after 1350°C, accompanied by defects such as stress and microcrack.

6.3.3.3.2 Coal-series kaolinite-epoxy resin E12 composite powder and properties of kaolinite ceramics prepared by selective laser sintering

The ceramic material used in this experiment is coal-series kaolinite. Its composition is shown in Table 6.6. The main compositions are Al_2O_3 and SiO_2 and the morphology and particle size distribution are shown in Figs. 6.35A and 6.36A, respectively, which are of irregular shapes with a D50 of 27.8 μm. The binder is bisphenol A type epoxy resin E12, and its morphology and particle size are shown in Figs. 6.35B and 6.36B, respectively, which are of irregular shapes with a D50 of 8.9 μm. The coal-series kaolinite powder and the epoxy resin E12 with a mass fraction of 15% (percentage in the total mass) are mixed by mechanical mixing method on a 3D powder mixer for 24 hours at a speed of 150 rpm, and the coal-series kaolinite-epoxy resin E12 composite powder is prepared.

As shown in Figs. 6.37, 6.38, and Table 6.7, the dimensional error, relative density, and flexural strength are used to characterize the properties of

TABLE 6.6 Composition of coal-series kaolinite powders.

Composition	SiO_2	Al_2O_3	Fe_2O_3	Others
Mass fraction (%)	51	46	0.8	2.2

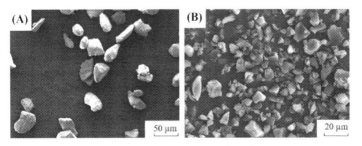

FIGURE 6.35 SEM images of raw material powders: (A) coal-series kaolinite; (B) epoxy resin.

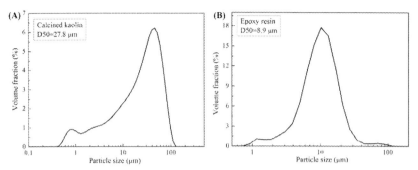

FIGURE 6.36 Particle size distribution: (A) coal-series kaolinite powder; (B) epoxy resin.

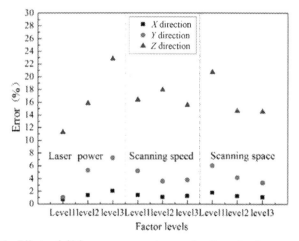

FIGURE 6.37 Effects of SLS process parameters on the dimensional accuracy of ceramic green parts.

the ceramic green body, respectively. As shown in Fig. 6.37, the dimensional error in the three directions increases with the increase of the laser power, but decreases with the increase of the scanning space, while the scanning speed has no significant effect on the dimensional error, and the dimensional error in the Z direction is obviously higher than that in the X and Y directions. When the laser scans the composite powder, the heat conduction between the powders softens the E12 in the heat-affected zone of the scanning region, thereby bonding the ceramic powder particles in the non-scanning region, causing secondary sintering and resulting in dimensional errors in the both X and Y directions. As the laser power gets higher or the scanning space gets lower, the laser energy density will get higher, and more heat will be transferred to the non-scanning area, so that the dimensional error becomes higher. The dimensional error in Z direction is high due to the

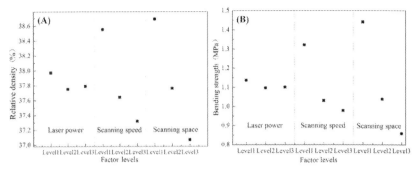

FIGURE 6.38 Effect of SLS process parameters on (A) relative density and (B) bending strength of ceramic green part.

"secondary sintering" on the upper and lower sides of the Z direction. Besides, when the laser scans the composite powder, the volatilized gas can only escape upward from the Z direction because the horizontal direction and the bottom of the rectangular samples are surrounded by unformed powder. At this time, an upward force is generated to push the ceramic particles upward, so that the amount of pores between the layers on Z direction increases, and the structure is loose, resulting in the increase of the dimensional error in the Z direction.

As shown in Fig. 6.38A and B, the relative density and bending strength of the ceramic green part change similarly. Among them, the laser power has little effect on the relative density and bending strength of the ceramic green part. However, the scanning speed and the scanning pitch have a significant influence on the relative density and bending strength of the ceramic green part. With the increase of scanning speed and scanning space, the relative density and the bending strength gradually decrease. This is because the increase of scanning speed and scanning space will lower the laser energy density, resulting in insufficient melting of the epoxy resin, which leads to incomplete wetting and bonding of ceramic particles by epoxy resin, so the relative density and bending strength of the ceramic green part decrease.

The best SLS process parameters could be determined by combining Table 6.7, Figs. 6.37 and 6.38: laser power of 5 W, scanning speed of 2000 mm/s, scanning space of 0.13 mm, and single-layer thickness of 0.15 mm. The SEM image of fractured ceramic green part is shown in Fig. 6.39. The ceramic particles are bonded by epoxy resin, and there are still a large number of pores among the particles. The dimensional error in Z direction is 10.43%, the relative density and the flexural strength are 37.89% and 0.984 MPa, respectively. When this set of parameters is selected, the bending strength of the green part is high, and the precision is good so that the postprocessing can be easily completed.

TABLE 6.7 Experimental results of SLS using coal-series kaolinite powder/E12 composite powder.

Laser power (W)	Scanning speed V (mm/s)	Scanning space (mm)	Dimensional error in Z direction (%)	Relative density (%)	Bending strength (MPa)
5	1800	0.11	12.41	39.76	1.830
5	2000	0.13	10.43	37.89	0.984
5	2200	0.15	11.00	36.27	0.594
7	1800	0.13	16.44	37.96	1.080
7	2000	0.15	12.51	37.01	0.920
7	2200	0.11	18.63	38.28	1.290
9	1800	0.15	20.19	37.93	1.060
9	2000	0.11	31.00	38.03	1.200
9	2200	0.13	16.98	37.42	1.050

FIGURE 6.39 SEM image of fractured ceramic green part.

The high-temperature sintering process is carried out after the ceramic green part is rubber discharged, and the XRD patterns of the porous coal-series kaolinite ceramics sintered at different temperatures are shown in Fig. 6.40. When the sintering temperature is 1350°C, the ceramic samples are composed of the mullite phase and the cristobalite phase. This is because when the coal-series kaolinite is decomposed to produce mullite phase during high-temperature sintering, and generate amorphous silica, and the amorphous silica is gradually converted into cristobalite at high-temperature, and the reaction equations are shown in Eqs. (6.5) and (6.6) [57]:

$$3(Al_2O_3 \cdot 2SiO_2)(s) \rightarrow 3Al_2O_3 \cdot 2SiO_2(s)(mullite) + 4SiO_2(s)(amorphous)$$

(6.5)

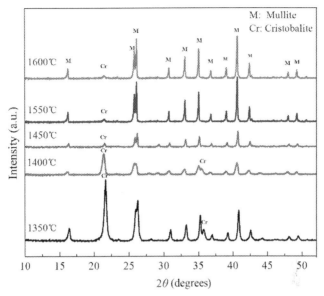

FIGURE 6.40 XRD pattern of porous coal-series kaolinite ceramics sintered at different temperatures.

$$SiO_2(s)(amorphous) \to SiO_2(s)(cristobalite) \quad (6.6)$$

With the increase of sintering temperature, the relative intensity of the mullite phase increases gradually, while the relative intensity of the cristobalite phase decreases gradually. When the temperature increases to 1450°C, the relative intensity of the cristobalite phase obviously decreases, indicating that the coal-series kaolinite green part has a strong mullitization reaction in the temperature range of 1350°C–1450°C, and it is basically completed at 1450°C. At this time, only a small amount of cristobalite remains in the sample, and the mullitization reaction formula of coal-series kaolinite is shown in Eq. (6.7) [58]:

$$x(3Al_2O_3 \cdot 2SiO_2)(s)(mullite) + ySiO_2(s)(cristobalite) \to z(mAl_2O_3 \cdot nSiO_2)(s)(mullite)$$
$$(6.7)$$

Fig. 6.41 shows the SEM images of fractured porous coal-series kaolinite ceramics sintered at different temperatures. As can be seen that the sintered sample has a 3D network skeleton, and the pores have irregular shapes and connect with each other, which is a typical porous structure. When the sintering temperature is 1350°C, there are a large number of fine grains in the coal-series kaolinite ceramic sample. As the temperature increases, some of the fine grains with higher activity gradually melt, and the cross-sectional morphology of the ceramic samples tends to be flat. When the sintering temperature becomes higher, part of the fine grains melts and fills into the spaces among the internal

FIGURE 6.41 SEM images of fractured porous coal-series kaolinite ceramics sintered at different temperatures: (A, B) 1350°C; (C, D) 1450°C; (E, F) 1550°C.

ceramic grains, and the rearrangement of the ceramic grains and the growth of the grains are both accelerated, so that the pores between the grains are eliminated continuously, forming sintering neck with large-area contact and high strength. At this time, the ceramic sample will undergo overall shrinkage, which is beneficial to the improvement of strength. When the sintering temperature increases to 1550°C, the fracture morphology of the sample is flat, and the number of through holes significantly decreases. The pore size distribution is wide, ranging from a few microns to 100 μm, indicating that the densification of the sample is relatively complete.

Fig. 6.42 shows the effect of sintering temperature on the shrinkage of porous coal-series kaolinite ceramics. As can be seen that the sample shrinks

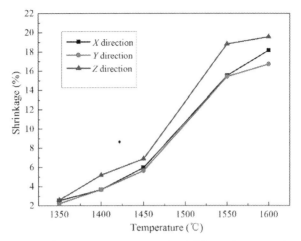

FIGURE 6.42 Effect of sintering temperature on the shrinkage of porous coal-series kaolinite ceramics.

in three directions of X, Y, and Z, and the shrinkage in the Z direction is obviously higher than the X and Y directions. Because the SLSed sample has loose structure and many pores between the layers in the Z direction, so the shrinkage in the high-temperature sintering stage is obviously higher than the other two directions. As the temperature increases, the shrinkage in all three directions increases continuously, but the changing rate of shrinkage increases firstly and then decreases. When the temperature increases from 1450°C to 1550°C, the shrinkage of porous coal-series kaolinite ceramics changes greatly from 6.88% to 18.84% in the Z direction. Because some fine grains in the ceramic sample melt and the rearrangement are severe when the samples are sintered in the range of 1450°C−1550°C, resulting in large shrinkage of the ceramic sample, but when the samples are sintered in the range of 1550°C−1600°C, the rearrangement of grains in the ceramic sample is relatively sufficient, and the changing rate of shrinkage decreases.

Fig. 6.43 shows the effect of sintering temperature on the porosity and bending strength of porous coal-series kaolinite ceramics. It can be seen that as the sintering temperature increases, the porosity of the porous coal-series kaolinite ceramics gradually decreases, while the bending strength increases gradually. As the sintering temperature increases, some of the fine grains with high activity are gradually melted to produce liquid phase and lower the viscosity [59], which is beneficial to the rearrangement and mass transfer of the grains, and promotes sintering, thereby increasing the densification of the material and resulting in increase of the strength. When the temperature increases from 1450°C to 1550°C, the porosity and bending strength of the porous coal-series kaolinite ceramics change greatly. The porosity decreases sharply from 44.55% to 27.49%, while the bending strength increases rapidly from 6.1 to 22.1 MPa.

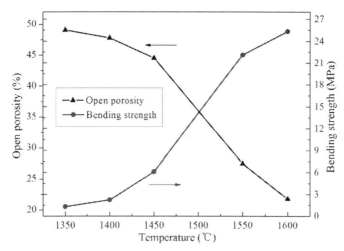

FIGURE 6.43 Effect of sintering temperature on the porosity and bending strength of coal-series kaolinite porous ceramics.

Because the raw materials of the coal-series kaolinite produce more liquid phase in this temperature range, and the degree of densification of porous kaolinite ceramics is remarkably improved. Besides, the mullite grains grow rapidly during this temperature period, and the growth is more complete, resulting in obvious increase in the strength of the porous coal-series kaolinite ceramics. As shown in Fig. 6.40, when the sintering temperature increases from 1450°C to 1550°C, the characteristic diffraction peak of mullite in the sintered sample becomes sharper, indicating that the mullite grains develop more completely. Considering the shrinkage, bending strength and porosity of porous coal-series kaolinite ceramics, the sintering temperature is chosen as 1450°C.

6.3.3.3.3 Cordierite-polymer composite powders and properties of cordierite ceramics prepared by selective laser sintering

The SEM image and particle size distribution of cordierite powders are shown in Fig. 6.44. The powders show an irregular shape with an average particle size of 38.6 μm, and the particle size distribution is not concentrated. The cordierite powders, polyvinyl alcohol, and nanosilica are mixed by mechanical mixing for 24 hours. The chemical composition of cordierite ceramic powders is shown in Table 6.8.

It is required first to consider the preheating temperature of the working chamber in the SLS-formed cordierite green part [38], so the differential thermal analysis (DSC) of the epoxy resin E12 is required. As shown in Fig. 6.45, the epoxy resin E12 has a glass transition temperature of 68.5°C. In order to make the binder to produce viscous flow after heating, and improve the wetting and bonding effect, the preheating temperature is

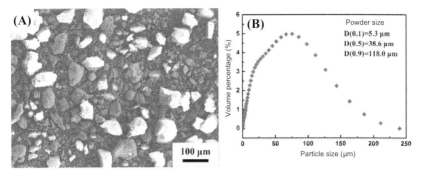

FIGURE 6.44 SEM image of and particle size of cordierite powder: (A) SEM image of cordierite powder; (B) particle size distribution.

TABLE 6.8 Composition of cordierite ceramic powders.

Composition	SiO_2	Al_2O_3	Fe_2O_3	MgO	Na_2O_3	Others impurities
Mass fraction (%)	50.4	34.58	0.46	13.1	0.032	1.428

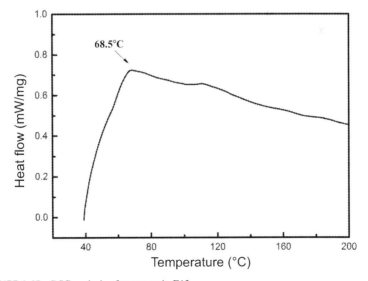

FIGURE 6.45 DSC analysis of epoxy resin E12.

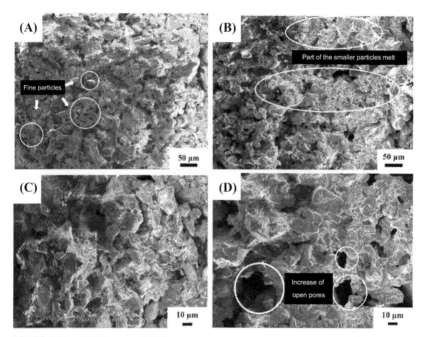

FIGURE 6.46 SEM images of SLSed cordierite green part and ceramics sintered at high-temperature: (A, C) the SEM images of the upper surface of SLSed cordierite green part; (B, D) the SEM images of the upper surface of cordierite ceramics sintered at 1350°C.

generally set below the glass transition temperature for 20°C, so the preheating temperature is set at 45°C.

Fig. 6.46A and C shows the SEM images of the upper surface of the SLSed cordierite green part. The laser power is 17 W and the scanning speed is 1500 mm/s in SLS process. There are a large number of pores on the surface of green parts, there are many closed pores and a small number of open pores, and their distribution is non-uniform. In addition, there are fine and loose particles on the surface. The SLS process relies on the thermal effects of epoxy E12 to complete the physical bond. Under the action of the laser, the epoxy resin E12 is melted to coat the high-melting cordierite ceramic particles, and the E12 is bonded to each other to form the sintering necks, so that the green part has original strength. Fig. 6.46B and D shows the SEM images of the upper surface of cordierite ceramic parts after sintering at 1350°C. The open pores on the surface of the part are more and concentrated, and the size of the open pores increases. Compared with the SLSed green part, during the high-temperature sintering process, due to the higher activity, the tiny ceramic particles melt and coat the unmelted cordierite powders to form a ceramic sintering neck, that is, the chemical bonding is completed through the liquid-phase sintering between the cordierite particles

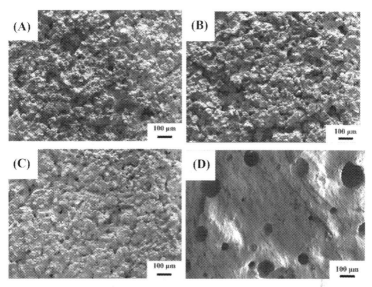

FIGURE 6.47 SEM images of cordierite ceramics sintered at different temperatures: (A) 1350°C; (B) 1375°C; (C) 1400°C; (D) 1450°C.

under high-temperature conditions. In addition, the epoxy resin decomposes to form pores in a large size.

Fig. 6.47 shows the SEM images of cordierite ceramics sintered at different temperatures. As the sintering temperature increases from 1350°C to 1400°C, the surface roughness of cordierite ceramic samples gradually decreases, and the amount of open pores gradually decreases. When the sintering temperature increases, the melt flow of the cordierite ceramic particles is more sufficient, and the bonding area increases. In addition, the shrinkage of the ceramics at high-temperature results in the decrease of porosity and increase of density, which is advantageous for the improvement of strength. When the sintering temperature reaches 1450°C, the microscopic surface of the sample is dense, and a large number of un-uniformly distributed closed pores with size of 50–100 μm appears because the temperature is close to the melting point of the cordierite ceramics (1460°C). At this time, the cordierite ceramic sample is melted and deformed, the gas originally existing in the open pores is exhausted from the surface to form circular pores.

Fig. 6.48 shows the XRD pattern of SLSed cordierite green part and cordierite ceramics sintered at different temperatures. The green part is mainly the μ-cordierite phase in the metastable state and α-cordierite phase in the high-temperature stable state, indicating the SLS process does not affect the phase composition of cordierite. When the sintering temperature is

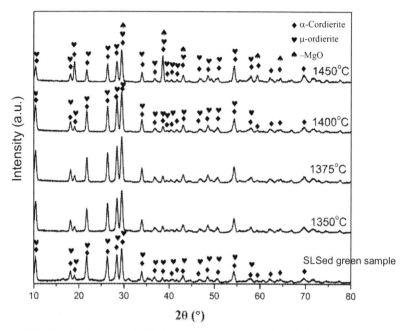

FIGURE 6.48 XRD pattern of SLSed cordierite green part and cordierite ceramics sintered at different temperatures.

below 1400°C, the low-temperature metastable μ-cordierite phase and the high-temperature stable α-cordierite phase could be smoothly obtained according to the process route established in the experiment, and as the sintering temperature increases, the low-temperature μ-cordierite phase gradually changes to the α-cordierite phase under the high-temperature stable state. However, when the sintering temperature reaches 1450°C, the second phase of MgO appears, indicating that the cordierite phase begins to decompose beyond the narrow synthesis temperature range [60].

According to Rice's research work [61], strength and porosity satisfies the following formula:

$$\sigma = \sigma_0 \exp(-bp) \tag{6.8}$$

where σ is the strength of the porous structure, σ_0 is the strength of the dense body, and b is the empirical constant determined by the pore characteristics, p is the porosity. It shows that the strength increases exponentially with the decrease of porosity. Fig. 6.49 shows the effect of sintering temperature on the porosity of cordierite ceramics.

As shown in Fig. 6.49, when the sintering temperature is lower than 1400°C, with the increase of sintering temperature, the apparent porosity and the open porosity slowly decrease, while they drastically decrease from

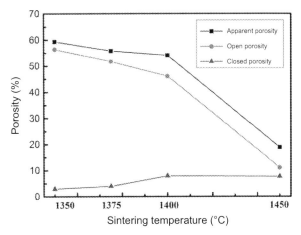

FIGURE 6.49 Effects of sintering temperature on porosity of cordierite ceramics.

1400°C to 1450°C. In addition, the closed porosity increases slowly as the sintering temperature increases, and remains substantially unchanged after reaching 1400°C. The decrease of the apparent porosity and open porosity proves that the porosity decrease with the increase of sintering temperature, and the increase of the closed porosity indicates that the increase of sintering temperature results in the sintering and melting of ceramic particles, and the open pores are filled to finally increase the density of the ceramic parts.

References

[1] S. Baklouti, J. Bouaziz, T. Chartier, et al., Binder burnout and evolution of the mechanical strength of dry-pressed ceramics containing poly (vinyl alcohol), J. Eur. Ceram. Soc. 21 (8) (2001) 1087–1092.

[2] M. Trunec, J. Klimke, Z.J. Shen, Transparent alumina ceramics densified by a combinational approach of spark plasma sintering and hot isostatic pressing, J. Eur. Ceram. Soc. 36 (16) (2016) 4333–4337.

[3] M. Jabbari, R. Bulatova, A.I.Y. Tok, et al., Ceramic tape casting: a review of current methods and trends with emphasis on rheological behavior and flow analysis, Mater. Sci. Eng. B 212 (2016) 39–61.

[4] S. Ananthakumar, A.R.R. Menon, K. Prabhakaran, et al., Rheology and packing characteristics of alumina extrusion using boehmite gel as a binder, Ceram. Int. 27 (2) (2001) 231–237.

[5] J. Liu, B. Zhang, C. Yan, et al., The effect of processing parameters on characteristics of selective laser sintering dental glass-ceramic powder, Rapid Prototyp. J. 16 (2) (2010) 138–145.

[6] N. Travitzky, A. Bonet, B. Dermeik, et al., Additive manufacturing of ceramic-based materials, Adv. Eng. Mater. 16 (6) (2014) 729–754.

[7] F. Doreau, C. Chaput, T. Chartier, Stereolithography for manufacturing ceramic parts, Adv. Eng. Mater. 2 (8) (2000) 493–496.
[8] D.A. Klosterman, R.P. Chartoff, N.R. Osborne, et al., Structural ceramic components via laminated object manufacturing, in: Proceedings of the International Conference on Rapid Product Development, Messe Stuttgart, Germany, 1996, pp. 247–256.
[9] L. Novakova-Marcincinova, I. Kuric, Basic and advanced materials for fused deposition modeling rapid prototyping technology, Manuf. Ind. Eng. 11 (1) (2012) 25–27.
[10] J. Ebert, E. Özkol, A. Zeichner, et al., Direct inkjet printing of dental prostheses made of zirconia, J. Dental Res. 88 (7) (2009) 673–676.
[11] H. Yves-Christian, W. Jan, M. Wilhelm, et al., Net shaped high performance oxide ceramic parts by selective laser melting, Phys. Procedia 5 (2010) 587–594.
[12] M.L. Griffith, J.W. Halloran, Freeform fabrication of ceramics via stereolithography, J. Am. Ceram. Soc. 79 (10) (1996) 2601–2608.
[13] C. Griggin, J. Daufenbach, C. Turner, Laminated object manufacturing of an extremely tough layered ceramic matrix composite, Final Phase I SBIR Report to the U. S. Army, Lone Peak Engineering, 1994.
[14] M.K. Agarwala, van Weeren, R., R. Vaidyanathan, et al., Structural ceramics by fused deposition of ceramics, in: Proceedings of the Solid Freeform Fabrication Symposium. Austin, Texas, The University of Texas at Austin, 1995, vol. 6, pp. 1–8.
[15] J. Yoo, M.J. Cima, S. Khanuja, et al., Structural ceramic components by AM, in: Solid Freeform Fabrication Symposium, 1993, pp. 40–50.
[16] I. Shishkovsky, I. Yadroitsev, P. Bertrand, et al., Alumina–zirconium ceramics synthesis by selective laser sintering/melting, Appl. Surf. Sci. 254 (4) (2007) 966–970.
[17] K. Subramanian, N. Vail, J. Barlow, et al., Selective laser sintering of alumina with polymer binders, Rapid Prototyp. J. 1 (2) (1995) 24–35.
[18] W. Zhou, D. Li, Z. Chen, et al., Research on SLA rapid prototyping characteristics of ceramic slurry and its engineering application, Aeronaut. Manuf. Technol. 8 (2010) 36–42.
[19] X. Li, Mechanism research and process optimization of 3DP forming technology, Tongji University, 2006.
[20] J. Will, R. Melcher, C. Treul, et al., Porous ceramic bone scaffolds for vascularized bone tissue regeneration, J. Mater. Sci. Mater. Med. 19 (8) (2008) 2781–2790.
[21] F.C. Fierz, F. Beckmann, M. Huser, et al., The morphology of anisotropic 3D-printed hydroxyapatite scaffolds, Biomaterials 29 (28) (2008) 3799–3806.
[22] G.A. Fielding, A. Bandyopadhyay, S. Bose, Effects of silica and zinc oxide doping on mechanical and biological properties of 3D printed tricalcium phosphate tissue engineering scaffolds, Dental Mater. 28 (2) (2012) 113–122.
[23] R. Melcher, S. Martins, N. Travitzky, et al., Fabrication of Al_2O_3-based composites by indirect 3D-printing, Mater. Lett. 60 (4) (2006) 572–575.
[24] J.P. Kruth, L. Froyen, J. Van Vaerenbergh, et al., Selective laser melting of iron-based powder, J. Mater. Process. Technol. 149 (1) (2004) 616–622.
[25] Y. Hagedorn, N. Balachandran, W. Meiners, et al., SLM of net-shaped high strength ceramics: new opportunities for producing dental restorations, in: Proceedings of the Solid Freeform Fabrication Symposium, Austin, TX, August 2011, pp. 8–10.
[26] Y. Wu, J. Du, K.L. Choy, et al., Laser densification of alumina powder beds generated using aerosol assisted spray deposition, J. Eur. Ceram. Soc. 27 (16) (2007) 4727–4735.
[27] Y. Shi, S. Liu, D. Zeng, et al., Laser Manufacturing Technology, Mechanical Industry Press, Beijing, 2011.

[28] S.J. Rodrigues, R.P. Chartoff, D.A. Klosterman, et al., Solid freeform fabrication of functional silicon nitride ceramics by laminated object manufacturing, in: Proceedings from Solid Freeform Fabrication Symposium, 2000.
[29] J. Deckers, J. Vleugels, J.-P. Kruth, Additive manufacturing of ceramics: a review, J. Ceram. Sci. Technol. 5 (4) (2014) 245–260.
[30] J.D. Cawley, A. Heuer, W.S. Newman, et al., Computer-aided manufacturing of laminated engineering materials, Am. Ceram. Soc. Bull. 75 (5) (1996) 75–79.
[31] I. Zein, D.W. Hutmacher, K.C. Tan, et al., Fused deposition modeling of novel scaffold architectures for tissue engineering applications, Biomaterials 23 (4) (2002) 1169–1185.
[32] J.N. Stuecker, J. Cesarano, D.A. Hirschfeld, Control of the viscous behavior of highly concentrated mullite suspensions for robocasting, J. Mater. Process. Technol. 142 (2) (2003) 318–325.
[33] M.C. Leu, L. Tang, B. Deuser, et al., Freeze-form extrusion fabrication of composite structures, in: Proceedings of the Solid Freeform Fabrication Symposium, Austin, TX, 2011, pp. 111–124.
[34] J.H. Liu, Y.S. Shi, Z.L. Lu, et al., Rapid manufacturing metal parts by laser sintering admixture of epoxy resin/iron powders, Adv. Eng. Mater. 8 (10) (2006) 988–994.
[35] D. Chen, J. Zhang, Basic Principles of Adhesive Bonding. Beijing: Science Press, 1992.
[36] L. Yin, X. Xu, Foundation of Adhesive Bonding and Adhesive. Aviation Industry Press, 1988.
[37] S. Kolosov, G. Vansteenkiste, N. Boudeau, et al., Homogeneity aspects in selective laser sintering (SLS), J. Mater. Process. Technol. 177 (1) (2006) 348–351.
[38] Y. Ren, Z. Liu, K. Yang, et al., Types and applications of zirconia materials, Chin. Ceram. 44 (4) (2008) 44–46.
[39] J. Langer, M.J. Hoffmann, O. Guillon, Electric field-assisted sintering in comgreen part with the hot pressing of yttria-stabilized zirconia, J. Am. Ceram. Soc. 94 (1) (2011) 24–31.
[40] M. Miao, Study on Al_2O_3 fiber reinforced cordierite matrix composites, Liaoning University of Technology, 2014.
[41] D. Kong, Research progress of coal series kaolinite and its application. Technol. Devel. Chem. Indus. 43 (7), 2014, 39–41.
[42] Z. Han, W. Cao, Z. Lin, et al., Research progress of selective laser sintering rapid prototyping technology for ceramic materials, J. Inorg. Mater. 19 (4) (2004) 705–713.
[43] K. Liu, Research on Ceramic Powder Laser Sintering/Cold Isostatic Pressing Composite Forming Technology, Huazhong University of Science and Technology, 2014.
[44] Z. Fan, N. Huang, Curing mechanism of selective laser sintering film-covered sand casting (core), J. Huazhong Univ. Sci. Technol. (Nat. Sci.) 29 (4) (2001) 60–62.
[45] Y. Shi, K. Liu, C. Li, et al., Laser selective sintering/cold isostatic pressing composite forming technology for zirconia parts, J. Mech. Eng. 50 (21) (2014) 118–123.
[46] Q. Deng, H. Zhang, Selective laser sintering of solid powders, Electr. Mach. 2 (1995) 32–35.
[47] K. Liu, Y. Shi, W. He, et al., Densification of alumina components via indirect selective laser sintering combined with isostatic pressing, Int. J. Adv. Manuf. Technol. 67 (9) (2013) 2511–2519.
[48] S. Xia, K. Liu, Study on preparation, SLS and post-process of ZrO_2 coated powder, in: International Conference on Progress in Additive Manufacturing, 2014, pp. 202–206.
[49] B.W. Xiong, H. Yu, Z.F. Xu, et al., Study on dual binders for fabricating SiC particulate preforms using selective laser sintering, Compos. Part. B 48 (5) (2013) 129–133.

[50] N.K. Vail, Preparation and characterization of microencapsulated, finely divided ceramic materials for selective laser sintering, University of Texas at Austin, 1994.
[51] K. Liu, Y. Shi, C. Li, et al., Indirect selective laser sintering of epoxy resin-Al_2O_3, ceramic powders combined with cold isostatic pressing, Ceram. Int. 40 (5) (2014) 7099−7106.
[52] K. Shahzad, J. Deckers, Z. Zhang, et al., Additive manufacturing of zirconia parts by indirect selective laser sintering, J. Eur. Ceram. Soc. 34 (1) (2014) 81−89.
[53] Q. Wei, P. Tang, J. Wu, et al., Microstructure and properties of selective laser sintered porous cordierite ceramics, J. Huazhong Univ. Sci. Technol. Nat. Sci. Ed. 44 (6) (2016) 46−51.
[54] K. Liu, H. Sun, Y. Shi, et al., Research on selective laser sintering of kaolin−epoxy resin ceramic powders combined with cold isostatic pressing and sintering, Ceram. Int. 42 (9) (2016) 10711−10718.
[55] Y. Liu, B. Zong, S. Peng, et al., Differential scanning calorimetry and phase characterization of ultrafine zirconium dioxide, Ind. Catal. (1)(1998) 55−59.
[56] H. Liu, Q. Xue, P. Zang, Study of the preparation of nano ZrO_2 powder by DSC, TGA and XRD techniques, Anal. Test. Technol. Instrum. 2 (1) (1996) 40−44.
[57] L. Xu, Study on Preparation of High Porosity Mullite Porous Ceramics by Solid Phase Sintering. Guangzhou: South China University of Technology, 2015.
[58] J. Zhang, Preparation and Growth Mechanism of Mullite Whiskers and their Application in Ceramic Toughening. China University of Geosciences (Wuhan) 2012.
[59] S. You, H. Zheng, D. Fu, et al., Structure and properties of porous ceramics synthesized by fly ash and anorthite, J. Chin. Ceram. Soc. 44 (12) (2016) 1718−1723.
[60] Y, Cao, Yufei. Study on the Preparation and Properties of Cordierite Honeycomb Ceramic Catalyst Carrier for Vehicle. Dalian Maritime University, 2013.
[61] R.W. Rice, Evaluation and extension of physical property-porosity models based on minimum solid area, J. Mater. Sci. 31 (1) (1996) 102−118.

Chapter 7

Application cases of additive manufacturing materials

7.1 Application case 1 of additive manufacturing polymer powder material

The samples used are powder material of PS granules, PA granules, and compatibilizer SMA. The forming equipment is the HRPS-II selective laser sintering (SLS) machine independently developed by the research group. The processing parameters are as follows: laser-scanning speed is 1700 mm/s, laser-scanning pitch is 0.10 mm, laser power is 12 W, laser energy density is 73.5 mJ/mm^2, and the preheating temperature is 100°C.

A large complex precision investment casting and sand mold are formed on the SLS machine, and the tensile strength is 41.3 MPa. The rapid-formed investment casting and sand mold are manufactured as the key components of large demands in China's aerospace, military, shipbuilding, automobile, machine tools, and other fields through investment precision casting and sand mold casting. By reducing the process flow, shortening the cycle, and decreasing the cost, the cost and cycle of casting process can be reduced by half, which greatly enhances the traditional casting process. The production process flowchart is shown in Fig. 7.1.

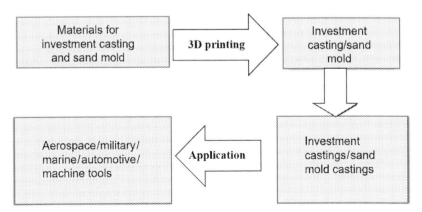

FIGURE 7.1 Production flowchart of AM/casting composite process.

The investment casting and sand mold formed by AM are featured by simple process, low cost and high precision, and the preheating process can reduce deformation. Huazhong University of Science and Technology and the University of Birmingham in the United Kingdom, Rolls-Royce Ltd in the United Kingdom, Airbus Corporation in France, European Space Agency, Tsinghua University, China Aviation Research Institute, Beijing Institute of Aeronautical Materials and other units jointly undertake the 7th Framework Program, "Casting of Large Ti Structures," Huazhong University of Science and Technology, is mainly engaged in the rapid manufacturing of wax molds for casting of large complex titanium alloy structural parts in the aviation field and the three-dimensional measurement technology of its castings, which provides Airbus with the casting wax molds needed to manufacture large-size titanium alloy aeronautical parts. Fig. 7.2 shows the wax mold and its castings provided by Huazhong University of Science and Technology for the Airbus for casting of large-size titanium alloy aeronautical parts, and Fig. 7.3 shows a large complex engine crankcase.

7.2 Application case 2 of additive manufacturing polymer powder material

This sample powder is made of nylon 12 granules by solvent precipitation method. Fig. 7.4 shows the morphology, particle size and distribution of the powder prepared by solvent precipitation method. The particles are close to the spherical shape, and the particle size is mainly concentrated between 40 and 70 μm. Such powder has good fluidity and satisfies the forming requirements of SLS.

The forming equipment is "HK P500" type SLS industrial grade 3D printer of Wuhan Huake 3D Technology Co., Ltd. The processing parameters are as follows: scanning speed is 4000 mm/s, laser power is 26 W, scanning pitch is 0.3 mm, and the powder leveling thickness is 0.1 mm.

To better serve the honorable disabled soldiers injured in the revolution and socially disabled groups, and meet their individualized and diversified needs, Hubei Rehabilitation Assistive Technology Center introduced the "HK P500" type SLS industrial grade 3D printer of Wuhan Huake 3D Technology Co., Ltd. at the end of 2015, and at the same time start cooperation with Huazhong University of Science and Technology to focus on the application of AM technology in the rehabilitation equipment industry. The rich 3D digital platform and advanced rehabilitation assistive is used in the center to design and manufacture the process. After careful research and scientific demonstration, the AM technology is applied to the research and development of rehabilitation assistive, and the products such as AM air-permeability prosthetic socket monolithic prosthetics, 3D printed spinal orthosis, and AM elastic bionic feet are developed as shown in Fig. 7.5. Staged results have been achieved after the actual use by patients.

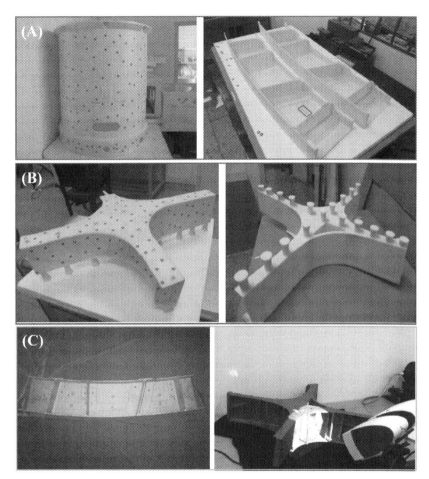

FIGURE 7.2 Wax mold and its castings provided manufactured for the Airbus for casting of large-size titanium alloy aeronautical parts: (A) wax mold for casting of aeronautical parts with an overall dimensions more than 1 m and a wall thickness only 3−4 mm; (B) wax mold of cross-joints for aeronautical use with an overall dimension more than 1 m and a complex internal structure; (C) aeronautical titanium alloy parts obtained from casting of large-sized complex wax mold.

Through SLS technology, the overall printing of short manufacturing cycle and intelligent process can be achieved through fast and accurate modeling. The buckle structure design can make the wearing convenient and flexible, and the fixing is firm and reliable. The mechanical structure design of the orthosis is designed to achieve local enhancement of the structure through special software system design, realizing a balance of functionality, breathability, aesthetics, and light weight.

FIGURE 7.3 Large complex engine crankcase.

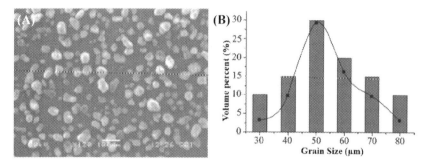

FIGURE 7.4 Morphology, grain size, and its distribution of powder prepared by solvent precipitation method: (A) grain morphology; (B) grain size and size distribution.

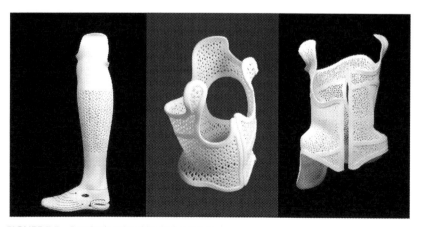

FIGURE 7.5 Prosthesis printed by SLS technology.

The final molded prosthesis has the following characteristics: diversified bionic design, personalized structural design, personalized functional design, standardized design, and manufacturing, which can meet actual needs.

7.3 Application case 3 of additive manufacturing polymer powder material

The large-scale cylinder sand sample is prepared by precoated sand and phenolic resin, and the precoated sand is prepared by thermal film coating. The original sand is scrubbed spherical sand, the granular shape is polygonal, the particle size is 74–149 μm, and the phenolic resin content is 4%, and Fig. 7.6 shows the DSC curve of precoated sand.

The forming equipment is the HRPS-II SLS sintering machine independently developed by the research group of Rapid Manufacturing Centre of HUST. The processing parameters are as follows: the preheating temperature is 50°C, the laser-scanning speed is 100 mm/s, the laser-scanning pitch is 0.10 mm, the laser power is 24 W, and the layered thickness is 0.25 mm.

The part having a certain initial strength is formed on an SLS machine, and then cured by heating in an oven to 170°C. The maximum strength after curing is 3.4 MPa, and the maximum bending strength is 5.43 MPa. During the forming process, when a new section appears, the sand should be preheated to about 90°C, so that the sand is slightly agglomerated and becomes the base of the part, which plays a fixed function. When the section becomes larger, the temperature can be kept at a temperature 5°C lower than the agglomerate temperature to avoid warpage.

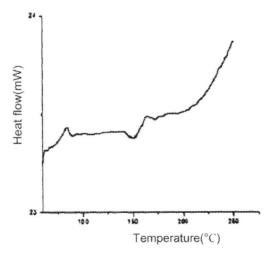

FIGURE 7.6 DSC curve of precoated sand.

654 Materials for Additive Manufacturing

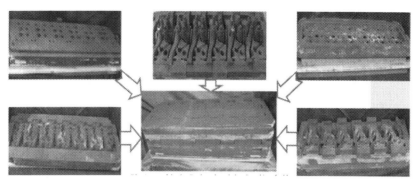

FIGURE 7.7 Large engine cylinder head sand core made with powder laser sintering (SLS) rapid manufacturing equipment (length × width × height: 1072.19 × 397.42 × 221.90 mm^3).

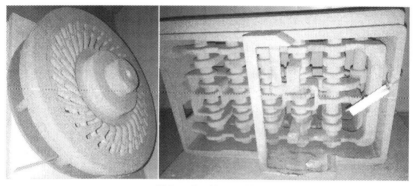

SLS sand mold assembly

FIGURE 7.8 Large cylinder sand mold of SLS additive manufacturing.

The large key components in aerospace, military, automotive and other fields formed by AM can shorten the design and manufacturing cycle and development process, reduce costs, and even form some very complex parts which cannot be formed by using traditional casting technologies.

As shown in Fig. 7.7, the components of the cylinder head sand core of six-cylinder engine cylinder more than one meter that meets the accuracy and strength requirements can be successfully formed and assembled in 1 week. Fig. 7.8 shows the impeller and riveter parts for the pump.

7.4 Application case 1 of additive manufacturing polymer wire material

Take the protected animal Great Spotted Woodpecker as the research object and scan its head with Micro-CT, and reconstruct the three-dimensional digital model of the skull based on the micro scanning data. Select the upper and lower beak structures and a biodegradable, high performance polylactic acid (PLA) of low shrinkage rate as the material, and mold the structures on a 3D printer based on FDM technology (Fig. 7.9). As a biodegradable material, PLA is recyclable. Its raw material has a wide range of lactic acid sources and can be obtained by fermentation of agricultural products such as corn and starch. PLA has high strength and good biocompatibility. Compared with the acrylonitrile-butadiene-styrene plastic commonly used in FDM process, PLA material is environmentally friendly and has low odor, which is suitable for indoor use, and the lower shrinkage rate ensures that warpage will not occur though the workbench is not heated while printing the model of large size.

PLA also has disadvantages such as poor toughness and low melt strength, which leads to difficulty in forming. In this example, the thermal stability of PLA is improved by adding a certain amount of chain extender ADR4370S. The TG curves of pure PLA and ADR chain extended and modified PLA are shown in Table 7.1, where, T_1 is the temperature at which the

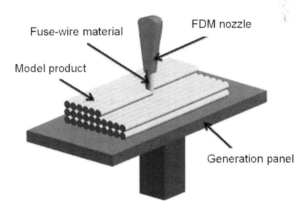

FIGURE 7.9 FDM desktop printer and the schematic diagram of its forming.

TABLE 7.1 TG data of pure PLA and ADR chain extended and modified PLA.

Sample formulation	T_1 (°C)	T_5 (°C)	T_{10} (°C)
1#	318.4	339.5	347.5
2#	325.7	341.8	349.0
3#	325.4	342.2	349.4
4#	326.7	342.6	349.6
5#	327.5	342.4	349.3

FIGURE 7.10 Three-dimensional model and physical map of the woodpecker axis in wrong direction.

mass loss is 1%, that is, the initial decomposition temperature, T_5 is the temperature at which the mass loss is 5%, and T_{10} is the temperature at which the mass loss is 10%. The T_1 of pure PLA is 318.4°C. After adding 0.2% chain extender ADR4370S, the T_1 of the modified PLA is 325.7°C, which is increased by about 7°C, and the T_5 and T_{10} are increased by about 2°C.

The MakerBot Replicator printer can form parts with maximum size up to 252 mm × 199 mm × 150 mm. When the process parameters are set as follows: the resolution of molded slice δ is 0.1 mm, the internal filling ratio is 100%, the nozzle moving speed is V, and the nozzle heating temperature is T, the surface of the finished product stacked along the long axis of the parallel beak is of high quality, and the degree of detail reduction is high.

The special head structure of the woodpecker is the result of its adjustment and adaptation to nature during the evolution process to absorb the pecking energy. It is of great significance to study the important role played by the woodpecker's beak in protecting it from impact damage. The FDM technology can quickly and completely form the bionic structure of the woodpecker's beak, which can provide the necessary foundation for the study of the performance test of the woodpecker's head and energy absorption mechanism in the environment with insufficient sample size (Fig. 7.10). It is expected to be applied to other activities such as popular science exhibitions and teaching demonstrations.

7.5 Application case 1 of polymer liquid materials for additive manufacturing

The raw materials such as alicyclic glycidyl ester, TDE-85; bisphenol A epoxy resin, E-51; epoxy acrylate, EA-612; alicyclic epoxy resin, ERL-4221; cyclohexyl dimethanol divinyl ether (CHDMDVE); polypropylene glycol diglycidyl ether diacrylate (PPGDEA); initiator triarylsulfonium hexafluoroantimonate salt (Ar3SSbF6), UV-6976; initiator benzoin dimethyl ether, and BDK (radical initiator) are used to prepare 3DPSL-1 type photopolymer in this case.

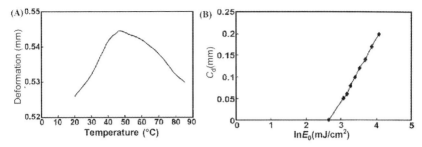

FIGURE 7.11 Thermomechanical performance analysis curve of photopolymer-cured product (A) and the relationship between natural logarithm of exposure energy and curing depth (B).

The photopolymer has a transmission depth (D_p) of 0.14 mm, a critical exposure amount (E_C) of 14.1 mJ/cm^2, and a viscosity of 359 mPa s at 30°C. At 25°C, the density of the 3DPSL-1 type photopolymer is 1.17 g/cm^3, the density after curing is 1.21 g/cm^3, and the volume shrinkage after curing is 3.31%. The sample band of photopolymer-cured product has a tensile strength of 21.0 MPa, an elastic modulus of 1108.2 MPa, an elongation at break of 10.6%, and a glass transition temperature of 47°C. The thermodynamic performance analysis curve and light curing curve of the cured photosensitive resin are shown in Fig. 7.11.

Use the HRPLA-I stereolithography rapid prototyping equipment, and set the operating parameters of the device as follows: $P = 95$ mW, $S = 0.08$ mm, $d = 0.10$ mm, $v = 4000$ mm/s and working temperature of photopolymer $= 30°C$ with reference to the measured critical exposure and Dp values. Compared with other AM technologies, SLA technology has the outstanding advantages of high precision, good surface quality and high utilization rate of raw materials (close to 100%), which is suitable for manufacturing parts with complex shapes and fine features. The photopolymer is applied as a printing material to a AM stereolithography rapid prototyping device to produce mobile phone and home appliance housings or models, and the production effect is good. The following illustration shows several application cases. Figs. 7.12–7.15 show several application examples.

7.6 Application case 1 of metal powder materials for additive manufacturing

7.6.1 Drum lid

This sample powder is made of S136 cast bar with a diameter of 30 mm by the air atomization method. Fig. 7.16 shows the morphology of the powder after aerosolization and its particle size and distribution. The powder is nearly spherical in shape and has an average particle size of 25 μm, which is

FIGURE 7.12 Automobile lamp.

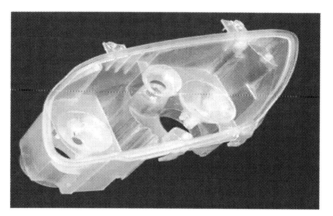

FIGURE 7.13 Automobile interior.

FIGURE 7.14 Mobile phone shell.

Application cases of additive manufacturing materials Chapter | 7 **659**

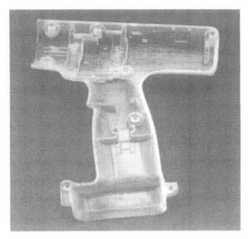

FIGURE 7.15 Hand saw housing.

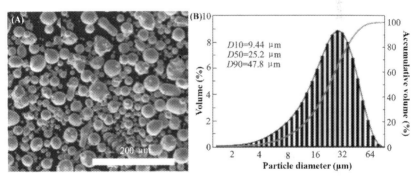

FIGURE 7.16 Micromorphology of S136 prepared by gas atomization (A) and particle size distribution (B).

TABLE 7.2 Chemical composition and content of S136 powder.

Element	C	Si	Mn	Cr	V	O	P	S	Fe
Wt (%)	0.29	0.80	0.56	13.67	0.31	0.034	0.01	–	Remaining

suitable for the forming requirements of SLM. Table 7.2 shows the chemical composition of the gas atomized powder, which is measured by direct-reading inductively coupled plasma emission spectrometer.

The forming equipment is the HRPM-II SLM system developed by the research group. The processing parameters selected are as follows: scanning

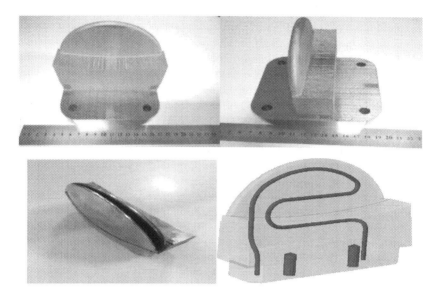

FIGURE 7.17 3D printed mold inserts.

speed is 650 mm/s, laser power is 250 W, scanning pitch is 0.06 mm, and paving layer thickness is 0.02 mm. Ninety-nine percent high purity argon gas is used during the forming process to prevent oxidation during the forming process.

The mold inserts formed by AM technology with a density of 99%, a hardness of 50 HRC and a surface roughness R_a of 15 μm are used. The conformal cooling waterway is designed such that the waterway is close to the surface of the product with a minimum distance to achieve sufficient cooling of the product, greatly improving the cooling efficiency and reducing the warpage of the product. The design for insert of the product and the waterway of the insert is shown in Fig. 7.17.

At the same time, the difference between the beryllium copper insert and the 3D printed insert in the test mode verification is compared. The cooling time and forming cycle are shown in Table 7.3. The plastic parts formed by the two schemes are deformed as shown in Fig. 7.18. It can be seen from the figure that the 3D printed inserts have significant improvement on the warpage of the product, and the gap between the two products is smaller and more uniform after assembly.

7.6.2 Lid and box

The mold inserts for another set of conformal cooling waterway are formed using S136 powder and AM forming process in Fig. 7.1, as shown in Fig. 7.19.

TABLE 7.3 Cooling time and forming cycle of two types of inserts.

Insert type	Injection molding and pressure maintain time (s)	Cooling time (s)	Production cycle (s)
3D printed inserts	7	16	40
Beryllium copper inserts	7	25	48

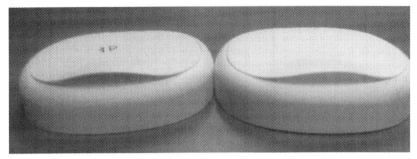

FIGURE 7.18 Product formed by 3D printed insert (left) and product formed by beryllium copper insert (right).

FIGURE 7.19 3D printed cooling jacket at the gate.

The insert makes the cooling waterway as close as possible to the cooling part, and the cooling effect is obvious. The scheme of hot runner company and AM design scheme is compared, as shown in Fig. 7.20.

The 3D printed gate cooling jacket can cool the gate well when forming the test mold, which can better ensure the mold temperature at the gate, and

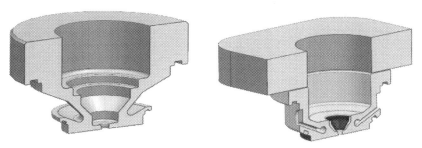

FIGURE 7.20 Cooling jacket at the gate of the hot runner company (*left*) and cooling jacket at the gate of the AM (*right*).

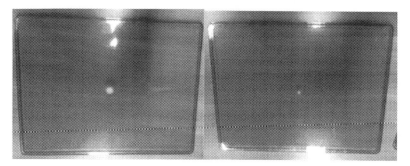

FIGURE 7.21 (*Left*) Products formed by the scheme of hot runner company and (*right*) products formed by AM.

the formed product can better meet the customer's requirements, especially for the production of transparent product, the specific effect of sample produced is shown in Fig. 7.21. Comparing the products formed by the cooling jacket provided by the hot runner company, it can be clearly found that the problem of blushing at the gate is well improved.

7.7 Application case 2 of metal powder materials for additive manufacturing

The denture sample is formed by Co-Cr alloy powder with an average particle size of 20 μm. The SLM forming process window has a laser power of 80−120 W, a scanning speed of 300−500 mm/s, and a scanning pitch of 0.04 mm. Within the above range, the sample of good quality can be formed by matching any laser power and scanning speed. The hardness is 476 ± 6 Hv, in the horizontal direction, the yield strength is 1142 MPa, the tensile strength is 1465 MPa and the elongation is 7.6%; in the vertical direction, the yield strength is 1002 MPa, the tensile strength is 1428 MPa, and the elongation is 10.5 %. The tensile strength of the sample is similar to that of the forged piece, and the elongation is similar to that of the casting. When using the EBSD technique, it is found that the part shows a preferred orientation approximating to the fiber texture in the direction close to [001]. A three-dimensional porous Co-Cr alloy implant is prepared on the basis of the above studies.

The accuracy of the denture mainly includes the wall thickness of inner crown and marginal adaptation. It is an important index to evaluate whether the denture meets the requirements of clinical application. Generally, the wall thickness of the inner crown is controlled between 0.3 and 0.5 mm. In the process of making the denture through the SLM process, the surface of the denture crown tends to have the adhered metal powder to be subsequently grinded and sandblasted. It has been experimentally found that the strength and wearing comfort of the denture is consistent with the design when the wall thickness of denture crown is set to 0.4 mm, as shown in Fig. 7.22.

Marginal adaptation refers to the vertical distance between the margin of the prosthesis and the cervical margin of the dental preparation. It reflects the accuracy and position of the prosthesis and is one of the main indicators for measuring the accuracy of the prosthesis. Because the position standard cannot be seen by the naked eye or is not easy to be detected by the probe, it is generally considered that the upper limit of the clinically acceptable margin difference is 100 μm. The marginal adaptation of the denture prosthesis and the matrix are mainly observed by the optical microscope. If the margin of the denture prosthesis and the matrix can be well overlapped after grinding, it can be said that the marginal adaptation of the denture prosthesis meets the requirements, otherwise it is considered as unqualified. Fig. 7.23 shows the coordination between the inner crown of the Co-Cr alloy denture made by SLM and the matrix. It can be seen that the margin of the inner crown of the denture completely coincides with the red line on the plaster model, indicating

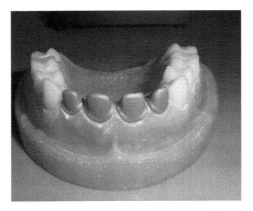

FIGURE 7.22 Coordination of SLM-formed Co-Cr alloy denture with matrix model.

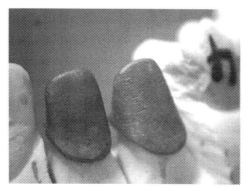

FIGURE 7.23 Coordination of the denture crown made by SLM with the matrix (OM).

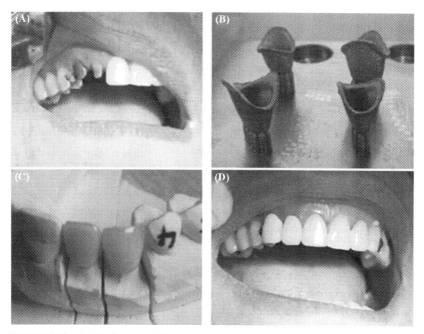

FIGURE 7.24 Application case of SLM-formed Co-Cr porcelain-fused denture prosthesis.

that the margin of the denture made by SLM has good adaptation and meets the clinical requirements.

On the basis of the previous research, the Co-Cr alloy porcelain-fused denture prosthesis made by SLM is finally successfully installed for a patient, as shown in Fig. 7.24. Fig. 7.24A shows the teeth that the patient needs to repair. The affected teeth obviously lose chewing function and seriously affect the aesthetics of the teeth. Fig. 7.24B shows a denture directly ordered for a patient using SLM technology. It can be seen that the outline is clear and there are no obvious defects. It takes only 2 hours from the acquisition of the data to the manufacture of the denture, while it usually takes about 1 week in traditional casting method. It can be seen that the denture manufactured by SLM greatly improves the manufacturing efficiency and saves a lot of treatment time for the patient. Fig. 7.24C shows that the SLM-formed denture fits to the plaster mold of the patient's teeth after baked porcelain. It can be seen that there is no crack on the denture of thick porcelain. The results of the previous study show that the metal-ceramic bonding strength can meet the international standards; at the same time, it can be seen that the denture and the plaster mold are closely matched. According to the doctor's judgment, the manufacturing precision and color are in full compliance with the clinical requirements. Fig. 7.24D shows the effect of wearing the denture prepared by SLM. It can be found that the color of denture well matches with the surrounding teeth, and the margin and the gums are tightly matched, which fully meets the clinical requirements. It is found that the gums are not inflamed after on-half year of use in the follow-up investigation to the patient, and the teeth are not stuffed with food residue; the patient is satisfactory to the use of dentures. It can be seen that the denture prepared by SLM technology not only has high precision, but also has low cost. It is estimated that the manufacturing cost is only 1/10 of that of traditional denture, and it is expected to be widely applied in the denture manufacturing industry.

7.8 Application case 1 of metal wire materials for additive manufacturing

The arc additive manufacturing technology is mainly applied to the patternless direct manufacturing of parts and the repair of damaged mechanical parts. Different wire materials and process parameters are selected according to different use occasions of parts. With 3D reverse technology and prototyping technology, the required machining model and related point cloud data can be obtained, so as to realize the intelligent repair of the failed and waste parts. The following introduces the additive manufacturing repair for a representative part.

The sample in the case uses the high-alloy steel flux-cored wire. The arc is stable during the test, the spatter lose rate is less than 2%, the forming and the welding process are good. The composition is shown in Table 7.4. The forming equipment is the KUKA robot and Funis CMT welder additive manufacturing

TABLE 7.4 Chemical composition of high-alloy steel flux-cored metal wire forming metal (%).

Element	C	Mn	Si	P	S	Cr	V
Wt (%)	0.22	0.85	0.19	0.010	0.012	7.90	0.25

platform built by the research group; the process parameters are set as follows: current: 140A, voltage: 14 V, and printing speed: 45 cm/min.

7.8.1 Printing of ultralarge (thin wall) parts

To verify the stability of the materials and parameters used in the printing of large parts, an ultralarge oval thin-walled part with a total length of 1200 mm, a total width of 1000 mm, and a total height of 720 mm is designed. The part has a draft angle of 5 degrees. Its continuous printing result is shown in Fig. 7.26. The actual size is shown in Table 7.5.

As shown in Fig. 7.25, the sidewalls of the ultralarge parts are layered uniformly without mixing layers, and the wall thickness is uniform, indicating that the thermal process in the entire printing process is uniform and unchanged. The test results show that the materials and parameters used are stable in the production of ultralarge parts.

TABLE 7.5 Forming dimensions of ultralarge (thin wall) parts.

Size	Printed parts
Total height (mm)	719.6
Total width (mm)	1001.8
Total length (mm)	1201.5
Thickness (mm)	5.2

FIGURE 7.25 Print results of ultralarge parts.

7.8.2 Multimaterial forging die forming

Forging die in the process of service not only the effect of force, but also by the influence of temperature. The service environment is complex and changeable, which requires high performance of the mold. The forging die itself not only needs to have good mechanical properties, such as high temperature strength and hardness, good toughness, and wear resistance, in the process of work also needs to have good thermal conductivity, heat resistance fatigue, thermal stability and oxidation resistance. In addition, the contact surface should also have good thermal wear and thermal melt loss. It is difficult for a single material forging die to fully meet the above requirements, which greatly affects the service life of the die. Therefore, the idea of multimaterial forging die manufacturing/reengineering of "base layer + transition layer + hard surface layer" is put forward here. The arc additive developed is used to manufacture specialty metal powder core wire and its forming equipment to realize the overall forming of multimaterial forging die (see Fig. 7.26). Among them, the base layer adopts MGY281 microslag powder core wire with high performance, and its hardness is 28HRC, which is a kind of crack resistant material with good plasticity. It has good yielding property under high pressure, which can effectively prevent the fracture failure of 5CrNiMo matrix under high pressure. The transition layer is made of MGY381 microslag powder core wire with high performance, and its hardness is 38HRC, which has high strength, hardness and good toughness. Meanwhile, it has a good combination with the substrate and matrix, and has a good transition effect in hardness and strength. The hard surface layer is made of MDY501 microslag high performance powder core wire, with a hardness of 54HRC, high hardness, strength, good thermal stability, fatigue resistance, wear resistance and good combination with the transition layer. The integrally formed multimaterial forging die has good toughness gradient and excellent combination between base and work layer. The performance of the forging die is significantly better than

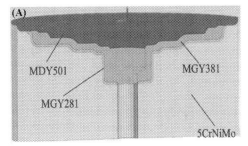

FIGURE 7.26 Multimaterial forging die made of arc additive: (A) schematic diagram; (B) real part figure.

the traditional 5CrNiMo forging die, and the service life is increased four to five times.

7.8.3 Hot forging die remanufacturing

Hot forging die is mainly used to bear the tension, pressure, bending, twisting and impact load of the key parts blank or part product production, the need for hot forging die variety, quantity. Car is the most prominent, usually a heavy truck forging more than 250 kinds of die, the need for forging nearly 1000 pairs. In the process of die forging production, hot forging die needs to bear the repeated effects of high temperature, high pressure and impact load, and often fails due to fast wear and thermal fatigue cracking. Traditional remanufacturing methods have problems such as high consumption of hot forging die steel, long remanufacturing cycle and low efficiency, resulting in high cost of hot forging die. Here forging die is implemented efficiently low cost manufacturing again (see Figs. 7.27 and 7.28), compared

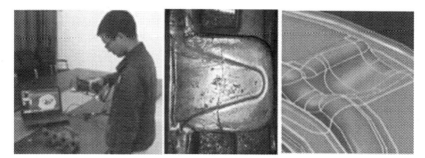

FIGURE 7.27 3D scanning and remanufacturing area design before mold remanufacturing.

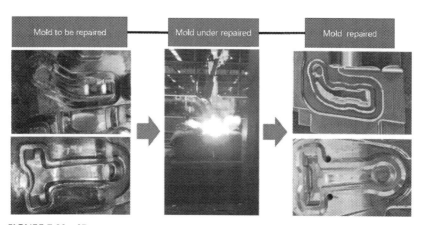

FIGURE 7.28 3D measurement and precision detection after mold repair.

with the traditional remanufacturing technology, material saving more than 50%, to improve the efficiency of remanufacturing by two to three times, the manufacturing cost will be reduced by more than 30%, the mold service life is increased by more than one time, obviously decrease the forging production cost, improve the overall profit and market competitiveness.

7.8.4 Part remanufacturing

The cylinder block of automobile engine will be partially damaged when used for a long time, which will lead to engine failure. To this end, the aluminum-silicon-magnesium alloy wire with a diameter of 1.2 mm was used for rapid manufacturing of the damaged part of the engine cylinder (see Fig. 7.29A), and the mechanical properties of the remanufacturing area reached the performance of 5356 welding wire. In addition, automobile engine automatic transmission oil road plate bottom around the bottom of the chamber is prone to crack, resulting in oil leakage, direct replacement is not only time-consuming and costly. To this end, the aluminum-silicon alloy wire with a diameter of 1.2mm was used to reproduce the crack in the oil circuit plate of the automatic transmission rapidly (see Fig. 7.29B), and the actual loading proved that the performance and life met the requirements (see Fig. 7.29C).

FIGURE 7.29 Automobile parts are remanufactured: (A) remanufactured rear engine head cylinder block; (B) remanufacturing the oil circuit board; (C) remanufacturing oil line board installation test.

7.9 Application case 2 of metal wire materials for additive manufacturing

WThe sample is made of low-alloy high-strength steel wire with the grade of YHJ507M. The composition of the wire is shown in Table 7.6. The KUKA robot and the Fronius CMT welder additive manufacturing platform built in by the research group are used to 3D print the multiaxial pipe structural members for construction, as shown in Figs. 7.30 and 7.31. The AM process parameters are shown in Table 7.7. It is verified by

TABLE 7.6 Chemical composition of metal wire YHJ507M formed metal (%).

C	Mn	Si	S	P	Cu	Cr
0.05–0.08	1.10–1.20	0.40–0.45	≤0.030	≤0.030	≤0.50	≤0.20

TABLE 7.7 Process parameters.

Current (A)	Voltage (V)	Speed of arc welding Torch (mm/s)	Protective gas and flow (L/min)	Wire feed rate (mm/s)
115–125	20–22	6–9	Argon-rich/20	70

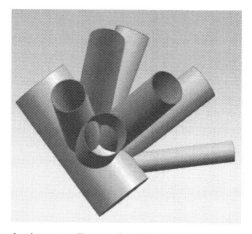

FIGURE 7.30 Simulated structure diagram of multiaxial piping system.

Application cases of additive manufacturing materials Chapter | 7   **671**

FIGURE 7.31 Structural parts of printed multiaxial piping system.

experiments that the variation of stacking melting height and melting width is the smallest, and the dimensional uniformity is the best under this parameter.

Multiaxial pipe joints are widely used in high-rise buildings, bridges, large offshore platforms and other engineering structures. At present, they are mainly manufactured by integral casting method. Because of the large wall thickness of the intersecting parts of each pipe, defects such as shrinkage cavity and shrinkage porosity are easily generated. The sample is directly formed by cold metal transition and layer-by-layer stacking, which provides a new solution for the traditional casting featured by high cost, low efficiency and multiple defects.

The experimental results show that the microstructure of multiaxial pipe structural members has a large number of small uniform ferrite and a small amount of pearlite; the fracture morphology is a large number of evenly distributed large and deep dimples, and the dimple size, uniformity and depth are greatly improved compared with the traditionally manufactured pipe joints; the tensile strength is 563 MPa, which is within the standard range; and the yield strength is about 437 MPa, which is higher than the standard value of 46%; the average elongation is 27.65%, which is higher than the standard value of 25.7%. In order to verify the accuracy and consistency of the test results, three sets of impact test samples are tested, the average impact absorption work is 147 J, and the impact absorption work of each sample is above 120 J, which is higher than the standard value of 145%. No cracks or pores are found in the penetration test. The test data indicates that the multiaxial pipe joints for arc additive manufacturing meet the requirements of construction steel for the microstructure and performance, and are of high dimensional accuracy.

7.10 Application case 3 of metal wire materials for additive manufacturing

This sample adopts the low-alloy high-strength steel metal wire with a grade of YHZ921A, the composition of the metal wire is shown in Table 7.8, the KUKA robot and the Fronius CMT welder additive manufacturing platform built by the research group are used to print out the double-arm shaft bracket of large ship, the process parameters are shown in Table 7.9, and the angle between the two arms is 52 degrees, as shown in Fig. 7.32.

The shaft bracket is used to support the high-speed rotating spiral propeller to ensure that the propeller works reliably and reduces the vibration generated when the propeller is working. Traditionally, the shaft bracket is manufactured by casting the propeller boss, the support arm and the cross arm separately, and then joining the three parts by means of tailor welding, so that the welded joint becomes a weak link and the production cycle is relatively long. The shaft bracket currently designed in China is generally bulky and cumbersome. Due to the large wall thickness, defects such as uneven composition, shrinkage cavity, shrinkage porosity, and pores are prone to occur in the casting method, while the shaft bracket manufactured by the AM method not

TABLE 7.8 Chemical composition of the formed metal (%).

C	Mn	Si	P	Ni	Cr	Mo	V
0.10	0.69	0.11	0.02	2.71	0.97	0.36	0.10

TABLE 7.9 Process parameters.

Current (A)	Voltage (V)	Speed of arc welding torch (mm s)	Protective gas and flow (L/min)	Wire feed rate (mm)
220–230	28–29	7	Argon-rich/20	70

FIGURE 7.32 Double-arm shaft bracket of a large ship.

only satisfies the requirements of use in performance, but also has no defects such as shrinkage cavity and shrinkage porosity inside. It is verified by experiments that the yield strength of the metal parts deposited by the metal wire is 530 MPa, which is higher than the standard value of 370 MPa; the tensile strength is 600 MPa, which is higher than the standard value of 470 MPa; the impact absorption at $-20°C$ is 70 J, which is higher than the standard value of 47 J.

7.11 Application case 1 of ceramic powder materials for additive manufacturing

The kaolinite-granulated powder with an average particle size of about 100 μm is used for manufacturing the sample, and the epoxy resin E06 is used as the binder to prepare kaolinite/E06 composite powder by mechanical mixing method. The kaolinite and E06 powder are uniformly mixed in a three-dimensional mixer at a mass ratio of 9:1 for 24 hours, and are taken out for SLS forming. The method is simple and convenient, and is suitable for the application and development of SLS forming of traditional ceramic products.

The forming equipment is the HRPS-IIIA system developed by the research group. The forming working chamber is 500 mm × 500 mm × 400 mm, the laser wavelength is 10.6 μm, and the spot is 0.2 mm. The processing parameters: laser power $P = 7.15$ W, scanning speed $V = 2000$ mm/s, scanning pitch $S = 0.1$ mm, single layer thickness $L = 0.15$ mm, and

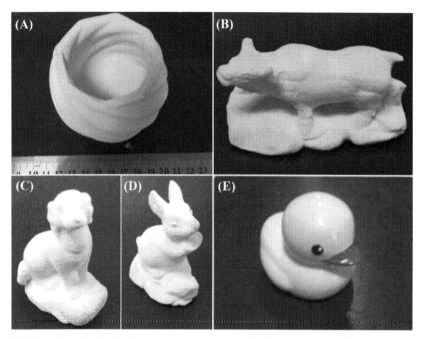

FIGURE 7.33 (A−D) Traditional ceramic parts prepared by SLS and (E) duck-like ceramic craft prepared by SLS/CIP after color painting and sintering ceramic green parts prepared by SLS.

preheating temperature of 55°C are selected in this experiment to form various ceramic parts as shown in Fig. 7.33A−D. In addition, the green parts prepared by SLS are subjected to cold isostatic pressing treatment under 200 MPa, color painting and high temperature sintering (1350°C) to produce a duck-like ceramic craft, as shown in Fig. 7.33E. The abovementioned technology makes the manufacture of ceramic crafts detached from the reliance on craftsmanship and traditional processing methods, further promoting the artistic creativity, so that people can participate in the manufacturing process of crafts, without having to master complicated skills.

7.12 Application case 2 of ceramic powder materials for additive manufacturing

The forming equipment used for the sample is SLS equipment independently developed by the rapid manufacturing center of Huazhong University of Science and Technology, and the forming size can reach 1700 mm × 1700 mm × 1700 mm. Aiming at the demand of integrated

FIGURE 7.34 SLS-shaped large-size SiC square mirror.

forming of large and complex ceramic parts, the equipment adopts multilayer and multipoint independent temperature control method combined with numerical simulation optimization to realize the uniform distribution of preheating temperature in a very large range of working fields. The distribution adopts the four-laser collaborative scanning method to meet the needs of a large range of work field scanning. Aiming at the difficulty of multilaser region stitching and scanning accuracy in a large range of work field, the automatic correction method of full-width multipoint vision measurement is adopted to realize the high-precision stitching of single-scan system and multiscan system. According to the requirements of processing technology and load conditions, the working area of multiple lasers is divided in real time to achieve higher manufacturing efficiency and quality.

A large SiC square mirror manufactured with the equipment is shown in Fig. 7.34. The square mirror is a key component in integrated circuit manufacturing equipment. The SiC composite powder is used by SLS forming and sintering to obtain ceramic parts with relative density greater than 95% and sintering shrinkage rate less than 2%. This method has solved the whole manufacturing problem of high performance large complex SiC ceramic parts, and is of great significance to promote the scientific and technological innovation of aerospace, electronic communication, electronic communication and other related industries.

7.13 Application case 3 of ceramic powder materials for additive materials

7.13.1 Manufacturing of cordierite ceramic parts

The cordierite powder with an average particle size of about 38.6 μm and irregular shape is used as the sample, and epoxy resin E12 is used

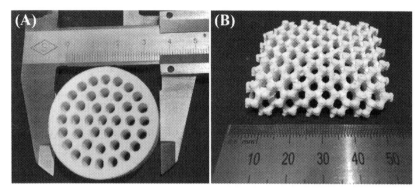

FIGURE 7.35 Cordierite parts with complex shape: (A) straight-hole honeycomb structure; (B) topology of spiral icositetrahedron.

as the binder to prepare cordierite/E12 composite powder by mechanical mixing method. Cordierite and E12 powder are uniformly mixed in a three-dimensional mixer for at a mass ratio of 9:1 24 hours, and are taken out for SLS forming. The processing parameters: laser power $P = 17$ W, scanning speed $V = 1500$ mm/s, scanning pitch $S = 0.15$ mm, single layer thickness $L = 0.15$ mm, and preheating temperature of 50°C are selected in this experiment to form porous cordierite ceramic green part. Then, the porous ceramic green part is subjected to degreasing and sintering (1400°C), and finally cordierite porous ceramics having a complicated pore structure are obtained, as shown in Fig. 7.35.

7.13.2 Manufacturing of Al_2O_3 ceramic parts

The Al_2O_3 granulated powder with an average particle size of 50–100 μm is used as the sample, and the epoxy resin E06 is used as the binder to form Al_2O_3/E06 composite powder by mechanical mixing method. The Al_2O_3 and E06 powder are uniformly mixed in a three-dimensional mixer at a mass ratio of 9.2:0.8 for 24 hours, and are taken out for SLS forming. The processing parameters: laser power $P = 5.5$ W, scanning speed $V = 1400$ mm/s, scanning pitch $S = 0.1$ mm, single layer thickness $L = 0.15$ mm, and preheating temperature of 55°C are selected in this experiment to form Al_2O_3 ceramics. Then, the green part is subjected to isostatic pressing operation, and the Al_2O_3 ceramic gear and the turbine disk are obtained by sintering at 1600°C for 4 hours, as shown in Fig. 7.36.

FIGURE 7.36 Aluminum parts with complex shape: (A) gear; (B) turbine disk.

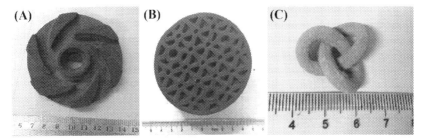

FIGURE 7.37 SiC ceramic parts with complex shape: (A) vane wheel; (B) mirror blank; (C) torsion ring.

7.13.3 Manufacturing of SiC ceramic parts

The SiC granulated powder with an average particle size of 100 μm is used as the sample and the epoxy resin E12 is used as the binder to prepare SiC/E12 composite powder by mechanical mixing method. The SiC and E12 powder are uniformly mixed in a three-dimensional mixer at a mass ratio of 8.5:1.5 for 24 hours, and are taken out for SLS forming. The processing parameters: laser power $P = 7$ W, scanning speed $V = 2000$ mm/s, scanning pitch $S = 0.15$ mm, single layer thickness $L = 0.15$ mm and preheating temperature of 35°C are selected in this experiment to form SiC ceramic parts, as shown in Fig. 7.37.

The use of SLS technology to manufacture fine ceramic or functional ceramic components with complex structures not only avoids costly and time-consuming secondary machining, but also eliminates the dependence of traditional slip casting on the mold, and truly realizes the high-precision manufacturing of ceramic parts with low cost and short cycle, which has broad application prospects in such fields as aerospace, automobile, and electronic information.

Chapter 8

Materials for four-dimensional printing

8.1 Definition of four-dimensional printing

With the development of social economics, science, and technology, the demands of high-end equipment for components are no longer limited to the conventional mechanical properties or functionalities but are expanded to the intelligent performances. The intelligent components, which can predictably change their shapes, properties, or functionalities, usually have complex, exquisite, and lightweight structures. It is complicated and even difficult to create these structures through traditional manufacturing technologies. Moreover, the utilization of composite materials and multiple materials are commonly required in the intelligent components to realize complex functionalities. The traditional manufacturing technologies are hard to meet both demands of intelligent components for structures and materials. In this context, an innovational additive manufacturing technology, namely four-dimensional (4D) printing, emerges based on the multidisciplinary integration of materials, machinery, mechanics, information, and so on. 4D printing provides effective solutions to the manufacturing of complex intelligent components.

4D printing was first proposed by Prof. Tibbits from Massachusetts Institute of Technology at the 2013 TED conference [1]. The researchers from this team prepared a one-dimensional rope-shape structure through additive manufacturing technology, which can automatically turn into a preset three-dimensional shape when it was put in water. This marked the start of 4D printing. Since then a study upsurge of 4D printing has been raised in academia and industry. The original definition of 4D printing refers to adding a fourth dimension, namely the time, to the 3D printing. That is, the 4D printing is a technology of using 3D printing technology to fabricate intelligent components, which can change their shapes with time under an external stimulus, such as heat, light, electricity, magnetism, PH, and so on. With the deepening study, the connotation of 4D printing has been further enriched. The new definition of 4D printing indicated that the changes of the 4D printing components under external stimuli are not only limited in shape changes but also include changes of properties and functionalities. Moreover, these changes can take place not only in the time dimension but also in the space dimension, and they should be controllable. Through the automatic adjustment of shapes, properties or functionalities under corresponding stimulus, 4D printing components can achieve a series of intelligent performances including self-adaption, self-assembly, self-repair, self-learning, etc. 4D printing is expected to promote innovative development of high-end intelligent equipment.

In traditional manufacturing technologies, the preparation of materials, design of structures, and realization of functionalities are three independent links. In contrast, 4D printing integrates the design of components' intelligent performances into the forming process, realizes the integrated manufacturing of materials, structures and functionalities, and simplifies the process from concept design to final products. This breakthrough technology is hopeful to bring great revolution to the traditional manufacturing technology as well as change the conventional design concepts and manners. It shows great application potential in aerospace, biomedicine, robotics, automobile, construction, etc. Thus, 4D printing has become the forefront and hot spot of research at home and abroad.

8.2 Research and development status of four-dimensional printing materials at home and abroad

4D printing technology enables the additive manufactured components to change the shapes, properties, or functionalities with time and space under certain stimuli. 4D printing materials play a key role in the realization of 4D printing. Both the smart materials (materials that can sense the change in environment and response to it by changing their shapes and/or chemical and physical properties) and the traditional materials (nonsmart materials) can be the 4D printing materials. The capacity of 4D printing components to achieve the above anticipated changes relies on the properties of the materials as well as their appropriate arrangement and synergy in the 3D space. At present, 4D printing materials has become the main bottleneck restricting the development of 4D printing technology as well as the difficulty in the breakthrough of 4D printing technology. To improve the performance of 4D printing components and expand the application fields of 4D printing, researches and developments of new materials are necessary. Currently, the 4D printing materials mainly include polymers and their composite materials, metals and their composite materials and ceramics and their composite materials. Next, the research and developing status of the above three groups of 4D printing materials will be introduced separately.

8.2.1 Polymers and their composite materials

Polymers and their composite materials are easy to be printed through additive manufacturing technologies. Moreover, they also have the advantages of low price and low density. Thus, polymers and their composites have become the most widely used 4D printing materials. At present, 4D printing polymers and their composite materials can be mainly divided into water-responsive, thermal-responsive, magneto-responsive, light-responsive and electro-responsive polymers and their composite materials.

8.2.1.1 Water-responsive polymers and their composite materials

Hydrogels are the most applied and studied 4D printing water-responsive polymers due to their high, extraordinary hydrophily which allow them to expand up to 200% of their original volume. In addition, the hydrogels also have high printability and are easy to be printed through direct ink writing (DIW). However, the shape changes of traditional hydrogels caused by water swelling are uncontrolled. To achieve controlled shape changes, the water swelling behaviors of the hydrogels must be programmed to endow them anisotropic swelling performance. In 2013, Tibbits et al. from the Massachusetts Institute of Technology in the United States succeeded in designing and preparing a composite polymer structure through coding the locations and shapes of a hydrogel and a rigid polymer [2]. When the composite polymer was put into the water, the hydrogel expanded, which caused the joint to fold in the other direction. The folding stopped when the rigid part hit another rigid part, and correspondingly, the composite polymer structure was stabilized in a new shape. Later, Jamal et al. from Johns Hopkins University in the United States prepared a bilayer material composed of photo-crosslinking polyethylene glycol (PEG) with varying molecular weights [3]. When put into the water, a predesigned shape change took place due to the differential swelling of the hydrogel layers. In 2016, Lewis et al. from Harvard University in the United States developed a hydrogel composite material mixed with cellulosic fibers [4]. The swelling performance of the hydrogel variated with the arrangements of the fibers. Through change of the orientation and amount of the fibers, they succeeded in programming the anisotropic swelling behavior, and thus controlled the shape change of this hydrogel composite material. In 2019, Song et al. from Huazhong University of Science and Technology developed a double network (DN) hydrogel composite material reinforced by carbon nanotubes (CNT), which was suitable for the DIW technology [5]. The addition of CNT improved the viscoelasticity of the DN hydrogel composite material and consequently enhanced its printability.

8.2.1.2 Thermoresponsive polymers and their composite materials

Thermoresponsive shape memory polymers (SMPs) and their composite materials are one group of materials that can return to their original shape upon heating. This phenomenon is called shape memory effect (SME). Due to the unique SME, low cost, high printability, and low density, thermoresponsive SMPs have been considered as one kind of promising 4D printing materials. And numerous studies focusing on the 4D printing technology of thermoresponsive SMPs and their composite materials have been reported. In 2016, Yang et al. from the University of Hong Kong prepared a 4D printing SMP filament through extrusion technology using the DiAPLEX MM-4520 pellets [6]. This SMP filaments could be printed through fused-deposition modeling (FDM) technology. The flower printed with this SMP could close themselves upon heating. Ge et al. from Singapore University of Technology and Design developed a 4D printing SMP, which could be printed through stereolithography appearance (SLA) technology [7]. This SMP was composed of methacrylate-based monomers and crosslinkers, photo initiator (2,4,6-trimethylbenzoyl), and photo absorbers (Sudan I and Rhodamine B). Using this SMP, they succeeded in the 4D printing of Eiffel tower, grippers, stents, and fowlers, which could change their shapes upon heating. Wu et al. from Georgia Institute of Technology in Atlanta prepared multishape active composites using multiple SMP fibers with different glass transition temperatures [8]. By tuning the volume fractions, locations and arrangement modes of the SMP fibers, bending deformation of those composites could be controlled. In 2019, Wu et al. from Huazhong University of Science and Technology developed a 4D printing SMP suitable for digital light processing (DLP) with tert-Butyl acrylate/1,6-hexanediol diacrylate (tBA/HDDA) networks [9]. After deformation under an external stress and a subsequent freeze-thaw treatment, this SMP could return to its original shape when it was placed in a 73°C water bath.

8.2.1.3 Magneto-responsive polymers and their composite materials

Magneto-responsive polymers and their composite materials are a group of materials that can change their shapes, properties or functionalities under an external magnetic field. At present, the main design and preparation method of magneto-responsive polymers and their composite materials is adding magnetic particles to the polymers and their composite materials. In 2017, Wei et al. from Harbin Institute of Technology developed a magneto-responsive composite material through adding iron oxide to the ultraviolet (UV) crosslinking poly (lactic acid)-based SMPs [10]. The architectures printed with this composite material could change their shapes in an alternating magnetic field. In the studies of Kim et al. from Massachusetts Institute of Technology in the United States, a 4D printing elastomer composite containing ferromagnetic microparticles was reported [11]. Through applying a magnetic field to the dispensing nozzle while printing, they reorient particles along the applied field program to impart patterned magnetic polarity to the printed filaments. Under an external magnetic field, different parts of this composite could make different responses; consequently a predesigned shape change could be produced. Later, Ze et al. from the Ohio State University Columbus in the United States developed a novel magnetic SMP composite by adding two types of magnetic particles into an amorphous SMP matrix [12]. The matrix softens via magnetic inductive heating of low-coercivity particles, and high-remanence particles with reprogrammable magnetization profiles drive the rapid and reversible shape change under actuation magnetic fields. Once cooled, the actuated shape can be locked.

8.2.1.4 Photo-responsive polymers and their composite materials

Photo-responsive polymers and their composite materials are a group of materials that can change their shape, properties or functionalities through physical or/and chemical reactions when exposed in a light. Liu et al. from North Carolina State University in the United States used a desktop printer to pattern inks of different light absorptivity as joints on otherwise homogeneous prestrained polymer sheets [13]. Printed ink on the surface of the polymer sheets discriminately absorbs light based on the wavelength of the light and the color of the ink. The absorbed light gradually heats the underlying polymer across the thickness of the sheet, which causes relief of strain to induce folding. These color patterns can be designed to absorb only specific wavelengths of light (or to absorb differently at the same wavelength using color hues), thereby providing control of sheet folding with respect to time and space. Wales from University of Nottingham in the United Kingdom developed a photochromic molecule, which can change from colorless to blue when exposed in the visible light and then return to colorless through the redox reaction when exposed in the oxygen [14]. Then they developed prepared a photochromic polymer material through adding these photochromic molecules into the bespoke polymeric ionic-liquid matrices.

8.2.2 Metals and their composite materials

As compared with the polymers and their composite materials, metals and their composite materials usually have better mechanical properties and can realize the integration of carrying and the shape, property or functionality changing. At present, 4D printing metals and their composite materials mainly include various shape memory alloys (SMAs) and their composites. SMAs are a group of alloys which can return from a deformed shape to their original shape under certain stimulus, such as temperature or magnetic variations. In addition, superelasticity (SE) is also a common shape memory performance in SMAs. That is, the alloy deformed under an applied stress can recovery its original shape after removing this external stress, with no need for any other stimulus. SMAs are one kind of smart materials with multiple functionalities, such sensing, driving, controlling and energy convert. Thus, SMAs have been widely studied in the 4D printing field. Currently, SMAs are mainly printed through powder-based additive manufacturing technologies, such as selective laser melting (SLM) and laser-engineered net shaping (LENS). At present, the SMA powders are mainly prepared through atomization method.

Among the current SMAs, Ni-Ti-based alloys are the most utilized SMAs in 4D printing thanks to their most excellent SME and SE (recovery strain up to 12%), high recovery stress (500−1000 MPa), high damping ratio (0.038 ± 0.004 in austenite, 0.002 ± 0.004 in martensite), outstanding corrosion and wear resistance, good mechanical properties, and good biocompatibility. Dadbakhsh et al. from KU Leuven in Belgium studied the influence of low laser parameters (LP, low laser power of 40 W with low laser scanning velocity of 160 mm/s) and high laser parameters (HP, high laser power of 250 W with high laser scanning velocity of 1100 mm/s) on the microstructures and martensitic transformation characteristics of SLM Ni-Ti alloy [15−16]. Both the above parameter combinations could fabricate components with relative densities of approximately 99%. Nevertheless, the HP-SLM samples mainly contained austenite at room temperature (RT) and exhibited SE due to the lower martensitic transformation temperatures, while the LP-SLM samples were mainly composed of martensite at RT and showed SME due to the higher martensitic transformation temperatures. They ascribed the higher martensitic transformation temperatures in LP-SLM samples to the precipitation of Ti-rich second phase as well as the lager grains caused by the lower cooling rate. In the research of Saedi et al. from University of Kentucky in the United States, it was also proved that fully dense Ni-Ti SMAs could be fabricated through either high laser power with high scanning velocity or low laser power with low scanning velocity [17]. However, a low laser power with a low scanning velocity was suggested to produce Ni-Ti SMAs since samples processed through low LP exhibited better SE response, lower mechanical hysteresis, higher hardness as compared to those fabricated through HP. The Ni-49.2Ti SMA created with a low laser power of 100 W and low scanning velocity of 125 mm/s showed almost perfect SE recovery strain of 5.77% after compressive deformation by 6.02% at $A_f + 10K$. This value is comparable to that in the conventional casting Ni-Ti SMAs. Furthermore, they also investigated the effect of solution treatment and subsequent aging treatments on the SE of a SLM Ni-49.2Ti alloy [18]. The stable SE recovery strain after 10 compressive cycles at $A_f + 15K$ was about 2.64% in the as-fabricated SLM samples. After a solution treatment at 1223K for 5.5 h, it was remarkably improved to 3.42% due to the homogenization of microstructure. Aging treatment at 623K for 18 h further improved the stable SE recovery strain in the solution treated SLM samples to 4.2%. They suggested that the precipitation strengthening caused by the precipitation of Ni_4Ti_3 particles was responsible for this SE enhancement. Marattukalam et al. from National Institute of Technology Karnataka in India investigated the effect of annealing treatment at 773K and 1273K on the microstructures and properties in LENS Ni-44.8Ti (in wt.%) alloy [19]. For the as-fabricated LENS samples, the recovery strain was about 7%−8% after compressive deformation by 10% at RT. Annealing treatment at 773K for 30 min further improved their recovery strain to 8.6%−10% due to the increase in volume fraction of B19′ martensite. However, after annealing treatment at 1273K for 30 min, the recovery strain in the SLM samples remarkably reduced to 3.8%−6.6% as a result of the stress relief effects. Lu et al. from South China University of Technology systematically investigated the variation of martensitic transformation characteristics and shape memory performance with the energy density (155−292 J/mm^3) in a SLM Ni-49.4Ti alloy [20]. As lowering the energy density, the martensitic transformation temperatures decreased, but the recovery strain at RT increased. The samples created with the energy density of 155 J/mm^3 showed a high stable SME recovery strain of 4.99% after ten tensile cycles at RT. Khoo et al. from Nanyang Technological University in Singapore fabricated a Ni-49.9Ti alloy through repetitive laser scanning [21]. These samples showed a SME recovery stain of 3.37% when tensile deformed at RT by about 5%. Through simple aging treatment at 673K for 5 min, their SME recovery strain was further improved to 3.59% due to the precipitation of fine Ni_4Ti_3 metastable phase.

Besides the Ni-Ti-based SMAs, some studies focusing on the 4D printing Cu-based SMAs have also been reported. In 2014, Mazzer et al. first succeeded in preparing a dense and fine Cu-11.85Al-3.2Ni-3Mn (in wt.%) alloy through SLM. SLM-simples with a grain size of 10−100 μm, a related density of 97%, no cracks, and single monoclinic martensite were obtained [22]. In 2016, Gustmann et al. further investigated the effect of processing parameters, scanning strategies, and additional remelting step on the microstructures and mechanical properties of the SLM Cu-11.85Al-3.2Ni-3Mn alloy [23−25]. The results showed that the optimum energy density is between 30 and 40 J/mm^3. Dense samples with a high relative density of 99% and full martensite could be obtained when the laser power and scanning velocity were above 300 W and 700 mm/s, respectively. Under compressive loading, the SLM samples exhibited high fracture strength (1515 ± 50 MPa) and fracture strain (18% ± 1.7%) comparable to the dense suction-cast samples (1500 ± 30 MPa and 16% ± 0.2%). Moreover, remelting the solidified layer could further improve the relative density, lower the surface roughness, and adjust the martensitic transformation temperatures of SLM samples without compromising their mechanical properties. In 2019, Tian et al. prepared a Cu-13.5Al-4Ni-0.5Ti (in wt.%) alloy through SLM successfully [26]. These SLM samples, which have an average grain size of 43 μm and a relative density up to 99.5%, showed higher elongation (about 7.6%) but lower ultimate tensile strength (about 541 MPa) than the casting samples.

8.2.3 Ceramics and their composite materials

Ceramics and their composite materials have stable physical and chemical properties as well as excellent wear and corrosion resistance. Thus, they show great application potential in aerospace, biomedicine, electronic communications, and so on. However, the traditional ceramics and their composite materials usually have high brittleness, and the additive manufactured ceramic structures are usually difficult to be deformed. To solve this limitation, Lu et al. from City University of Hong Kong developed an elastomeric poly (dimethylsiloxane) matrix nanocomposites that could be printed, deformed, and then transformed into silicon oxycarbide matrix, making the growth of complex ceramic origami and 4D-printed ceramic structures possible [27]. These nanocomposites could be printed through DIW to create the ceramic precursors, which were soft and can be stretched beyond three times their initial length. They succeeded in the 4D printing of ceramic with this material using the following method: First, the well-designed ceramic precursor was printed on a substrate that was prestretched by a homemade stretch device. In this case, some elastic energy was previously stored in the printed the ceramic precursor. After releasing of prestretch, the ceramic precursor deformed, forming the predesigned 4D printing structures. Finally, the deformed ceramic precursor transformed into ceramic after a heat treatment.

8.3 Research progress of our team on four-dimensional printing materials

8.3.1 Cu-Al-Ni-based shape memory alloys

At present, Ni-Ti and Cu-based alloys are two types of the most widely used SMAs. Although Ni-Ti SMAs show better SME and SE than Cu-based SMAs, they are only used at a low operating temperature not exceeding 100°C [28], and they have poor machinability [29]. In contrast, Cu-based SMAs can reach higher transformation temperatures and are preferential candidates for high-temperature applications (such as in thermal actuators and thermal sensors) up to 240°C due to their superelastic or pseudoelastic properties, two-way SME, and high damping ability [30]. Moreover, the excellent processing performance of Cu-based SMAs enables the manufacture of complex-shaped parts. Cu-based SMAs are currently divided into two series, namely Cu-Al-Ni-based and Cu-Zn-Al-based, whose martensite transformation temperatures can be adjusted within a wide range. Since the transformation temperature of Cu-based SMAs is sensitive to the variation of composition, it can be tailored by adjusting the ratios of elements in the alloy to satisfy various temperature requirements in different application environments. More significantly, Cu-Al-Ni-based alloys exhibits a higher martensite transformation temperature than do Cu-Zn-Al-based alloys. For example, Sugimoto et al. developed a Cu-Al-Ni-Mn alloy with good thermodynamic stability and antiaging behavior at temperatures above 100°C and this alloy showed the capability for high-temperature applications [31].

However, Cu-Al-Ni SMAs are brittle in nature due to the intergranular cracking, which is caused by the high elastic anisotropy. The large elastic anisotropy makes the intergranular elasticity and plastic deformation of the

Cu-based SMAs extremely uncoordinated, which causes the grain boundary stress concentration and intergranular cracking. Currently, grain refinement and plasticity enhancement are two main methods to improve the mechanical properties of Cu-based SMAs. Researchers found that the addition of B, Ce, and V elements could result in the grain refinement of Cu-Zn-Al-based alloys, whereas the addition of Ti and Zr elements could refine the grains of Cu-Al-Ni-based alloys. In addition, it is also effective to improve the performance of Cu-Al-Ni-based alloys through rapid solidification methods such as melt-spinning, melt extraction, spray forming, and SLM.

As an additive manufacturing technique, SLM can create bulk parts by melting specific predefined areas of a powder bed layer-by-layer. The processing of a thin powder layer (20–50 μm) on massive substrate plates in combination with small laser spot diameters (approximately 100 μm) results in high intrinsic cooling rates of $2.13 - 2.97 \times 10^6$ °C/s. Fine grain sizes within approximately one micron can be obtained because of the high cooling rates during SLM. If the parts prepared by SLM have a high relative density, the yield and ultimate tensile strength are comparable to (even better than) those of conventional manufactured samples. Recently, the investigations about SLM technology of SMAs mainly focused in Ni-Ti-based alloys. Only few studies were performed in Cu-based alloys. Gustmann et al. studied the effect of SLM parameters on the microstructures and mechanical properties of Cu-11.85Al-3.2Ni-3Mn SMAs and obtained nearly dense samples with higher mechanical properties than those of the casting counterparts [23–25]. The SLM alloys have an average ultimate tensile strength of 620 ± 50 MPa, an elongation of $8.2\% \pm 0.9\%$ at RT, and an average hardness of 245 ± 20 HV. However, the mechanical properties of Cu-based SMAs still need to be improved, and there is a lack of studies on their high-temperature performances.

This study investigated the SLM of a new Cu-based SMA, Cu-13.5Al-4Ni-0.5Ti. According to the phase diagram of Cu-Al-Ni ternary alloy, the alloy with 4 wt.% Ni and 13.5 wt.% Al exhibits a good SME. The addition of Ti element is expected to improve the material strength because of the grain refinement effect. The objectives are to obtain the optimal SLM processing parameters to fabricate fully dense samples ($\rho \geq 99\%$), clarify the phase formation, microstructures and tensile properties at RT and 200°C, and understand their performance characteristic and difference from casting alloys. The findings provide a basic insight on SLM-fabricating high-performance Cu-based SMAs with tailored microstructure and elevated mechanical properties.

8.3.1.1 Experimental

8.3.1.1.1 Preparation of Cu-Al-Ni-Ti alloy powder

The experimental material was gas-atomized spherical Cu-13.5Al-4Ni-0.5Ti powder that was prepared by Powder Metallurgy Research Institute of Central South University. The chemical composition of the powder was shown in Table 8.1. The density (7.11 g/cm^3) of the alloy powder was determined by a fully automatic

TABLE 8.1 Chemical composition of experimental Cu-Al-Ni-Ti powder.

Element	Al	Ni	Ti	Others	Cu
Mass fraction (wt.%)	13.5	4	0.5	<0.1	>81.9

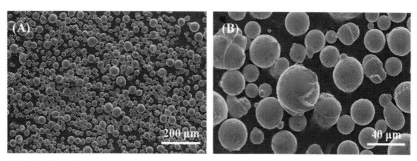

FIGURE 8.1 Micromorphology of Cu-13.5Al-4Ni-0.5Ti powder under scanning electron microscopy (SEM).

true density analyzer (AccuPyc 1330, United States). The powder particles showed a nearly spherical shapes and smooth surfaces as shown in Fig. 8.1. Small satellite particles were observed to adhere to the surface of large particles. The powder size was 19.8–46.7 μm with an average value of 30.5 μm (Fig. 8.2).

8.3.1.1.2 Fabrication of Cu-Al-Ni-Ti alloy

The samples were fabricated by an SLM 125HL machine (Solutions GmbH, Germany). The SLM machine was equipped with a 400-W Gaussian beam fiber laser. The laser beam diameter ($\Phi_{90\%}$) is approximately 80 μm. The process was performed in an argon atmosphere (approximately 500 ppm oxygen). Based on previous extensive SLM forming experiments, the laser power and scanning speed have greater effect on the formed parts than the layer thickness and scan spacing. Therefore, the laser power and scanning speed were chosen as the experimental variables of SLM process optimization experiments. Some cube samples (in $8 \times 8 \times 5$ mm^3) for process optimization were fabricated by SLM. The processing parameters with fixed layer thickness (0.04 mm), fixed scanning spacing (0.09 mm), varying laser power (250–340 W) and varying scanning speed (600–900 mm/s) were chosen. The long bidirectional scanning strategy was used, as shown in Fig. 8.3. To reduce the residual thermal stress during SLM, a 67 degrees rotation of the scanning direction between two consecutive layers N and N + 1 was set [32]. The substrate was preheated to 200°C before processing. Fig. 8.4A illustrates the Cu-13.5Al-4Ni-0.5Ti alloy cubes ($8 \times 8 \times 5$ mm^3) fabricated by SLM

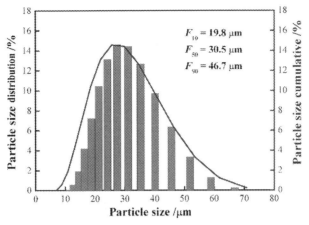

FIGURE 8.2 Particle size distribution of Cu-13.5Al-4Ni-0.5Ti powder.

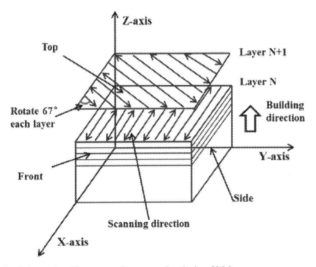

FIGURE 8.3 Schematic of laser scanning strategies during SLM.

using various processing parameters. In addition, the hardness testing samples (cubes of $8 \times 8 \times 5$ mm^3) and tensile samples were fabricated with the optimized processing parameter (maximum relative density), as shown in Fig. 8.4B.

8.3.1.1.3 Characterization of Cu-Al-Ni-Ti alloy

The density of the cube samples was measured using a high-precision balance (Sartorius MC210P, Germany), whose measurement principle is Archimedes method. The phase identification was performed by X-ray

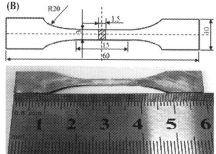

FIGURE 8.4 The Cu-13.5Al-4Ni-0.5Ti samples fabricated by SLM: (A) cubes (8 × 8 × 5 mm³) fabricated by various processing parameters (layer thickness: 0.04 mm, scanning space: 0.09 mm); (B) tensile samples (laser power: 310 W, scanning speed: 800 mm/s, layer thickness: 0.04 mm, scanning space: 0.09 mm).

diffraction (XRD) (XRD-7000S, Shimadzu, Japan) with a Cu target in a conventional X-ray tube, which operated at an accelerating voltage of 40 kV and an electron beam current of 30 mA. The SLM samples for microstructure observations were cut from the substrate by wire-electrode cutting, ground and polished with automatic grinding and polishing machine (Ecomet300/ Automet300, Buehler, United States), and chemically etched with a solution (vol. 50% H_2O and vol. 50% HNO_3) at RT for approximately 10 s. The microstructures were investigated by optical microscopy (OM VHX-1000C, Keyence, Japan), scanning electron microscopy (SEM JSM-7600F, Japan) and transmission electron microscopy (TEM JEM-2100, Japan).

The Vickers microhardness of the samples was determined by a hardness test machine (430-SVD, Wilson Hardness, United States) with a 5 kg load and 30 s dwell time. Five points were chosen and measured at different positions of each surface to obtain an average value. The tensile samples were cut to the required dimension by a CNC machine (WDW3200, China), as shown in Fig. 8.4B. Because the working environment of Cu-Al-Ni alloys at high temperature is mainly at 100−200°C [33], the tensile tests were performed on a mechanical testing machine (Zwick/Roell Z020, Zwick, Germany) at RT and 200°C. The tests were performed using a cross speed of 0.5 mm/min and a constant tension load. The strain was recorded by a laser extensometer. The fracture morphology of the tensile samples was observed by SEM (JSM-7600F, Japan).

8.3.1.2 Results and discussion

8.3.1.2.1 Effect of SLM processing parameters on relative density

Fig. 8.5 illustrates the relative densities of Cu-13.5Al-4Ni-0.5Ti SMAs fabricated by SLM with various laser powers and scanning speeds. When the scanning speed is 600−700 mm/s (Fig. 8.5A, part A), the density

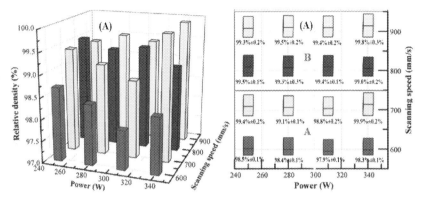

FIGURE 8.5 (A) The relationship between the laser power and the scanning speed and the sample density; (B) Top view of Fig. 8.5(A): Part A is the density value of the sample at the lower scanning speed (600–700 mm/s); Part B is the density value of the sample at the higher scanning speed (800–900 mm/s).

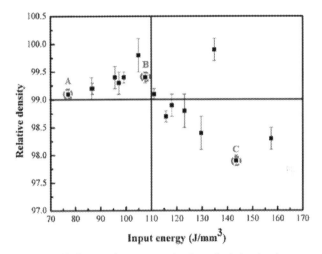

FIGURE 8.6 Relationship between input energy density and relative density.

substantially decreases with the increase in laser power but increases with the increase in scanning speed. The combination of the processing parameter, a high laser power and a low scanning speed easily makes the metal solution boil in molten pools due to an extremely high temperature, which causes the splashing of small metal droplets and finally the formation of small pores in the samples and a lower density [34]. When the scanning speed is improved to 800–900 mm/s (Fig. 8.5B, part B), the relative density exceeds 99%. In this case, the density only slightly varies with the laser power and scanning speed. Thus, the laser energy is enough and appropriate to completely melt powder to form a continuous molten pool. The input energy density is usually used to evaluate the effects of the processing parameters on the density of SLM samples, which can be described as [35]:

$$E = \frac{P}{HDV} \quad (8.1)$$

here, E is the input energy density (in J/mm^3), P is the laser power (in W), V is the scanning speed (in mm/s), H is the scanning space (in mm), and D is the layer thickness (in mm). Fig. 8.6 illustrates the relationship between the input energy density and the relative density of SLM-fabricated samples.

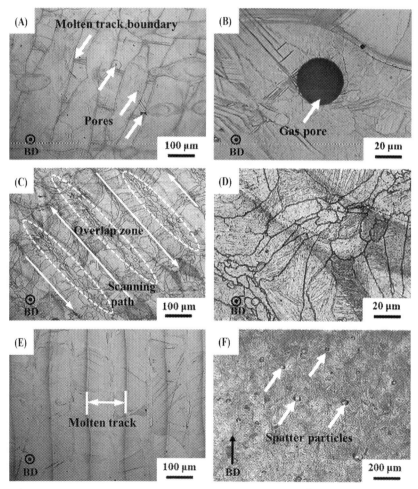

FIGURE 8.7 Micro morphologies by OM: (A), (B) Point A (77 J/mm^3, 250 W and 900 mm/s); (C), (D) Point B (107 J/mm^3, 310 W and 800 mm/s); (E), (F) Point C (147 J/mm^3, 310 W and 600 mm/s).

When the input energy density is 77–110 J/mm^3, the relative density exceeds 99% and slightly increases with the increase in energy density. However, the relative density significantly decreases with the increase in energy density when the input energy density is over 110 J/mm^3.

Fig. 8.7 displays the morphology in SLM-fabricated samples at three input laser energy densities values of points A, B, and C in Fig. 8.6. When the input energy density is extremely low, a discontinuous spread of melting tracks is formed because some of the powder has not been completely melted [34]. Meanwhile, the irregular boundaries of molten pools are produced even without fusion overlapping, as shown in Fig. 8.7A. Consequently, irregular micropores result from the insufficient melt (Fig. 8.7A), and a small amount of round gas pores (Fig. 8.7B) is found. With the increase in input energy density, the melting tracks become more continuous and fully overlap one another, as shown in Fig. 8.7C. No pores are found on the overlapping boundaries (Fig. 8.7D). Fig. 8.7E shows that the samples that were fabricated with a very high energy density have nearly linear overlapping boundaries, and macrocracks are generated because of excessive residual stress. Moreover, many micrometal balls can be found, as shown in Fig. 8.7F. The extremely high energy density can result in molten pool boiling due to the high metal temperatures and splashing of small powder particles [36]. In addition, the large laser energy density input produces large spheroidal particles that separate from one another, as shown in Fig. 8.8, which will create more void (Fig. 8.8C) formation during the powder-spreading process. The voids will finally cause many pores to form in the SLM-fabricated sample, which decreases the density of the sample. Simultaneously, spheroidization will increase the molten pool surface roughness, affect the powder spreading and followup melt spread and accumulation, and reduce the

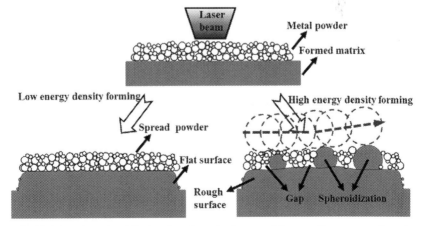

FIGURE 8.8 After spheroidization the schematic diagram of the spreading process of powder during SLM.

relative density. Based on the density analysis of the test samples fabricated with different parameters, when the laser energy density is approximately 110 J/mm^3 (the power is 310 W, the scanning speed is 800 mm/s, the layer thickness is 0.04 mm, and the scanning space is 0.09 mm), the sample has the highest density and best formability.

8.3.1.2.2 Phases and microstructures

Fig. 8.9 shows the phase composition of the Cu-Al-Ni alloy from the phase diagram and XRD measurements. The theoretical solidification pathway of Cu-13.5Al-4Ni alloy is: $L \rightarrow L + \beta \rightarrow \beta \rightarrow \alpha + NiAl + \gamma_2$. Thus, during the slow cooling process, the Cu-13.5Al-4Ni alloy is composed of α, NiAl and γ_2 phases at RT, where α is a solid solution with the face-centered cubic (FCC) structure of the Al element incorporated into Cu element, NiAl is an intermetallic compound, and γ_2 is a solid solution based on an electron compound [29]. However, the XRD results in Fig. 8.9B indicate only the β_1' phase (the martensite phases that inherit the composition and order of austenite) in the SLM-fabricated Cu-13.5Al-4Ni-0.5Ti samples, which is attributed to the rapid cooling and solidification process of SLM. Because the cooling time is insufficient, the transition of β_1 phase into α phase and γ_2 phase can be suppressed. When the cooling rate is greater than 5–6°C/min, the eutectoid transition is suppressed [29]. The cooling rate of SLM is up to $2.13 \times 10^6 - 2.97 \times 10^{6\circ}$C/s [37]. Therefore, the solidification route at a high cooling rate is: $L \rightarrow L + \beta \rightarrow \beta \rightarrow \beta_1 \rightarrow \beta_1'$. Commonly, the direct casting Cu-Al-Ni SMAs must be quenched to complete the thermoelastic martensitic transformation. Therefore, more thermoelastic martensite (β_1') remain at RT,

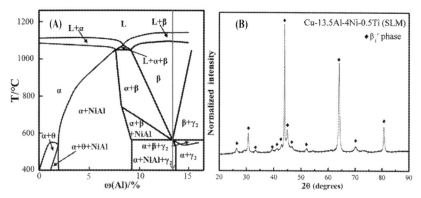

FIGURE 8.9 Phase composition of Cu-Al-Ni alloys: (A) Cu-Al-Ni (4% Ni fraction) ternary phase diagram (The red arrow shows the solidification path and phase transition of Cu-13.5Al-4Ni); (B) X-ray diffraction patterns of SLM-fabricated Cu-13.5Al-4Ni-0.5Ti samples (laser power: 310 W; scanning speed: 800 mm/s; layer thickness: 0.04 mm; scanning space: 0.09 mm; laser energy density: 107 J/mm^3). (For interpretation of the references to color in this figure legend, the reader is referred to the Web version of this article.)

which improves the shape memory properties of SLM-fabricated Cu-13.5Al-4Ni-0.5Ti SMAs.

Fig. 8.10 illustrates the SEM micro morphologies of the SLM samples (laser power: 310 W; scanning speed: 800 mm/s; layer thickness: 0.04 mm; scanning space: 0.09 mm; and laser energy density: 107 J/mm^3). Fig. 8.10A and B show the grain morphology on the X−Y plane. The SLM Cu-13.5Al-4Ni-0.5Ti sample exhibits a "bimodal" grain size distribution with equiaxed grains having diameters in the range of 10−30 μm in the regions where the melt tracks overlap. In the center of the melting tracks, the grains are elongated perpendicular to the scanning direction and have typical sizes of approximately 30−80 μm. Fig. 8.10C with a high magnification in Fig. 8.10B presents a typical lath martensite and in the internal cross grain growth, which is consistent with the XRD test. The average grain size can be estimated as follows [38]:

$$d = 2\sqrt{\frac{S}{\pi N}} \quad (8.2)$$

where S is the area of a certain circle, N is the effective number of grains inside this circle, and d is the average diameter of grains. According to the

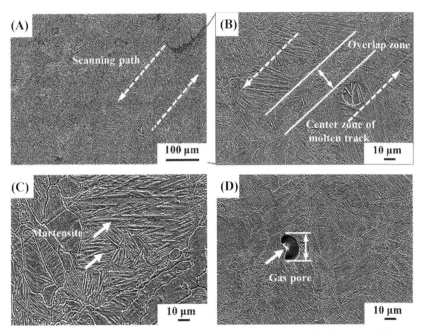

FIGURE 8.10 SEM microstructure of the Cu-13.5Al-4Ni-0.5Ti alloy fabricated by SLM: (A) Microstructure on X−Y plane; (B) High-magnification microstructure of the circle area marked in (A); (C) Martensite morphology; (D) Micropore.

high-magnification morphology of grains, as shown in Fig. 8.10D, the average diameter of grains calculated by Eq. (8.2) is approximately 43 μm. Saud et al. fabricated a Cu-11.9Al-4Ni-0.7Ti SMA by a conventional casting method with an average grain size of 400 μm [39]. The grain size of SLM-fabricated samples is only one-tenth of that of the casting samples, which is the combined effect of the high cooling rate during SLM and the grain refining of the Ti element. Fig. 8.10D shows a micropore with the diameter of approximately 10 μm. The reason is the fast cooling and solidification speed during SLM processing, so the gas carried in molten pools is too late to overflow [34]. The presence of these pores may prevent the SLM samples from achieving full density. However, the so-called X-phase (Cu_2TiAl) described in the reference [40] is not observed by SEM, which we will discuss next. The result of TEM will prove the existence of the X-phase.

To prove the existence of the X-phase (Cu_2TiAl), TEM is applied. Fig. 8.11A shows the brightfield image of the SLM-fabricated sample (laser power: 310 W; scanning speed: 800 mm/s; layer thickness: 0.04 mm; scanning space: 0.09 mm; laser energy density: 107 J/mm^3), which indicates that the X-phase (Cu_2TiAl) is randomly distributed on the surface of the matrix. Fig. 8.11B and C present the selected area diffraction pattern (SADP) of Fig. 8.11A. The X-phase (Cu_2TiAl) and twin martensite can be identified from SADP. The X-phase is granular with a size of 20–50 nm, which indicates that the SLM does not affect the structure of the precipitates but merely causes size refinement. The X-phase is dispersed on the substrate and hinders the growth of the grains, resulting in grain refinement [40]. Thus, the Ti element in the alloy can refine the grains.

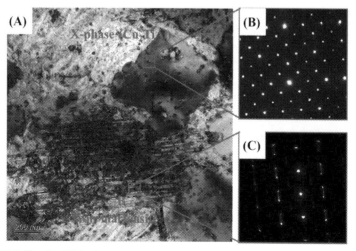

FIGURE 8.11 (A) TEM microstructure of the Cu-13.5Al-4Ni-0.5Ti alloy fabricated by SLM; B, C Corresponding selecte7d area diffraction pattern (SADP) data.

8.3.1.2.3 Mechanical properties

Fig. 8.12 shows the hardness values of the Cu-13.5Al-4Ni-0.5Ti alloy fabricated by SLM (laser power: 310 W; scanning speed: 800 mm/s; layer thickness: 0.04 mm; scanning space: 0.09 mm; laser energy density: 107 J/mm^3). The hardness values at different measurement points are similar, which indicates the uniform composition distribution in the sample. The average hardness on the X−Z plane (289.1 ± 16.9 HV, Fig. 8.12B) is slightly higher than that on the X−Y plane (267.1 ± 24.2 HV, Fig. 8.12B) because the X−Z plane has a slightly higher molten pool boundary density than the X−Y plane. However, there is no obvious difference in phases between X−Y and X−Z planes. Silva et al. fabricated a Cu-11.85Al-3.2Ni-3Mn SMA alloy by SLM (SLM Solutions GmbH) with an average hardness of 249.3 HV, which exceeds the casting standard by approximately 20.9 HV because of grain refinement [41]. The average grain size of the Cu-11.85Al-3.2Ni-3Mn alloy exceeds 123 mm, which is three times higher than that of the alloy fabricated in this study. Therefore, the grain refinement effect of the Ti element contributes to the high hardness of the SLM-fabricated Cu-13.5Al-4Ni-0.5Ti alloy. The excessive cooling rate of SLM also inhibits the precipitation of the brittle g2 phase and increases the hardness of the alloy.

Fig. 8.13 illustrates the RT stress-strain curves of the SLM-fabricated Cu-13.5Al-4Ni-0.5Ti SMAs for three process parameters: (A) 77 J/mm^3 (250 W and 900 mm/s); (B) 107 J/mm^3 (310 W and 800 mm/s); and (C) 147 J/mm^3 (310 W and 600 mm/s). The average ultimate tensile strength of the SLM-fabricated samples for the three process parameters are 420 ± 16, 541 ± 26 and 476 ± 14 MPa, respectively. The average ultimate elongation of the SLM-fabricated samples for the three process parameters are 5.72% ± 0.31%, 7.63% ± 0.39%, and 6.32% ± 0.23%, respectively. The results show that the SLM-fabricated sample has the best tensile properties at moderate energy densities (107 J/mm^3, 310 W and 800 mm/s), which is consistent with the microscopic analysis results of the forming quality in

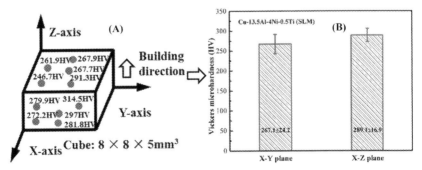

FIGURE 8.12 (A) Microhardness test specimen of the Cu-13.5Al-4Ni-0.5Ti alloy fabricated by SLM; (B) Average hardness of X−Y and X−Z planes.

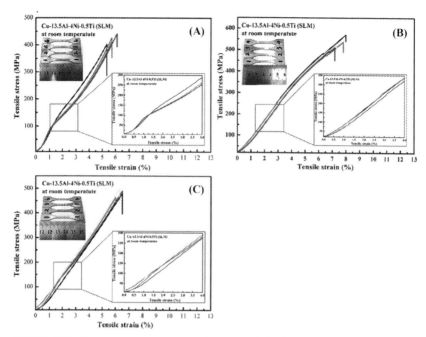

FIGURE 8.13 Room temperature stress-strain curves of the SLM-fabricated Cu-13.5Al-4Ni-0.5Ti SMAs under three process parameters: (A) 77 J/mm^3 (250 W and 900 mm/s); (B) 107 J/mm^3 (310 W and 800 mm/s); (C) 147 J/mm^3 (310 W and 600 mm/s).

Fig. 8.7. At low energy density, the powders are not fully melted to form continuous melting tracks and overlap zones, which results in the formation of voids and reduces the mechanical properties of the alloy. When the laser energy is too high, the temperature of the molten pool is too high to cause the spatter of the molten pool to form pores, which reduces the mechanical properties of the alloy. Compared with the as-cast Cu-11.9Al-4Ni-0.7Ti SMAs [39], the ultimate tensile strength decreases by 23%, and the elongation increases by 163%. The high elongation is related to the grain refinement and phase composition in the alloy. The grain size of the as-cast Cu-11.9Al-4Ni-0.7Ti by Saud et al. is approximately 400 μm [39], which is approximately 9−10 times larger than that of the alloy fabricated in this study. Additionally, the absence of the brittle γ_2 phase in samples should benefit the ductility improvement of SLM-fabricated Cu-13.5Al-4Ni-0.5Ti alloys. However, the ultimate tensile strength is 79 MPa lower than that of the as-cast Cu-11.9Al-4Ni-0.7Ti SMAs, which may be attributed to the presence of pores in Fig. 8.10D. The reason is that the stress easily concentrates where pores are under tensile stress or compressive stress, so cracks are first formed from the pores and finally decreases the ultimate strength value.

Fig. 8.14 illustrates the RT tensile fracture morphologies of the SLM-fabricated Cu-13.5Al-4Ni-0.5Ti alloys. The macroscopic fracture

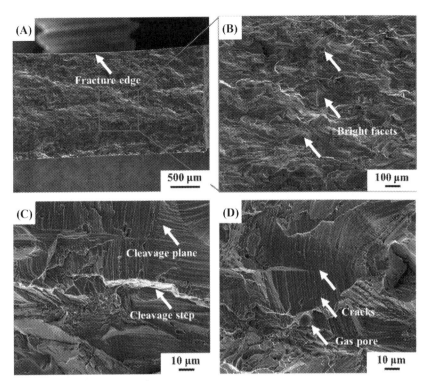

FIGURE 8.14 Room temperature tensile fracture morphologies of the SLM-fabricated Cu-13.5Al-4Ni-0.5Ti alloys (laser power: 310 W; scanning speed: 800 mm/s; layer thickness: 0.04 mm; scanning space: 0.09 mm): (A) SEM at low magnification; and (B), (C), and (D) SEM at high magnification.

(Fig. 8.14A) is composed of many facets with irregular orientations, from which radial stripes (Fig. 8.14B) and bright facets (Fig. 8.14B) can be observed. Thus, the macroscopic fracture features show that the tensile samples have no obvious plastic deformation and exhibit a brittle fracture. Moreover, many cleavage steps, cleavage planes and river-like patterns can be found on the microscopic fracture, as shown in Fig. 8.14C. This appearance also indicates that the tensile samples are a cleavage fracture. Remarkably, gas pores (Fig. 8.14D) are observed in the fracture, around which there is a crack distribution. Thus, the cracks first appear at the pores due to stress concentration in the tension process. Then, the crack propagations interconnect to form steps, which will merge or disappear during propagation and eventually form a river-like pattern.

Fig. 8.15 illustrates the differential scanning calorimetry (DSC) curve and tensile curves at 200°C of the SLM-fabricated Cu-13.5Al-4Ni-0.5Ti alloys (laser power: 310 W; scanning speed: 800 mm/s, layer thickness: 0.04 mm; scanning space: 0.09 mm; laser energy density: 107 J/mm^3). Fig. 8.15A

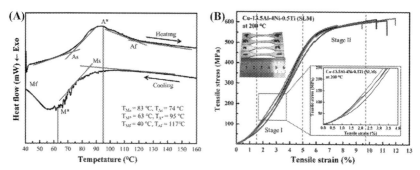

FIGURE 8.15 (A) DSC curve of the SLM Cu-13.5Al-4Ni-0.5Ti sample; (B) Tensile curves of the SLM Cu-13.5Al-4Ni-0.5Ti samples at 200°C.

shows the martensitic transformation starting temperature of the alloy is approximately 83°C, the end temperature of the martensite transformation is approximately 40°C, the starting temperature of the austenite transformation is approximately 63°C, and the end temperature transition temperature of the austenite is approximately 117°C, which indicates that the alloy is in a martensite state at RT and has high-temperature application potential. Because the working environment of Cu-Al-Ni alloys at high temperature is mainly at 100°C–200°C [42], the tensile tests were performed at 200°C, as shown in Fig. 8.15B. The average ultimate tensile strength is 611 ± 9 MPa, which is nearly 70 MPa higher than that obtained at RT, whereas the average ultimate elongation is $10.78\% \pm 1.87\%$, which is nearly 3.15% higher than that obtained at RT. The strength is improved because the tensile samples are in the austenitic state at 200°C. The austenitic parent phase is a bcc structure, which is not prone to plastic slip under external force, whereas martensite is prone to twin deformation and martensite reorientation, so the strength of the parent phase is higher than that of the martensite phase. Fig. 8.15B shows that the samples undergo two stages of deformation during the stretching process at 200°C. Stage I is the elastic deformation stage, where the sample is in the austenite state. In stage II, a stress-induced martensitic transformation occurs after the stress reaches a certain level, and a strain platform appears in the curve. Then, martensite deformation and brittle fracture occur after a certain degree of deformation. Thus, the plasticity of the alloy increases at 200°C. However, the stress-induced martensite is not stable. When the external force disappears, the martensite transforms into the austenite parent phase.

Fig. 8.16 shows the high temperature (200°C) tensile fracture morphologies of the SLM-fabricated Cu-13.5Al-4Ni-0.5Ti alloys. Apparent cleavage characteristics are observed on the fracture surface, as shown in Fig. 8.16A. However, unlike the RT tensile, there are obvious fibrous regions and dimples in the high-magnification SEM morphologies, as shown in Fig. 8.16B, C, and D.

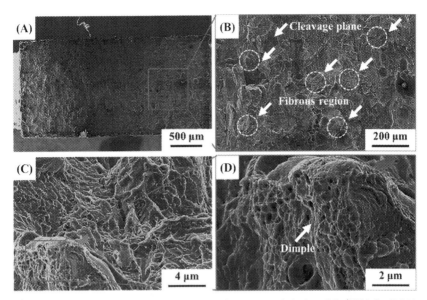

FIGURE 8.16 High temperature (200°C) tensile fracture morphologies of the SLM Cu-13.5Al-4Ni-0.5Ti alloys: (A) SEM at low magnification; (B), (C), and (D) SEM at high magnification.

As a result, the fracture is characterized by the combination of brittle fracture and ductile fracture, so the strength and plasticity of the alloy are enhanced at 200°C, and there is obvious yielding. The study shows that the SLM-fabricated Cu-13.5Al-4Ni-0.5Ti alloys have good mechanical properties and application potential at high temperatures.

8.3.1.3 Conclusion

The Cu-13.5Al-4Ni-0.5Ti copper-based SMA was fabricated by SLM from prealloy powder. The effect of the SLM processing parameters on the relative density was investigated, and the optimized parameters were determined. In addition, the microstructure, phase composition, and tensile properties at RT and 200°C of the SLM Cu-13.5Al-4Ni-0.5Ti alloy were studied. The main findings and conclusions can be summarized as follows.

1. When the scanning speed is 800–900 mm/s, the relative density is more than 99%, and the density slightly varies with the laser power and scanning speed. The alloy density is relatively high when the input energy density is approximately 110 J/mm^3. When the energy density is too low, a discontinuous spread of melting tracks will be formed because the powder is not completely melted. Meanwhile, the irregular boundaries of molten pools are produced; there may even be no fusion overlapping. Consequently, there are irregular micropores as a result of insufficient melt and a small amount of round gas pores. A high input laser energy

density will produce large, separate spheroidal particles, which form more voids during the powder-spreading process. Finally, the voids cause many pores to form in the SLM sample and decrease the density of the sample.
2. Due to the rapid cooling rate during the SLM process, the α-phase and γ_2-phase are suppressed, and the β_1'-phase and Cu^2TiAl-phase are generated in the SLM Cu-13.5Al-4Ni-0.5Ti alloy. Elongated strip-like grains traverse the melting tracks. Large-size grains are found at the center of the melting tracks, whereas the grain size in the overlap zone is obviously refined. The average grain size is approximately 43 μm.
3. The hardness values of the SLM-fabricated Cu-13.5Al-4Ni-0.5Ti alloy are 267.1−289.1 HV. The SLM-fabricated sample has the best tensile properties at moderate energy densities (107 J/mm^3, 310 W and 800 mm/s). The average ultimate tensile strength and elongation are 541 ± 26 MPa and 7.63% ± 0.39% at RT, respectively. The elongation is approximately 5.41% higher than that of as-cast Cu-11.9Al-4Ni-0.7Ti SMAs. It should be related to the grain refinement, which is the combined effect of the high cooling rate during SLM and grain refining of the Ti element. The average ultimate tensile strength is 611 ± 9 MPa at 200°C, which increases by nearly 70 MPa compared to that obtained at RT. Accordingly, the elongation is 10.78% ± 1.87%, which increases by nearly 3.15% compared to that obtained at RT. The parent phase with the bcc structure is higher strength than martensite, and the stress-induced martensite transformation at 200°C increases the plasticity of the alloy.

8.3.2 Cu-Zn-Al-based shape memory alloys

At present, Cu-Al-Ni-based and Cu-Zn-Al-based SMAs are the two most used Cu-based SMAs. However, the current studies about the SLM technology of Cu-based SMAs mainly focuses on Cu-Al-Ni-based SMAs. The studies referring to the SLM Cu-Zn-Al-based SMAs are still hardly reported. Cu-Zn-Al-based alloys have good plasticity and processability. Thus, the Cu-Zn-Al-based SMAs was fabricated through SLM technology in this work. The chemical composition of the raw Cu-Zn-Al-based SMAs was designed to Cu-25.5Zn-4Al-0.6Mn. The Al content of 4 wt.% was selected to avoid the deterioration of SME and the increase of brittleness caused by the too high and too low Al content. The Zn content of 25.5 wt.% was selected to obtain a Zn equivalent between 36−48 at.%. In addition, it is known that the addition of Mn is favorable for the grain refinement, the reduction of quenched-in vacancies. Mn can suppress the migration of vacancies, Zn and Al atoms during aging, and thus decrease the tendency of martensite stabilization. Therefore, 0.6 wt.% Mn was also added in this Cu-Zn-Al-based alloy. The effect of laser energy density, w, which is defined as the energy level of absorbed by volumetric unit of powder, on the microstructures and mechanical properties are investigated.

8.3.2.1 Experimental

8.3.2.1.1 Preparation of Cu-Zn-Al-Mn alloy powder

The experimental material was gas-atomized spherical Cu-25.5Zn-4Al-0.6Mn powder. Fig. 8.17 showed the particle size distribution and micromorphology under SEM of Cu-25.5Zn-4Al-0.6Mn powder. The average powder particle size was 34.4 μm, and the powder particles showed nearly spherical shapes.

8.3.2.1.2 Fabrication of Cu-Zn-Al-Mn alloy

The samples were fabricated by a SLM HKM125 machine of Wuhan Huake 3D Technology Co., Ltd. The Cu-13.5Al-4Ni-0.5Ti powder was dried at 80°C for 12 h in a vacuum drying oven before printing. The SLM process was performed in an argon atmosphere (the oxygen content maintained a value lower than 0.03%). The processing parameters with fixed layer thickness (0.04 mm), fixed scanning spacing (0.09 mm), varying laser power (250, 300, and 350 W) and varying scanning speed (400, 500, 600, 700, and 800 mm/s) were chosen. Fifteen groups of processing parameters were performed through the combination of different laser power and different scanning speed. The long bidirectional scanning strategy was used, as shown in Fig. 8.18. To reduce the residual thermal stress during SLM, a 67 degrees rotation of the scanning direction between two consecutive layers N and N + 1 was set. The substrate was preheated to 100°C before printing. After printing, the SLM samples were cut off from the substrate using a filament cutter.

8.3.2.1.3 Characterization of Cu-Zn-Al-Mn alloy

The density of the cube was measured using a fully automatic true density analyzer (Micromeritics Instrument Corp., United States), whose measurement principle is Archimedes method. The true density of the powder was 7.8342 g/cm³, which was considered as the density of a 100% dense cube. The relative density, θ, of the SLM cube was calculated by the following formula:

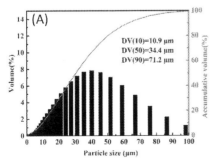

FIGURE 8.17 (A) Particle size distribution and (B) micromorphology under SEM of Cu-13.5Al-4Ni-0.5Ti powder.

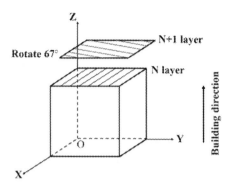

FIGURE 8.18 Schematic of laser scanning strategies during SLM.

$$\theta = \frac{\text{density of SLM samples}}{\text{true density of powder}} \times 100\% \qquad (8.3)$$

The microstructures were investigated by SEM (JSM-7600F, Japan). The phase identification was performed by XRD (X'Pert3 Powder, PANalytical B.V., Netherlands) with a Cu target in a conventional X-ray tube. The scanning speed was 10 degrees/min. The Vickers microhardness of the samples was determined by a hardness test machine (TMVS-1) with a 200 g load and 10 s dwell time. Five points were chosen and measured at different positions of each surface to obtain an average value. The tensile tests were performed on a mechanical testing machine (Zwick/Roell Z020, Zwick, Germany). The tests were performed using a cross speed of 1 mm/min.

8.3.2.2 Results and discussion

8.3.2.2.1 Effect of SLM processing parameters on relative density

Fig. 8.19 showed the relative densities of SLM Cu-25.5Zn-4Al-0.6Mn SMAs fabricated through different laser energy density. The highest relative density of the SLM Cu-25.5Zn-4Al-0.6Mn SMAs was about 96.96%. As raising the laser energy density, the relative density of the SLM Cu-25.5Zn-4Al-0.6Mn SMAs first increased and then decreased. The reason was that when the laser energy density was too low, the powders cannot be melted completely, and it was difficult to form a continuous molten pool due to the high viscosity of the liquid at a low temperature. In this case, small pores were produced, which resulted in a low relative density. When the laser energy density increased, the powder was melted completed, the viscosity of the liquid decreased, and continuous molten pool was formed. Thus, the relative density of the SLM samples increased. However, when the laser energy density was too high, it easily made the metal solution boil in molten pools due to an extremely high temperature, which causes the splashing of small metal droplets and finally the formation of small pores in the samples. This

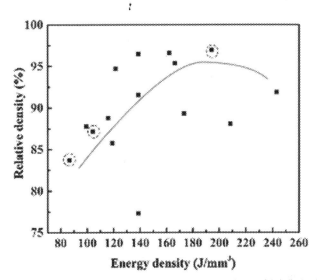

FIGURE 8.19 Relative density of SLM Cu-25.5Zn-4Al-0.6Mn SMAs fabricated with different laser energy density.

was the reason why the relative density decreased when the laser energy density was above 200 J/mm^3. Fig. 8.20 showed the optical micrographs of SLM Cu-25.5Zn-4Al-0.6Mn SMAs fabricated with the laser energy densities of 86.8, 104.2 and 194.4 J/mm^3, respectively. Correspondingly, their relative densities were 83.73%, 87.18%, and 96.96%. The results showed that when the laser energy density was no higher than 200 J/mm^3, both the numbers and sizes of the pores in the SLM samples deceased with increasing the laser energy density. It is feasible to reduce the porosity and increase the relative density of the SLM samples through optimizing the laser energy density.

8.3.2.2.2 Phases and microstructures

Fig. 8.21 showed the XRD patterns of the Cu-25.5Zn-4Al-0.6Mn powder as well as the SLM Cu-25.5Zn-4Al-0.6Mn alloy fabricated with different laser energy density. Only the peaks of martensite existed in the XRD patterns of the raw powder as shown in Fig. 8.21A, indicating that there was only martensite phase in the raw powder. Differently, strong peaks of the α phase and weak peaks of the β phase appeared in the XRD pattern of the SLM sample fabricated with a laser energy density of 86.8 J/mm^3 (Fig. 8.21B), showing that many α phase and a few β phase existed. Here, the α phase is a FCC solid solution based on Cu, which β phase with a body-centered cubic (BCC) solid solution based on CuZn. As further raising the laser energy density, the peaks of β phase became weaker and completely disappeared when the laser energy density was 194.4 J/mm^3. Fig. 8.22 showed the vertical section of

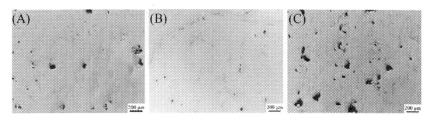

FIGURE 8.20 Optical micrographs of SLM Cu-25.5Zn-4Al-0.6Mn SMAs fabricated with laser energy densities of (A) 86.8 J/mm^3, (B) 104.2 J/mm^3, and (C) 194.4 J/mm^3, respectively.

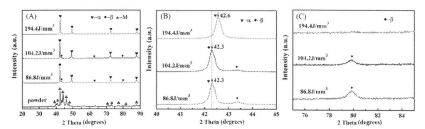

FIGURE 8.21 XRD patterns of the Cu-25.5Zn-4Al-0.6Mn powder as well as the SLM Cu-25.5Zn-4Al-0.6Mn alloy fabricated with different laser energy density. (A) XRD patterns between 20 and 90 degrees; (B) XRD patterns between 40 and 45 degrees; (C) XRD patterns between 75 and 85 degrees.

the ternary phase diagram of Cu-Zn-Al when the Al content was 4 wt.%. As indicated by the red arrow, many β phase as well as a few α and γ phases can be obtained in the Cu-25.5Zn-4Al alloy after equilibrium solidification. However, because the melting point of Zn (419.5°C) was much lower than those of Cu (1083.4°C) and Al (660°C), significant preferential evaporation of Zn will take place during SLM. As a result, the practical chemical compositions of the SLM sample will shift to the left. This was the reason why many α phase and a few β phase appeared in the SLM samples. With the increase of the laser energy density, the preferential evaporation of Zn increased, and thus the fraction of β phase decreased. On the other hand, the result in Fig. 8.21 indicated that the whole peaks of the α phase shifted to the right as raising the laser energy density, showing that its crystalline interplanar spacing, d, decreased according to the Bragg's law ($2d\sin\theta = n\lambda$). As known, the solution of Zn in Cu will result in an increase in the crystalline interplanar spacing due to the much bigger size of Zn atom that of Cu atom. With the increase of the laser energy density, the amount of Zn atom solutioned in the Cu decreased thanks to the enhanced evaporation of Zn. Therefore, the crystalline interplanar spacing of α phase decreased.

Fig. 8.23 showed the optical micrographs of the X-Z plane of the SLM Cu-25.5Zn-4Al-0.6Mn sample fabricated with a laser energy density of 194.4 J/mm^3. Obvious outlines of the molten pools were observed. And some pored existed at the boundaries of the molten pools (Fig. 8.23A). In

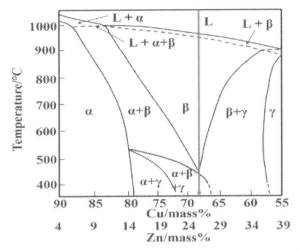

FIGURE 8.22 Vertical section of ternary phase diagram of Cu-Zn-Al when the Al content is 4 wt.%.

FIGURE 8.23 Optical micrographs under different magnifications of SLM Cu-25.5Zn-4Al-0.6Mn sample fabricated with a laser energy density of 194.4 J/mm^3. (A) under 100x magnification; (B) under 500x magnification.

addition, the grain size of this SLM sample was about 15 μm (Fig. 8.23B), whereas the grain size of the cast Cu-Zn-Al-based SMAs was commonly about 500 μm. Obviously, SLM technology can greatly refine the grains. This was ascribed to the high cooling rate during SLM, which caused a big undercooling, thus enhanced the driving force for the nucleation and suppressed the growth of grains. In the backscattered electron images of the SLM Cu-25.5Zn-4Al-0.6Mn sample fabricated with a laser energy density of 194.4 J/mm^3, obvious lamellar structures were observed as shown in Fig. 8.24. Because there were only peaks of α phase existed in XRD pattern of the SLM Cu-25.5Zn-4Al-0.6Mn sample fabricated with a laser energy density of 194.4 J/mm^3 and no brightness difference was observed in its

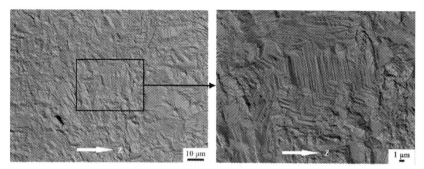

FIGURE 8.24 Backscattered electron images of SLM Cu-25.5Zn-4Al-0.6Mn sample fabricated with a laser energy density of 194.4 J/mm^3.

backscattered electron image, the lamellar structures was determined to be α phase. In cast Cu-Zn-Al-based SMAs, α phase usually have a long ribbon shape due to the low cooling rate during solidification. However, strong Marangoni convection occurs in the small molten pools during SLM because of the high gradients of temperature and viscosity. Moreover, the formed layer will be reheated when the next layer is being scanning by the laser. Thus, the microstructures in the SLM samples are significantly different from that of the cast.

8.3.2.2.3 Mechanical properties

Fig. 8.25 showed the ultimate tensile strength of the SLM Cu-25.5Zn-4Al-0.6Mn samples fabricated with different laser energy density. When the laser energy density was 86.8 J/mm^3, the ultimate tensile strength was 82.2 MPa. When increasing the laser energy density to 104.2 J/mm^3, the ultimate tensile strength slightly increased to 95.0 MPa. When further increasing the laser energy density to 194.4 J/mm^3, the ultimate tensile strength remarkably increased to 328.3 MPa, which was higher than that of the cast Cu-Zn-Al-Mn alloy. The higher amounts of pores in the SLM samples fabricated with low laser energy density were responsible for their lower ultimate tensile strength. The higher ultimate tensile strength in the SLM samples than in the cast one was ascribed to the finer grains. Lots of dimples were observed in the fracture micrographs of the SLM Cu-25.5Zn-4Al-0.6Mn sample fabricated with a laser energy density of 194.4 J/mm^3 (Fig. 8.26), indicating that it was plastic fracture. However, it should be noted that its relative density still did not reach 100%. And some unmelted powder existed (Fig. 8.26A), which were speculated to be the locations of fracture sources.

As raising the laser energy density, the microhardness of the SLM Cu-25.5Zn-4Al-0.6Mn samples decreased (Fig. 8.27). The reason was that with the increase of the laser energy density, the high-hardness brittle β

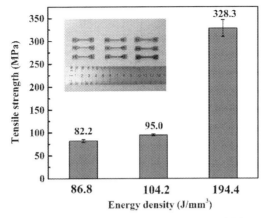

FIGURE 8.25 Ultimate tensile strength of the SLM Cu-25.5Zn-4Al-0.6Mn samples fabricated with different laser energy density.

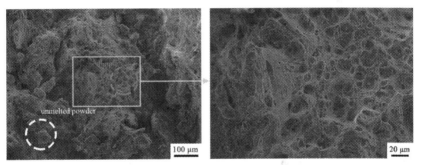

FIGURE 8.26 SEM micrographs of fractures of SLM Cu-25.5Zn-4Al-0.6Mn sample fabricated with a laser energy density of 194.4 J/mm^3.

phase decreased and even disappeared. Correspondingly, the α phase with low hardness and good plasticity increased.

8.3.2.3 Conclusion

1. The laser energy density greatly affected the relative density of the SLM Cu-25.5Zn-4Al-0.6Mn SMAs. That is, the relative density first increased and then decreased as raising the laser energy density. A highest relative density of 96.96% was obtained.
2. The phases in the SLM Cu-25.5Zn-4Al-0.6Mn SMAs were different from that in the raw powders. Because great preferential evaporation of Zn took place during SLM, the practical Zn content was lower than the designed one. Lots of α phase and small amount of β phase existed in the SLM Cu-25.5Zn-4Al-0.6Mn SMAs. As raising the laser energy density, the β phase decreased and completely disappeared.

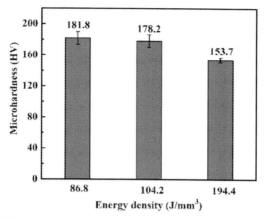

FIGURE 8.27 Microhardness of the SLM Cu-25.5Zn-4Al-0.6Mn samples fabricated with different laser energy density.

3. Overlapping molten pools were observed in the micrographs of the SLM Cu-25.5Zn-4Al-0.6Mn SMAs. The α phase in the SLM Cu-25.5Zn-4Al-0.6Mn SMAs showed a lamellar structure due to the complexity of SLM process.
4. As raising the laser energy density was from 86.8 to 194.4 J/mm^3, the ultimate tensile strength increased from 82.2 to 328.3 MPa thanks to the decrease of pores, but the microhardness decreased from 181.8 to 153.7 HV due to the decrease of the high-hardness β phase.

8.3.3 Double network hydrogel reinforced by carbon nanotubes

The CNTs are thin cylinders of carbon that bonded together in unique honeycomb nanostructure. The CNTs are commonly divided by several various forms based on cylindrical concentric planes on layer wall, that is, single wall, double wall, and multiwall [43]. The CNTs also has excellent properties as reinforcing materials for the improvement of polymer performances. There are three advantages of using CNTs to reinforce the polymer matrix; good dispersive ability, strong interfacial network with polymers, and superior collaborative alignment with the polymer matrix. The advantages of CNTs reinforcement within the polymer structures are made they exist in a wide variety of applications such as; thermoplastics, liquid crystalline, thermosetting polymer, and water-soluble of polymers. Based on the literature, the term improvement performances of polymer by CNTs reinforce is represented with toughness [44], tensile strength-modulus [45], thermal-electrical conductivity [46], and temperature of glass transition [47]. However, that is quietly rare of discussion about the effect of CNTs reinforcement in case of SME and printability.

On the other hand, the syntheses of hydrogels with two different properties, that is, the SME and self-healing, have been reported [48]. These hydrogels are known as the DN hydrogels. DN hydrogels are very promising materials and classified as soft and tough. As the articles have reported, DN hydrogels synthesized by combining a network of PEG as the physical network and polyvinyl alcohol (PVA) as the self-healable. The combined of network polymer has become one of the popular studies due to their unique properties [48]. The structural properties of DN gels can improve by nanoparticle reinforcement. Additionally, the PVA is the suitable polymer that has considered candidates as matrix agent for CNTs as reported by researchers [49] and became second network structure of DN hydrogels. The PVA also has contributed an important role in SME due to their crystalline microdomains will activating that phenomena [48]. In the first network, the PEG contributes not only to recovery or self-healable phenomenon but also as the modification of materials to let the hydrophobicity and functionality of hydrogels structure. In addition, the PEG composites with CNTs reinforcement also has reported [50], whereby indicates this material is the best candidate to collaborate with PVA. In addition, there are several methods to fabricate the hydrogel, but 3D printing is one of the best methods to fabricate the high complex object. The DIW is the most suitable printing process for colloidal gels that is based on the injection of gel precursor. The gel precursor must have a good viscoelasticity response control. In this case, the flow deposition nozzle will be adjusted to facilitate shape retention in layer production. In addition, the viscoelasticity properties can be approached by rheology analysis, and this result is closely correlated to printability. The printability is a part of the feasibility study of material prior to use for the printing process.

Ultimately, in this work, we first syntheses and actualizes the first network of PEG with acryloyl chloride, dichloromethane, diethyl ether, and triethylamine to get PEG-diacrylate. Then the second network of PVA-CNTs composites was synthesized incorporates by PEG of the first network, through relief of ammonium persulfate (AP) as a crosslinker agent. The influence of (multiwalled) CNTs within DN hydrogels structure on the shape memory phenomena and printability were investigated. This research seeks to understand the effect of CNTs on the shape memory phenomena and printability of PEG-PVA DN hydrogels composite, which to our knowledge has not been observed previously in the article.

8.3.3.1 Experimental

8.3.3.1.1 Raw materials of double network hydrogel reinforced by carbon nanotubes

PEG (MW = 1800–2200) and AP (98%) were purchased from Tianjin Tian Li Chemical Reagents Co. Ltd., China. PVA (MW = 146000–186000) were produced and purchased from Sigma-Aldrich, Missouri, the United States, and

CNT ($d = 8-15$ nm, and $l = 50$ μm) were produced and purchased from Chengdu Organic Chemicals Co. Ltd, China. Dichloromethane, diethyl ether, triethylamine (99%), and acryloyl chloride (96%) were purchased from Sinopharm Chemical Reagent Co., Ltd, Shanghai, China, Damao Chemical Reagent Factory, Tianjin, China, and Aladdin Industrial Corp., Shanghai, China.

8.3.3.1.2 Preparation of double network hydrogel reinforced by carbon nanotubes

The PEG with acrylate compounds attached to two chain ends were used to synthesize the model DN hydrogel system by forming a covalently cross-linked network. The first network of DN hydrogels was prepared by dissolving the PEG (powders, 10% m/w) in dichloromethane. A fourfold excess of triethylamine was then added to the PEG-solution, followed by the addition of a fourfold excess of acryloyl chloride at RT. The mixture was vigorously stirred in an ice bath for 6 h as shown in Fig. 8.28A. The solutions were then treated with 500 mL of diethyl ether, which resulted in the precipitation of a slurry of PEG-diacrylate that was separated in a vacuum (Fig. 8.28B) and dried for 28 h at 37°C. To synthesize the DN system hydrogels, another network must be prepared after the first network of PEG-diacrylate was obtained.

To prepare the second network and synthesize the CNT-reinforced composite, homogeneous PVA solutions were obtained by dissolving 10% m/w of PVA powders into distilled water at 90°C. The PVA solutions were vigorously stirred in a closed system to avoid water evaporation. Then, after stirring for 15 min, CNT powders were added to the PVA solutions; this procedure successfully distributed the CNTs into the polymers. Next, the first PEG-diacrylate network and AP were simultaneously added; AP acted as the catalyst to promote the crosslinking reaction. In addition, the weight ratio between PEG and PVA was 70:30 due to the beneficial properties of such DN hydrogels reported in previous articles. The DN hydrogel composite

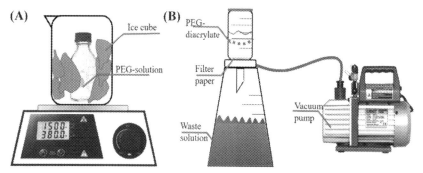

FIGURE 8.28 Schematic diagram of synthetic process of PEG: (A) stir in ice bath; and (B) filtration in vacuum.

solutions were vigorously stirred at 90°C for 30 min. Finally, the polymer gels were poured into a mold and put into an oven at 60°C for 5 h to complete the hydrogel crosslinking process (Fig. 8.29). To complete the process for shape memory phenomenon, the gels must be programmed through a freezing and thawing process. In addition, hydrogel composites with different contents of CNTs (0%, 0.125%, 0.25%, and 0.5%) were prepared to investigate the effect of the CNT content on the properties of this hydrogel composite.

8.3.3.1.3 Characterization of double network hydrogel reinforced by carbon nanotubes

The structural morphology and elemental mapping were studied by comparison with literature reports and obtained via SEM-energy dispersive spectroscopy (Thermo Scientific FEI Quanta 200) and the Fourier transform infrared spectroscopy (FTIR, Vertex 70 Bruker). The FTIR analysis characteristic IR bands samples were measured with wavelength from 400 to 4000 cm^{-1}.

Thermogravimetric analysis (TGA) and DSC were used to measure the thermal properties of the DN hydrogel composite. The samples (5 mg) were prepared for TGA analysis using a PerkinElmer Pyris 1 at a heating rate of 10°C/min to trace the weight loss in the range from 25°C to 800°C under nitrogen. Additionally, the hydrogel composite samples were prepared under nitrogen for DSC analysis (PerkinElmer DSC8000) at a heating rate of 5°C/min to trace heat flow from −50°C to 100°C. To measure the water content in the hydrogel, a comparison between the heat absorbed with pure water (334 J/g), can be used to obtain the nonfreezable water content (NBW) from the equation below using the free water content (FW) and the equilibrium water content (EWC).

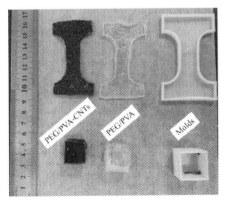

FIGURE 8.29 PEG-PVA hydrogel composites with and without CNTs and their molds used in crosslinking process.

$$FW = \left(\frac{\text{Enthalpy of melting endotherm}}{\text{Enthalpy of melting pure water}}\right) \times 100 \qquad (8.4)$$

$$NBW = FW - EWC \qquad (8.5)$$

The shape memory behavior of DN hydrogel composite was investigated based on the shape recovery process, which was recorded with a digital camera and monitored with a protractor application.

The swelling behavior measurements and gel fraction analysis were adapted from a previously published method. The swelling behavior was measured by the swelling ratio of perfectly dried 1 g samples after immersion in 500 mL of distilled water at 37°C for 6 d. The samples were removed from the immersion bath and weighed on days one and six. The swelling ratio and gel fraction were calculated according to the equation below, where W_s is the weight of the swollen sample, W_d is the weight of the dried sample before immersion, and W_r is the weight of the dried sample after immersion.

$$\text{Swelling Ratio}(\%) = \frac{(W_s - W_d)}{W_d} \times 100 \qquad (8.6)$$

$$\text{Gel Fraction}(\%) = \frac{(W_r - W_d)}{W_d} \times 100 \qquad (8.7)$$

XRD (PANalytical X'Pert3 Powder X-ray diffractometer) images were obtained to determine the crystallinity of the hydrogel composite samples at RT (25°C). The hydrogel composite samples (0.3 × 0.3 × 0.3 mm^3) were prepared and placed into the holder, and data were obtained using 2θ values from 5 to 50 degrees, a scan analysis rate of 5°C/min, 40 kV and 40 mA.

Viscoelasticity was obtained through rheological testing via a TA instruments HR-2 rheometer. The testing parameters that were used in this isothermal rheology method were carried out with the sample holder ($d = 25$ mm) at a constant 1 Hz frequency, 25°C and 2% stress-strain.

8.3.3.2 Results and discussion

8.3.3.2.1 Thermal properties

In this study, the samples without treatment were chosen for observation by TGA-DSC that determines the freezing point by the onset temperature data and water content analysis. Therefore, the freezing point temperature is directly linked to shape memory phenomenon on hydrogels as well as water content. In this case, the water content will act as a provider that surround the thermal transition mechanism to shape memory phenomenon. As shown in Fig. 8.30A, the TGA curve contains three different zones based on mass losses. The first zone represents water release from the polymer hydrogel structure as the temperature increases. The water content was released, and

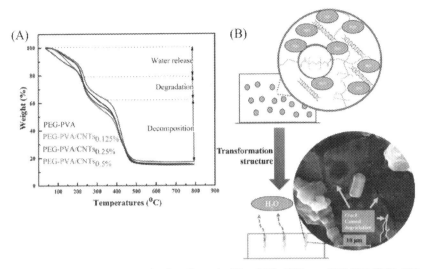

FIGURE 8.30 Thermal properties by effect of adding 0.5% CNTs on PEG-PVA/CNTs DN hydrogels composite. (A) TGA curve analysis; (B) an illustration of weight loss phenomenon on samples.

the weight of the hydrogel decreased from 150°C to 200°C. This zone was set as the EWC data. In the second zone between 200°C and 350°C, the degradation of the hydrogel occurs mainly. The last zone starting at 350°C corresponds to the decomposition of the hydrogels through significant weight loss reached 65% of the total.

Theoretically, the analysis properties of water contained in hydrogels are used thermal analysis [51]. As mentioned earlier, the hydrogel is a hydrophilic polymer chain that absorbs a considerable amount of water. In this case, the hydrogel properties are strongly influenced by the EWC. The water content observed during thermal analysis studies are described as water molecules in hydrophilic zones [52]. These water zones represent three different forms: nonfreezable bound water, freezable bound water, and FW. Nonfreezable bound water is described as a high-density hydrogen-bound compound with an indiscernible transition phase. The freezable water is described as high-intensity water interspersed with molecules in the hydrogel polymer structure. Unlike the other zones, FW does not take part in hydrogen bonding even with the hydrogel polymer structure molecules.

The other thermal analysis of samples was studied through DSC analysis as depicted in Fig. 8.31A, the property of water was monitored and shows the data calculated using Eqs. (8.4 and 8.5) as the result in Table 8.2. In the present study, freezable water will decrease as the CNT content addition, and the FW having the lowest. In this case, the lower FW

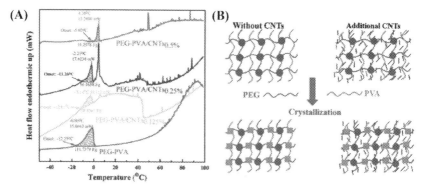

FIGURE 8.31 Thermal properties by effect of adding 0.5% CNTs on PEG-PVA/CNTs DN hydrogels composite: (A) DSC curve with water content analysis; (B) an illustration of PVA phenomenon at onset temperatures.

TABLE 8.2 Water content analysis of hydrogel composite samples.

Properties	–	+CNTs$_{0.125\%}$	+CNTs$_{0.25\%}$	+CNTs$_{0.5\%}$
Onset crystallization of freezing water	−12.29°C	−12.01°C	−11.26°C	−5.02°C
Equilibrium water content	33.89%	20.94%	20.66%	14.31%
Freezable FW content	32.86%	17.30%	16.51%	5.36%
Nonfreezable water content	1.53%	3.64%	4.15%	8.95%

indicates a low melting endotherm and decreasing crosslinking. Subsequently, in the water evaporation stages on thermal condition is consisted of fast and slow stages. The less of FW contains has indicated the length of slow evaporation stages that means the water has longer times trapped within hydrogels structure.

On the other hand, the onset temperatures of water crystallization increased as the content of the CNTs added, which indicated that the water content decreased. Based on DSC result, the data analysis led to the choice of −20°C as freezing point treatment for obtaining PVA crystallization (Fig. 8.30B) as the programming of the shape memory. The temperature selection is needed lower than onset crystallization data because the sample used for treatment dimensionally bigger than DSC analysis (5 mg). Our consideration is strengthened by some other literature [53–54] that summarized the optimum freezing treatment to produce crystallization on PVA located at −20°C.

8.3.3.2.2 Shape memory effect

SME is represented as technologically important materials of stimuli responsive that capable to response lies in the shape change [55]. Literally, The SME does not demonstrate the effect by themselves without activation. As seen in Fig. 8.32, in details about the SME procedure, after synthesized the polymer is formed into their first initials or permanent structure (shape A). Hereinafter, the polymer is deformed purposefully and fixed into second temporary structure (shape B), this section called programming. During application of an external stimulus, the polymer recovers into permanent structure (shape B to A). Presumably, the hydrogels are required in order to withstand temporary physical deformation and then activated the shape memory by external stimulus to obtain shape-change structure.

The SME on hydrogels has two specific structural transformation requirements such as shape memory behavior by reversible thermal transition and a crosslinking structure to maintain the initials-temporary structures [55]. The first requirement is the water molecular mobility for shape fixing and recovery as illustrated in Fig. 8.32. In addition, the crosslinking of hydrogel depends on the contribution of polymer composition, concentration and molecular weight. Consequently, the SME on hydrogels has driven by two specific reasons such as crosslinking and water diffusion mobility as depicted in Fig. 8.32. The hydrophilic properties of hydrogel have made the specific liquid absorption phenomenon that enables to swell in water by mobility of diffusion.

8.3.3.2.3 Swelling

PVA-based hydrogels are classified as the temperature sensitive shape memory materials with hydrophilic network structure. The term of hydrophilic is exposure to water as medium activator the shape memory phenomenon of those material by water induced through immersion and obtain swelling as reported [48,56,57]. Regardless of PVA crystalline, the swelling phenomenon is contributed equilibrium to the SME of hydrogels due the functional ability to manipulate structure by water absorbed. Theoretically, the swelling properties of hydrogels depend on water

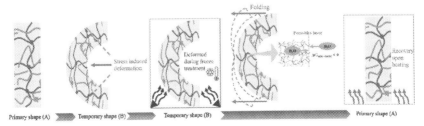

FIGURE 8.32 Illustration of SME procedure with chronological of the transformative structure.

diffusion. The water plays as a plasticizer in the network system of hydrogels, whereby identifies the interaction between polymer and water. Based on principles of the thermodynamic polymer, the interaction between polymer and water can be described as a free energy of mixing (ΔG_{mix}) and crosslinked polymer as the elastic free energy ($\Delta G_{elastic}$). In detail, one of the energy systems of swelling phenomena between hydrogels (crosslinked) and water interaction are illustrated through Eq. (8.8).

$$\Delta G_{system} = \Delta G_{mix} + \Delta G_{elastic} \qquad (8.8)$$

As depicted in Fig. 8.33B, the early day of immersing hydrogels, may occur the $\Delta G_{mix} + \Delta G_{elastic} < 0$ that means the equilibrium reaction will shift to product and swollen started by water diffusing. During the immersion of hydrogels, the energy state of swelling reaction shifts into the improvement until it equals to zero ($\Delta G_{mix} + \Delta G_{elastic} = 0$). The zero-equilibrium state of energy interaction will be driven to discontinue of the swelling phenomena. In this present study, we are used the highest content of CNTs (0.5%) to observe the SME and swelling ability. As shown in Fig. 8.33A, the swelling properties were monitored and calculated using Eqs. (8.6 and 8.7). Our analysis supports a model already proposed in the literature [48], which describes the lower percentage of swelling ratio and the higher percentage of gel fraction were indicated of obtaining higher polymer gels density. As expected from the comparative study, Fig. 8.33A shows an increase in the number of freeze/thaw treatment cycles resulted in a higher network arrangement of the porous structure on the surface of the hydrogel. The phenomenon of freeze/thaw treatment cycles increased the polymer network density in the sample that contained

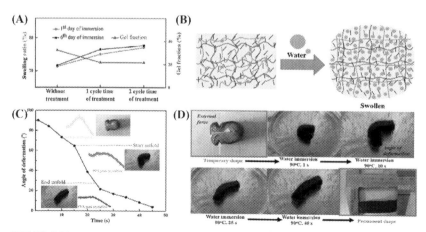

FIGURE 8.33 Transformative phenomenon of PEG-PVA/CNTs$_{0.5\%}$ hydrogel composites. (A) Swelling ratio at equilibrium state and gel fraction; (B) illustration swollen ability; (C) graph of deformed angle as a time function; (D) photos showing SME under water at 90°C.

CNTs; in this case, the swelling ratio increased as the gel fraction decreased which is affected to increasing hydrogel porous structure.

As depicted in Fig. 8.33D that shows the shape memory phenomenon of hydrogels composite through the thermoresponsive under water. The deformation of the sample under an external force was to program the crystallinity during the freeze-thaw process. The figure shows that the temporary deformation was stable at RT due to highly crystalline microdomains of PVA. Furthermore, when the hydrogel composites were immersed in hot water at 90°C, the deformation angle decreased rapidly (Fig. 8.33C) from 15 to 25 s (64.5−21 degrees) and recovered completely after 40 s to the primary structure. These phenomena occur due to the interruption of the crystalline PVA at 90°C. It is expected that the PVA plays an important role in this SME phenomenon due to the impact of physical crosslinks on storage deformation. Essentially, the SME phenomenon is related not only crystalline PVA but also the swelling capabilities and gel fraction of the folding structure. In this case, the CNTs content has successfully maintained the SME without significantly reducing the DN hydrogels performance.

8.3.3.2.4 Printability

Recently, 3D printing extrusion-based method called DIW has emerged. This method is a material fabrication method controlled by a computer to manufacture a 3D designed object through an ink deposition nozzle. The prospectus materials for DIW are needed to possible flow through the tight nozzle and maintain its structure due to the viscoelasticity properties, as confirmed by literature [58]. The viscoelasticity of gel precursor for the DIW is closely linked with rheological analysis and strongly correlated with printability. Our approach is to observe the nanoparticulate by CNTs to provide interparticle forces and improves rheological properties that has good yield strength to resist deformation after printing. In this case, our hypothesis is summarized in that suspension gels must have good viscoelasticity with sufficient yield stress to support layer arranging during the printing process. The viscoelasticity analysis was performed directly after synthesizing and chemically crosslinking the hydrogel composites. The sample was observed after determining loss modulus (G'') as a time sweep function at a constant frequency and temperature. Fig. 8.34A shows the viscoelasticity behavior of the hydrogel composites with different amounts of CNTs added. The graph depicted an improving viscoelasticity while increasing CNTs contain; in this case, the present study matches with expectations from the previous study [59], which shows nanoparticulate additions will increase viscoelasticity. According to rheological theory [60], flows of complex fluids are affected by time dependency due to viscoelastic behavior and our trend of graph results show that the hydrogel composites are liquid-like. The liquid-like (2) Effect of CNT on swelling behavior of DN hydrogel describes the hydrogels solution will behave like liquid in a certain time. The term of certain time is identified as how long hydrogel solutions

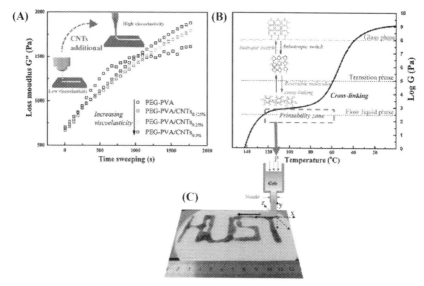

FIGURE 8.34 PEG-PVA/CNTs DN hydrogels composite. (A) Rheology properties of comparison CNTs content by loss modulus as the time sweeping function and resistivity properties. (B) Schematic of illustration DIW method incorporates gels transformation graph through rheology effect properties of relaxation stress modulus. (C) Manually trial of HUST letters by medical injection ($d = 0.3$ cm).

are possible to print or deposition by DIW before they change to the solidified structure. Based on time sweep analysis on rheology, the DN hydrogels composite are identified that feasible to print during 16 min for constructing 3D model object. The time-dependent of rheology was linked to deformation of printing process in order to construct the structure. Our rheology observation method based on the graph study as depicted in Fig. 8.34B that printability zone is appropriated after chemically crosslinked. On the other hand, as shown in Fig. 8.34C, we are trying manually via medical injection to construct HUST words by PEG-PVA/CNTs$_{0.5\%}$ hydrogel. In this study, the analysis and monitoring of the gel samples were performed to further establish their printing feasibility (at the RT, 25°C). In our study, the CNTs has successfully improve the printability as well as viscoelasticity, and the crosslinked network polymer also contributed equally. Our material is capable to fulfill the printability for DIW by improvement the viscoelastic utilized CNTs content.

Hereinafter, the PEG-PVA/CNTs DN hydrogels composite is proven to have a good viscoelasticity for printability requirement. The exploration of PEG-PVA/CNT DN hydrogel composites for further development of additive manufacturing through processing parameters such as scanning speed, gel filament size, and nozzle spacing and height relative to the holder is therefore feasible. The hydrogel composites are ready to further develop constructing 3D objects by DIW printer as a 3D printing method. In addition, the PEG-PVA/CNTs DN hydrogels composite not only possible to do 3D printing method but also 4D printing method, whereby need the time function as the changeable structure to fulfill their criteria. There is no crucial difference in case of processing between 3D and 4D printing, but the main difference is located on the materials uses are needed a capability of the changeable structure. Based on our results, this prospective material has shape memory behavior which has a possibility to 4D printing development with their parameters.

8.3.3.3 Conclusion

In this study, CNTs were successfully distributed within a DN hydrogel of PEG/PVA by optimizing physical mixing and allowed chemical-physical crosslinks. The thermal analysis properties of DN hydrogel composite are observed by TGA and DSC to analyze water content and consider the freezing temperature treatment for optimizing physical crosslink of PVA. The PVA crystalline as the physical crosslink has optimized by the freeze at $-20°C$ and thaw at $25°C$. The freeze-thaw method of treatment has successfully constructed temporary shape and recover to their initial shape when activated as shape memory behavior. The SME can be activated at $90°C$ by water induced. In addition, the CNTs content has successfully maintained the shape memory performance of hydrogel and improved the viscoelasticity as well as printability. Finally, the PEG-PVA/CNTs of DN hydrogel composite is appropriate for further development of 3D/4D printing strategy and has advantages in customizable structure folding and predictability due to freeze-thaw treatment as a programming method. We believe that the concept of 3D/4D printing through our material is worthy of additional exploration in additive manufacturing due to the applications of shape memory hydrogels in a variety of technological fields.

8.3.4 Acrylate-based shape memory polymer

SPMs are representative stimuli responsive polymers, can recover from specific temporary shapes to their original shapes when exposed to external stimuli such as temperature, light, electricity, magnetism, and moisture [61–63]. This process of shape recovery is defined as SME. There have been extensive investigations performed in the field of SMPs which turn out to be a kind of prospective smart materials. Compared to other shape memory materials like SMAs, the deformation performance of SMPs can be tailored more easily with much lower material and processing costs. Besides, SMPs also show low density and activated temperature, high elastic deformation, as well as promising biocompatibility and biodegradability. Precisely because of these outstanding performances, SMPs present a huge application potential in aerospace domains such as deployable structures, hinge, folding wing, as well as the biomedical fields, including self-tightening sutures, fasteners, and drug delivery carriers.

Currently, the manufacturing of SMPs still heavily relies upon traditional processes such as injection molding, casting, and extrusion, which are difficult to fabricate highly personalized and customized SMP parts with intricate structures. Additive manufacturing technology refers to a class of techniques that fabricate parts through adding materials via a layer-by-layer manner according to their 3D computer-aided design models without any tooling or molds. Theoretically speaking, it has the capability of quickly building any complex-shaped and high-precision parts. When smart materials are used in additive manufacturing, it adds the fourth dimension of time into the 3D space coordinates, where the shape, property, or functionality of additive manufactured structures can respond in an intended manner from external environment stimuli. This is the 4D printing [64].

SMPs, the most extensively studied smart materials, have been fabricated primarily via polyjet and FDM. Despite polyjet technology can provide the printed parts with high accuracy, it is subjected to high equipment cost and low-strength parts. In addition, it is relatively difficult to develop customizable materials due to the viscosity limitation for the equipment. FDM technology is mainly used for the fabrication of thermoplastic polymer filaments. But, the chemical structure of thermoplastic SMPs is physically crosslinking, resulting in their shape recovery properties worse in contrast with those of thermosetting SMPs. It dramatically degrades shape-memory properties of the printed thermoplastic SMPs. Consequently, continuously more researchers are eager to fabricate thermosetting SMPs to achieve 4D printing and attain excellent shape memory performance. The vat photosensitive is a superior alternative, which is a mature additive manufacturing technology based on UV curing of liquid photosensitive resin and has the advantage of excellent surface finish and high resolution. This technology can be classified into two typical categories, namely, SLA and DLP based on a laser beam or a projector used as light source. Miao et al. developed smart soybean oil epoxidized acylate inks printed by a modified SLA technology called photolithographic- and stereolithographic-tandem, and successfully manufactured 4D smart scaffolds for biomedical applications [65]. In contrast with SLA, DLP has a faster fabrication rate because the liquid photosensitive resin is solidified into one layer each time. Choong et al. optimized the energy density and curing depths during DLP process and successfully printed a complex SMP bucky ball with remarkable shape memory behavior [66]. Huang et al. fabricated 4D shape changing hydrogels via the ultrafast DLP printing, and the printed hydrogels showed excellent shape memory performance [67]. Until now, most of works have focused on the shape memory properties of several available photosensitive SMPs, and printable shape memory photosensitive resins for the vat photopolymerization are still lacking.

Therefore, in this work, a new 4D printable acrylate-based photosensitive resin is designed and prepared for DLP. The resin is comprised of tBA,

HDDA and 2,4,6-trimethylbenzoyldiphenyl phosphine oxide (TPO) serving as the photo initiator. SMP parts with tBA/HDDA networks are synthesized through in-situ polymerization under the UV light, and exhibit SME during the deformation-recovery testing. The thermal stability, crystallization performance and chemical structure of both cured and uncured resins are systematically investigated. The shape memory properties in terms of shape recovery ratio (R_r), shape fixity ratio (R_f) and cycle life of DLP-printed parts with different crosslinker concentrations are also unveiled. This work is beneficial to broaden the variety of materials for 4D printing and provide valuable guidance for the optimization of shape memory properties of other SMPs.

8.3.4.1 Experimental

8.3.4.1.1 Preparation of acrylate-based shape memory polymer

The resins for DLP are composed of tBA, HDDA, and TPO. More specifically, the tBA is used for a monomer and HDDA serves as a crosslinker. The monomer (soft component) and the crosslinker (hard component) make up a dual segment, which frequently exists in the structure of SMPs. When heated above T_g, the soft component enables highly elastic strain into a temporary shape and the hard component is able to maintain thermally stable. The TPO, a kind of canary yellow powder, acts as UV photoinitiator producing free radicals, which rapidly initiates the polymerization of monomer and crosslinker.

Commercial tBA monomer was mixed with HDDA and TPO. They were all obtained from Shan Dong Yinglang Chemical Co., Ltd and were utilized as received without any purification. Photosensitive resins with five mass fractions of HDDA ranging from 10 wt.% to 50 wt.% were prepared. The mass and composition proportion of these photosensitive resins prepared are listed in Table 8.3, and 1 g TPO was added to each 100 g tBA/HDDA solution.

Taking the preparation process of 10 wt.% HDDA as an example, 10 g HDDA was first slowly poured into 90 g tBA to get a mixed solution, followed by the addition of 1 g TPO and mixture for 3 min with ultrasonic vibration, and then the homogeneous, sediment free and transparent liquid with low viscosity was obtained. Therefore, the prepared photosensitive resins were very suitable for UV curing. The preparation process of the other four ratios of photosensitive resins was identical with that of the resin containing 10 wt.% HDDA. The schematic diagram of preparation process of photosensitive resin and the synthesized crosslinked network is shown in Fig. 8.35.

8.3.4.1.2 Fabrication of acrylate-based shape memory polymer

A desk top Moonray DLP printer (Soonsolid, China) was used to make the photosensitive resin cured under a projection UV light. The wavelength of the UV

TABLE 8.3 The mass and composition proportion of the five kinds of photosensitive resins.

Material	The mass fraction of HDDA				
	10 wt.% (g)	20 wt.% (g)	30 wt.% (g)	40 wt.% (g)	50 wt.% (g)
tBA	90	80	70	60	50
HDDA	10	20	30	40	50
TPO	1	1	1	1	1

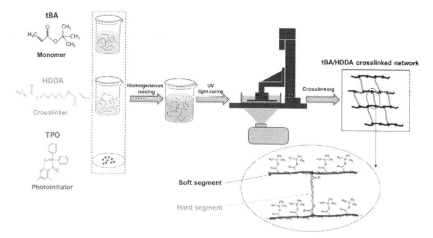

FIGURE 8.35 Schematic illustration of the preparation process of the tBA/HDDA crosslinked network made up of tBA monomers (*soft segments*) and HDDA crosslinkers (*hard segments*) via digital light processing (DLP).

light is 405 nm, and the power density is 1250–1450 μW/cm^2, which are both intrinsic parameters of this printer and could not be adjusted. The layer thickness was set to be 50 μm and the exposure time for each layer was set as 30 s.

8.3.4.1.3 Characterization of digital light processing-fabricated acrylate-based shape memory polymer

The thermal stability and crystallization behavior of the DLP-fabricated specimens were analyzed by a TG/DSC synchronous thermal analyzer (STA449F3, Netzsch, Germany). About 5 mg testing samples cut off from the DLP-fabricated specimens were placed in an alumina crucible and heated from 30°C to 600°C at a speed of 10°C/min in a pure nitrogen atmosphere.

The polymorphism characteristics of the DLP-printed parts were studied by an X'pert3 power X-ray diffractometer (PANalytical B.V., Netherlands) with a PIXcel versatile matrix detector and the diffraction angle of 2θ varied from 10 to 70 degrees at a scanning rate of 3 degrees/min under the Cu Kα radiation.

Fourier transform infrared spectroscopy (FTIR, VERTEX 70 Bruker, Germany) was carried out for chemical structure analysis of the photosensitive resins (both uncured and cured) with different compositional proportions.

Dynamic mechanical analysis (DMA) was carried out with a Perkin Elmer Diamond DMA apparatus to evaluate the dynamic thermal-mechanical properties of the DLP-printed specimens. Measurements were done in compression mode at a frequency of 1 Hz and with a heating rate of 5°C/min in the temperature range of 30°C–200°C. The test samples were processed into cylindrical shapes with the dimensions of Φ10 mm × 2 mm.

Shape R_f and shape R_r are two important parameters reflecting shape memory properties. The R_f can indicate the ability of the SMPs to fix the mechanical deformation applied in the programming of process, while the R_r demonstrates the ability of the material to recover to its original shape. They can be determined from Eqs. (8.9 and 8.10), respectively.

$$R_f = \frac{\varepsilon_{unload}}{\varepsilon_{load}} \times 100\% \tag{8.9}$$

$$R_r = \frac{\varepsilon_r}{\varepsilon_{unload}} \times 100\% = \frac{\varepsilon_{unload} - \varepsilon_{final}}{\varepsilon_{unload}} \times 100\% \tag{8.10}$$

here, ε_{load} is the strain applied at a temperature above T_g, while ε_{unload} is the strain measured when the external load is removed after cooling to a temperature below T_g, and ε_{final} is the strain measured when the DLP-printed SMPs have recovered to its original shape. As a result, ε_r, the deformation ratio during the recovery process when the SMPs are reheated, can be also expressed as ε_{unload} minus ε_{final}.

The R_f and R_r were tested by an advanced and precise TA ARES-G2 DMA instrument. Measurements were done in a force-controlled mode and samples were printed into rectangular bars with the size of 20 mm × 10 mm × 1 mm. The testing procedures were performed in the following five steps: (1) Heating, where the samples were heated from 30°C to 10°C above T_g at a rate of 3°C/min and then thermally insulated for 10 min; (2) deformation, where the samples were exerted an external force at a rate of 0.1 N/min up to an applied strain (ε_{load}) around 10%; (3) cooling, where the external force remaining unchanged, the samples were cooled to 30°C at the speed of 10°C/min; (4) fixing, where the external force was removed; and (5) recovery, where the samples were reheated to 10°C above T_g. The strain at the end of each step was recorded and applied to Eqs. (8.9 and 8.10) to get the values of R_f and R_r.

Fold-deploy tests were conducted in the following four steps (as shown in Fig. 8.36): (1) A circular strip specimen with a diameter of 25 mm and a thickness of 1 mm was printed by DLP and then cut a small gap (original shape); (2) the strip was immersed into a hot water bath (the temperature was set as T_1) with a certain temperature exceeding T_g about 20°C for 5 min and exerted an external force to get a straight bar (temporary shape); (3) the deformed specimen was fixed by quickly immersed in cold water (the temperature was set as T_2) at RT (about 25°C) and then the external force was removed; (4) it was put back into the hot water bath and recovered to the original shape. All printed SMPs with different HDDA contents went through these four steps and the time spent of the fourth step was recorded to characterize the shape recovery rate.

8.3.4.2 Results and discussion

8.3.4.2.1 Thermal stability and crystallization behavior

The thermogravimetric curves of the DLP-printed SMPs are shown in Fig. 8.37, it is obviously found that the thermal stability of the DLP-printed SMPs enhances with the increase of the mass fraction of HDDA. At 230°C, the DLP-printed parts begin to undergo thermal decomposition and the second thermal decomposition occurred at 400°C. When the temperature increases from 100°C to 230°C, there is a slight decrease in mass due to the evaporation of volatile compositions, and it shows an inverse proportion relationship between the content of HDDA and the evaporation loss. In view of this result, it is suggested that the ambient temperature should not exceed 200°C during the whole printing process and working condition. It is crucial that the volatile substances should be cleaned up.

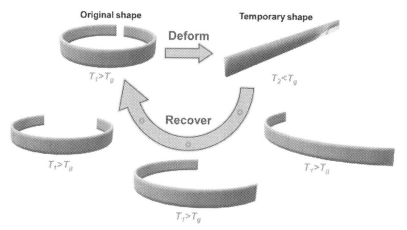

FIGURE 8.36 Scheme diagram of deformation and recovery process in the fold-deploy tests of DLP-printed SMPs.

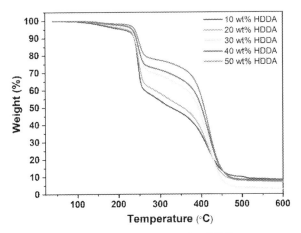

FIGURE 8.37 Thermogravimetric curves of the DLP-printed SMPs.

Fig. 8.38 demonstrates the DSC curves of the SMPs with the mass fraction of HDDA from 10 to 50 wt.%. The endothermic peak reflecting crystal melting is not observed, which clearly confirms that the crosslinked networks of the DLP-printed SMPs do not contain crystalline phase. In addition, there is only one single step on each curve, which indicates that the SMPs are amorphous taking T_g as the critical characteristic parameter (the characterization of T_g is discussed in the latter section).

The phase identification of DLP-printed SMPs are analyzed by XRD in Fig. 8.39. Not crystal diffraction peak but amorphous diffuse peak around $2\theta = 18$ degrees is observed, which indicates that these DLP-printed SMPs are amorphous. This result is in complete agreement with DSC measurements. The correlation is related to the rapid cure rate of free radical polymerization, leading to the DLP-printed SMPs incompetently aligning into an ordered crystal structure over a long range. While the presence of diffraction peaks indicates that the internal atoms have a certain degree of regularity in a small range. In the five XRD curves illustrated in Fig. 8.39, the diffraction intensity of each sample was normalized to the scale of 0–1. It is found that the five diffuse peaks have approximately the same intensity and width, so the increase in the content of the crosslinking agent does not change the order of the small-scale arrangement of internal atoms. The most remarkable result emerging from the curves is that the highest point of amorphous peak moves slightly toward high angle direction when the proportion of HDDA increases. This result indicates that the spacing of orderly arrangement of atoms gradually decreases with the increase of the content of HDDA, which can be attributed to the fact that deeper crosslinking will bring about tighter molecular chains, resulting in a shorter distance between short-range ordered molecular chains.

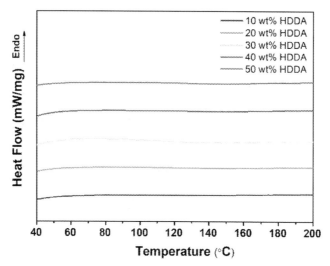

FIGURE 8.38 DSC curves of the DLP-printed SMPs.

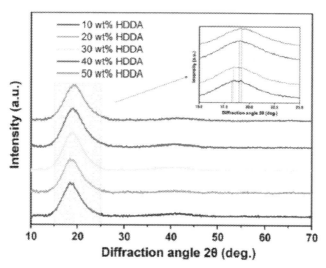

FIGURE 8.39 XRD curves of the DLP-printed SMPs.

8.3.4.2.2 Chemical structure analysis

FTIR analysis was conducted to verify the conversion of the photosensitive in the DLP process. The FTIR spectra of uncured resins is shown in Fig. 8.40A, absorption peaks at 2940 and 2863 cm^{-1} could be assigned to the symmetric stretching vibration of CH_3 and CH_2, respectively. The absorption peak corresponding to C=C stretching vibration at 1636 cm^{-1}

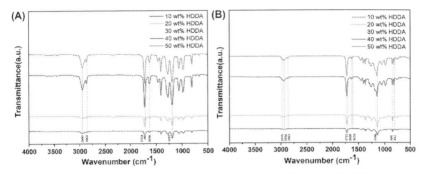

FIGURE 8.40 FTIR spectra of (A) uncured resins and (B) cured resins with different proportions of compositions.

decreases obviously as shown in Fig. 8.40B, which can be attributed to the addition polymerization of double bonds contained in both tBA and HDDA. The consumption of C=C has a significant influence on the oscillation of C−O−C, which results in the peak representing to C−O−C stretching vibration shifted from 1192 to 1148 cm^{-1}. For the same reason, the peak at 2940 cm^{-1} indicating the oscillation of CH$_3$ splits into two peaks locating at 2975 and 2936 cm^{-1}. The peak at 1724 cm^{-1} can be ascribed to the stretching vibration of the C=O, changing slightly to 1722 cm^{-1} maybe owing to the inductive effect, which originates from the electron-donating group adding to the C atom connected to the C=O double bond. It is clear that the intensity of absorption peaks enhances with the increase of HDDA content.

8.3.4.2.3 Shape memory properties

Compared with DSC measurements, DMA is commonly utilized to characterize the T_g of polymers due to its higher accuracy. This is because the thermal effect of polymers is obscure, while the storage modulus of polymers can vary by several orders of magnitude when a glass transition occurs. The T_g is determined by the temperature, at which the tan δ curves obtained by DMA reached the peak value, as shown in Fig. 8.41A. The T_gs of DLP-printed SMPs are shown in Fig. 8.40B. The T_g for the SMP with 10 wt% HDDA is 48.3°C, and the T_g approximately linearly increases with per additional 10 wt.% HDDA. More precisely, the mass fraction of 30 wt.% seems to be a dividing point. When the mass fraction is less than 30 wt.%, the T_g increases by about 5°C for per additional 10 wt.% HDDA, while the increment of T_g is around 10°C when the mass fraction is over 30 wt.%. As shown in Fig. 8.41A, the peak height decreases and the peak becomes wider with increasing the mass fraction of HDDA, which results from the fact that the crosslinking degree of SMPs increases with the increasing concentration of HDDA, causing an increment of storage modulus and much difficulty in

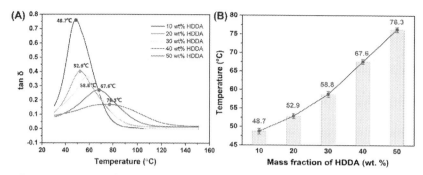

FIGURE 8.41 (A) Tan δ of SMP with the HDDA mass fractions of 10, 20, 30, 40 and 50 wt.%, (B) The glass transition temperature (T_g) as a function of the mass fraction of HDDA.

the movement of molecular networks. Thus, these SMPs show relatively high viscosity in a wider temperature range.

The storage modulus [i.e., elastic modulus (E)] reflects the amount of energy that a polymer reserve. The storage modulus is named as E_g when the SMP is in its glassy phase and defined as E_r when changing to the rubber phase. Previous works have confirmed that the storage modulus of a SMP ought to undergo two to three orders of magnitude decline when the SMP is heated from low temperature ($<T_g$, glassy phase) to high temperature ($>T_g$, rubbery phase), and then the modulus value reaches a plateau after such a descent. These typical characteristics are the most critical guidelines for the SMPs design. The drop of E indicates the significant movement of molecular chain and the rubbery plateau arises from the restriction of chain slippage at a longer length scale. In this work, the storage modulus of the printed SMPs with different HDDA proportions as a function of temperature range between 25°C and 130°C is shown in Fig. 8.42. It is evident that the storage modulus of the DLP-printed samples decreases drastically as the temperature increases, which is totally consistent with the judgment basis of SMPs as aforementioned. The E_g of 10 wt.% HDDA sample at 25°C is 927 MPa, which increases to 1.48×10^3 MPa when the HDDA concentration is 50 wt.%, and the higher HDDA mass fraction generally tends to show a higher storage modulus with minor deviations occurring in the SMP with 30 wt.% HDDA. The curve of SMP with 30 wt.% HDDA is the steepest in the decline stage, indicating that its modulus deceases fastest with the increase of temperature.

The released energy can be quantitatively reflected by the reduced storage modulus when the SMP undergoes a transition from the glassy phase to the rubbery phase. The released energy of FDM-printed SMPs was 1.65×10^3 MPa ($E_g - E_r$) from the reference. In our work, SMP with 50 wt.% crosslinker has the highest storage modulus, and the declined storage modulus is 1.47×10^3 MPa which is 12.2% less than that of FDM-printed SMPs. This result should be mainly due to the different mechanisms

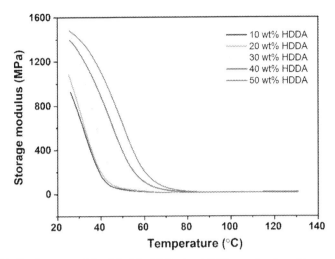

FIGURE 8.42 Storage modulus (E) of the DLP-printed SMPs as a function of temperature.

of achieving 4D printing through the two technologies. FDM technology imposes prestrain during the forming process, so the driving signal can be inserted into materials in advance. As a result, the released energy is particularly dependent on the process parameters. But the SMP parts are programmed by applying an external force after DLP process in our study. Therefore, the external force plays a key role in the released energy, and more released energy will be obtained as the external force enhances within the range of elastic strength. It should be noted that E_r is solely related to the performance of the material owing to the disappearance of driving force (internal or external). It is obvious that E_r for DLP-printed parts (4.35 MPa) gets an improvement of 33.4% compared to the FDM-printed ones (3.26 MPa), proving better mechanical properties of the DLP-printed tBA/HDDA. This should be ascribed to the crosslinked network structure existing in thermosetting polymers, which reflects the superiority of DLP in comparison with FDM in terms of the fabrication of thermosetting SMP.

Fig. 8.43A shows the shape fixities of DLP-printed SMPs with the mass fraction of HDDA ranging from 10 to 50 wt.% under the different thermomechanical cycle numbers. The shape fixity ratio is more than 91% and declines slightly with increasing the cycle number during the initial four cycles. In the first cycle, the DLP-printed SMPs with the 10, 20, and 30 wt.% HDDA show the comparatively high shape fixity values of 92%, 97%, and 96%, respectively. It is obvious that the DLP-printed SMPs with higher concentrations of HDDA exhibit a shorter cycle life, and fail to work continuously after seven cycles, five cycles, and three cycles for SMPs with 30, 40, and 50 wt.% HDDA, respectively. This result can be explained by

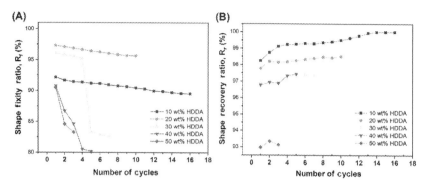

FIGURE 8.43 Effect of increasing mass fraction of HDDA on (A) shape fixity ratio (R_f) and (B) shape recovery ratio (R_r) of the DLP-printed SMPs under consecutive thermomechanical cycles.

the fact that the higher proportion of HDDA possesses a higher crosslink degree, leading to the lower mobility of polymer molecular chains. The SMP with 20 wt.% HDDA shows the highest R_f, and its cycle life (10 cycles) is superior to that of the SMP with 30 wt.% HDDA (7 cycles) but relatively shorter compared with that of the SMP with 10 wt.% HDDA (16 cycles). Fig. 8.43B shows the outstanding shape recovery properties dependent on the chemical composition of tBA/HDDA network. The DLP-printed SMPs with different proportions of HDDA tend to obtain a higher shape recovery ratio when undergoing more cycles, and the SMPs with lower HDDA contents have stronger shape recovery performance. This result can be ascribed to the following two reasons. Firstly, repeated cyclic deformation makes the molecular structure more flexible and more conducive to recover the original shape. Secondly, the lower concentration of crosslinkers facilitates a comparatively flexible polymer network. As a result, the R_r of SMP with 10 wt.% HDDA is up to 100% after 14 thermomechanical cycles. All the DLP-printed SMPs achieve considerably high shape recovery ratio of over 97% except that the one of SMP with 10 wt.% HDDA is around 93%.

Previous investigation respect to 4D printing has been done on the SLA-printed tBA/DEGDA network [68]. The results showed that the shape memory properties of this photosensitive resin exhibited desirable shape memory performance with 100% full recovery and an outstanding durability of 22 cycles. The SLA-printed SMPs with 10, 20, and 30 wt.% crosslinker got the shape fixity ratios of 85%, 95%, and 94%, respectively, which are all lower than 92%, 97%, and 96% of the DLP-printed SMPs with the HDDA contents of 10, 20, and 30 wt.% respectively in our present work. This probably results from the stronger flexibility of C—O bond existing in DEGDA, which makes it more prone to mobile when external force removed, thus giving rise to relatively lower shape fixity ratio but higher shape recovery ratio. Although this work missed to achieve 22 thermomechanical cycles that was

TABLE 8.4 Comparison of the cycle life between this work and the referred work.

	The concentration of crosslinker				
	10 wt.%	20 wt.%	30 wt.%	40 wt.%	50 wt.%
Cycle life of this work	16	10	7	5	3
Cycle life of the referred work	22	8	6	1	1

a bright spot of the referred work, the cycle life snatches a victory if the concentration of crosslinker exceeds 20 wt%, as shown in Table 8.4

In a previous work [69], a polyurethane acrylate was synthesized by compounding epoxy acrylate and isobornyl acrylate, and the resin was fabricated by SLA to achieve 4D printing. The results showed that the shape memory polyurethane fabricated by SLA held excellent endurance (16 cycles), shape fixity ratio (96% ± 1%) and shape recovery ratio (100% ± 1%). The variation trends of R_f and R_r under different cycle numbers in this work are consistent with the results in their work, and a comparable cycle life (16 cycles) and shape recovery ratios (100% after the 14th cycle) are obtained. The high shape fixity values of 97% appear in the initial three cycles of the 20 wt.% crosslinker SMP, which are better than those of the SLA-printed SMP prepared.

The entire process of fold-deploy test for the SMP with 20 wt.% HDDA is demonstrated in Fig. 8.44. The sample was immersed into a 73°C (T_g = 52.9°C) water bath and completed the recovery process with 11 s, and the one with 50 wt.% HDDA just had a 7 s of recovery time, indicating a fairly good shape recovery rate. Fig. 8.45 clearly shows that the SMPs with more than 20 wt.% HDDA have a shorter recovery time, namely, faster recovery rate, because the more HDDA content leads to higher storage modulus and more elastic potential energy released during the shape recovery process.

8.3.4.3 Conclusion

In present work, a novel SMP consisting of tBA/HDDA crosslinked network has been synthesized and printed by DLP technology to successfully achieve 4D printing, broadening the range of materials that can be used in the area of 4D printing. This work systematically investigated the thermal stability, crystallization behavior, chemical structure changes both cured and uncured, and shape memory properties of DLP-printed parts with different HDDA contents. The major results and findings are as follows:

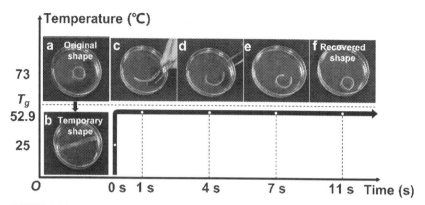

FIGURE 8.44 The entire shape recovery process of the SMP with 20 wt.% HDDA in a 73°C water bath.

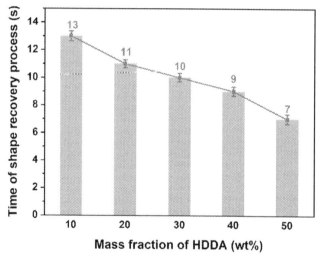

FIGURE 8.45 The time spent of shape recovery process when deformed samples were reheated back to above T_g.

1. The thermal stability slightly enhances with the increasing content of crosslinker, and the working temperature of the DLP-printed SMPs is optimal not to exceed 200°C.
2. The DLP-printed SMPs are amorphous, which is consistently confirmed by the testing results of both DSC and XRD. More specifically, no crystallizing exothermic peak appears in the DSC curves and no spiculate crystalline peak is observed in the XRD diagrams. The distance between short-range ordered molecular chains decrease as the crosslinker concentration increases, leading to the diffuse peak in XRD curves shift to right.

3. The addition reaction of double bonds in tBA and HDDA results in the rapid solidification of the resins, leading to the evident decline in C=C peak stretching vibration at 1636 cm^{-1}, and the shift of the characteristic peaks of C–O–C and C=O. The aforementioned changes are identical in SMPs with different concentrations of HDDA, and the higher the content of HDDA is, the stronger the absorption peak presents.
4. The T_gs and moduli of SMPs can be tunable by adjusting the concentration of crosslinker. The T_g of 76.3°C for the DLP-printed SMP with 50 wt.% HDDA is the highest, and the T_g gets a roughly linear reduction as the concentration of HDDA decreases. The T_g decreases to 48.7°C when the HDDA proportion is 10 wt.%. At the temperature of 25°C, the storage modulus of the SMP with 50 wt.% crosslinker is as high as 1.48 × 103 MPa.
5. The SMP with 10 wt.% HDDA possesses the best cycling endurance (16 cycles) and a shape recovery ratio of 100% after undergoing 14 cycles. The 20 wt.% one exhibits an outstanding shape fixity ratio of 96% ± 1% but relatively shorter cycle life (10 cycles). The DLP-printed SMP with 50 wt.% HDDA has the fastest recovery rate owing to its high storage modulus. Therefore, it can be concluded that the DLP-printed SMPs prepared in this work indicate exhibit shape memory properties.

There is an urgent need for developing new SMPs, which are photosensitive and manufacturable for SLA or DLP to achieve 4D printing. This study will set the stage for further research into this field. Although 4D printing technology is still in its infancy, it will grow to maturity under continuous research in the future.

References

[1] S. Tibbits, 4D Printing: multi-material shape change, Architect. Des. 84 (1) (2014) 116–121.
[2] D. Raviv, W. Zhao, C. Mcknelly, et al., Active printed materials for complex self-evolving deformations, Sci. Rep. (2014) 7422.
[3] M. Jamal, S.S. Kadam, R. Xiao, et al., Bio-origami hydrogel scaffolds composed of photo-crosslinked PEG bilayers, Adv. Healthc. Mater. 2 (8) (2013) 1142–1150.
[4] G.A. Sydney, E.A. Matsumoto, R.G. Nuzzo, et al., Biominetic 4D printing, Nat. Mater. 15 (2016) 413–418.
[5] N. Nurly, Q. Yan, B. Song, et al., Effect of carbon nanotubes reinforcement on the polyvinyl alcohol-polyethylene glycol double-network hydrogel composites: a general approach to shape memory and printability, Eur. Polym. J. 110 (2019) 114–122.
[6] Y. Yang, Y.H. Chen, Y. Wei, et al., 3D printing of shape memory polymer for functional part fabrication, Int. J. Adv. Manuf. Technol. 84 (2016) 9–12.
[7] Q. Ge, A.H. Sakhaei, H. Lee, et al., Multimaterial 4D printing with tailorable shape memory polymers, Sci. Rep. 6 (2016) 31110.
[8] J.T. Wu, C. Yuan, Z. Ding, et al., Multi-shape active composites by 3D printing of digital shape memory polymers, Sci. Rep. 13 (2016) 24224.

[9] H.Z. Wu, P. Chen, C.Z. Yan, et al., Four-dimensional printing of a novel acrylate-based shape memory polymer using digital light processing, Mater. Des. 171 (2019) 107704.
[10] H. Wei, Q. Zhang, Y. Yao, et al., Direct-write fabrication of 4D active shape-changing structures based on a shape memory polymer and its nanocomposite, ACS Appl. Mater. Interfaces 9 (1) (2017) 876–883.
[11] Y. Kim, H. Yuk, R. Zhao, et al., Printing ferromagnetic domains for untethered fast-transforming soft materials, Nature 558 (7709) (2018) 274–279.
[12] Q.J. Ze, X. Kuang, J. Wong, et al., Magnetic shape memory polymers with integrated multifunctional shape manipulations, Adv. Mater. (2019).
[13] Y. Liu, B. Shaw, M.D. Dickey, et al., Sequential self-folding of polymer sheets, Sci. Adv. 3 (3) (2017) e1602417.
[14] D.J. Wales, Q. Cao, K. Kastner, et al., 3D-printable photochromic molecular materials for reversible information storage, Adv. Mater. (2018) 1800159.
[15] S. Dadbakhsh, M. Speirs, J.P. Kruth, et al., Effect of SLM parameters on transformation temperatures of shape memory nickel titanium parts, Adv. Eng. Mater. 16 (9) (2014) 1140–1146.
[16] S. Dadbakhsh, M. Speirs, J.P. Kruth, et al., Influence of SLM on shape memory and compression behaviour of Ni-Ti scaffolds, CIRP Ann. Manuf. Technol. 64 (2015) 209–212.
[17] S. Saedi, N.S. Moghaddam, A. Amerinatanzi, et al., On the effects of selective laser melting process parameters on microstructure and thermomechanical response of Ni-rich Ni-Ti, Acta Mater. 144 (2018) 552–560.
[18] S. Saedi, A.S. Turabi, M.T. Andari, et al., The influence of heat treatment on the thermomechanical response of Ni-rich Ni-Ti alloys manufactured by selective laser melting, J. Alloy. Compd. 677 (2016) 204–210.
[19] J.J. Marattukalam, V.K. Balla, M. Das, et al., Effect of heat treatment on microstructure, corrosion, and shape memory characteristics of laser deposited NiTi alloy, J. Alloy. Compd. 744 (2018) 337–346.
[20] H.Z. Lu, C. Yang, X. Luo, et al., Ultrahigh-performance Ti-Ni shape memory alloy by 4D printing, Mater. Sci. Eng. A 763 (2019) 138166.
[21] Z.X. Khoo, J. An, C.K. Chua, et al., Effect of heat treatment on repetitively scanned SLM Ni-Ti shape memory alloy, Materials 12 (1) (2019) 1–19.
[22] E.M. Mazzer, C.S. Kiminami, P. Gargarella, et al., Atomization and selective laser melting of a Cu-Al-Ni-Mn shape memory alloy, Mater. Sci. Forum 802 (2014) 343–348.
[23] T. Gustmann, J.M. Dos Santos, P. Gargarella, et al., Properties of Cu-based shape memory-alloys prepared by selective laser melting, Shape Mem. Super Elast. 3 (1) (2017) 24–36.
[24] T. Gustmann, A. Neves, U. Kühn, et al., Influence of processing parameters on the fabrication of a Cu-Al-Ni-Mn shape-memory alloy by selective laser melting, Addit. Manuf. 11 (2016) 23–31.
[25] T. Gustmann, H. Schwab, U. Kühn, et al., Process optimization, microstructures and mechanical properties of a Cu-based shape memory alloy fabricated by selective laser melting, Mater. Des. 153 (2018) 129–138.
[26] J. Tian, W.Z. Zhu, Q.S. Wei, et al., Process optimization, microstructures and mechanical properties of a Cu-based shape memory alloy fabricated by selective laser melting, J. Alloy. Compd. 785 (2019) 754–764.
[27] G. Liu, Y. Zhao, G. Wu, et al., Origami and 4D printing of elastomer-derived ceramic structures, Sci. Adv. 4 (8) (2018) eaat0641.

[28] U. Sari, Influences of 2.5wt.% Mn addition on the microstructure and mechanical properties of Cu-Al-Ni shape memory alloys, Int. J. Miner. Metall. Mater. 17 (2010) 192–198.

[29] X.J. Liu, R. Kaimuma, I. Qnuma, et al., Phase equilibria in the Cu-rich portion of the Cu-Al binary system, J. Alloy. Compd. 264 (2000) 201–208.

[30] M. Elahinia, M. Elahiniaa, N.S. Moghaddama, et al., Additive manufacturing of Ni-Ti-Hf high temperature shape memory alloy, Scr. Mater. 145 (2018) 90–94.

[31] J.V. Humbeeck, R. Stalmans, M. Chandrasekaran, et al., On the stability of shape memory alloys, Eng. Asp. Shape Mem. Alloy. 35 (5) (1990) 96–105.

[32] L. Ma, H. Bin, Temperature and stress analysis and simulation in fractal scanning-based laser sintering, Int. J. Adv. Manuf. Technol. 34 (9–10) (2007) 898–903.

[33] C.A. Biffi, A. Tuissi, High-temperature shape memory alloys, J. Mater. Eng. Perform. 23 (10) (2014) 3727–3735.

[34] L. Thijs, K. Kempen, J.P. Kruth, et al., Fine-structured aluminium products with controllable texture by selective laser melting of pre-alloyed AlSi10Mg powder, Acta Mater. 61 (5) (2013) 1809–1819.

[35] W. Xu, M. Brandt, S. Sun, et al., Additive manufacturing of strong and ductile Ti-6Al-4V by selective laser melting via in situ martensite decomposition, Acta Mater. 85 (2015) 74–84.

[36] M. Ma, Z. Wang, M. Gao, et al., Layer thickness dependence of performance in high-power selective laser melting of 1Cr18Ni9Ti stainless steel, J. Mater. Process. Technol. 215 (2015) 142–150.

[37] Y. Li, D. Gu, Parametric analysis of thermal behavior during selective laser melting additive manufacturing of aluminum alloy powder, Mater. Des. 63 (2014) 856–867.

[38] P. Gargarella, C.S. Kiminami, E.M. Mazzer, et al., Phase formation, thermal stability and mechanical properties of a Cu-Al-Ni-Mn shape memory alloy prepared by selective laser melting, Mater. Res. 18 (Suppl. 1) (2015) 35–38.

[39] S.N. Saud, E. Hamzah, T. Abubakar, et al., Influence of Ti additions on the martensitic phase transformation and mechanical properties of Cu-Al-Ni shape memory alloys, J. Therm. Anal. Calorim. 118 (1) (2014) 111–122.

[40] J. Dutkiewicz, T. Czeppe, J. Morgiel, Effect of titanium on structure and martensitic transformation in rapidly solidified Cu-Al-Ni-Mn-Ti alloys, Mater. Sci. Eng. A 273–275 (3) (1999) 703–707.

[41] D. Silva, M. Romero, G. Piter, et al., Laser surface re-melting of a Cu-Al-Ni-Mn shape memory alloy, Mater. Sci. Eng. A 661 (2016) 61–67.

[42] J. Zhang, L.W. Zhou, D.H. Jiang et al., High temperature shape memory alloys, Precious Met. 21 (2001) 96–101.

[43] B. Massoumi, P. Jafarpour, M. Jaymand, et al., Functionalized multiwalled carbon nanotubes as reinforcing agents for poly (vinyl alcohol) and poly (vinyl alcohol)/starch nanocomposites: synthesis, characterization and properties, Polym. Int. 64 (5) (2015) 689–695.

[44] B.X. Yang, J.H. Shi, X. Li, et al., Mechanical reinforcement of poly (1-butene) using polypropylene-grafted multiwalled carbon nanotubes, J. Appl. Polym. Sci. 113 (2) (2009) 1165–1172.

[45] J.Y. Kim, S.I. Han, S.H. Kim, Crystallization behaviors and mechanical properties of poly (ethylene 2,6-naphthalate)/ multiwall carbon nanotube nanocomposites, Polym. Eng. Sci. 47 (11) (2007) 1715–1723.

[46] B.R. Sankapal, K. Setyowati, J. Chen, et al., Electrical properties of air-stable, iodine-doped carbon-nanotube-polymer composites, Appl. Phys. Lett. 91 (17) (2007) 173103.

[47] J.Q. Pham, C.A. Mitchell, J.L. Bahr, et al., Glass transition of polymer/single-walled carbon nanotube composite films, J. Polym. Sci. Part. B (Polym. Phys.) 41 (24) (2003) 3339–3345.

[48] G. Li, H. Zhang, D. Fortin, et al., Poly (vinyl alcohol)-poly (ethylene glycol) double-network hydrogel: a general approach to shape memory and self-healing functionalities, Langmuir 31 (2015) 11709–11716.

[49] E. Choi, J. Shin, Z. Khaleel, et al., Synthesis of electroconductive hydrogel films by an electro-controlled click reaction and their application to drug delivery systems, Polym. Chem. 6 (2015) 4473–4478.

[50] K. Shah, D. Vasileva, A. Karadaghy, et al., Development and characterization of polyethylene glycol-carbon nanotube hydrogel composite, J. Solut. Chem. 3 (40) (2015) 7950–7962.

[51] M.B. Tork, N.H. Nejad, S. Ghalehbagh, et al., In situ green synthesis of silver nanoparticles/chitosan/poly vinyl alcohol/ polyethylene glycol hydrogel nanocomposite for novel finishing of nasal tampons, J. Ind. Text. 45 (2014) 1399–1416.

[52] A. Yamada-Nosaka, K. Ishikiryama, M. Todoki, et al., 1H-NMR studies on water in methacrylate hydrogels. I, J. Appl. Polym. Sci. 39 (1990) 2443–2452.

[53] M.K. Lee, H. Bae, S. Lee, et al., Freezing/thawing processing of PVA in the preparation of structured microspheres for protein drug delivery, Macromol. Res. 19 (2) (2011) 130–136.

[54] C.M. Hassan, N.A. Peppas, Structure and morphology of freeze/thawed PVA hydrogels, Macromolecules 33 (7) (2000) 2000.

[55] T. Xie, Recent advances in polymer shape memory, Polymer 52 (2011) 4985–5000.

[56] N.A. Peppas, P. Bures, W. Leobandung, et al., Hydrogels in pharmaceutical formulations, Eur. J. Pharm. Biopharm. 50 (1) (2000) 27–46.

[57] H. Du, J. Zhang, Solvent induced shape recovery of shape memory polymer based on chemically cross-linked poly (vinyl alcohol), Soft Matter 6 (14) (2010) 3370.

[58] J. Lewis, Direct ink Writ, . 3D Funct. Mater. 16 (17) (2006) 2193–2204.

[59] N. Yang, Microrheology and microstructure of poly (vinyl alcohol)-based physical gels, Univ. West. Ontario Dr. Phil. (2011).

[60] H. Zhang, D. Han, Q. Yan, et al., Light-healable hard hydrogels through photothermally induced melting-crystallization phase transition, J. Mater. Chem. A 2 (33) (2014) 13373–13379.

[61] H. Meng, G. Li, A review of stimuli-responsive shape memory polymer composites, Polymer 54 (9) (2013) 2199–2221.

[62] J. Leng, X. Lan, Y. Liu, et al., Shape-memory polymers and their composites: stimulus methods and applications, Prog. Mater. Sci. 56 (7) (2011) 1077–1135.

[63] J. Hu, Y. Zhu, H. Huang, et al., Recent advances in shape–memory polymers: structure, mechanism, functionality, modeling and applications, Prog. Polym. Sci. 37 (12) (2012) 1720–1763.

[64] F. Momeni, S.M. Hassani, N.X. Liu, et al., A review of 4D printing, Mater. Des. 122 (2017) 42–79.

[65] S. Miao, H. Cui, M. Nowicki, et al., Stereolithographic 4D bioprinting of multi-responsive architectures for neural engineering, Adv. Biosyst. (2018) 1800101.

[66] Y.Y.C. Choong, S. Maleksaeedi, H. Eng, et al., Curing characteristics of shape memory polymers in 3D projection and laser stereolithography, Virtual Phys. Prototyp. 12 (1) (2017) 77–84.

[67] L. Huang, R. Jiang, J. Wu, et al., Ultrafast digital printing toward 4D shape changing materials, Adv. Mater. 29 (7) (2016) 1605390.
[68] Y.Y.C. Choong, S. Maleksaeedi, H. Eng, et al., 4D printing of high-performance shape memory polymer using stereolithography, Mater. Des. 126 (2017) 219–225.
[69] T. Zhao, R. Yu, X. Li, et al., 4D printing of shape memory polyurethane via stereolithography, Eur. Polym. J. 101 (2018) 120–126.

Index

Note: Page numbers followed by "*f*" and "*t*" refer to figures and tables, respectively.

A

ABS. *See* Acrylonitrile–butadiene–styrene (ABS)
Absorption
　coefficient, 411–412, 413*f*
　of laser by metal powder, 415–419
　of laser energy by metal, 412–415
　　laser polarization effects, 414–415
　　laser wavelength effects, 412
　　metal properties effect, 412
　　metal surface roughness effects, 414
　　metal temperature effects, 412–413
　rate
　　of mixed powder to laser, 418, 418*t*
　　powder particle size, 418
Acceleration of photosensitizer, 235
Acoustic vibration, 27
Acrylate-based SPM, 721–735
　characterization of digital light processing-fabricated, 724–726
　chemical structure analysis, 728–729
　fabrication, 723–724
　preparation, 723, 724*f*
　shape memory properties, 729–733
　thermal stability and crystallization behavior, 726–727
Acrylic copolymers
　hydration method of, 400
　polymerization method of, 400
　water-soluble support material, 399–400
Acrylonitrile–butadiene–styrene (ABS), 50, 184, 185*t*
　filament, 373–378
　　performance indexes, 377*t*
　　preparation, 376–378
　　selection of ABS resin brand, 373–376
　plastic, 654–655
　powder material
　　characteristics of ABS materials, 184–185
　　sintering performance of ABS powder, 186–188
　terpolymer, 374*f*
Acryloyl chloride, 711–712
Additive manufacturing (AM), 9, 191, 597. *See also* Four-dimensional printing (4D printing)
　air-permeability prosthetic socket monolithic prosthetics, 650
　extrusion-based method, 719–720
　FeCrMoVC tool steel, 5–6
　forming mechanisms of metal materials
　　absorption of laser by metal powder, 415–419
　　absorption of laser energy by metal, 412–415
　　dynamics and stability of melting pool, 436–439
　　laser energy transfer, 409–411
　　temperature, stress and strain fields in SLM forming process, 419–436
　polymer powder material, 649–654
　　DSC curve of precoated sand, 653*f*
　　large complex engine crankcase, 652*f*
　　large cylinder sand mold of SLS additive manufacturing, 654*f*
　　large engine cylinder head sand core, 654*f*
　　morphology, particle size, and distribution of powder, 652*f*
　　production flowchart of AM/casting composite process, 649*f*
　　prosthesis printed by SLS technology, 652*f*
　　wax mold and castings, 651*f*
　polymer wire material, 654–656
　　TG data of pure PLA and ADR chain, 655*t*
　　3D model and physical map of woodpecker, 656*f*

741

Additive manufacturing (AM) (*Continued*)
  preparation of ceramic parts, 598t
  and principle of ceramic materials, 597–606
    FDM of ceramic filaments, 604–605
    LOM and ceramic sheets, 602–603
    SL and ceramic slurry, 599–600
    SLM technology and ceramic powders, 601–602
    SLS technology and ceramic powders, 605–606
    3D printing and ceramic powders, 600–601
  technologies, 10–11, 679–680, 722
  and materials, 1–3
  for metal materials and principles, 403–409, 404f
Additives, 59
Aermet 100 steel, 5–6
Aerospace, 685–686
Agglomeration temperature, 41–42, 83
Aggregation structure, 41–49
  density of sintered parts, 44–45
  dimensional accuracy of sintered parts, 47–49
  fractured surface shape of sintered parts, 46–47
  mechanical properties of sintered parts, 45–46
  sintering temperature window, 41–44
Aging of nylon 12, 77–79
AISI 420 stainless steel
  effects of grain refinement on tensile strength, 569–570
  mechanical properties, 567–569
  microscopic defects on tensile strength, 570–575
  microstructure, 564–567, 565f, 566f
Alcohol, 263–264
  alcohol-water solution, 59
  selection of, 268–269
Alicyclic epoxy acrylate, 349–359
  SLA properties of resin, 354–359
  synthesis of, 351–354
Alicyclic epoxy compounds, 266f
Alicyclic glycidyl ester, 656
α,α-diethoxy acetophenone (DEAP), 214
Aluminum alloy materials, 6
Aluminum oxide (Al_2O_3), 7
  Al_2O_3-polymer composite powder, 617–619
  ceramic parts manufacturing, 676

Aluminum powder, 119
  content effects
    on dimensional accuracy of SLS parts, 125–126
    on mechanical properties, 123–125, 123f, 124f
  particle size effects on properties of SLS parts, 127–129, 128f
Aluminum-based alloys, 5–6
Aluminum-silicon alloy wire, 669
AM. *See* Additive manufacturing (AM)
Amine, 248–249
Ammonium persulfate (AP), 711
Amorphous polymers, 34, 41, 48f, 61, 163
  selection of, 53–54
  SLS mathematical model of, 16–18
Anisotropic swelling performance, 681
Antioxidants, 51, 55–57
  on antiaging properties of nylon 12 powder, 76t
  on mechanical properties of nylon 12 SLS samples, 78t
AP. *See* Ammonium persulfate (AP)
APDL. *See* AYSYS parameterized language (APDL)
Apparent density, 439–441, 440t, 448f
Application cases
  additive manufacturing polymer powder material, 649–654
  additive manufacturing polymer wire material, 654–656
Arc additive manufacturing technology, 665
Arc fuse deposition forming, 575–579
Aryl hexafluoroantimonate iodonium salt (DPI-SbF$_6$), 198–200
Austenitic parent phase, 699–700
Automobile engine, 669
Average heat flux of laser, 424
AYSYS parameterized language (APDL), 466–473

B

Backscattered electron images, 706–708, 708f
Balling effect, 26
BCC. *See* Body-centered cubic (BCC)
Beer–Lambert law, 27
Benzoin dimethyl ether, 205–206
Benzophenone, initiation mechanism of, 200f
Benzyl alcohol
  acceleration effect on OXT221, 239
  accelerator in cationic SLA, 230–239
  selection of, 238–239

β-carboxyethyl acrylate (β-CEA), 323–324
"Bimodal" grain size distribution, 695–696
Biomedicine, 685–686
Bis(3-ethyl-3-oxetanyl)methyl group ether, 260
    acceleration effect on, 239
    infrared spectroscopy analysis, 233–234
    SLA reaction mechanism, 260–264
Blending modification, 396
Body-centered cubic (BCC), 705–706
Bond jetting, 1–2
Bonding
    mechanism, 389–391
    model, 389
    process, 391–392
Break-away support materials, 389–395
    bonding mechanism, 389–391
    bonding model, 389
    bonding process, 391–392
    material factors, 389–391
    preparation of, 392–395
Bulk density
    of powder, 60, 60*t*
    of pure nylon 12 powder and nylon 12/PTW, 143*t*
Bulk strength of materials, 40–41

C

CAD. *See* Computer-aided design (CAD)
Calcium stearate, 57
Carbamate acrylate analysis, 301–305
Carbon depleted zone, 571–573
Carbon fiber (CF), 152
Carbon fiber/nylon 12 composite powder material, 150–162
    preparation and characterization of
        characteristics of composite powder, 153–157
        preparation process of composite powder, 152–153
        surface treatment of fiber powder, 151–152
        SLS properties of carbon fiber/nylon 12 composite powder, 158–162
Carbon nanotubes (CNT), 681, 713*f*
    DN hydrogel reinforced by, 710–721
Carbon-rich zone, 571–573
Cationic initiator, 198–200
Cationic photoinitiated polymerization mechanism, 207–213
Cationic SLA, 230–239
Cationic solid materials, 264–278
    contact angle, 272–273
    injection parameters, 269–272
    maximum spreading factor, 272–273
    preparation of solid material resin, 269
    printing stability, 274–275
    raw materials for, 265–269
    SLA properties of, 275–278
    spreading time, 272–273
    sputtering coefficient, 272–273
Center for Technology Transfers in Ceramics (CTTC), 1–2
Ceramic
    and composite materials, 685–686
    direct deposition, 1–2
    filaments, 604–605
    materials, 597
        AM technology and principle, 597–606
        forming mechanism of ceramics, 606–609
        preparation for SLS, 609–645
    sheets, 602–603
    slurry, 599–600
Ceramic powder(s). *See also* Composite powder
    for additive manufacturing, 673–675
    manufacturing of Al_2O_3 ceramic parts, 676
    manufacturing of cordierite ceramic parts, 675–676
    manufacturing of SiC ceramic parts, 677
    SLS-shaped large-size SiC square mirror, 675*f*
    traditional ceramic parts prepared by SLS, 674*f*
    and binders, 609–612
    SLM technology and, 601–602
    SLS technology and, 605–606
    3D printing and, 600–601
Cerium salt, 248–249
CF. *See* Carbon fiber (CF)
Chain-stopping antioxidants. *See* Free radical trapping agents
Charge-transfer complexes (CTCs), 248–249
CHDMDVE. *See* Cyclohexyl dimethanol divinyl ether (CHDMDVE)
Chemical reaction mechanism of shrinkage of free-radical system, 206–207
Chemical structure analysis, 728–729
CIP. *See* Cold isostatic pressing (CIP)
Cladding track, 548

CMT. See Cold metal transfer (CMT)
CNC. See Computerized numerical control (CNC)
CNT. See Carbon nanotubes (CNT)
Co-Cr alloy, 559f, 560f
  mechanical properties
    hardness, 561
    tensile properties, 561–564
  microstructure, 559–561
  porcelain-fused denture prosthesis, 664–665
Coal-series kaolinite-epoxy resin E12 composite powder, 633–640
Cold isostatic pressing (CIP), 619
Cold metal transfer (CMT), 407
Commercial tBA monomer, 723
Complex parts, 381, 405–406
Composite materials, 3
Composite powder(s)
  Al_2O_3-polymer, 617–619
  characteristics of, 153–157
    melting/crystallization of composite powder, 154–156
    microscopic shape of powder, 154
    particle size and particle size distribution, 153–154
    thermal weight loss analysis of composite powder, 156–157
  coal-series kaolinite-epoxy resin E12, 633–640
  cordierite-polymer, 640–645
  kaolinite-polymer, 630–645
  material of nylon 12, 91–162
    carbon fiber/nylon 12 composite powder material, 150–162
    nanosilica/nylon 12 composite powder materials, 98–111
    of nylon 12 coated aluminum, 111–129
    nylon 12/copper composite powder material, 130–137
    of nylon 12/PTW, 138–150
    of nylon 12/rectorite, 92–98
  materials, 9–49
    aggregation structure, 41–49
    bulk strength of materials, 40–41
    particle size, 26–29
    particle size distribution, 29–31
    shape of powder particles, 31–33
    on SLS processing, 24–49
    surface tension, 24–26
    viscosity, 33–40

noncoal series kaolinite-epoxy resin E06, 630–633
  preparation process of, 152–153
  sintered materials, 94
Composition on laser absorption rate, 416–418
  absorption rate of mixed powder to laser, 418
Compound initiator, 201
Computer-aided design (CAD), 403–404, 604
Computerized numerical control (CNC), 606
Contact angle, 272–273, 326
Continuous laser scanning line, 12
Cooling directly through cooling water, 66–67
Cooling method and speed, 64–69
  cooling directly through cooling water, 66–67
  cooling oil temperature by supplying cooling water in jacket, 67, 67f
  cooling outside kettle and distillation cooling, 68–69, 68f
  natural cooling, 64–66
Copolymerization modification, 397
Copper PA, 4–5
Cordierite ceramic parts, manufacturing of, 675–676
  aluminum parts of complex shape, 677f
  cordierite parts of complex shape, 676f
Cordierite powder, 675–676
Cordierite-polymer composite powders, 640–645
Corrosion resistance, 454–455
Coupling
  agent, 58
  mode, 415–416
Crack, 497–498, 497f
Crank-Nicholson form, 17–18
Creep theory, 429–430
Crystal orientation, 552
Crystal shrinkage, 83
Crystalline polymers, 4–5, 41, 43, 61
  selection of, 54–55
  SLS mathematical model of, 18–19, 19f
Crystallinity (CI), 144–145
Crystallization, 80
  behaviors, 576, 726–727
  enthalpy, 66–67
  exotherm process, 155–156
  and melting characteristics of nylon 12/copper composite powder, 135–136
  performance, 143–145

CTCs. *See* Charge-transfer complexes (CTCs)
CTTC. *See* Center for Technology Transfers in Ceramics (CTTC)
Cu-13.5Al-4Ni-0.5Ti copper-based SMA, 701–702
Cu-Al-Ni-based SMAs, 686–702
  experimental
    characterization of Cu-Al-Ni-Ti alloy, 689–690
    fabrication of Cu-Al-Ni-Ti alloy, 688–689
    preparation of Cu-Al-Ni-Ti alloy powder, 687–688
  results and discussion
    mechanical properties, 697–701, 697f, 698f
    phases and microstructures, 694–696
    effect of SLM processing parameters on relative density, 690–694
Cu-Al-Ni-Ti alloy
  characterization, 689–690
  fabrication, 688–689
    Cu-13.5Al-4Ni-0.5Ti samples fabricated, 690f
    laser scanning strategies, 689f
  powder, 687–688, 688f, 688t, 689f
Cu-Zn-Al-based SMAs, 702–710
  characterization of Cu-Zn-Al-Mn alloy, 703–704
  effect of SLM processing parameters on relative density, 704–705
  fabrication of Cu-Zn-Al-Mn alloy, 703
  mechanical properties, 708–709
  phases and microstructures, 705–708, 707f
  preparation of Cu-Zn-Al-Mn alloy powder, 703
Cu-Zn-Al-Mn alloy
  characterization, 703–704
  fabrication, 703
  powder, 703, 703f
Cubic packing powder bed, 21, 21f
Curing shrinkage of photopolymer, 225–228
Cyclohexyl dimethanol divinyl ether (CHDMDVE), 656

D

D-nanosilica/PA12, 109–110, 109t
DEAP. *See* α,α-diethoxy acetophenone (DEAP)
Decarburization, 571, 572f
Defoamer, 269

Deformation process, 21–22, 22f
Densification
  mechanism, 494–495
  shrinkage, 83
Density, 555
  and mechanical properties of formed entities, 447–450
  of sintered parts, 44–45, 44f
Desk top Moonray DLP printer, 723–724
Di-*n*-butylamine, 296
DiAPLEX MM-4520 pellets, 681–682
Diaryl hexafluorophosphate iodonium salt (DPI-PF$_6$), 198–200
Diazonium salt, 208–209
Dichloromethane, 711–712
Die forging production, 668–669
Die swelling, 365
Diethyl ether, 711–712
Diethylene glycol-water system, 62
Differential scanning calorimetry (DSC), 36–37, 37f, 43f, 81f, 699–700, 700f, 728f
  analysis of powder, 119
  curves, 105f
    thermal properties data obtained from, 106t
Differential thermal analysis (DSC), 640–642
Diffusion theory, 389–390
Digital light processing (DLP), 681–682, 728–729
  DLP-fabricated acrylate-based SPM, 724–726
    deformation and recovery process in fold-deploy tests, 726f
Digital medical AM materials, 2–3
Diluent monomers, 191, 197–198, 197t
Dimensional accuracy, 167, 167f, 186, 186t, 529–545
  dimensional changes at processing positions, 534–537
    collaboration diagram of machining parts, 536f
    dimensions of part, 536t
    scanning galvanometer, 535f
  of HIPS sintered parts, 174t
  machining error of XY plane, 529–534
  of PC sintered parts, 181f
  of sintered parts, 47–49, 49t
  of SLS parts, 125–126
  Z-axis accuracy, 537–545
3,4-Dimethoxybenzyl alcohol, 268–269
Dimethoxyethane (DME), 253–255

746 Index

Dimethyl adipate (3,4-epoxycyclohexylmethyl), 210−212, 211f
Dimethylolpropionic acid (DMPA), 307
Direct ink writing (DIW), 681, 719−720
Discrete-layer-superposition, 403−404
Dispersant, 57
Dissolution
 precipitation method, 614−617
 temperature, 63−64, 64t
DIW. See Direct ink writing (DIW)
DLP. See Digital light processing (DLP)
DMA. See Dynamic mechanical analysis (DMA)
DME. See Dimethoxyethane (DME)
DMPA. See Dimethylolpropionic acid (DMPA)
Double network (DN), 681
 hydrogel reinforcement by CNTs, 710−721
  characterization, 713−714
  preparation, 712−713
  printability, 719−721
  raw materials, 711−712
  SME, 717
  swelling, 717−719
  thermal properties, 714−716, 715f, 716f
DPI-PF$_6$. See Diaryl hexafluorophosphate iodonium salt (DPI-PF$_6$)
DPI-SbF$_6$. See Aryl hexafluoroantimonate iodonium salt (DPI-SbF$_6$)
DPIOC/TEA system, 249−251
Drum lid, 657−660
DSC. See Differential scanning calorimetry (DSC); Differential thermal analysis (DSC)
DuraForm GF, 4−5
Dye-onium salt photosensitive system, 252−259
Dynamic mechanical analysis (DMA), 725
Dynamics of melting pool, 437
 melt flow inside melting pool, 438f

E

EA. See Epoxy acrylate resin (EA)
Elastic deformation stage, 584−585
Elastic hysteresis, 370
Elastomeric poly (dimethylsiloxane) matrix nanocomposites, 685−686
Electric resistivity of metals, 412, 412t
Electrical vector, 414
Electromagnetic vibration, 27

Electron beam manufacturing, 403
Electronic communications, 685−686
Embrittlement temperatures, 50
 of thermoplastic resins, 50t
Energy
 balance between laser and metal action, 410−411
 dispersion rate, 20
 energy-carrying beam, 407
 spectrum analysis of powder, 117
Enthalpy, 424
 value, 97
Eosin, 252−253
Epoxy acrylate, 340−349, 733
Epoxy acrylate resin (EA), 343
Epoxy compound, 208−209, 208f
Epoxy resin, 166−167, 167f, 173−174
 epoxy resin E06, 673
Equilibrium water content (EWC), 713−714
ESEM photos, 590−592
Ethanol, 52, 63
2-(2-Ethoxyethoxy) ethyl acrylate, 284
2-Ethyl-9,10-dimethoxy anthracene, 235
EWC. See Equilibrium water content (EWC)
Exposure, amount of, 218
Extrusion process, 364

F

Fabrication
 of acrylate-based SPM, 723−724
 of Cu-Zn-Al-Mn alloy, 703
Face-centered cubic (FCC), 694−695
FDM. See Fused deposition modeling (FDM)
Feeding section, 365−366
Ferrocenium salt, 210
Fiber powder, surface treatment of, 151−152
Film coating method, 111
Finite element method, 430
 numerical simulation method, 419
 for transient heat conduction, 421−422
Finite element model, 425, 425f
 of powder and substrate, 467
First-order ordinary differential equations, 421
Flattening tests, 584−587, 586t
Flow theory, 429−430
Fold-deploy tests, 726
Forging die, 667−668
Formation mechanism of warping deformation, 225−227

Index 747

Four-dimensional printing (4D printing), 679–680. *See also* Additive manufacturing (AM)
  elastomer composite, 682–683
  materials
    acrylate-based SPM, 721–735
    Cu-Al-Ni-based SMAs, 686–702
    Cu-Zn-Al-based SMAs, 702–710
    DN hydrogel reinforced by CNTs, 710–721
  research and development status, 680–686
Fourier transform infrared spectroscopy (FTIR), 100–101, 101*f*, 713, 725
  spectra of uncured resins and cured resins, 729*f*
Fracture morphology, 558–559, 671–672
Fracture surfaces, 579–582
  shape of sintered parts, 46–47
Fraunhofer diffraction theory, 27
Free radical trapping agents, 55–57
Free water content (FW content), 713–714
Free-radical
  initiator, 200
  photoinitiated polymerization mechanism, 204–207
  system, 206–207, 284–287
    preparation of physical materials, 285
    principles for selecting raw materials, 284–285
    properties of free-radical solid materials, 285–287
Freezable water, 715
Freeze-thaw process, 719
Frenkel's two-liquid-drop model, 20–21, 20*f*, 35
Frenkel's viscous flow mechanism, 21
FTIR. *See* Fourier transform infrared spectroscopy (FTIR)
Fumed silica, 82
Functional precision, 362
Funis CMT welder additive manufacturing, 665–666
Fused deposition modeling (FDM), 361, 365–366, 597–598, 681–682
  of ceramic filaments, 604–605
  modeling materials in, 373–387
  polymer materials
    analysis of material modeling process, 361–366
    performance requirements for, 371–372
    thermodynamic transformation of polymer processing, 366–371
  principle, 361
  support materials in, 387–400
Fusing section, 365–366
FW content. *See* Free water content (FW content)

G

Galerkin method, 421–422
γ-aminopropyltriethoxysilane (KH-550), 58–59
Gas atomized powder, 657–659
Gaussian beam, 11–12, 437
Gaussian distribution, 423–424
Gaussian energy distribution, 566–567
Gel fraction analysis, 714
General purpose polystyrene (GPPS), 392–393
Geometric precision, 362
Glass transition temperature, 25–26, 36, 37*t*, 42–43, 657
Gold (Au), 405–406
GPPS. *See* General purpose polystyrene (GPPS)
Grain refinement effect, 697
  on tensile strength, 569–570

H

HA. *See* Hydroxyapatite (HA)
Hanging drop method, 25
Hardness, 561
HAZ. *See* Heat affected zone (HAZ)
HBVE. *See* Hydroxybutyl vinyl ether (HBVE)
HDDA, 197–198, 330, 723, 729–730, 730*f*, 732*f*
HEA. *See* Hydroxyethyl acrylate (HEA)
Heat affected zone (HAZ), 432–434
Heat transfer, 504–505, 504*f*
  control equation, 16–17
Heat transfer in polymer powder materials, 15–16
Heat treatment
  process and properties, 588–592
  temperature, 590, 591*f*
Heat-generating agent, 576–577
Heating process of polymer powder materials by laser
  characteristics of laser input energy, 11–13
  interaction between laser and polymer powder materials, 13–15
  SLS mathematical model
    of amorphous polymers, 16–18

748 Index

Heating process of polymer powder materials by laser (*Continued*)
  of crystalline polymer, 18–19
  transfer of heat in polymer powder materials, 15–16
Heterogeneous nucleation, 70, 550–551
Hexafluorophosphate, 210–212
High impact polystyrene (HIPS), 3–4, 168, 170*t*, 393
  powder material, 168–176
    postprocessing on properties of SLS parts of, 172–176
    properties of SLS parts of, 169–171, 171*t*
    SLS process characteristics of, 168–169
  powder sintered impeller, 177*f*
High-magnification microstructure, 500, 501*f*
HIP. *See* Hot isostatic pressing (HIP)
HIPS. *See* High impact polystyrene (HIPS)
"HK P500" type SLS industrial grade 3D printer, 650
Hole, 498–499, 499*f*
Home and abroad, 4D printing materials at, 680–686
  ceramics and composite materials, 685–686
  metals and composite materials, 683–685
  polymers and composite materials, 680–683
Homogeneous nucleation, 70
Hot forging die remanufacturing, 668–669
  automobile parts, 669*f*
  3D measurement and precision detection, 668*f*
  3D scanning and remanufacturing area design, 668*f*
Hot isostatic pressing (HIP), 498
HRPLA-I stereolithography rapid prototyping equipment, 657
Humectants, 324–325
Hybrid polymerization system, 213–214
Hybrid system solid materials, 279–283
  preparation of, 280
  properties of, 280–283
Hydration method of acrylic copolymers, 400
Hydrogels, 681, 715
Hydroxyapatite (HA), 5
  decomposers, 55–57
Hydroxybutyl vinyl ether (HBVE), 265–267
Hydroxycyclohexylacetophenone, 200*f*
Hydroxyethyl acrylate (HEA), 295
Hygroscopicity, 372
Hysteresis effect, 370

I

Incremental tangent stiffness method, 431–432
Induction period, 261–262
Infrared spectroscopy analysis
  of polyethylene glycol diacrylate, 321–322
  of UVR6105 and OXT221, 233–234
Infrared spectrum, 101–102
Inkjet printing, 1–2
Inorganic powders, 57
Instantaneous elastic deformation, 366–368
Intercalation mechanism in SLS process, 96–98
Interface bonding with nylon 12, 126–127
Interpenetrating network (IPN), 204
Interpolation function, 421
Inverted solvent polarity dependence, 255–257
Iodonium salt, 209–210, 240–259
  absorption peaks of, 249*t*
  dye-onium salt photosensitive system, 252–259
  iodonium salt-amine composite system, 248–251
  synthesis of, 241–247
IPN. *See* Interpenetrating network (IPN)
Iron-based alloys, 5–6, 455
Isobornyl acrylate, 733

K

Kaolinite, 609–611
  kaolinite-granulated powder, 673
  kaolinite-polymer composite powder, 630–645
KUKA robot, 665–666

L

316 L stainless steel
  mechanical properties, 552–553
  microstructure, 546–552, 550*f*
Lambert's law, 411
Laminated objected manufacturing (LOM), 597–598
  and ceramic sheets, 602–603
Laser. *See also* Selective laser melting (SLM); Selective laser sintering (SLS)
  absorption by metal powder, 415–419

Index **749**

absorption rate, 416–418, 417t, 418t
absorptivity, 418
beam, 15–16, 436–437
engineering net shaping, 403
heating process of polymer powder materials by, 11–19
input energy characteristics, 11–13, 13f
and metal action, 409–410
method, 27
polarization effects, 414–415
and polymer powder materials, 13–15
power, 489–490, 525–527
  density, 409
wavelength effects, 412
Laser energy
  absorption by metal, 412–415
  density, 38f, 39, 109–110, 708–709, 710f
  transfer
    changes in physical state caused by laser and metal action, 409–410
    energy balance between laser and metal action, 410–411
Laser sintering
  performance, 145–146
    single-layer laser scanning photo, 146f
  process, 15, 98
  properties of nylon 12 PTW composite
    analysis of impact section morphology, 148–149
    crystallization performance, 143–145
    laser sintering performance, 145–146
    mechanical properties of SLS parts, 147–148, 147t
    powder spreading performance, 143
    thermal stability of composite powder, 150
Laser-engineered net shaping (LENS), 683–684
Laser-melted metal powder, 443
Layered silicate clay, 93
LCST. *See* Lower critical solution temperature (LCST)
LENS. *See* Laser-engineered net shaping (LENS)
Lewis acid, 208–209
Lid and box, 660–662
  cooling jacket at gate of hot runner company, 662f
  product formation by scheme of hot runner company, 662f
Light intensity, 217–218, 319
  loss, 218

Light wave, 414
Linear energy density, 474–480
Linear shrinkage rate, 225
Liquid metals, 500
LOM. *See* Laminated objected manufacturing (LOM)
Low-magnification morphology, 555–557, 563–564
Low-temperature pulverization, 31–32, 33f, 50–51, 84f, 98–99
Low-viscosity urethane acrylate, 293–305
Lower critical solution temperature (LCST), 324
Lubricant, 57

M

Machining error of XY plane, 529–534
  process parameters on XY plane accuracy, 529–530
  of thin-walled parts, 530–532
  3D scanning measurement of manufactured parts, 533–534
Magnesium stearate, 57
Magneto-responsive polymers and composite materials, 682–683
MakerBot Replicator printer, 656
MAN5004 laser diffraction particle size analyzer, 102
Marginal adaptation, 663–664
Mark–Houwink empirical equation, 35–36
Martensitic transformation, 553–554, 684–685, 699–700
Mathematical model of heat transfer in SLM forming process, 422–424
  thermophysical parameters
    of pure iron, 423t
    of pure nickel, 423t
Maximum spreading factor, 272–273, 326
MDY501 miniature slag flux-core wire, 581t
Mechanical mixing method, 613
Mechanical property, 75
  indicators, 383–384
Melt flow rate, 35
Melt solidification shrinkage, 83
Melt spheroidization, 495–496, 495f
Melt viscosity, 371
Melting
  and crystallization characteristics of nylon 12 on SLS process, 79–83
  fillers on sintering, 83t

Melting (*Continued*)
  nucleating agent on preheating
    temperature, 82*t*
  and crystallization of nylon 12,
    104–106
  process, 79
  section, 365–366
  temperature, 371–372
Melting/crystallization of composite powder,
    154–156
Melting/solidification process, 559–561
Metal inert-gas welding (MIG), 407
Metal materials, 5–7
  AM forming mechanisms of metal
    materials, 409–439
  AM technologies for metal materials and
    principles, 403–409
  metal powder for SLM, 439–455
  metal wire for arc fuse deposition forming,
    575–579
  microstructure and properties of WAAM,
    579–593
  properties and microstructure characteristics
    of metal powder, 455–575
Metal powder
  absorption of laser, 415–419
    composition on laser absorption rate,
      416–418
    coupling mode, 415–416
    powder particle size on absorption rate,
      418
    power melting process on absorption
      rate, 419
    transmission depth of laser, 416
  for additive manufacturing, 662–665
    chemical composition and content of
      S136 powder, 659*t*
    cooling time and forming cycle of two
      types of inserts, 661*t*
    coordination of denture crown, 664*f*
    coordination of SLM-formed Co-Cr alloy
      denture, 663*f*
    drum lid, 657–660
    lid and box, 660–662
    metallurgical characteristics of SLM
      metal powder, 455–507
    micromorphology of S136 prepared by
      gas atomization, 659*f*
    microstructure characteristics and
      mechanical properties, 545–575
    product formed by 3D printed insert,
      661*f*
    SLM-formed Co-Cr porcelain-fused
      denture prosthesis, 664*f*
    surface roughness and dimensional
      accuracy of formed parts, 507–545
    3D printed cooling jacket at gate, 661*f*
    3D printed mold inserts, 660*f*
  for SLM, 439–455
    metal and alloy powder materials for
      AM, 454–455
    powder oxygen content effects on
      formability, 452–453
    powder particle size effects on
      formability, 439–450
    powder sphericity effects on formability,
      450–452
    thermal properties of, 468–469, 468*t*
Metal wire for arc fuse deposition forming,
    575–579
  characterization of metal wire properties,
    578–579
  design and preparation technology of wire
    materials, 575–577
Metal wire materials for additive
    manufacturing, 665–673
  chemical composition
    of formed metal, 672*t*
    of high-alloy steel flux-cored metal wire,
      665*t*
    of metal wire YHJ507M formed metal,
      670*t*
  hot forging die remanufacturing, 668–669
  multimaterial forging die forming,
    667–668
  part remanufacturing, 669
  printing of ultralarge (thin wall) parts, 666
  process parameters, 670*t*, 672*t*
  simulated structure diagram of multiaxial
    piping system, 670*f*
  structural parts of printed multiaxial piping
    system, 671*f*
Metal-based flux core wire material
    preparation equipment, 576–577
Metal-based wire, 576–577
Metal(s)
  and alloy powder materials for AM,
    454–455
  and composite materials, 683–685
  droplets, 457–458
  properties effect, 412
  surface roughness effects, 414
  temperature effects, 412–413
Metallic copper, 135

Index 751

Metallurgical characteristics of SLM metal powder
  organization characteristics, 499–507
    surface energy spectrum analysis results, 506f
    typical microstructure photo of SLM, 505f
  pores, 486–499
  spheroidization, 455–485
Methyl methacrylate, 399
MGY281 miniature slag flux-core wire, 578–579, 580t
Micro-size spheroidization analysis, 461–464
Micron-sized polymer powder materials, 61
Microscopic
  defects on tensile strength, 570–575
  morphology, 450–451, 456–457, 548–550, 582
  shape of powder, 114–117, 154
Microstructures
  morphology, 554–555
  phases and, 694–696, 705–708
Mie scattering theory, 27
MIG. *See* Metal inert-gas welding (MIG)
Modeling materials. *See also* Support materials
  in FDM, 373–387
    ABS filament, 373–378
    nylon filaments, 386–387
    polycarbonate, 383–386
    polylactic acid filament, 378–382
Modified SLA forming materials, 329–359
  alicyclic epoxy acrylate, 349–359
  epoxy acrylate, 340–349
  nano-SiO_2 modified SLA, 330–340
Moisture, 263–264
Molecular weight, 34–35
Monomers, 265–267
Montmorillonite, 93
Multiaxial pipe joint component
  analysis of microstructure, 582–584
  mechanical properties of forming components, 584–587
    impact test at room temperature, 587, 587t
    tensile and flattening tests, 584–587
  scanning electron microscopy analysis of fracture surfaces, 579–582
Multimaterial forging die forming, 667–668

N

Nano indentation test, 571, 572f
Nano-SiO_2 modified SLA, 330–340
  forming material, 331–340
  SiO_2 surface treatment, 330
Nanofiller/nylon composite powder, 98–99
Nanoparticles, 91–92
Nanoreinforced materials, 138
Nanosilica
  effects
    on mechanical properties of nylon 12 SLS parts, 108–111
    on melting and crystallization of nylon 12, 104–106
    on thermal stability of nylon 12, 107–108
  surface modification of, 99
Nanosilica/nylon 12 composite powder materials, 98–111
  analysis of powder characteristics, 102
  dispersion of nanosilica in nylon 12 matrix, 103–104
  interface bonding between nanosilica and nylon 12, 100–102, 100f
  preparation of nanosilica/nylon 12 composite powder
    preparation process of powder, 99–100
  surface modification of nanosilica, 99
Nantokite, 75
Natural cooling, 64–66, 65f, 66f
NBW. *See* Nonfreezable water content (NBW)
Necking stage, 584–585
Neutralizer, 317–319
Newton-Raphson, 429
Newtonian fluids, 33–34
Ni-Ti-based alloys, 684–685
Nickel-based alloys, 5–6, 455
Non-equilibrium solidification process, 567
Non-Newtonian fluids, 33–34
Noncoal series kaolinite-epoxy resin E06, 630–633
Nonfreezable bound water, 715
Nonfreezable water content (NBW), 713–714
Nonspherical powder particles, 85–86, 86f
Nucleation during powder precipitation, 70–72, 71t
Nylon, 51, 56t, 614–615
  filaments, 386–387
  process flow for preparing nylon 12 powder, 52f
Nylon 12 coated aluminum, composite powder material of, 111–129
  aluminum powder particle size effects on properties of SLS parts, 127–129

Nylon 12 coated aluminum, composite powder material of (*Continued*)
  dispersion state of aluminum powder particles, 126–127
  preparation and characterization of
    characterization of powder materials, 112–119
    preparation process of, 112
    SLS process characteristics, 119–126
Nylon 12, 89, 97
  crystallization rate, 80
  granules, 650
  powder, 52
  powder materials, 61–91
    cooling method and speed, 64–69
    dissolution temperature, 63–64
    nucleation during powder precipitation, 70–72
    postprocessing of nylon powder, 73
    selection of solvents, 62–63
    stirring, 69–70
    thermal history on preparation of powder, 73
  powder SLS parts
    physical and mechanical properties, 89, 90t
    precision of parts, 89–91
    on SLS process, 79–83
  SLS process characteristics of nylon 12 powder, 73–89
Nylon 12/copper composite powder material, 130–137
  preparation and characterization of, 130–132
  SLS-forming process of, 132–137
Nylon 12/PTW, 138–150
  laser sintering properties of nylon 12 PTW composite, 143–150
  preparation of nylon 12/PTW composite powder, 139
  properties of nylon 12/PTW composite powder, 139–143
Nylon 12/rectorite composite powder material, 92–98
  intercalation mechanism in SLS process, 96–98
  preparation, 93–94
  properties of SLS parts, 94–96
Nylon 12/rectorite composite powder sintered materials
  preparation of composite powder sintered materials, 94
  preparation of OREC, 93–94

O

Oligomers, 191, 265–267
  low-viscosity urethane acrylate, 293–305
  polyethylene glycol diacrylate, 320–323
  polypropylene glycol diglycidylether diacrylate, 288–293
  of SLA, 305–306
    solid materials, 288–329
    support materials by, 323–329
  waterborne urethane acrylate, 306–319
Optical microscopy (OM), 689–690, 692f
  microscopic morphology, 574–575
Optimum laser energy density, 109–110
Organic clay, 97
Organic rectorite (OREC), 93
  preparation of, 93–94
  SEM pictures of, 94f
Output spectrum, 217
Over curing depth, 223–224
Overlapping ratio, 515–521, 518f
  overlapped melting channel, 516f
  overlapping of melting channel, 520f
  theoretical values of overlapping ratio, 517t
Oxidation resistance, 454–455
Oxidized carbon fiber, 152
OXT221. *See* Bis(3-ethyl-3-oxetanyl)methyl group ether
Oxygen content of metal powder, 461–464

P

P value, 343–344, 351–352
p-methoxyphenol, 290
PA. *See* Polyamide (PA)
PANalytical X'Pert3 Powder X-ray diffractometer, 714
Particle size, 30f, 112–114, 113t, 153–154, 153t
  distribution, 112–114, 153–154
  principle, 29–30
    on SLS processing, 30–31
  principle, 26–27
  of pure nylon 12 and composite powder, 141t
  on SLS processing, 27–29
PC. *See* Polycarbonate (PC)
PE. *See* Polyethylene (PE)
PEG. *See* Polyethylene glycol (PEG)
PEG-PVA/CNTs DN hydrogels composite, 720–721
PEK. *See* Polyetheretherketone (PEK)

Index

Perkin Elmer DSC-7 differential scanning calorimeter, 79
Photo-responsive polymers and composite materials, 683
Photochromic molecule, 683
Photoinitiator, 195–196, 198–201, 325
  iodonium salt and, 241–247
  selection of, 267–268
Photolithographic-tandem, 722
Photopolymer, 195–196, 195t, 657
  characteristic parameters of, 215–216
  characteristics of materials, 218–228
  curing shrinkage of, 225–228
  SLA properties of, 218–224
  toughened photopolymer material of epoxy acrylate, 340–349
Photosensitive resin, 723, 724t
Photosensitizer additive, 201
Photothermal effect, 409–410
Pigment, 269
Pinch instability theory, 438–439
PLA. *See* Polylactic acid (PLA)
PLAGA. *See* Polylactide-glycolide (PLAGA)
Plastic deformation stage, 584–585
Plastically deformable matrix, 161
PLLA. *See* Poly-L-lactic acid (PLLA)
PLS nanocomposite. *See* Polymer/layered silicate nanocomposite (PLS nanocomposite)
PMMA. *See* Polymethyl-methacrylate (PMMA)
Polarization, 414
Poly-L-lactic acid (PLLA), 5
Polyacrylate, 52–53
Polyamide (PA), 3–4
Polycaprolactone polyol, 212–213, 212f
Polycarbonate (PC), 3–4, 53–54, 54t, 176, 383–386
  powder, 29
  powder material, 176–184
    SLS process and properties, 178–182
    sintered parts
      density and mechanical properties, 180t
      postprocessing effects on properties, 182–184
Polyetheretherketone (PEK), 5
Polyethylene (PE), 4–5, 50
Polyethylene glycol (PEG), 295, 681, 712–713, 712f
  diacrylate, 320–323
    analysis of product of, 321–323
    synthesis experiment, 320
    synthesis reaction formula of, 321
Polylactic acid (PLA), 654–655
  characteristics, 378
  filament, 378–382
  performance indicators, 382t
  preparation, 378–382
Polylactide-glycolide (PLAGA), 5
Polymer materials, 3–5
  for AM
    ABS powder material, 184–188
    composite powder material of nylon 12, 91–162
    nylon 12 powder materials, 61–91
    PC powder material, 176–184
    preparation, composition, and characterization of polymers, 49–61
    SLS processing mechanism of polymer, 9–49
    styrene-based amorphous polymer powder materials, 162–176
  liquid materials for additive manufacturing, 656–657
    automobile interior, 658f
    automobile lamp, 658f
    hand saw housing, 659f
    mobile phone shell, 658f
    thermomechanical performance analysis curve, 657f
  matrix materials
    selection of amorphous polymers, 53–54
    selection of crystalline polymers, 54–55
Polymer powder materials
  by laser, 11–19
  SLS mechanism, 19–24
    Frenkel's two-liquid-drop model, 20–21
    sintered cube model, 21–24
Polymer/layered silicate nanocomposite (PLS nanocomposite), 92
Polymer(s)
  and composite materials, 680–683
    magneto-responsive, 682–683
    photo-responsive, 683
    thermoresponsive, 681–682
    water-responsive, 681
  and composite powder materials
    characterization of SLS polymer materials, 59–61
    composition of SLS polymer materials, 53–59
    low-temperature pulverization, 50–51
    preparation, 49–53
    solvent precipitation, 51–53

Polymer(s) (*Continued*)
  melt intercalation, 96–98
  molecular chain, 97
  photoinitiation system, 201
  thermodynamic transformation, 366–371
Polymerization
  inhibitor, 343
  kinetics, 215
  method of acrylic copolymers, 400
Polymethyl-methacrylate (PMMA), 3–4, 53
Polymorphism, 725
Polyoxymethylene (POM), 54–55
Polypropylene (PP), 4–5, 50
Polypropylene glycol diglycidyl ether diacrylate (PPGDEA), 288–293, 656
  influencing factors, 289–291
  infrared spectrum of product, 291–292
  properties of product, 292–293
  reaction mechanism, 288–289
  synthesis of oligomers, 288
Polystyrene (PS), 3–4, 9
  powder materials, 163
Polyvinyl alcohol (PVA), 5, 395–396, 711
  PVA-based hydrogels, 717–718
  water-soluble support material, 396–399
POM. *See* Polyoxymethylene (POM)
Pores, 486–499
  classification of pores and formation mechanism, 494–499
    crack, 497–498
    hole, 498–499
    melt spheroidization, 495–496
    scanning line, 496–497, 496f
  factors affecting porosity
    forming process, 489–494
    powder characteristics, 486–489
Postprocessing
  effects on properties
    of PC sintered parts, 182–184
    of SLS parts, 174–176
    enhanced parts after, 178f
  method, 436
  of nylon powder, 73
  of SAN powder material, 163–165
  selection and use of reinforced resins, 172–174
Postreaction modification, 397
Potassium titanate whisker (PTW), 139, 140f
Powder
  bed, 498
  materials characterization

  analysis on thermal weight loss of powder, 117–118
  differential scanning calorimetry analysis of powder, 119
  energy spectrum analysis of powder, 117
  microscopic shape of powder, 114–117
  particle size and particle size distribution, 112–114
  powder-based additive manufacturing technologies, 683–684
  powder-based metal raw materials, 408
  powder-spreading process, 693–694, 693f
  properties, 461–464
    analysis of surface energy spectrum, 462f
    sphericity effects on formability, 450–452
    single-pass single-layer melting pool, 451f
    surface microscopic morphology of block parts, 452f
  spheroidization, 85
  spreading performance, 119–121, 143, 158
  whiteness, 61, 61t
Powder oxygen content effects on formability, 452–453
  characteristics of test powder, 453t
  surface microscopic morphology of block parts, 454f
Powder particle
  shape
    principle, 31–32
    on SLS processing, 33
  size and shape of, 59–60, 60f
  size effects on formability
    and apparent density, 439–441
    density and mechanical properties of formed entities, 447–450
    on single-pass forming shape, 441–443
    on single-pass scanning line width, 443–446
    on surface scanning, 446–447
  size on absorption rate, 418, 419t
Power
  density spectrum, 217–218
  melting process on absorption rate, 419
  and speed, 511–512, 511f
PowerScanner-STD, 533–534
PP. *See* Polypropylene (PP)
PPGDEA. *See* Polypropylene glycol diglycidyl ether diacrylate (PPGDEA)
Precision extrusion, 362
Preheating
  method, 435, 650

Index 755

temperature, 121–122, 122t
  of composite powder, 146t
Prepolymers. *See* Oligomers
Printability, 711, 719–721
  PEG-PVA/CNTs DN hydrogels composite, 720f
Printing of ultralarge (thin wall) parts, 666.
  See also Additive manufacturing (AM); Four-dimensional printing (4D printing)
  forming dimensions of ultralarge (thin wall) parts, 666t
  multimaterial forging die, 667f
  print results of ultralarge parts, 666f
Process parameters on XY plane accuracy, 529–530, 530f, 530t
Processed layer thickness, 527–529
Product precision, 362
Prophylactic antioxidants. *See* Hydroperoxide decomposers
Prototyping technology, 665
PS. *See* Polystyrene (PS)
PTW. *See* Potassium titanate whisker (PTW)
PVA. *See* Polyvinyl alcohol (PVA)

Q
Quasi-cleavage fracture, 573–574

R
R&D. *See* Research and development (R&D)
Raw materials of DN hydrogel reinforced by CNTs, 711–712
Reactive diluents. *See* Diluent monomers
Rectorite, 92
Reinforced resins, 172–174
Relative density, 45–46
Rene series, 6
Research and development (R&D), 2–3
Residual stress distribution, 432–434, 433f
Resin materials, SLA for, 202–203
Resin parameters, 338
  of photopolymer, 223–224
Reversible thermal transition, 717
Reynolds number (*Re*), 271–272
Rubber duck ceramic craft, 673–674

S
SADP. *See* Selected area diffraction pattern (SADP)
SAN. *See* Styrene–acrylonitrile (SAN)

Scanning
  line interval, 480–481
  pictures of Al/PA, 115, 115f, 116f, 117f
  pitch, 491–493, 515–516, 517t
  SEM-energy dispersive spectroscopy, 713
  shape of low-temperature brittle, 103–104, 104f
  speed, 490–491, 490f
    effects, 469–470, 469f, 470f
  strategy, 436, 513–515
Scanning electron microscopy (SEM), 102, 102f, 447, 449f, 476f, 478f, 479f, 689–690, 695f, 699f, 709f
  analysis of fracture surfaces, 579–582
Screw extrusion process, 361–365
Screw rotational speed, 363–364
SE. *See* Superelasticity (SE)
Sedimentation method, 27
Selected area diffraction pattern (SADP), 696
Selective laser melting (SLM), 551, 597–598, 683–684, 701f, 704f
  and ceramic powders, 601–602
  equipment, 403
  metal powder for, 439–455
  processing parameters effect on relative density, 690–694, 691f, 704–705, 705f
  SLM forming process, temperature, stress and strain fields, 419–436
    temperature field, 420–429
    thermal stress and thermal strain field, 429–436
  SLM metal powder, metallurgical characteristics of, 455–507
  technology, 403–406, 404f, 507–545, 510f
    lightweight porous parts, 406f
    personalized artificial prosthesis, 407f
Selective laser sintering (SLS), 9, 10f, 403, 605–606, 649
  forming mechanism of ceramics, 606–609
  mathematical model
    of amorphous polymers, 16–18
    of crystalline polymer, 18–19
  mechanism of polymer powder materials, 19–24
  parts of nylon 12/rectorite composite powder, 94–96
    mechanical properties, 94–95, 95t
    thermal properties, 95–96, 96t
  parts SAN powder material, 165–167
    dimensional accuracy, 167
    mechanical properties, 165–167, 165t

Selective laser sintering (SLS) (*Continued*)
  polymer materials, 53–59
    additives, 59
    bulk density of powder, 60
    dispersant, 57
    lubricant, 57
    powder whiteness, 61
    selection and surface treatment of filler, 57–59
    selection of polymer matrix materials, 53–55
    size and shape of powder particles, 59–60
    stabilizer, 55–57
  preparation of ceramic materials for, 609–645
    ceramic powders and binders, 609–612
    preparation of composite, 612–617
    properties of, 617–645
  process and properties of PC powder material, 178–182
  processing mechanism of polymer, 9–49
    heating process of polymer powder materials, 11–19
    mechanism of SLS of polymer powder materials, 19–24
    properties of polymer and composite powder materials, 24–49
  properties of carbon fiber/nylon 12 composite powder
    mechanical properties of sintered parts, 158–159
    powder spreading performance, 158
    sectional shape of sintered part, 159–162
  SLS-forming process
    crystallization and melting characteristics, 135–136
    interface formation between nylon 12 and copper, 133–135
    of nylon 12/copper composite powder and parts properties, 132–137
    properties of sintered parts, 136–137
SEM. *See* Scanning electron microscopy (SEM)
Semicrystalline polymer, 25–26
Sensitization mechanism, 214
Shaft bracket, 672–673, 673*f*
Shape fixity ratio, 722–723
Shape memory
  materials, 717–718
  phenomenon, 714–715
  properties, 729–733

  cycle life between this work and referred work, 733*t*
  entire shape recovery process of SMP, 734*f*
  time spent of shape recovery process, 734*f*
Shape memory alloys (SMAs), 683–684
Shape memory effect (SME), 681–682, 717, 717*f*
Shape memory polymers (SMPs), 681–682
Shape recovery ratio, 722–723
Shearing test of wire arc addictive manufacturing-manufactured high-alloy steel, 592–593, 593*f*
Shrinkage
  coefficient, 277–278
  deformation of nylon 12, 83–89
    mixture of powders of particle sizes, 87*t*
    powder particle size on preheating temperature, 87*t*
SiC, 7
  ceramic parts, 677, 677*f*
  square mirror, 675
Side surface roughness of part, 522–529
  laser power, 525–527
  processed layer thickness, 527–529
  tilt angle, 522–525
Sieving method, 27
Silane coupling agent, 58
Silica (SiO_2), 72
Silicon nitride (Si_3N_4), 7
Single layer, surface roughness of, 510–521
  overlapping ratio, 515–521
  power and speed, 511–512
  scanning strategy, 513–515
Single pass, surface roughness of, 508–510
Single-channel multilayer scanning process, 432
Single-layer scanning of agglomerated powder, 88, 88*f*
Single-pass
  forming shape, 441–443, 441*f*
  scanning line, 441–442
  scanning line width, 443–446
  single-layer scanning test, 451
Sintered cube model, 21–24, 28
Sintered nylon 12/OREC composite material, 95
Sintered parts, 47*f*
  density of, 44–45, 44*f*
  dimensional accuracy of, 47–49
  fractured surface shape of, 46–47

Index   757

mechanical properties of, 45−46
mechanical properties of, 158−159
properties of, 136−137
sectional shape of, 159−162
Sintering, 19−20
  materials, 24
  performance of ABS powder
    dimensional accuracy, 186
    mechanical properties, 186−188, 186*t*
    process, 86
    temperature window, 41−44
Size distribution. *See* Particle size distribution
Skorohod model, 22
SL. *See* Stereolithography (SL)
SLA. *See* Stereolithography apparatus (SLA)
SLM. *See* Selective laser melting (SLM)
SLS. *See* Selective laser sintering (SLS)
SLS process characteristics. *See also* Selective laser sintering (SLS)
  of HIPS powder material, 168−169
  of nylon 12 coated aluminum composite powder
    aluminum powder content effects on dimensional accuracy, 125−126
    aluminum powder content effects on mechanical properties, 123−125
    powder spreading performance, 119−121
    preheating temperature, 121−122
  of nylon 12 powder, 73−89
    melting and crystallization characteristics, 79−83
    shrinkage and warping, 83−89
    thermal oxygen stability, 73−79
Smart materials, 680, 683−684, 721−722
SMAs. *See* Shape memory alloys (SMAs)
SME. *See* Shape memory effect (SME)
SMPs. *See* Shape memory polymers (SMPs)
Solid materials, SLA, 228−287
  benzyl alcohol accelerator in cationic SLA, 230−239
  cationic solid materials, 264−278
  free-radical system, 284−287
  hybrid system solid materials, 279−283
  iodonium salt, 240−259
  trimethylene oxide, 260−264
Solid-liquid interface, 503
Solidification, 694−695
  morphology of single-melting channel, 477−479
Solvent
  evaporation, 613−614

precipitation method, 31−32, 32*f*, 51−53, 82
coating method, 130
selection, 62−63
moisture contents on powder particle size, 63*t*
Specific heat capacity, 423
Spherical powder particles, 85−86, 86*f*
Spheroidization, 455−485, 458*f*, 459*f*, 465*f*, 466*f*, 480*f*
  analysis of, 456−460
  influencing factors
    forming atmosphere, 464−465
    forming process parameters, 466−485
    powder properties, 461−464
  SLM forming, 444*f*
  wetting of the melting pool and substrate, 443*f*
Spray drying method, 31−32, 32*f*
Sputtering coefficient, 272−273
Square mirror, 675
SR256. *See* 2-(2-Ethoxyethoxy) ethyl acrylate
SR306. *See* Tripropylene glycol diacrylate
SR454. *See* Triethyl oxide trimethylolpropane triacrylate
Stability of melting pool, 438−439
Stabilizer, 55−57
Stainless steel powder, 446−447, 483−484
Step-by-step reaction mechanism, 257−258
Stereolithographic-tandem, 722
Stereolithography (SL), 597−598
  and ceramic slurry, 599−600
Stereolithography apparatus (SLA), 191−228, 681−682
  characteristic of ultraviolet light source, 216−218
  exposure methods, 191
  material, 195−203
  modified SLA forming materials, 329−359
  oligomers in, 288−329
  photopolymer
    characteristic parameters of, 215−216
    characteristics of materials, 218−228
  reaction mechanism, 203−214
  resins, 195−196
  on solid materials, 228−287
Stereolithography appearance. *See* Stereolithography apparatus (SLA)
Stirring, 69−70, 70*t*
Storage modulus, 730, 731*f*
Stress-induced martensitic transformation, 699−700

Stress-strain relationship equations, 429–432
  stress and strain of thermal elastoplastic, 429–430
  thermal elastoplastic finite element method, 430–432
Styrene-based amorphous polymer powder materials, 162–176
  HIPS powder material, 168–176
  polystyrene powder materials, 163
  styrene–acrylonitrile copolymer powder material, 163–167
Styrene–acrylonitrile (SAN), 163
  copolymer powder material, 163–167
    preparation and characterization, 163
    properties of SLS parts SAN powder material, 165–167
    SLS-forming and postprocessing, 163–165
Sulfonium salt, 209–210
Superelasticity (SE), 683–684
Support materials, 387–389. *See also* Modeling materials
  in FDM, 387 400
    break-away support materials, 389–395
    water-soluble support materials, 395–400
  by oligomers, 323–329
Surface modification of nanosilica, 99
Surface morphology, 513, 514*f*, 519*f*
Surface roughness, 507–529, 508*f*, 509*f*, 512*f*
  data of scanning methods, 515*t*
  side surface roughness of part, 522–529
  of single layer, 510–521
  of single pass, 508–510
Surface scanning, 446–447, 446*f*
Surface tension, 269–271
  principle, 24–25
  on SLS process, 25–26
Surface treatment
  of fiber powder, 151–152
  of filler, 57–59
Surfactant, 269
Swelling, 717–719, 718*f*
  behavior measurements, 714

T

tBA/HDDA. *See* Tert-Butyl acrylate/1, 6-hexanediol diacrylate (tBA/HDDA)
TCP. *See* Tricalcium phosphate (TCP)
TDI. *See* Toluene diisocyanate (TDI)
TEM. *See* Transmission electron microscopy (TEM)
Temperature, 34
  control, 63–64
  field, 426*f*, 427*f*
    analysis of molten pool in SLM forming process, 425–429
    finite element method for transient heat conduction, 421–422
    mathematical model of heat transfer in SLM forming process, 422–424
    three-dimensional transient temperature field, 420–421
  shrinkage, 83
  temperature-sensitive material, 324
Tensile
  properties, 561–564, 562*t*, 568*f*
  strength
    curve, 450, 450*f*
    grain refinement effects on, 569–570
    microscopic defects on, 570–575
    test, 584–587, 586*t*, 592
Terpenoid, 201, 201*f*
Tert-Butyl acrylate/1, 6-hexanediol diacrylate (tBA/HDDA), 681–682
TG/DSC synchronous thermal analyzer, 724
TGA. *See* Thermogravimetric analysis (TGA)
Thermal actuators, 686
Thermal conductivity, 14–15
  of alloy material, 423
Thermal elastoplastic
  finite element method, 430–432
  stress and strain of, 429–430
Thermal oxygen stability
  antiaging properties of nylon 12 powder, 77*t*
  of nylon 12 powder on SLS process, 73–79
Thermal sensors, 686
Thermal stability, 726–727
  of composite powder, 150
  of nylon 12, 107–108
Thermal stress and strain field, 429–436
  analysis, 432–435
  equations of stress-strain relationship, 429–432
  method of reducing thermal stress and strain, 435–436
Thermal weight loss
  analysis of composite powder, 156–157
  of powder, 117–118
Thermo-mechanical characteristic curve, 366
Thermodynamic polymer, 717–718

Index 759

Thermodynamic transformation of polymer processing, 366−371
Thermogravimetric (TG)
  curve, 74, 726, 727f
  temperatures, 107−108, 107t
  of NPA12, 118t
Thermogravimetric analysis (TGA), 713−714
Thermooxidative aging mechanism, 74−75
Thermoresponsive polymers and composite materials, 681−682
Thin-walled parts, machining error of, 530−532, 531f, 533f
Three-dimension (3D)
  computer-aided design models, 722
  measurement technology, 650
  printed spinal orthosis, 650
  reverse technology, 665
  scanning measurement of manufactured parts, 533−534
  solid parts, 9−10, 403−404
  transient temperature field, 420−421
Three dimensional printing (3DP), 600−601
Ti6Al4V alloy
  mechanical properties, 555−559
    density, 555
  microstructure, 553−555
TIG. See Tungsten inert-gas welding (TIG)
Tilt angle, 522−525, 523f, 526f
Titanium and titanium alloys, 455
Titanium silicide (Ti_3SiC_2), 7
Titanium-based alloys, 5−7
Toluene diisocyanate (TDI), 295
Toughened photopolymer material of epoxy acrylate, 340−349
TPO. See 2,4,6-Trimethylbenzoyldiphenyl phosphine oxide (TPO)
TPS-SbF_6. See Triaryl hexafluoroantimonate sulfonium salt (TPS-SbF_6)
Transient heat conduction, 421−422
Transient temperature field, 420
Transmission depth of laser, 416, 416f
Transmission electron microscopy (TEM), 689−690, 696f
Triallylamine, 205−206
Triaryl hexafluoroantimonate sulfonium salt (TPS-SbF_6), 198−200, 199f
Tricalcium phosphate (TCP), 7
Triethyl oxide trimethylolpropane triacrylate, 284
Triethylamine, 711−712
2,4,6-Trimethylbenzoyldiphenyl phosphine oxide (TPO), 722−723

Trimethylene oxide, 260−264
  experimental part, 260
  OXT221 SLA reaction mechanism, 260−261
Tripropylene glycol diacrylate, 284
Tungsten inert-gas welding (TIG), 407
Two-dimension (2D)
  manufacturing, 403−404. See also Additive manufacturing (AM)
  transient temperature field finite element, 421

U

Ultraviolet (UV), 191, 682−683
  curing materials, 192t
  illuminance meter, 217−218
  light source, 216−218
  photoinitiator, 723
Urethane acrylate
  analysis of carbamate acrylate, 301−305
  low-viscosity, 293−305
  principle of preparation of, 297
  synthesis conditions of, 297−301
  synthesis experiment, 295−297
UV. See Ultraviolet (UV)
UVR6105
  acceleration of photosensitizer on, 235
  infrared spectroscopy analysis, 233−234

V

Van der Waals force, 24−25
Vaporization, 410, 571
Vat photosensitive, 722
Vickers microhardness, 690
Vinokurov empirical relationship, 467−468
Vinyl ether, 265−267
Viscoelasticity of polymers, 369−371
Viscosity, 269−271
  principle, 33−35
    molecular weight, 34−35
    temperature, 34
  on SLS processing, 35−40
  viscosity-average molecular weight and melt flow rate, 36t
Volume deformation, 23
Volume shrinkage, 47−48
  rate, 225

W

WAAM. See Wire and arc additive manufacture (WAAM)

Warpage factor, 277–278
Warping deformation of nylon 12, 83–89
Water content analysis, 714–715, 716t
Water diffusion mobility, 717
Water-atomized powder, 486, 487f
Water-responsive polymers
  and composite materials, 681
Water-soluble support materials, 395–400
Waterborne polyurethane acrylate, 309–312
Waterborne urethane acrylate, 306–319
  analysis of test method, 307–308
  hydrophilic properties of, 312–315
  SLA speed of, 316–319
  synthesis experiment, 307
  waterborne polyurethane acrylate, 309–312
Weber number (*We*), 271–272
Wetting agents, 324–325
Whisker-reinforced materials, 138
Wire and arc additive manufacture (WAAM), 407–409, 408f, 579
  microstructure and properties of WAAM of multiaxial pipe joint component, 579–587
  and performance of typical repaired components, 588–593
Wire materials, design and preparation technology of, 575–577

X

X-ray diffraction (XRD), 689–690
  curves of DLP-printed SMPs, 728f
  images, 714
  patterns, 705–706, 706f
X'pert3 power X-ray diffractometer, 725
XRD. *See* X-ray diffraction (XRD)

Y

Yagi-Kun model, 15
Young's modulus, 567–569

Z

Z-axis accuracy, 537–545
  height measurement values
    of optimized tower parts, 545t
    of tower part, 543t
  powder layer thickness of powder, 540t
  powder leveling thickness results, 541f
  profile of SLM-formed part, 542f
  SLM-formed parts, 544f
Zero-equilibrium state, 718–719
Zero-shear viscosity, 33–34
Zirconium dioxide (ZrO_2)
  ZrO_2-E06 composite powder, 620–622
  ZrO_2-nylon 12 composite powder, 625–629
  ZrO_2-polymer composite powder, 620–629
  ZrO_2-stearic acid composite powder, 623–625

Printed in the United States
By Bookmasters